D0207105

# Preface

This *Student's Solutions Manual* contains selected solutions and strategies for solving typical exercises in the text, *Fundamentals of College Algebra, Eleventh Edition*, by Earl W. Swokowski and Jeffery A. Cole.

In each exercise set, all odd-numbered solutions are included, with an emphasis on the solutions of the applied "word" problems. In the review exercise section at the end of each chapter, a mixture of odd-numbered and even-numbered solutions are included. For the discussion exercises at the end of each chapter, all odd-numbered solutions are included. I have tried to illustrate enough solutions so that the student will be able to obtain an understanding of all types of problems in each section.

A significant number of today's students are involved in various outside activities, and find it difficult, if not impossible, to attend all class sessions. This manual should help meet the needs of these students. In addition, it is my hope that this manual's solutions will enhance the understanding of all readers of the material and provide insights to solving other exercises.

I would appreciate any feedback concerning errors, solution correctness, solution style, or manual style—comments from students using previous editions have greatly strengthened the text's supplements as well as the text itself. These and any other comments may be sent directly to me at the address below or in care of the publisher: Thomson Higher Education, 10 Davis Drive, Belmont, CA 94002.

I would like to thank: Mike Rosenborg and Sudhir Goel for accuracy checking; Joan Cole, my wife, for proofing various features of the manual; Brian Morris and the late George Morris, of Scientific Illustrators, for creating the mathematically precise art package; and Sally Lifland and Gail Magin, of Lifland et al., Bookmakers, for assembling the final manuscript. I dedicate this book to my children, Becky and Brad.

Jeffery A. Cole
Anoka-Ramsey Community College
11200 Mississippi Blvd. NW
Coon Rapids, MN 55433

# Table of Contents


**1 Fundamental Concepts of Algebra** .......... 1

1.1 Real Numbers • .......... 1
1.2 Exponents and Radicals • .......... 3
1.3 Algebraic Expressions • .......... 7
1.4 Fractional Expressions • .......... 11
*Chapter 1 Review Exercises* • .......... 16
*Chapter 1 Discussion Exercises* • .......... 20

**2 Equations and Inequalities** .......... 21

2.1 Equations • .......... 21
2.2 Applied Problems • .......... 24
2.3 Quadratic Equations • .......... 26
2.4 Complex Numbers • .......... 32
2.5 Other Types of Equations • .......... 35
2.6 Inequalities • .......... 41
2.7 More on Inequalities • .......... 45
*Chapter 2 Review Exercises* • .......... 49
*Chapter 2 Discussion Exercises* • .......... 55

**3 Functions and Graphs** .......... 57

3.1 Rectangular Coordinate Systems • .......... 57
3.2 Graphs of Equations • .......... 60
3.3 Lines • .......... 68
3.4 Definition of Function • .......... 73
3.5 Graphs of Functions • .......... 81
3.6 Quadratic Functions • .......... 91
3.7 Operations on Functions • .......... 97
*Chapter 3 Review Exercises* • .......... 104
*Chapter 3 Discussion Exercises* • .......... 112

**4 Polynomial and Rational Functions** .......... 115

4.1 Polynomial Functions of Degree Greater Than 2 • .......... 115
4.2 Properties of Division • .......... 123
4.3 Zeros of Polynomials • .......... 128
4.4 Complex and Rational Zeros of Polynomials • .......... 134
4.5 Rational Functions • .......... 139
4.6 Variation • .......... 153
*Chapter 4 Review Exercises* • .......... 156
*Chapter 4 Discussion Exercises* • .......... 161

Student Solutions Manual

for
Swokowski and Cole's

# Fundamentals of College Algebra

Eleventh Edition

**Jeffery A. Cole**
*Anoka-Ramsey Community College*

THOMSON
BROOKS/COLE

Australia • Canada • Mexico • Singapore • Spain • United Kingdom • United States

Printed in the United States of America
1 2 3 4 5 6 7 08 07 06 05 04

Printer: Thomson/West

0-534-46459-9

For more information about our products, contact us at:
**Thomson Learning Academic Resource Center**
**1-800-423-0563**

For permission to use material from this text or product, submit a request online at **http://www.thomsonrights.com.**
Any additional questions about permissions can be submitted by email to **thomsonrights@thomson.com.**

Cover photo: Getty Images

**Thomson Higher Education**
**10 Davis Drive**
**Belmont, CA 94002-3098**
**USA**

**Asia (including India)**
Thomson Learning
5 Shenton Way
#01-01 UIC Building
Singapore 068808

**Australia/New Zealand**
Thomson Learning Australia
102 Dodds Street
Southbank, Victoria 3006
Australia

**Canada**
Thomson Nelson
1120 Birchmount Road
Toronto, Ontario M1K 5G4
Canada

**UK/Europe/Middle East/Africa**
Thomson Learning
High Holborn House
50–51 Bedford Road
London WC1R 4LR
United Kingdom

**Latin America**
Thomson Learning
Seneca, 53
Colonia Polanco
11560 Mexico
D.F. Mexico

**Spain (including Portugal)**
Thomson Paraninfo
Calle Magallanes, 25
28015 Madrid, Spain

# Table of Contents


**5 Inverse, Exponential, and Logarithmic Functions ..... 163**

5.1 Inverse Functions • ..... 163
5.2 Exponential Functions • ..... 166
5.3 The Natural Exponential Function • ..... 174
5.4 Logarithmic Functions • ..... 179
5.5 Properties of Logarithms • ..... 186
5.6 Exponential and Logarithmic Equations • ..... 190
*Chapter 5 Review Exercises* • ..... 197
*Chapter 5 Discussion Exercises* • ..... 202

**6 Systems of Equations and Inequalities ..... 205**

6.1 Systems of Equations • ..... 205
6.2 Systems of Linear Equations in Two Variables • ..... 211
6.3 Systems of Inequalities • ..... 214
6.4 Linear Programming • ..... 221
6.5 Systems of Linear Equations in More Than Two Variables • ..... 226
6.6 The Algebra of Matrices • ..... 233
6.7 The Inverse of a Matrix • ..... 238
6.8 Determinants • ..... 242
6.9 Properties of Determinants • ..... 245
6.10 Partial Fractions • ..... 249
*Chapter 6 Review Exercises* • ..... 253
*Chapter 6 Discussion Exercises* • ..... 259

**7 Sequences, Series, and Probability ..... 261**

7.1 Infinite Sequences and Summation Notation • ..... 261
7.2 Arithmetic Sequences • ..... 267
7.3 Geometric Sequences • ..... 270
7.4 Mathematical Induction • ..... 274
7.5 The Binomial Theorem • ..... 279
7.6 Permutations • ..... 282
7.7 Distinguishable Permutations and Combinations • ..... 284
7.8 Probability • ..... 286
*Chapter 7 Review Exercises* • ..... 291
*Chapter 7 Discussion Exercises* • ..... 297

**8 Topics from Analytic Geometry ..... 301**

8.1 Parabolas • ..... 301
8.2 Ellipses • ..... 305
8.3 Hyperbolas • ..... 312
*Chapter 8 Review Exercises* • ..... 319
*Chapter 8 Discussion Exercises* • ..... 322

# Notations

The following notations are used in the manual.

| | |
|---|---|
| *Note:* | { Notes to the student pertaining to hints on solutions, common mistakes, or conventions to follow. } |
| { } | { comments to the reader are in braces } |
| LS | { Left Side of an equation } |
| RS | { Right Side of an equation } |
| $\approx$ | { approximately equal to } |
| $\Rightarrow$ | { implies, next equation, logically follows } |
| $\Leftrightarrow$ | { if and only if, is equivalent to } |
| • | { bullet, used to separate problem statement from solution or explanation } |
| ★ | { used to identify the answer to the problem } |
| § | { *section* references } |
| $\forall$ | { For all, i.e., $\forall x$ means "for all $x$". } |
| $\mathbb{R} - \{a\}$ | { The set of all real numbers except $a$. } |
| $\therefore$ | { therefore } |
| QI–QIV | { quadrants I, II, III, IV } |

# To the Student

This manual is a text supplement and should be read along *with* the text. Read all exercise solutions in this manual since explanations of concepts are given and then appear in subsequent solutions. All concepts necessary to solve a particular problem are not reviewed for every exercise. If you are having difficulty with a previously covered concept, look back to the section where it was covered for more complete help. The writing style I have used in this manual reflects the way I explain concepts to my own students. It is not as mathematically precise as that of the text, including phrases such as "goes down" or "touches and turns around." My students have told me that these terms help them understand difficult concepts with ease.

Lengthier explanations and more steps are given for the more difficult problems. Additional information that my students have found helpful is included—see page 36. The guidelines given in the text are followed for some solutions—see page 142.

In the review sections, the solutions are somewhat abbreviated since more detailed solutions were given in previous sections. However, this is not true for the word problems in these sections since they are unique. In easier groups of exercises, representative solutions are shown. Occasionally, alternate solutions are also given.

All figures have been plotted using computer software, offering a high degree of precision. The calculator graphs are from various TI screens. When possible, each piece of art was made with the same scale to show a realistic and consistent graph.

This manual was done using EXP: *The Scientific Word Processor.* I have used a variety of display formats for the mathematical equations, including centering, vertical alignment, and flushing text to the right. I hope that these make reading and comprehending the material easier for you.

# Chapter 1: Fundamental Concepts of Algebra

## 1.1 Exercises

[1] (a) Since $x$ and $y$ have opposite signs, the product $xy$ is negative.

(b) Since $x^2 > 0$ and $y > 0$, $x^2y > 0$.

(c) Since $x < 0$ {$x$ is negative} and $y > 0$ {$y$ is positive}, $\frac{x}{y}$ is negative.

Thus, $\frac{x}{y} + x$ is the sum of two negatives, which is *negative.*

(d) Since $y > 0$ and $x < 0$, $y - x > 0$.

[3] (a) Since $-7$ is to the left of $-4$ on a coordinate line, $-7 \boxed{<} -4$.

(b) Using a calculator, we see that $\frac{\pi}{2} \approx 1.5708$. Hence, $\frac{\pi}{2} \boxed{>} 1.57$.

(c) $\sqrt{225} \boxed{=} 15$ *Note:* $\sqrt{225} \neq \pm 15$

[5] (a) Since $\frac{1}{11} = 0.\overline{09}$, $\frac{1}{11} \boxed{>} 0.09$. (b) Since $\frac{2}{3} = 0.\overline{6}$, $\frac{2}{3} \boxed{>} 0.6666$.

(c) Since $\frac{22}{7} = 3.\overline{142857}$ and $\pi \approx 3.141593$, $\frac{22}{7} \boxed{>} \pi$.

[7] (a) "$x$ is negative" is equivalent to $x < 0$. We symbolize this by writing

$$x \text{ is negative} \iff x < 0.$$

(b) $y$ is nonnegative $\iff$ $y \geq 0$

(c) $q$ is less than or equal to $\pi$ $\iff$ $q \leq \pi$

(d) $d$ is between 4 and 2 $\iff$ $2 < d < 4$

(e) $t$ is not less than 5 $\iff$ $t \geq 5$

(f) The negative of $z$ is not greater than 3 $\iff$ $-z \leq 3$

(g) The quotient of $p$ and $q$ is at most 7 $\iff$ $\frac{p}{q} \leq 7$

(h) The reciprocal of $w$ is at least 9 $\iff$ $\frac{1}{w} \geq 9$

(i) The absolute value of $x$ is greater than 7 $\iff$ $|x| > 7$

*Note:* An informal definition of absolute value that may be helpful is

$$|\,\textit{something}\,| = \begin{cases} \textit{itself} & \text{if } \textit{itself} \text{ is positive or zero} \\ -(\textit{itself}) & \text{if } \textit{itself} \text{ is negative} \end{cases}$$

[9] (a) $|-3-2| = |-5| = -(-5)$ {since $-5 < 0$} $= 5$

(b) $|-5| - |2| = -(-5) - 2 = 5 - 2 = 3$

(c) $|7| + |-4| = 7 + [-(-4)] = 7 + 4 = 11$

[11] (a) $(-5)\,|3-6| = (-5)\,|-3| = (-5)[-(-3)] = (-5)(3) = -15$

(b) $|-6|/(-2) = -(-6)/(-2) = 6/(-2) = -3$

(c) $|-7| + |4| = -(-7) + 4 = 7 + 4 = 11$

[13] (a) Since $(4 - \pi)$ is positive, $|4 - \pi| = 4 - \pi$.

(b) Since $(\pi - 4)$ is negative, $|\pi - 4| = -(\pi - 4) = 4 - \pi$.

(c) Since $(\sqrt{2} - 1.5)$ is negative, $|\sqrt{2} - 1.5| = -(\sqrt{2} - 1.5) = 1.5 - \sqrt{2}$.

**15** (a) $d(A,\ B) = |\,7-3\,| = |\,4\,| = 4$ (b) $d(B,\ C) = |\,-5-7\,| = |\,-12\,| = 12$

(c) $d(C,\ B) = d(B,\ C) = 12$ (d) $d(A,\ C) = |\,-5-3\,| = |\,-8\,| = 8$

**17** (a) $d(A,\ B) = |\,1-(-9)\,| = |\,10\,| = 10$ (b) $d(B,\ C) = |\,10-1\,| = |\,9\,| = 9$

(c) $d(C,\ B) = d(B,\ C) = 9$ (d) $d(A,\ C) = |\,10-(-9)\,| = |\,19\,| = 19$

*Note:* Exer. 19–24: Since $|\,a\,| = |\,-a\,|$, the answers could have a different form.

For example, $|\,-3-x\,| \geq 8$ is equivalent to $|\,x+3\,| \geq 8$.

**19** $A = x$ and $B = 7$, so $d(A,\ B) = |\,7-x\,|$.

Thus, "$d(A,\ B)$ is less than 5" can be written as $|\,7-x\,| < 5$.

**21** $d(A,\ B) = |\,-3-x\,| \ \Rightarrow\ |\,-3-x\,| \geq 8$

**23** $d(A,\ B) = |\,x-4\,| \ \Rightarrow\ |\,x-4\,| \leq 3$

**25** Pick an arbitrary value for $x$ that is less than $-3$, say $-5$.

Since $3+(-5) = -2$ is negative, we conclude that if $x < -3$, then $3+x$ is negative.

Hence, $|\,3+x\,| = -(3+x) = -x-3$.

**27** If $x < 2$, then $2-x > 0$, and $|\,2-x\,| = 2-x$.

**29** If $a < b$, then $a-b < 0$, and $|\,a-b\,| = -(a-b) = b-a$.

**31** Since $x^2+4 > 0$ for every $x$, $|\,x^2+4\,| = x^2+4$.

**33** $\text{LS} = \dfrac{ab+ac}{a} = \frac{ab}{a} + \frac{ac}{a} = b+c$ $\boxed{\neq}$ RS $(b+ac)$.

**35** $\text{LS} = \dfrac{b+c}{a} = \frac{b}{a} + \frac{c}{a}$ $\boxed{=}$ RS.

**37** $\text{LS} = (a \div b) \div c = \frac{a}{b} \cdot \frac{1}{c} = \frac{a}{bc}$. $\text{RS} = a \div (b \div c) = a \div \frac{b}{c} = a \cdot \frac{c}{b} = \frac{ac}{b}$. LS $\boxed{\neq}$ RS

**39** $\text{LS} = \dfrac{a-b}{b-a} = \dfrac{-(b-a)}{b-a} = -1$ $\boxed{=}$ RS.

**41** (a) On the TI-83 Plus, the absolute value feature is choice 1 under MATH, NUM.

Enter abs($3.2^2 - \sqrt{\ }(3.15)$). On the TI-86, abs is F5 under MATH, NUM.

$\left|3.2^2 - \sqrt{3.15}\right| \approx 8.4652$

(b) $\sqrt{(15.6-1.5)^2 + (4.3-5.4)^2} \approx 14.1428$

**43** (a) $\dfrac{1.2 \times 10^3}{3.1 \times 10^2 + 1.52 \times 10^3} \approx 0.6557 = 6.557 \times 10^{-1}$

(b) $(1.23 \times 10^{-4}) + \sqrt{4.5 \times 10^3} \approx 67.08 = 6.708 \times 10^1$

**45** Construct a right triangle with sides of lengths $\sqrt{2}$ and 1. The hypotenuse will have length $\sqrt{(\sqrt{2})^2 + 1^2} = \sqrt{3}$. Next construct a right triangle with sides of lengths $\sqrt{3}$ and $\sqrt{2}$. The hypotenuse will have length $\sqrt{(\sqrt{3})^2 + (\sqrt{2})^2} = \sqrt{5}$.

**47** The large rectangle has area = width × length = $a(b+c)$. The sum of the areas of the two small rectangles is $ab+ac$. Since the areas are the same, we have $a(b+c) = ab+ac$.

49 (a) Since the decimal point is 5 places to the right of the first nonzero digit,

$$427{,}000 = 4.27 \times 10^5.$$

(b) Since the decimal point is 8 places to the left of the first nonzero digit,

$$0.000\ 000\ 098 = 9.8 \times 10^{-8}.$$

51 (a) Moving the decimal point 5 places to the right, we have $8.3 \times 10^5 = 830{,}000$.

(b) Moving the decimal point 12 places to the left, we have

$$2.9 \times 10^{-12} = 0.000\ 000\ 000\ 002\ 9.$$

53 Since the decimal point is 24 places to the left of the first nonzero digit,

$$0.000\ 000\ 000\ 000\ 000\ 000\ 000\ 001\ 7 = 1.7 \times 10^{-24}.$$

55 It is helpful to write the units of any fraction, and then "cancel" those units to determine the units of the final answer.

$$\frac{186{,}000 \text{ miles}}{\text{second}} \cdot \frac{60 \text{ seconds}}{1 \text{ minute}} \cdot \frac{60 \text{ minutes}}{1 \text{ hour}} \cdot \frac{24 \text{ hours}}{1 \text{ day}} \cdot \frac{365 \text{ days}}{1 \text{ year}} \cdot 1 \text{ year} \approx 5.87 \times 10^{12} \text{ mi}$$

57 $$\frac{\dfrac{1.01 \text{ grams}}{\text{mole}}}{\dfrac{6.02 \times 10^{23} \text{ atoms}}{\text{mole}}} \cdot 1 \text{ atom} = \frac{1.01 \text{ grams}}{6.02 \times 10^{23}} \approx 0.1678 \times 10^{-23} \text{ g} = 1.678 \times 10^{-24} \text{ g}$$

59 $$\frac{24 \text{ frames}}{\text{second}} \cdot \frac{60 \text{ seconds}}{1 \text{ minute}} \cdot \frac{60 \text{ minutes}}{1 \text{ hour}} \cdot 48 \text{ hours} = 4.1472 \times 10^6 \text{ frames}$$

61 (a) $1 \text{ ft}^2 = 144 \text{ in}^2$, so the force on one square foot of a wall is

$$144 \text{ in}^2 \times 1.4 \text{ lb/in}^2 = 201.6 \text{ lb}.$$

(b) The area of the wall is $40 \times 8 = 320 \text{ ft}^2$, or $320 \text{ ft}^2 \times 144 \text{ in}^2/\text{ft}^2 = 46{,}080 \text{ in}^2$.

The total force is $46{,}080 \text{ in}^2 \times 1.4 \text{ lb/in}^2 = 64{,}512$ lb.

Converting to tons, we have $64{,}512 \text{ lb}/(2000 \text{ lb/ton}) = 32.256$ tons.

## 1.2 Exercises

1 $(-\frac{2}{3})^4 = (-\frac{2}{3}) \cdot (-\frac{2}{3}) \cdot (-\frac{2}{3}) \cdot (-\frac{2}{3}) = \frac{16}{81}$

*Note:* Do not confuse $(-x)^4$ and $-x^4$ since $(-x)^4 = x^4$ and $-x^4$ is the negative of $x^4$.

3 $\frac{2^{-3}}{3^{-2}} = \frac{3^2}{2^3} = \frac{9}{8}$ *Note:* Remember that negative exponents don't necessarily equate to negative results—that is, $2^{-3} = \frac{1}{2^3} = \frac{1}{8}$, not $-\frac{1}{8}$.

5 $-2^4 + 3^{-1} = -16 + \frac{1}{3} = -\frac{48}{3} + \frac{1}{3} = \frac{-47}{3}$

7 $16^{-3/4} = 1/16^{3/4} = 1/(\sqrt[4]{16})^3 = 1/2^3 = \frac{1}{8}$

9 $(-0.008)^{2/3} = (\sqrt[3]{-0.008})^2 = (-0.2)^2 = 0.04 = \frac{4}{100} = \frac{1}{25}$

[11] $(\frac{1}{2}x^4)(16x^5) = (\frac{1}{2} \cdot 16)x^{4+5} = 8x^9$

[13] A common mistake is to write $x^3x^2 = x^6$, and another is to write $(x^2)^3 = x^5$.

The following solution illustrates the proper use of the exponent rules.

$$\frac{(2x^3)(3x^2)}{(x^2)^3} = \frac{(2 \cdot 3)x^{3+2}}{x^{2 \cdot 3}} = \frac{6x^5}{x^6} = 6x^{5-6} = 6x^{-1} = \frac{6}{x}$$

[15] $(\frac{1}{6}a^5)(-3a^2)(4a^7) = \frac{1}{6} \cdot (-3) \cdot 4 \cdot a^{5+2+7} = -2a^{14}$

[17] $\frac{(6x^3)^2}{(2x^2)^3} \cdot (3x^2)^0 =$

$\frac{6^2x^{3 \cdot 2}}{2^3x^{2 \cdot 3}} \cdot 1$ {an expression raised to the zero power is equal to 1} $= \frac{36x^6}{8x^6} = \frac{36}{8} = \frac{9}{2}$

[19] $(3u^7v^3)(4u^4v^{-5}) = 12u^{7+4}v^{3+(-5)} = 12u^{11}v^{-2} = \frac{12u^{11}}{v^2}$

[21] $(8x^4y^{-3})(\frac{1}{2}x^{-5}y^2) = 4x^{-1}y^{-1} = \frac{4}{xy}$

[23] $(\frac{1}{3}x^4y^{-3})^{-2} = (\frac{1}{3})^{-2}(x^4)^{-2}(y^{-3})^{-2} = (\frac{3}{1})^2x^{-8}y^6 = 3^2x^{-8}y^6 = \frac{9y^6}{x^8}$

[25] $(3y^3)^4(4y^2)^{-3} = 81y^{12} \cdot 4^{-3}y^{-6} = 81y^6 \cdot \frac{1}{64} = \frac{81}{64}y^6$

[27] $(-2r^4s^{-3})^{-2} = (-2)^{-2}r^{-8}s^6 = \frac{s^6}{(-2)^2r^8} = \frac{s^6}{4r^8}$

[29] $(5x^2y^{-3})(4x^{-5}y^4) = 20x^{-3}y = \frac{20y}{x^3}$

[31] $\left(\frac{3x^5y^4}{x^0y^{-3}}\right)^2$ {remember that $x^0 = 1$} $= \frac{9x^{10}y^8}{y^{-6}} = 9x^{10}y^{8+6} = 9x^{10}y^{14}$

[33] $(4a^{3/2})(2a^{1/2}) = 4 \cdot 2a^{(3/2)+(1/2)} = 8a^{4/2} = 8a^2$

[35] $(3x^{5/6})(8x^{2/3}) = 3 \cdot 8x^{(5/6)+(4/6)} = 24x^{9/6} = 24x^{3/2}$

[37] $(27a^6)^{-2/3} = 27^{-2/3}a^{-12/3} = \frac{a^{-4}}{27^{2/3}} = \frac{1}{(\sqrt[3]{27})^2a^4} = \frac{1}{3^2a^4} = \frac{1}{9a^4}$

[39] $(8x^{-2/3})x^{1/6} = 8x^{(-4/6)+(1/6)} = 8x^{-3/6} = \frac{8}{x^{1/2}}$

[41] $\left(\frac{-8x^3}{y^{-6}}\right)^{2/3} = \frac{(-8)^{2/3}(x^3)^{2/3}}{(y^{-6})^{2/3}} = \frac{(\sqrt[3]{-8})^2\,x^{(3)(2/3)}}{y^{(-6)(2/3)}} = \frac{(-2)^2x^2}{y^{-4}} = \frac{4x^2}{y^{-4}} = 4x^2y^4$

[43] $\left(\frac{x^6}{9y^{-4}}\right)^{-1/2} = \frac{x^{-3}}{9^{-1/2}y^2} = \frac{9^{1/2}x^{-3}}{y^2} = \frac{3}{x^3y^2}$

[45] $\frac{(x^6y^3)^{-1/3}}{(x^4y^2)^{-1/2}} = \frac{(x^6)^{-1/3}(y^3)^{-1/3}}{(x^4)^{-1/2}(y^2)^{-1/2}} = \frac{x^{-2}y^{-1}}{x^{-2}y^{-1}} = 1$

**47** $\sqrt[4]{x^3} = (x^3)^{1/4} = x^{3\cdot(1/4)} = x^{3/4}$

**49** $\sqrt[3]{(a+b)^2} = \left[(a+b)^2\right]^{1/3} = (a+b)^{2/3}$

**51** $\sqrt{x^2+y^2} = (x^2+y^2)^{1/2}$ *Note:* $\sqrt{x^2+y^2} \neq x+y$

**53** (a) $4x^{3/2} = 4x^1x^{1/2} = 4x\sqrt{x}$

(b) $(4x)^{3/2} = (4x)^1(4x)^{1/2} = (4x)^1\,4^{1/2}x^{1/2} = 4x\cdot 2\cdot x^{1/2} = 8x\sqrt{x}$

**55** (a) $8 - y^{1/3} = 8 - \sqrt[3]{y}$ (b) $(8-y)^{1/3} = \sqrt[3]{8-y}$

**57** $\sqrt{81} = \sqrt{9^2} = 9$

**59** $\sqrt[5]{-64} = \sqrt[5]{-32}\,\sqrt[5]{2} = \sqrt[5]{(-2)^5}\,\sqrt[5]{2} = -2\sqrt[5]{2}$

**61** In the denominator, you would like to have $\sqrt[3]{2^3}$. How do you get it? Multiply by $\sqrt[3]{2^2}$, or, equivalently, $\sqrt[3]{4}$. Of course, we have to multiply the numerator by the same value so that we don't change the value of the given fraction.

$$\frac{1}{\sqrt[3]{2}} = \frac{1}{\sqrt[3]{2}}\cdot\frac{\sqrt[3]{4}}{\sqrt[3]{4}} = \frac{\sqrt[3]{4}}{\sqrt[3]{2\cdot 4}} = \frac{\sqrt[3]{4}}{\sqrt[3]{8}} = \frac{\sqrt[3]{4}}{2} = \tfrac{1}{2}\sqrt[3]{4}$$

**63** $\sqrt{9x^{-4}y^6} = (9x^{-4}y^6)^{1/2} = 9^{1/2}(x^{-4})^{1/2}(y^6)^{1/2} = 3x^{-2}y^3 = \dfrac{3y^3}{x^2}$

**65** $\sqrt[3]{8a^6b^{-3}} = 2a^2b^{-1} = \dfrac{2a^2}{b}$

*Note:* For exercises similar to those in 67–74, pick a multiplier that will make all of the exponents of the terms in the denominator a multiple of the index.

**67** The index is 2. Choose the multiplier to be $\sqrt{2y}$ so that the denominator contains only terms with even exponents. $\sqrt{\dfrac{3x}{2y^3}} = \sqrt{\dfrac{3x}{2y^3}}\cdot\dfrac{\sqrt{2y}}{\sqrt{2y}} = \sqrt{\dfrac{6xy}{4y^4}} = \dfrac{\sqrt{6xy}}{2y^2}$, or $\dfrac{1}{2y^2}\sqrt{6xy}$

**69** The index is 3. Choose the multiplier to be $\sqrt[3]{3x^2}$ so that the denominator contains only terms with exponents that are multiples of 3.

$$\sqrt[3]{\frac{2x^4y^4}{9x}} = \sqrt[3]{\frac{2x^4y^4}{9x}}\cdot\frac{\sqrt[3]{3x^2}}{\sqrt[3]{3x^2}} = \frac{\sqrt[3]{6x^6y^4}}{\sqrt[3]{27x^3}} = \frac{\sqrt[3]{x^6y^3}\,\sqrt[3]{6y}}{3x} = \frac{x^2y\,\sqrt[3]{6y}}{3x} = \frac{xy}{3}\sqrt[3]{6y}$$

**71** The index is 4. Choose the multiplier to be $\sqrt[4]{3x^2}$ so that the denominator contains only terms with exponents that are multiples of 4.

$$\sqrt[4]{\frac{5x^8y^3}{27x^2}} = \sqrt[4]{\frac{5x^8y^3}{27x^2}}\cdot\frac{\sqrt[4]{3x^2}}{\sqrt[4]{3x^2}} = \frac{\sqrt[4]{15x^{10}y^3}}{\sqrt[4]{81x^4}} = \frac{\sqrt[4]{x^8}\,\sqrt[4]{15x^2y^3}}{3x} = \frac{x^2\,\sqrt[4]{15x^2y^3}}{3x}$$

$$= \tfrac{x}{3}\sqrt[4]{15x^2y^3}$$

[73] The index is 5. Choose the multiplier to be $\sqrt[5]{4x^2}$ so that the denominator contains only terms with exponents that are multiples of 5.

$$\sqrt[5]{\frac{5x^7y^2}{8x^3}} = \sqrt[5]{\frac{5x^7y^2}{8x^3}} \cdot \frac{\sqrt[5]{4x^2}}{\sqrt[5]{4x^2}} = \frac{\sqrt[5]{x^5}\,\sqrt[5]{20x^4y^2}}{2x} = \frac{x\,\sqrt[5]{20x^4y^2}}{2x} = \frac{1}{2}\sqrt[5]{20x^4y^2}$$

[75] $\sqrt[4]{(3x^5y^{-2})^4} = 3x^5y^{-2} = \dfrac{3x^5}{y^2}$

[77] $\sqrt[5]{\dfrac{8x^3}{y^4}}\,\sqrt[5]{\dfrac{4x^4}{y^2}} = \sqrt[5]{\dfrac{8x^3}{y^4}}\,\sqrt[5]{\dfrac{4x^4}{y^2}} \cdot \dfrac{\sqrt[5]{y^4}}{\sqrt[5]{y^4}} = \dfrac{\sqrt[5]{32x^5}\,\sqrt[5]{x^2y^4}}{\sqrt[5]{y^{10}}} = \dfrac{\sqrt[5]{32x^5}\,\sqrt[5]{x^2y^4}}{y^2} = \dfrac{2x}{y^2}\sqrt[5]{x^2y^4}$

[79] $\sqrt[3]{3t^4v^2}\,\sqrt[3]{-9t^{-1}v^4} = \sqrt[3]{-27t^3v^6} = -3tv^2$

[81] $\sqrt{x^6y^4} = \sqrt{(x^3)^2(y^2)^2} = \sqrt{(x^3)^2}\,\sqrt{(y^2)^2} = |\,x^3\,|\,|\,y^2\,| = |\,x^3\,|\,y^2$ since $y^2$ is always nonnegative. *Note:* $|\,x^3\,|$ could be written as $x^2\,|\,x\,|$.

[83] $\sqrt[4]{x^8(y-1)^{12}} = \sqrt[4]{(x^2)^4((y-1)^3)^4} = |\,x^2\,|\,|\,(y-1)^3\,| = x^2\,|\,(y-1)^3\,|$, or $x^2(y-1)^2\,|\,(y-1)\,|$

[85] $(a^r)^2 = a^{2r}$ $\boxed{\neq}$ $a^{(r^2)}$ since $2r \neq r^2$ for *all* values of $r$; for example, let $r = 1$.

[87] $(ab)^{xy} = a^{xy}b^{xy}$ $\boxed{\neq}$ $a^xb^y$ for *all* values of $x$ and $y$; for example, let $x = 1$ and $y = 2$.

[89] $\sqrt[n]{\dfrac{1}{c}} = \left(\dfrac{1}{c}\right)^{1/n} = \dfrac{1^{1/n}}{c^{1/n}}$ $\boxed{=}$ $\dfrac{1}{\sqrt[n]{c}}$

[91] (a) $(-3)^{2/5} = [(-3)^2]^{1/5} = 9^{1/5} \approx 1.5518$

(b) $(-5)^{4/3} = [(-5)^4]^{1/3} = 625^{1/3} \approx 8.5499$

[93] (a) $\sqrt{\pi + 1} \approx 2.0351$ (b) $\sqrt[3]{15.1} + 5^{1/4} \approx 3.9670$

[95] $\$200(1.04)^{180} \approx \$232{,}825.78$

[97] $W = 230$ kg $\Rightarrow$ $L = 0.46\sqrt[3]{W} = 0.46\sqrt[3]{230} \approx 2.82$ m

[99] $b = 75$ and $w = 180$ $\Rightarrow$ $W = \dfrac{w}{\sqrt[3]{b-35}} = \dfrac{180}{\sqrt[3]{75-35}} \approx 52.6.$

$b = 120$ and $w = 250$ $\Rightarrow$ $W = \dfrac{w}{\sqrt[3]{b-35}} = \dfrac{250}{\sqrt[3]{120-35}} \approx 56.9.$

It is interesting to note that the 75-kg lifter can lift 2.4 times his/her body weight and the 120-kg lifter can lift approximately 2.08 times his/her body weight, but *the formula ranks the 120-kg lifter as the superior lifter.*

[101] $W = 0.1166h^{1.7}$

| Height | 64 | 65 | 66 | 67 | 68 | 69 | 70 | 71 |
|---|---|---|---|---|---|---|---|---|
| Weight | 137 | 141 | 145 | 148 | 152 | 156 | 160 | 164 |

| Height | 72 | 73 | 74 | 75 | 76 | 77 | 78 | 79 |
|---|---|---|---|---|---|---|---|---|
| Weight | 168 | 172 | 176 | 180 | 184 | 188 | 192 | 196 |

## 1.3 Exercises

[1] $(3x^3+4x^2-7x+1)+(9x^3-4x^2-6x)=12x^3-13x+1$

[3] $$\begin{aligned}(4x^3+5x-3)-(3x^3+2x^2+5x-7) &= 4x^3+5x-3-3x^3-2x^2-5x+7\\ &= (4x^3-3x^3)-2x^2+(5x-5x)+(-3+7)\\ &= x^3-2x^2+4\end{aligned}$$

[5] $$\begin{aligned}(2x+5)(3x-7) &= (2x)(3x)+(2x)(-7)+(5)(3x)+(5)(-7)\\ &= 6x^2-14x+15x-35\\ &= 6x^2+x-35\end{aligned}$$

[7] $$\begin{aligned}(5x+7y)(3x+2y) &= (5x)(3x)+(5x)(2y)+(7y)(3x)+(7y)(2y)\\ &= 15x^2+10xy+21xy+14y^2 = 15x^2+31xy+14y^2\end{aligned}$$

[9] $(2u+3)(u-4)+4u(u-2)=(2u^2-5u-12)+(4u^2-8u)=6u^2-13u-12$

[11] $$\begin{aligned}(3x+5)(2x^2+9x-5) &= 3x(2x^2+9x-5)+5(2x^2+9x-5)\\ &= (6x^3+27x^2-15x)+(10x^2+45x-25)\\ &= 6x^3+37x^2+30x-25\end{aligned}$$

[13] $$\begin{aligned}(t^2+2t-5)(3t^2-t+2) &= t^2(3t^2-t+2)+2t(3t^2-t+2)+(-5)(3t^2-t+2)\\ &= (3t^4-t^3+2t^2)+(6t^3-2t^2+4t)+(-15t^2+5t-10)\\ &= 3t^4+5t^3-15t^2+9t-10\end{aligned}$$

[15] $$\begin{aligned}(x+1)(2x^2-2)(x^3+5) &= 2\,[(x+1)(x^2-1)](x^3+5)\\ &= 2(x^3+x^2-x-1)(x^3+5)\\ &= 2(x^6+x^5-x^4+4x^3+5x^2-5x-5)\\ &= 2x^6+2x^5-2x^4+8x^3+10x^2-10x-10\end{aligned}$$

[17] $\dfrac{8x^2y^3-10x^3y}{2x^2y}=\dfrac{8x^2y^3}{2x^2y}-\dfrac{10x^3y}{2x^2y}=4y^2-5x$

[19] $\dfrac{3u^3v^4-2u^5v^2+(u^2v^2)^2}{u^3v^2}=\dfrac{3u^3v^4}{u^3v^2}-\dfrac{2u^5v^2}{u^3v^2}+\dfrac{u^4v^4}{u^3v^2}=3v^2-2u^2+uv^2$

[21] We recognize this product as the difference of two squares.

$$(2x+3y)(2x-3y)=(2x)^2-(3y)^2=4x^2-9y^2$$

[23] $(x^2+2y)(x^2-2y)=(x^2)^2-(2y)^2=x^4-4y^2$

[25] $(x^2+9)(x^2-4)=x^4-4x^2+9x^2-36=x^4+5x^2-36$

[27] $(3x+2y)^2=(3x)^2+2(3x)(2y)+(2y)^2=9x^2+12xy+4y^2$

[29] $(x^2-3y^2)^2=(x^2)^2-2(x^2)(3y^2)+(3y^2)^2=x^4-6x^2y^2+9y^4$

[31] We could expand $(x+2)^2$ and $(x-2)^2$ and then multiply the resulting expressions, but the following solution is simpler.

$$\begin{aligned}(x+2)^2(x-2)^2 &= [(x+2)(x-2)]^2 && \{a^2b^2=(ab)^2\} \\ &= (x^2-4)^2 && \{\text{difference of two squares}\} \\ &= (x^2)^2-2(x^2)(4)+(4)^2 && \{\text{square of a binomial}\} \\ &= x^4-8x^2+16 && \{\text{simplify}\}\end{aligned}$$

[33] $(\sqrt{x}+\sqrt{y})(\sqrt{x}-\sqrt{y})=(\sqrt{x})^2-(\sqrt{y})^2=x-y$

[35] $(x^{1/3}-y^{1/3})(x^{2/3}+x^{1/3}y^{1/3}+y^{2/3})$

$$\begin{aligned}&= x^{1/3}(x^{2/3}+x^{1/3}y^{1/3}+y^{2/3})-y^{1/3}(x^{2/3}+x^{1/3}y^{1/3}+y^{2/3}) \\ &= x+x^{2/3}y^{1/3}+x^{1/3}y^{2/3}-x^{2/3}y^{1/3}-x^{1/3}y^{2/3}-y=x-y\end{aligned}$$

This exercise illustrates how the difference of *any* two terms can be factored as the difference of cubes. Another example of this concept is

$$x-5=(\sqrt[3]{x}-\sqrt[3]{5})(\sqrt[3]{x^2}+\sqrt[3]{5x}+\sqrt[3]{25}).$$

[37] Use Product Formula (3) on text page 36.

$$(x-2y)^3=(x)^3-3(x)^2(2y)+3(x)(2y)^2-(2y)^3=x^3-6x^2y+12xy^2-8y^3$$

[39] $$\begin{aligned}(2x+3y)^3 &= (2x)^3+3(2x)^2(3y)+3(2x)(3y)^2+(3y)^3 \\ &= 8x^3+3(4x^2)(3y)+3(2x)(9y^2)+27y^3 \\ &= 8x^3+36x^2y+54xy^2+27y^3\end{aligned}$$

*Note:* Exer. 41–44: Treat these as "the sum of the squares plus twice the product of all possible pairs of terms," that is,

$$(x+y+z)^2=x^2+y^2+z^2+2xy+2xz+2yz.$$

[41] $(a+b-c)^2=[a+b+(-c)]^2=a^2+b^2+c^2+2ab-2ac-2bc$

[43] $$\begin{aligned}(2x+y-3z)^2 &= (2x)^2+(y)^2+(-3z)^2+2(2x)(y)+2(2x)(-3z)+2(y)(-3z) \\ &= 4x^2+y^2+9z^2+4xy-12xz-6yz\end{aligned}$$

[45] Always factor out the greatest common factor first. $rs+4st=s(r+4t)$

[47] The greatest common factor is $3a^2b$. $3a^2b^2-6a^2b=3a^2b(b-2)$

[49] The greatest common factor is $3x^2y^2$. $3x^2y^3-9x^3y^2=3x^2y^2(y-3x)$

[51] $\{\text{gcf}=5x^3y^2\}$ $15x^3y^5-25x^4y^2+10x^6y^4=5x^3y^2(3y^3-5x+2x^3y^2)$

[53] We recognize this as a trinomial that may be able to be factored into the product of two binomials. $8x^2-53x-21=(8x+3)(x-7)$

[55] The factors for $x^2+3x+4$ would have to be of the form $(x+\_)$ and $(x+\_)$. The factors of 4 are 1 & 4 and 2 & 2, but their sums are 5 and 4, respectively.

Thus, $x^2+3x+4$ is *irreducible.*

[57] $6x^2+7x-20=(3x-4)(2x+5)$

59 $12x^2 - 29x + 15 = (3x - 5)(4x - 3)$

61 $4x^2 - 20x + 25 = (2x - 5)(2x - 5) = (2x - 5)^2$

63 $25z^2 + 30z + 9 = (5z + 3)(5z + 3) = (5z + 3)^2$

65 $45x^2 + 38xy + 8y^2 = (5x + 2y)(9x + 4y)$

67 $36r^2 - 25t^2 = (6r)^2 - (5t)^2 = (6r + 5t)(6r - 5t)$

69 $z^4 - 64w^2 = (z^2)^2 - (8w)^2 = (z^2 + 8w)(z^2 - 8w)$

71 $x^4 - 4x^2 = x^2(x^2 - 4) = x^2(x^2 - 2^2) = x^2(x + 2)(x - 2)$

73 $x^2 + 25$ is irreducible. *Note:* A common mistake is to confuse the *sum* of two squares with the *difference* of two squares.

75 $75x^2 - 48y^2 = 3(25x^2 - 16y^2) = 3\,[(5x)^2 - (4y)^2] = 3(5x + 4y)(5x - 4y)$

77 We recognize this expression as the sum of two cubes.

$$\begin{aligned} 64x^3 + 27 &= (4x)^3 + (3)^3 \\ &= (4x + 3)\,[(4x)^2 - (4x)(3) + (3)^2] \\ &= (4x + 3)(16x^2 - 12x + 9) \end{aligned}$$

79 We recognize this expression as the difference of two cubes.

$$\begin{aligned} 64x^3 - y^6 &= (4x)^3 - (y^2)^3 \\ &= (4x - y^2)\,[(4x)^2 + (4x)(y^2) + (y^2)^2] \\ &= (4x - y^2)(16x^2 + 4xy^2 + y^4) \end{aligned}$$

81 We recognize this expression as the sum of two cubes.

$$\begin{aligned} 343x^3 + y^9 &= (7x)^3 + (y^3)^3 \\ &= (7x + y^3)\,[(7x)^2 - (7x)(y^3) + (y^3)^2] \\ &= (7x + y^3)(49x^2 - 7xy^3 + y^6) \end{aligned}$$

83 We recognize this expression as the difference of two cubes.

$$\begin{aligned} 125 - 27x^3 &= (5)^3 - (3x)^3 \\ &= (5 - 3x)\,[(5)^2 + (5)(3x) + (3x)^2] \\ &= (5 - 3x)(25 + 15x + 9x^2) \end{aligned}$$

85 Since there are more than 3 terms, we will try to factor by grouping first.

$$\begin{aligned} 2ax - 6bx + ay - 3by &= 2x(a - 3b) + y(a - 3b) \\ &= (2x + y)(a - 3b) && \{\text{factor out } (a - 3b)\} \end{aligned}$$

87

$$\begin{aligned} 3x^3 + 3x^2 - 27x - 27 &= 3(x^3 + x^2 - 9x - 9) && \{\textbf{gcf} = 3\} \\ &= 3[x^2(x + 1) - 9(x + 1)] && \{\text{factor by grouping}\} \\ &= 3(x^2 - 9)(x + 1) && \{\text{factor out } (x + 1)\} \\ &= 3(x + 3)(x - 3)(x + 1) && \{\text{difference of two squares}\} \end{aligned}$$

89 Since there are more than 3 terms, we will try to factor by grouping first.

$$x^4 + 2x^3 - x - 2 = x^3(x+2) - 1(x+2) = (x^3-1)(x+2)$$

Now recognize $x^3 - 1$ as the difference of two cubes.

$$(x^3-1)(x+2) = [(x-1)(x^2+x+1)](x+2) = (x-1)(x+2)(x^2+x+1)$$

91 $a^3 - a^2b + ab^2 - b^3 = a^2(a-b) + b^2(a-b) = (a^2+b^2)(a-b)$

93 We could treat this expression as the difference of two squares or the difference of two cubes. Factoring as the difference of two squares and then as the sum and difference of two cubes leads to the following:

$$\begin{aligned} a^6 - b^6 &= (a^3)^2 - (b^3)^2 \\ &= (a^3+b^3)(a^3-b^3) \\ &= (a+b)(a-b)(a^2-ab+b^2)(a^2+ab+b^2) \end{aligned}$$

95 We might first try to factor this expression by grouping since it has more than 3 terms, but this would prove to be unsuccessful. Instead, we will group the terms containing $x$ and the constant term together, and then proceed as in Example 10(c).

$$x^2 + 4x + 4 - 9y^2 = (x+2)^2 - (3y)^2 = (x + 2 + 3y)(x + 2 - 3y)$$

97 We will group the terms containing $y$ and the constant term together, and then proceed as in Example 10(c).

$$\begin{aligned} y^2 - x^2 + 8y + 16 &= (y^2 + 8y + 16) - x^2 \\ &= (y+4)^2 - (x)^2 \\ &= (y+4+x)(y+4-x) \end{aligned}$$

99 We should first note that one of the two variable terms, $y^6$, is the square of the other, $y^3$. Thus, we may treat this expression as a simple trinomial that can be factored into the product of two binomials.

$$y^6 + 7y^3 - 8 = (y^3+8)(y^3-1) = (y+2)(y^2-2y+4)(y-1)(y^2+y+1)$$

101 
$$\begin{aligned} x^{16} - 1 = (x^8+1)(x^8-1) &= (x^8+1)(x^4+1)(x^4-1) \\ &= (x^8+1)(x^4+1)(x^2+1)(x^2-1) \\ &= (x^8+1)(x^4+1)(x^2+1)(x+1)(x-1) \end{aligned}$$

103 The dimensions of I are $(x)$ and $(x-y)$. The area of I is $(x-y)x$, and the area of II is $(x-y)y$. The area $A = \underline{x^2 - y^2} = (x-y)x + (x-y)y = \underline{(x-y)(x+y)}.$

105 (a) For the 25-year-old female, use

$C_f = 66.5 + 13.8w + 5h - 6.8y$ with $w = 59$, $h = 163$, and $y = 25$.

$C_f = 66.5 + 13.8(59) + 5(163) - 6.8(25) = 1525.7$ calories

(continued)

For the 55-year-old male, use

$C_m = 655 + 9.6w + 1.9h - 4.7y$ with $w = 75$, $h = 178$, and $y = 55$.

$C_m = 655 + 9.6(75) + 1.9(178) - 4.7(55) = 1454.7$ calories

(b) As people age they require fewer calories. The coefficients of $w$ and $h$ are positive because large people require more calories.

## 1.4 Exercises

[1] $\dfrac{3}{50} + \dfrac{7}{30} = \dfrac{3}{2 \cdot 5^2} + \dfrac{7}{2 \cdot 3 \cdot 5} = \dfrac{3 \cdot 3 + 7 \cdot 5}{2 \cdot 3 \cdot 5^2} = \dfrac{9 + 35}{2 \cdot 3 \cdot 5^2} = \dfrac{44}{2 \cdot 3 \cdot 5^2} = \dfrac{22}{3 \cdot 5^2} = \dfrac{22}{75}$

[3] $\dfrac{5}{24} - \dfrac{3}{20} = \dfrac{5}{2^3 \cdot 3} - \dfrac{3}{2^2 \cdot 5} = \dfrac{5 \cdot 5 - 3(2 \cdot 3)}{2^3 \cdot 3 \cdot 5} = \dfrac{25 - 18}{2^3 \cdot 3 \cdot 5} = \dfrac{7}{2^3 \cdot 3 \cdot 5} = \dfrac{7}{120}$

[5] $\dfrac{2x^2 + 7x + 3}{2x^2 - 7x - 4} = \dfrac{(2x+1)(x+3)}{(2x+1)(x-4)} = \dfrac{x+3}{x-4}$

[7] $\dfrac{y^2 - 25}{y^3 - 125} = \dfrac{(y+5)(y-5)}{(y-5)(y^2+5y+25)} = \dfrac{y+5}{y^2+5y+25}$

[9] $\dfrac{12 + r - r^2}{r^3 + 3r^2} = \dfrac{(3+r)(4-r)}{r^2(r+3)} = \dfrac{4-r}{r^2}$

[11] $\dfrac{9x^2 - 4}{3x^2 - 5x + 2} \cdot \dfrac{9x^4 - 6x^3 + 4x^2}{27x^4 + 8x} = \dfrac{(3x+2)(3x-2)}{(3x-2)(x-1)} \cdot \dfrac{x^2(9x^2 - 6x + 4)}{x(3x+2)(9x^2 - 6x + 4)} = \dfrac{x}{x-1}$

[13] $\dfrac{5a^2 + 12a + 4}{a^4 - 16} \div \dfrac{25a^2 + 20a + 4}{a^2 - 2a} = \dfrac{(5a+2)(a+2)}{(a^2+4)(a+2)(a-2)} \cdot \dfrac{a(a-2)}{(5a+2)(5a+2)} =$

$\dfrac{a}{(a^2+4)(5a+2)}$

[15] $\dfrac{6}{x^2 - 4} - \dfrac{3x}{x^2 - 4} = \dfrac{6 - 3x}{x^2 - 4} = \dfrac{3(2-x)}{(x+2)(x-2)} = \dfrac{-3}{x+2}$. Since $2 - x = -(x - 2)$, we canceled the two factors, $2 - x$ and $x - 2$, and replaced them with $-1$. In general, you may do this whenever you encounter factors of the form $a - b$ and $b - a$ in the numerator and the denominator, respectively, of a fractional expression.

[17] $\dfrac{2}{3s+1} - \dfrac{9}{(3s+1)^2} = \dfrac{2(3s+1)}{(3s+1)^2} - \dfrac{9}{(3s+1)^2} = \dfrac{6s + 2 - 9}{(3s+1)^2} = \dfrac{6s - 7}{(3s+1)^2}$

[19] $\dfrac{2}{x} + \dfrac{3x+1}{x^2} - \dfrac{x-2}{x^3} = \dfrac{2x^2}{x^3} + \dfrac{(3x+1)x}{x^3} - \dfrac{(x-2)}{x^3} = \dfrac{2x^2 + 3x^2 + x - x + 2}{x^3} = \dfrac{5x^2 + 2}{x^3}$

**21** $\dfrac{3t}{t+2}+\dfrac{5t}{t-2}-\dfrac{40}{t^2-4} = \dfrac{3t}{t+2}+\dfrac{5t}{t-2}-\dfrac{40}{(t+2)(t-2)}$

$$= \frac{3t(t-2)}{(t+2)(t-2)}+\frac{5t(t+2)}{(t+2)(t-2)}-\frac{40}{(t+2)(t-2)}$$

$$= \frac{3t^2-6t+5t^2+10t-40}{(t+2)(t-2)}$$

$$= \frac{8t^2+4t-40}{(t+2)(t-2)}$$

$$= \frac{4(2t+5)(t-2)}{(t+2)(t-2)} = \frac{4(2t+5)}{t+2}$$

**23** $\dfrac{4x}{3x-4}+\dfrac{8}{3x^2-4x}+\dfrac{2}{x} = \dfrac{4x(x)+8+2(3x-4)}{x(3x-4)} = \dfrac{4x^2+6x}{x(3x-4)} = \dfrac{2x(2x+3)}{x(3x-4)} = \dfrac{2(2x+3)}{3x-4}$

**25** $\dfrac{2x}{x+2}-\dfrac{8}{x^2+2x}+\dfrac{3}{x} = \dfrac{2x(x)-8+3(x+2)}{x(x+2)} = \dfrac{2x^2+3x-2}{x(x+2)} = \dfrac{(2x-1)(x+2)}{x(x+2)} = \dfrac{2x-1}{x}$

**27** $\dfrac{p^4+3p^3-8p-24}{p^3-2p^2-9p+18} = \dfrac{p^3(p+3)-8(p+3)}{p^2(p-2)-9(p-2)}$ {factor by grouping}

$= \dfrac{(p^3-8)(p+3)}{(p^2-9)(p-2)}$ {factor out $p+3$; factor out $p-2$}

$= \dfrac{(p-2)(p^2+2p+4)(p+3)}{(p+3)(p-3)(p-2)}$ {difference of two cubes; difference of two squares}

$= \dfrac{p^2+2p+4}{p-3}$ {cancel $p+3$ and $p-2$}

**29** $3+\dfrac{5}{u}+\dfrac{2u}{3u+1} = \dfrac{3u(3u+1)+5(3u+1)+2u(u)}{u(3u+1)}$ {common denominator}

$= \dfrac{9u^2+3u+15u+5+2u^2}{u(3u+1)}$ {multiply terms}

$= \dfrac{11u^2+18u+5}{u(3u+1)}$ {add like terms}

**31** $\dfrac{2x+1}{x^2+4x+4}-\dfrac{6x}{x^2-4}+\dfrac{3}{x-2} = \dfrac{(2x+1)(x-2)-6x(x+2)+3(x^2+4x+4)}{(x+2)^2(x-2)} =$

$$\frac{-x^2-3x+10}{(x+2)^2(x-2)} = -\frac{x^2+3x-10}{(x+2)^2(x-2)} = -\frac{(x+5)(x-2)}{(x+2)^2(x-2)} = -\frac{x+5}{(x+2)^2}$$

**33** The lcd of the entire expression is $ab$.

Thus, we will multiply both the numerator and denominator by $ab$.

$$\frac{\frac{b}{a}-\frac{a}{b}}{\frac{1}{a}-\frac{1}{b}} = \frac{\left(\frac{b}{a}-\frac{a}{b}\right)\cdot ab}{\left(\frac{1}{a}-\frac{1}{b}\right)\cdot ab} = \frac{b^2-a^2}{b-a} = \frac{(b+a)(b-a)}{b-a} = a+b$$

**35** The lcd of the entire expression is $x^2y^2$.

Thus, we will multiply both the numerator and denominator by $x^2y^2$.

$$\frac{\dfrac{x}{y^2}-\dfrac{y}{x^2}}{\dfrac{1}{y^2}-\dfrac{1}{x^2}}=\frac{\left(\dfrac{x}{y^2}-\dfrac{y}{x^2}\right)\cdot x^2y^2}{\left(\dfrac{1}{y^2}-\dfrac{1}{x^2}\right)\cdot x^2y^2}=\frac{x^3-y^3}{x^2-y^2}=\frac{(x-y)(x^2+xy+y^2)}{(x+y)(x-y)}=\frac{x^2+xy+y^2}{x+y}$$

**37** The lcd of the entire expression is $xy$.

Thus, we will multiply both the numerator and denominator by $xy$.

$$\frac{y^{-1}+x^{-1}}{(xy)^{-1}}=\frac{\frac{1}{y}+\frac{1}{x}}{\frac{1}{xy}}=\frac{\left(\frac{1}{y}+\frac{1}{x}\right)\cdot xy}{\left(\frac{1}{xy}\right)\cdot xy}=\frac{x+y}{1}=x+y$$

**39** $$\frac{\dfrac{5}{x+1}+\dfrac{2x}{x+3}}{\dfrac{x}{x+1}+\dfrac{7}{x+3}}=\frac{\dfrac{5(x+3)+2x(x+1)}{(x+1)(x+3)}}{\dfrac{x(x+3)+7(x+1)}{(x+1)(x+3)}}=\frac{5x+15+2x^2+2x}{x^2+3x+7x+7}=\frac{2x^2+7x+15}{x^2+10x+7}$$

**41** $$\frac{\dfrac{3}{x-1}-\dfrac{3}{a-1}}{x-a}=\frac{\dfrac{3(a-1)-3(x-1)}{(x-1)(a-1)}}{x-a}=\frac{3a-3x}{(x-1)(a-1)(x-a)}=\frac{3(a-x)}{(x-1)(a-1)(x-a)}=$$
$$-\frac{3}{(x-1)(a-1)}$$

**43** $$\frac{(x+h)^2-3(x+h)-(x^2-3x)}{h}=\frac{2xh+h^2-3h}{h}=\frac{h(2x+h-3)}{h}=2x+h-3$$

**45** $$\frac{\dfrac{1}{(x+h)^3}-\dfrac{1}{x^3}}{h}=\frac{\dfrac{x^3-(x+h)^3}{(x+h)^3x^3}}{h}=\frac{x^3-(x+h)^3}{hx^3(x+h)^3}=$$
$$\frac{[x-(x+h)][x^2+x(x+h)+(x+h)^2]}{hx^3(x+h)^3}=\frac{-h(3x^2+3xh+h^2)}{hx^3(x+h)^3}=-\frac{3x^2+3xh+h^2}{x^3(x+h)^3}$$

**47** $$\frac{\dfrac{4}{3x+3h-1}-\dfrac{4}{3x-1}}{h}=\frac{\dfrac{4(3x-1)-4(3x+3h-1)}{(3x+3h-1)(3x-1)}}{h}=\frac{12x-4-12x-12h+4}{h(3x+3h-1)(3x-1)}=$$
$$\frac{-12h}{h(3x+3h-1)(3x-1)}=\frac{-12}{(3x+3h-1)(3x-1)}$$

**49** The conjugate of $\sqrt{t}-5$ is $\sqrt{t}+5$. Multiply the numerator and the denominator by the conjugate of the denominator. This will eliminate the radical in the denominator.

$$\frac{\sqrt{t}+5}{\sqrt{t}-5}=\frac{\sqrt{t}+5}{\sqrt{t}-5}\cdot\frac{\sqrt{t}+5}{\sqrt{t}+5}=\frac{(\sqrt{t})^2+2\cdot 5\sqrt{t}+5^2}{(\sqrt{t})^2-5^2}=\frac{t+10\sqrt{t}+25}{t-25}$$

[51] $\dfrac{81x^2-16y^2}{3\sqrt{x}-2\sqrt{y}} = \dfrac{81x^2-16y^2}{3\sqrt{x}-2\sqrt{y}} \cdot \dfrac{3\sqrt{x}+2\sqrt{y}}{3\sqrt{x}+2\sqrt{y}}$

$= \dfrac{(9x+4y)(9x-4y)(3\sqrt{x}+2\sqrt{y})}{9x-4y} = (9x+4y)(3\sqrt{x}+2\sqrt{y})$

[53] We must recognize $\sqrt[3]{a}-\sqrt[3]{b}$ as the first factor of the product formula for the difference of two cubes, $x^3-y^3=(x-y)(x^2+xy+y^2)$. The second factor is then $\left(\sqrt[3]{a}\right)^2+\left(\sqrt[3]{a}\right)\left(\sqrt[3]{b}\right)+\left(\sqrt[3]{b}\right)^2=\sqrt[3]{a^2}+\sqrt[3]{ab}+\sqrt[3]{b^2}$.

$$\frac{1}{\sqrt[3]{a}-\sqrt[3]{b}} = \frac{1}{\sqrt[3]{a}-\sqrt[3]{b}} \cdot \frac{\sqrt[3]{a^2}+\sqrt[3]{ab}+\sqrt[3]{b^2}}{\sqrt[3]{a^2}+\sqrt[3]{ab}+\sqrt[3]{b^2}} = \frac{\sqrt[3]{a^2}+\sqrt[3]{ab}+\sqrt[3]{b^2}}{a-b}$$

[55] $\dfrac{\sqrt{a}-\sqrt{b}}{a^2-b^2} = \dfrac{\sqrt{a}-\sqrt{b}}{a^2-b^2} \cdot \dfrac{\sqrt{a}+\sqrt{b}}{\sqrt{a}+\sqrt{b}} = \dfrac{a-b}{(a+b)(a-b)(\sqrt{a}+\sqrt{b})} = \dfrac{1}{(a+b)(\sqrt{a}+\sqrt{b})}$

[57] $\dfrac{\sqrt{2(x+h)+1}-\sqrt{2x+1}}{h} = \dfrac{\sqrt{2(x+h)+1}-\sqrt{2x+1}}{h} \cdot \dfrac{\sqrt{2(x+h)+1}+\sqrt{2x+1}}{\sqrt{2(x+h)+1}+\sqrt{2x+1}}$

$= \dfrac{(2x+2h+1)-(2x+1)}{h(\sqrt{2(x+h)+1}+\sqrt{2x+1})} = \dfrac{2}{\sqrt{2(x+h)+1}+\sqrt{2x+1}}$

[59] $\dfrac{\sqrt{1-x-h}-\sqrt{1-x}}{h} = \dfrac{\sqrt{1-x-h}-\sqrt{1-x}}{h} \cdot \dfrac{\sqrt{1-x-h}+\sqrt{1-x}}{\sqrt{1-x-h}+\sqrt{1-x}}$

$= \dfrac{(1-x-h)-(1-x)}{h(\sqrt{1-x-h}+\sqrt{1-x})} = \dfrac{-1}{\sqrt{1-x-h}+\sqrt{1-x}}$

[61] $\dfrac{4x^2-x+5}{x^{2/3}} = \dfrac{4x^2}{x^{2/3}} - \dfrac{x}{x^{2/3}} + \dfrac{5}{x^{2/3}} = 4x^{4/3} - x^{1/3} + 5x^{-2/3}$

[63] $\dfrac{(x^2+2)^2}{x^5} = \dfrac{x^4+4x^2+4}{x^5} = \dfrac{x^4}{x^5} + \dfrac{4x^2}{x^5} + \dfrac{4}{x^5} = x^{-1}+4x^{-3}+4x^{-5}$

*Note:* Exercises 65–82 are worked using the factoring concept given as the third method of simplification in Example 9.

[65] The *smallest* exponent that appears on the factor $x$ is $-3$.

$$x^{-3}+x^2 \text{ \{factor out } x^{-3}\} = x^{-3}(1+x^{2-(-3)}) = x^{-3}(1+x^5) = \frac{1+x^5}{x^3}$$

[67] $x^{-1/2}-x^{3/2} = x^{-1/2}(1-x^2) = \dfrac{1-x^2}{x^{1/2}}$

[69] $(2x^2-3x+1)(4)(3x+2)^3(3)+(3x+2)^4(4x-3) =$

$(3x+2)^3[12(2x^2-3x+1)+(3x+2)(4x-3)] = (3x+2)^3(36x^2-37x+6)$

[71] The smallest exponent that appears on the factor $(x^2-4)$ is $-\frac{1}{2}$ and the smallest exponent that appears on the factor $(2x+1)$ is 2. Thus, we will factor out $(x^2-4)^{-1/2}(2x+1)^2$.

(continued)

$$(x^2-4)^{1/2}(3)(2x+1)^2(2)+(2x+1)^3(\tfrac{1}{2})(x^2-4)^{-1/2}(2x)$$
$$=(x^2-4)^{-1/2}(2x+1)^2[6(x^2-4)+x(2x+1)]$$

If you are unsure of this factoring, it is easy to visually check at this stage by merely multiplying the expression—that is, we mentally add the exponents on the factor $(x^2-4)$, $-\frac{1}{2}$ and 1, and we get $\frac{1}{2}$, which is the exponent we started with.

Proceeding, we may simplify as follows:

$$=\frac{(2x+1)^2(8x^2+x-24)}{(x^2-4)^{1/2}}$$

**73** $(3x+1)^6(\frac{1}{2})(2x-5)^{-1/2}(2)+(2x-5)^{1/2}(6)(3x+1)^5(3)$

$=(3x+1)^5(2x-5)^{-1/2}[(3x+1)+18(2x-5)]$ {factor out $(3x+1)^5(2x-5)^{-1/2}$}

$$=\frac{(3x+1)^5(3x+1+36x-90)}{(2x-5)^{1/2}}=\frac{(3x+1)^5(39x-89)}{(2x-5)^{1/2}}$$

**75** $\dfrac{(6x+1)^3(27x^2+2)-(9x^3+2x)(3)(6x+1)^2(6)}{(6x+1)^6}=$

$$\frac{(6x+1)^2[(6x+1)(27x^2+2)-18(9x^3+2x)]}{(6x+1)^6}=\frac{27x^2-24x+2}{(6x+1)^4}$$

**77** $\dfrac{(x^2+2)^3(2x)-x^2(3)(x^2+2)^2(2x)}{[(x^2+2)^3]^2}=\dfrac{(x^2+2)^2(2x)[(x^2+2)^1-x^2(3)]}{(x^2+2)^6}=$

$$\frac{2x(x^2+2-3x^2)}{(x^2+2)^4}=\frac{2x(2-2x^2)}{(x^2+2)^4}=\frac{4x(1-x^2)}{(x^2+2)^4}$$

**79** $\dfrac{(x^2+4)^{1/3}(3)-(3x)(\frac{1}{3})(x^2+4)^{-2/3}(2x)}{[(x^2+4)^{1/3}]^2}=\dfrac{(x^2+4)^{-2/3}[3(x^2+4)-2x^2]}{(x^2+4)^{2/3}}=\dfrac{x^2+12}{(x^2+4)^{4/3}}$

**81** $\dfrac{(4x^2+9)^{1/2}(2)-(2x+3)(\frac{1}{2})(4x^2+9)^{-1/2}(8x)}{[(4x^2+9)^{1/2}]^2}=$

$$\frac{(4x^2+9)^{-1/2}[2(4x^2+9)-4x(2x+3)]}{(4x^2+9)}=\frac{18-12x}{(4x^2+9)^{3/2}}=\frac{6(3-2x)}{(4x^2+9)^{3/2}}$$

**83** Table $Y_1=\dfrac{113x^3+280x^2-150x}{22x^3+77x^2-100x-350}$ and $Y_2=\dfrac{3x}{2x+7}+\dfrac{4x^2}{1.1x^2-5}$.

| $x$ | $Y_1$ | $Y_2$ |
|---|---|---|
| 1 | −0.6923 | −0.6923 |
| 2 | −26.12 | −26.12 |
| 3 | 8.0392 | 8.0392 |
| 4 | 5.8794 | 5.8794 |
| 5 | 5.3268 | 5.3268 |

The values for $Y_1$ and $Y_2$ agree. Therefore, the two expressions might be equal.

## Chapter 1 Review Exercises

1 (a) $(\frac{2}{3})(-\frac{5}{8}) = -\frac{1}{3}\cdot\frac{5}{4} = -\frac{5}{12}$ (b) $\frac{3}{4}+\frac{6}{5} = \frac{15}{20}+\frac{24}{20} = \frac{39}{20}$

(c) $\frac{5}{8}-\frac{6}{7} = \frac{35}{56}-\frac{48}{56} = -\frac{13}{56}$ (d) $\frac{3}{4}\div\frac{6}{5} = \frac{3}{4}\cdot\frac{5}{6} = \frac{1}{4}\cdot\frac{5}{2} = \frac{5}{8}$

3 (a) $x$ is negative $\Leftrightarrow$ $x < 0$ (b) $a$ is between $\frac{1}{2}$ and $\frac{1}{3}$ $\Leftrightarrow$ $\frac{1}{3} < a < \frac{1}{2}$

(c) The absolute value of $x$ is not greater than 4 $\Leftrightarrow$ $|\,x\,| \le 4$

4 (c) $|\,3^{-1} - 2^{-1}\,| = \left|\frac{1}{3}-\frac{1}{2}\right| = \left|\frac{2}{6}-\frac{3}{6}\right| = \left|-\frac{1}{6}\right| = -(-\frac{1}{6}) = \frac{1}{6}$

5 (a) $d(A, C) = |\,-3-(-8)\,| = |\,5\,| = 5$

(b) $d(C, A) = d(A, C) = 5$

(c) $d(B, C) = |\,-3-4\,| = |\,-7\,| = -(-7) = 7$

6 (a) $(x+y)^2 = x^2 + 2xy + y^2 \boxed{\neq} x^2 + y^2$ for every nonzero $x$ and nonzero $y$.

(b) $\frac{1}{\sqrt{x+y}} = \frac{1}{\sqrt{x}} + \frac{1}{\sqrt{y}}$ is not true if $x = y = 1$ {we need only find one set of values of the variables for which the expression is false}.

(c) $\frac{1}{\sqrt{c}-\sqrt{d}} = \frac{1}{\sqrt{c}-\sqrt{d}}\cdot\frac{\sqrt{c}+\sqrt{d}}{\sqrt{c}+\sqrt{d}} \boxed{=} \frac{\sqrt{c}+\sqrt{d}}{c-d}$

7 (a) $93{,}700{,}000{,}000 = 9.37 \times 10^{10}$ (b) $0.000\,004\,02 = 4.02 \times 10^{-6}$

9 If $x \le -3$, then $x + 3 \le 0$, and $|\,x+3\,| = -(x+3) = -x-3$.

10 If $2 < x < 3$, then $x - 2 > 0$ {$x-2$ is positive} and $x - 3 < 0$ {$x-3$ is negative}. Thus, $(x-2)(x-3) < 0$ {positive times negative is negative}, and since the absolute value of an expression that is negative is the negative of the expression,

$|\,(x-2)(x-3)\,| = -(x-2)(x-3)$, or, equivalently, $(2-x)(x-3)$.

11 $-3^2 + 2^0 + 27^{-2/3} = -9 + 1 + \frac{1}{(\sqrt[3]{27})^2} = -8 + \frac{1}{3^2} = -\frac{72}{9} + \frac{1}{9} = \frac{-71}{9}$

13 $(3a^2b)^2(2ab^3) = (9a^4b^2)(2ab^3) = 18a^5b^5$

15 $\frac{(3x^2y^{-3})^{-2}}{x^{-5}y} = \frac{3^{-2}x^{-4}y^6}{x^{-5}y} = \frac{x^5y^5}{3^2x^4} = \frac{xy^5}{9}$

17 $(-2p^2q)^3\left(\frac{p}{4q^2}\right)^2 = (-8p^6q^3)\left(\frac{p^2}{16q^4}\right) = -\frac{p^8}{2q}$

18 $c^{-4/3}c^{3/2}c^{1/6} = c^{-8/6}c^{9/6}c^{1/6} = c^{(-8+9+1)/6} = c^{2/6} = c^{1/3}$

19 $\left(\frac{xy^{-1}}{\sqrt{z}}\right)^4 \div \left(\frac{x^{1/3}y^2}{z}\right)^3 = \frac{x^4y^{-4}}{z^2}\cdot\frac{z^3}{xy^6} = \frac{x^3z}{y^{10}}$

21 $\left[(a^{2/3}b^{-2})^3\right]^{-1} = (a^2b^{-6})^{-1} = a^{-2}b^6 = \frac{b^6}{a^2}$

[23] $\dfrac{r^{-1}+s^{-1}}{(rs)^{-1}}=\left(\frac{1}{r}+\frac{1}{s}\right)\div\frac{1}{rs}=\left(\frac{1}{r}+\frac{1}{s}\right)\cdot rs=s+r$

[24] Do not expand $(u+v)^3$ since it can be combined with $(u+v)^{-2}$.

$$(u+v)^3(u+v)^{-2}=(u+v)^1=u+v$$

[25] $s^{5/2}s^{-4/3}s^{-1/6}=s^{(15-8-1)/6}=s^{6/6}=s^1=s$

[27] $\sqrt[3]{(x^4y^{-1})^6}=(x^4y^{-1})^{6/3}=(x^4y^{-1})^2=x^8y^{-2}=\dfrac{x^8}{y^2}$

[29] Since $\sqrt[3]{4}=\sqrt[3]{2^2}$, we need to multiply the numerator and the denominator by $\sqrt[3]{2}$ to obtain a cube in the radicand of the denominator. $\dfrac{1}{\sqrt[3]{4}}=\dfrac{1}{\sqrt[3]{4}}\cdot\dfrac{\sqrt[3]{2}}{\sqrt[3]{2}}=\dfrac{\sqrt[3]{2}}{\sqrt[3]{8}}=\frac{1}{2}\sqrt[3]{2}$

[31] $\sqrt[3]{4x^2y}\,\sqrt[3]{2x^5y^2}=\sqrt[3]{8x^7y^3}=\sqrt[3]{8x^6y^3}\,\sqrt[3]{x}=2x^2y\,\sqrt[3]{x}$

[32] $\sqrt[4]{(-4a^3b^2c)^2}=\sqrt[4]{16a^6b^4c^2}=\sqrt[4]{2^4a^4b^4}\,\sqrt[4]{a^2c^2}=2ab\,\sqrt[4]{(ac)^2}=2ab\,\sqrt{ac}$

[33] $\dfrac{1}{\sqrt{t}}\left(\dfrac{1}{\sqrt{t}}-1\right)=\dfrac{1}{\sqrt{t}}\left(\dfrac{1}{\sqrt{t}}-\dfrac{\sqrt{t}}{\sqrt{t}}\right)=\dfrac{1}{\sqrt{t}}\left(\dfrac{1-\sqrt{t}}{\sqrt{t}}\right)=\dfrac{1-\sqrt{t}}{t}$

[35] $\dfrac{\sqrt{12x^4y}}{\sqrt{3x^2y^5}}=\sqrt{\dfrac{12x^4y}{3x^2y^5}}=\sqrt{\dfrac{4x^2}{y^4}}=\dfrac{2x}{y^2}$

[37] $\sqrt[3]{\dfrac{1}{2\pi^2}}=\dfrac{1}{\sqrt[3]{2\pi^2}}\cdot\dfrac{\sqrt[3]{4\pi}}{\sqrt[3]{4\pi}}=\dfrac{\sqrt[3]{4\pi}}{\sqrt[3]{8\pi^3}}=\dfrac{\sqrt[3]{4\pi}}{2\pi}$, or $\dfrac{1}{2\pi}\sqrt[3]{4\pi}$

[39] $\dfrac{1-\sqrt{x}}{1+\sqrt{x}}=\dfrac{1-\sqrt{x}}{1+\sqrt{x}}\cdot\dfrac{1-\sqrt{x}}{1-\sqrt{x}}=\dfrac{1-2\sqrt{x}+x}{1-x}$

[40] $\dfrac{1}{\sqrt{a}+\sqrt{a-2}}=\dfrac{1}{\sqrt{a}+\sqrt{a-2}}\cdot\dfrac{\sqrt{a}-\sqrt{a-2}}{\sqrt{a}-\sqrt{a-2}}=\dfrac{\sqrt{a}-\sqrt{a-2}}{a-(a-2)}=\dfrac{\sqrt{a}-\sqrt{a-2}}{2}$

[41] $\dfrac{81x^2-y^2}{3\sqrt{x}+\sqrt{y}}=\dfrac{81x^2-y^2}{3\sqrt{x}+\sqrt{y}}\cdot\dfrac{3\sqrt{x}-\sqrt{y}}{3\sqrt{x}-\sqrt{y}}=\dfrac{(9x+y)(9x-y)(3\sqrt{x}-\sqrt{y})}{9x-y}=$

$(9x+y)(3\sqrt{x}-\sqrt{y})$

[43] $(3x^3-4x^2+x-7)+(x^4-2x^3+3x^2+5)=x^4+x^3-x^2+x-2$

[45] $(x+4)(x+3)-(2x-1)(x-5)=(x^2+7x+12)-(2x^2-11x+5)=-x^2+18x+7$

[47]
$$\begin{aligned}(3y^3-2y^2+y+4)(y^2-3)&=(3y^3-2y^2+y+4)y^2+(3y^3-2y^2+y+4)(-3)\\&=(3y^5-2y^4+y^3+4y^2)+(-9y^3+6y^2-3y-12)\\&=3y^5-2y^4-8y^3+10y^2-3y-12\end{aligned}$$

[49] $(a-b)(a^3+a^2b+ab^2+b^3)=(a^4+a^3b+a^2b^2+ab^3)-(a^3b+a^2b^2+ab^3+b^4)=$

$a^4-b^4$

[50] $\dfrac{9p^4q^3-6p^2q^4+5p^3q^2}{3p^2q^2}=\dfrac{9p^4q^3}{3p^2q^2}-\dfrac{6p^2q^4}{3p^2q^2}+\dfrac{5p^3q^2}{3p^2q^2}=3p^2q-2q^2+\frac{5}{3}p$

[51] $(3a-5b)(2a+7b)=6a^2+11ab-35b^2$

[53] $(13a^2 + 4b)(13a^2 - 4b) = (13a^2)^2 - (4b^2)^2 = 169a^4 - 16b^2$

[54] $$(a^3 - a^2)^2 = (a^3)^2 - 2(a^3)(a^2) + (a^2)^2 = a^6 - 2a^5 + a^4$$

Alternatively, we could factor out $a^2$ first.

$$(a^3 - a^2)^2 = \left[a^2(a-1)\right]^2 = (a^2)^2(a-1)^2 = a^4(a^2 - 2a + 1) = a^6 - 2a^5 + a^4$$

[55] $(2a + b)^3 = (2a)^3 + 3(2a)^2(b) + 3(2a)(b)^2 + (b)^3 = 8a^3 + 12a^2b + 6ab^2 + b^3$

[57] $(3x + 2y)^2(3x - 2y)^2 = \left[(3x + 2y)(3x - 2y)\right]^2 = (9x^2 - 4y^2)^2 = 81x^4 - 72x^2y^2 + 16y^4$

[59] $60xw + 70w = 10w(6x + 7)$

[61] $28x^2 + 4x - 9 = (14x + 9)(2x - 1)$

[63] $2wy + 3yx - 8wz - 12zx = y(2w + 3x) - 4z(2w + 3x) = (y - 4z)(2w + 3x)$

[64] $2c^3 - 12c^2 + 3c - 18 = 2c^2(c - 6) + 3(c - 6) = (2c^2 + 3)(c - 6)$

[65] $8x^3 + 64y^3 = 8(x^3 + 8y^3) = 8\left[(x)^3 + (2y)^3\right] = 8(x + 2y)(x^2 - 2xy + 4y^2)$

[67] $$\begin{aligned} p^8 - q^8 &= (p^4)^2 - (q^4)^2 \\ &= (p^4 + q^4)(p^4 - q^4) \\ &= (p^4 + q^4)(p^2 + q^2)(p^2 - q^2) \\ &= (p^4 + q^4)(p^2 + q^2)(p + q)(p - q) \end{aligned}$$

[69] $w^6 + 1 = (w^2)^3 + (1)^3 = (w^2 + 1)(w^4 - w^2 + 1)$, which cannot be factored any further.

[71] $x^2 + 36$ is irreducible

[72] $$\begin{aligned} x^2 - 49y^2 - 14x + 49 &= (x^2 - 14x + 49) - 49y^2 \\ &= (x - 7)^2 - (7y)^2 \\ &= (x - 7 + 7y)(x - 7 - 7y) \end{aligned}$$

[73] $$\begin{aligned} x^5 - 4x^3 + 8x^2 - 32 &= x^3(x^2 - 4) + 8(x^2 - 4) \\ &= (x^3 + 8)(x^2 - 4) \\ &= \left[(x + 2)(x^2 - 2x + 4)\right]\left[(x + 2)(x - 2)\right] \\ &= (x - 2)(x + 2)^2(x^2 - 2x + 4) \end{aligned}$$

[75] $\dfrac{6x^2 - 7x - 5}{4x^2 + 4x + 1} = \dfrac{(3x - 5)(2x + 1)}{(2x + 1)(2x + 1)} = \dfrac{3x - 5}{2x + 1}$

[76] $\dfrac{r^3 - t^3}{r^2 - t^2} = \dfrac{(r - t)(r^2 + rt + t^2)}{(r + t)(r - t)} = \dfrac{r^2 + rt + t^2}{r + t}$

[77] $\dfrac{6x^2 - 5x - 6}{x^2 - 4} \div \dfrac{2x^2 - 3x}{x + 2} = \dfrac{(3x + 2)(2x - 3)}{(x + 2)(x - 2)} \cdot \dfrac{x + 2}{x(2x - 3)} = \dfrac{3x + 2}{x(x - 2)}$

[79] $\frac{7}{x+2}+\frac{3x}{(x+2)^2}-\frac{5}{x}=\frac{7(x)(x+2)+3x(x)-5(x+2)^2}{x(x+2)^2}$

$=\frac{7x^2+14x+3x^2-5x^2-20x-20}{x(x+2)^2}=\frac{5x^2-6x-20}{x(x+2)^2}$

[80] $\frac{x+x^{-2}}{1+x^{-2}}=\frac{x+\frac{1}{x^2}}{1+\frac{1}{x^2}}=\frac{\left(x+\frac{1}{x^2}\right)\cdot x^2}{\left(1+\frac{1}{x^2}\right)\cdot x^2}=\frac{x^3+1}{x^2+1}$. We could factor the numerator,

but since it doesn't lead to a reduction of the fraction, we leave it in this form.

[81] $\frac{1}{x}-\frac{2}{x^2+x}-\frac{3}{x+3}=\frac{1(x+1)(x+3)-2(x+3)-3x(x+1)}{x(x+1)(x+3)}$

$=\frac{x^2+4x+3-2x-6-3x^2-3x}{x(x+1)(x+3)}=\frac{-2x^2-x-3}{x(x+1)(x+3)}$

[83] $\frac{x+2-\frac{3}{x+4}}{\frac{x}{x+4}+\frac{1}{x+4}}=\frac{\frac{(x+2)(x+4)-3}{x+4}}{\frac{x+1}{x+4}}=\frac{x^2+6x+5}{x+1}=\frac{(x+1)(x+5)}{x+1}=x+5$

[85] $(x^2+1)^{3/2}(4)(x+5)^3+(x+5)^4(\frac{3}{2})(x^2+1)^{1/2}(2x)$

$=(x^2+1)^{1/2}(x+5)^3[4(x^2+1)+3x(x+5)]$

$=(x^2+1)^{1/2}(x+5)^3(7x^2+15x+4)$

[86] $\frac{(4-x^2)(\frac{1}{3})(6x+1)^{-2/3}(6)-(6x+1)^{1/3}(-2x)}{(4-x^2)^2}=\frac{2(6x+1)^{-2/3}[(4-x^2)+x(6x+1)]}{(4-x^2)^2}$

$=\frac{2[4-x^2+6x^2+x]}{(6x+1)^{2/3}(4-x^2)^2}$

$=\frac{2(5x^2+x+4)}{(6x+1)^{2/3}(4-x^2)^2}$

[87] (5.5 liters)$\left(10^6\ \frac{\text{mm}^3}{\text{liter}}\right)\left(5\times 10^6\ \frac{\text{cells}}{\text{mm}^3}\right)=2.75\times 10^{13}$ red blood cells {27.5 *trillion*}

[88] $\frac{\text{70 (or 90) beats}}{\text{minute}}\cdot\frac{\text{60 minutes}}{\text{1 hour}}\cdot\frac{\text{24 hours}}{\text{1 day}}\cdot\frac{\text{365 days}}{\text{1 year}}\cdot 80\text{ years}=$

$2.94336\times 10^9$ (or $3.78432\times 10^9$) beats

[89] $h=86$ cm and $w=13$ kg $\Rightarrow$

$S=(0.007184)w^{0.425}h^{0.725}=(0.007184)(13)^{0.425}(86)^{0.725}\approx 0.54\text{ m}^2.$

[90] $p=40$ dyne/cm$^2$ and $v=60$ cm$^3$ $\Rightarrow$ $c=pv^{-1.4}=40(60)^{-1.4}\approx 0.13$ dyne-cm.

## Chapter 1 Discussion Exercises

**1** 1 gallon $\approx 0.13368$ ft$^3$ is a conversion factor that would help. The volume of the tank is 10,000 gallons $\approx 1336.8$ ft$^3$. Use $V = \frac{4}{3}\pi r^3$ to determine the radius. $1336.8 = \frac{4}{3}\pi r^3 \Rightarrow r^3 = \frac{1002.6}{\pi} \Rightarrow r \approx 6.83375$ ft. Then use $S = 4\pi r^2$ to find the surface area. $S = 4\pi(6.83375)^2 \approx 586.85$ ft$^2$.

**3** We first need to determine the term that needs to be added and subtracted. Since $25 = \underline{5}^2$, it makes sense to add and subtract $2 \cdot \underline{5}x = 10x$. Then we will obtain the square of a binomial—i.e., $(x^2 + 10x + 25) - 10x = (x+5)^2 - 10x$. We can now factor this expression as the difference of two squares,

$$(x+5)^2 - 10x = (x+5)^2 - (\sqrt{10x})^2 = (x + 5 + \sqrt{10x})(x + 5 - \sqrt{10x}).$$

**5** Try $\dfrac{3x^2 - 4x + 7}{8x^2 + 9x - 100}$ with $x = 10^3$, $10^4$, and $10^5$. You get approximately 0.374, 0.3749, and 0.37499. The numbers seem to be getting closer to 0.375, which is the decimal representation for $\frac{3}{8}$, which is the ratio of the coefficients of the $x^2$ terms. In general, the quotients of this form get close to the ratio of leading coefficients as $x$ gets larger.

**7** Follow the algebraic simplification given.

| | |
|---|---|
| 1 Write down his/her age | Denote the age with $x$. |
| 2 Multiply it by 2 | $2x$ |
| 3 Add 5 | $2x + 5$ |
| 4 Multiply this sum by 50 | $50(2x+5) = 100x + 250$ |
| 5 Subtract 365 | $(100x + 250) - 365 = 100x - 115$ |
| 6 Add his/her height (in inches) | $100x - 115 + y$, where $y$ is the height |
| 7 Add 115 | $100x - 115 + y + 115 = 100x + y$ |

As a specific example, suppose the age is 21 and the height is 68. The number obtained by following the steps is $100x + y = 2168$ and we can see that the first two digits of the result equal the age and the last two digits equal the height.

# Chapter 2: Equations and Inequalities

## *2.1 Exercises*

1 $-3x+4=-1 \;\Rightarrow\; -3x=-5 \;\Rightarrow\; x=\frac{5}{3}$

3 $4x-3=-5x+6 \;\Rightarrow\; 4x+5x=6+3 \;\Rightarrow\; 9x=9 \;\Rightarrow\; x=1$

5
| | |
|---|---|
| $4(2y+5)=3(5y-2)$ | given |
| $8y+20=15y-6$ | multiply terms |
| $26=7y$ | get constants on one side, variables on the other |
| $y=\frac{26}{7}$ | solve for $y$ |

7
| | |
|---|---|
| $\frac{1}{5}x+2=3-\frac{2}{7}x$ | given |
| $(\frac{1}{5}x+2)\cdot 35=(3-\frac{2}{7}x)\cdot 35$ | multiply by the lcd, 35 |
| $7x+70=105-10x$ | simplify |
| $17x=35$ | simplify |
| $x=\frac{35}{17}$ | solve for $x$ |

9 Decimal values can be thought of as equivalent rationals, i.e., $0.3=\frac{3}{10}$. We then multiply all terms by the lcd, 10 in this case, just as we multiplied by 35 in Exercise 7. A common mistake is to multiply 10 times a term such as $0.3(x+1)$ and get $3(10x+10)$. However, this is just like multiplying 10 times $ab$, which is $10ab$. Thus, $10[0.3(x+1)]=10(0.3)(x+1)=3(x+1)$. With that in mind, we proceed with the problem: $[0.3(3+2x)+1.2x=3.2]\cdot 10 \;\Rightarrow\; 3(3+2x)+12x=32 \;\Rightarrow$

$$9+6x+12x=32 \;\Rightarrow\; 18x=23 \;\Rightarrow\; x=\frac{23}{18}$$

11 $\left[\frac{3+5x}{5}=\frac{4-x}{7}\right]\cdot 35 \;\Rightarrow\; 21+35x=20-5x \;\Rightarrow\; 40x=-1 \;\Rightarrow\; x=-\frac{1}{40}$

*Note:* You may have solved equations such as those in Exercise 11 using a process called *cross-multiplication* in the past. This method is sufficient for problems of the form $\frac{P}{Q}=\frac{R}{S}$, but the guidelines for solving an equation containing rational expressions (page 64 in the text) apply to rational equations of a more complex form.

13 The lcd is $4(4x+1)$ and we need to remember that $x$ cannot equal $-\frac{1}{4}$.

$\left[\frac{13+2x}{4x+1}=\frac{3}{4}\right]\cdot 4(4x+1) \;\Rightarrow\; 4(13+2x)=3(4x+1) \;\Rightarrow\; 52+8x=12x+3 \;\Rightarrow$

$49=4x \;\Rightarrow\; x=\frac{49}{4}$. Since $x\neq -\frac{1}{4}$, $x=\frac{49}{4}$ is a solution.

15 The lcd is $x$ and we need to remember that $x$ cannot equal 0.

$\left[8-\frac{5}{x}=2+\frac{3}{x}\right]\cdot x \;\Rightarrow\; 8x-5=2x+3 \;\Rightarrow\; 6x=8 \;\Rightarrow\; x=\frac{4}{3}$

[17] $(3x-2)^2 = (x-5)(9x+4) \Rightarrow 9x^2 - 12x + 4 = 9x^2 - 41x - 20 \Rightarrow$

$29x = -24 \Rightarrow x = -\frac{24}{29}$

[19] $(5x-7)(2x+1) - 10x(x-4) = 0 \Rightarrow 10x^2 - 9x - 7 - 10x^2 + 40x = 0 \Rightarrow$

$31x = 7 \Rightarrow x = \frac{7}{31}$

[21] $\left[\frac{3x+1}{6x-2} = \frac{2x+5}{4x-13}\right] \cdot (6x-2)(4x-13) \Rightarrow 12x^2 - 35x - 13 = 12x^2 + 26x - 10 \Rightarrow$

$-3 = 61x \Rightarrow x = -\frac{3}{61}$ {note that $x \neq \frac{1}{3}, \frac{13}{4}$}

[23] $\left[\frac{2}{5} + \frac{4}{10x+5} = \frac{7}{2x+1}\right] \cdot 5(2x+1) \Rightarrow (4x+2) + 4 = 35 \Rightarrow 4x = 29 \Rightarrow x = \frac{29}{4}$

[25] Since $2x - 4 = 2(x-2)$ and $3x - 6 = 3(x-2)$, the lcd is $2 \cdot 3 \cdot 5(x-2) = 30(x-2)$.

$\left[\frac{3}{2x-4} - \frac{5}{3x-6} = \frac{3}{5}\right] \cdot 30(x-2) \Rightarrow 3(15) - 5(10) = 3(6)(x-2) \Rightarrow$

$18x = 31 \Rightarrow x = \frac{31}{18}$

[27] $2 - \frac{5}{3x-7} = 2 \Rightarrow \frac{5}{3x-7} = 0 \Rightarrow$ *no solution* since the numerator is never 0.

[29] $\frac{1}{2x-1} = \frac{4}{8x-4} \Rightarrow \frac{1}{2x-1} = \frac{4}{4(2x-1)} \Rightarrow \frac{1}{2x-1} = \frac{1}{2x-1}$. This is an identity,

and the solutions consist of every number in the domains of the given expressions.

Thus, the solutions are all real numbers except $\frac{1}{2}$, which we denote by $\mathbb{R} - \{\frac{1}{2}\}$.

[31] $\left[\frac{7}{y^2-4} - \frac{4}{y+2} = \frac{5}{y-2}\right] \cdot (y+2)(y-2) \Rightarrow 7 - 4(y-2) = 5(y+2) \Rightarrow$

$5 = 9y \Rightarrow y = \frac{5}{9}$

[33] $(x+3)^3 - (3x-1)^2 = x^3 + 4 \Rightarrow (x^3 + 9x^2 + 27x + 27) - (9x^2 - 6x + 1) =$

$x^3 + 4 \Rightarrow 33x = -22 \Rightarrow x = -\frac{2}{3}$

[35] $\left[\frac{9x}{3x-1} = 2 + \frac{3}{3x-1}\right] \cdot (3x-1) \Rightarrow 9x = 2(3x-1) + 3 \Rightarrow 9x = 6x + 1 \Rightarrow$

$3x = 1 \Rightarrow x = \frac{1}{3}$, which is not in the domain of the given expressions. **No solution**

[37] $\left[\frac{1}{x+4} + \frac{3}{x-4} = \frac{3x+8}{x^2-16}\right] \cdot (x+4)(x-4) \Rightarrow x - 4 + 3(x+4) = 3x + 8 \Rightarrow$

$4x + 8 = 3x + 8 \Rightarrow x = 0$

[39] $\left[\frac{4}{x+2} + \frac{1}{x-2} = \frac{5x-6}{x^2-4}\right] \cdot (x+2)(x-2) \Rightarrow 4(x-2) + x + 2 = 5x - 6 \Rightarrow$

$0 = 0$, indicating an identity. The solution is $\mathbb{R} - \{\pm 2\}$.

[41] $\left[\frac{2}{2x+1} - \frac{3}{2x-1} = \frac{-2x+7}{4x^2-1}\right] \cdot (2x+1)(2x-1) \Rightarrow$

$2(2x-1) - 3(2x+1) = -2x + 7 \Rightarrow -2x - 5 = -2x + 7 \Rightarrow$

$-5 = 7$, a contradiction. **No solution**

[43] $\left[\dfrac{5}{2x+3}+\dfrac{4}{2x-3}=\dfrac{14x+3}{4x^2-9}\right]\cdot(2x+3)(2x-3) \Rightarrow$

$5(2x-3)+4(2x+3)=14x+3 \Rightarrow 18x-3=14x+3 \Rightarrow 4x=6 \Rightarrow x=\frac{3}{2}$,

which is not in the domain of the given expressions. **No solution**

*Note:* For Exercises 45–50, we must show that LS = RS.

[45] $\text{LS}=(4x-3)^2-16x^2=(16x^2-24x+9)-16x^2=9-24x=\text{RS}$

[47] $\text{LS}=\dfrac{x^2-9}{x+3}=\dfrac{(x+3)(x-3)}{x+3}=x-3=\text{RS}$, $\forall x$ except $x=-3$.

[49] $\text{LS}=\dfrac{3x^2+8}{x}=\dfrac{3x^2}{x}+\frac{8}{x}=\frac{8}{x}+3x=\text{RS}$

[51] Substituting $-2$ for $x$ in $4x+1+2c=5c-3x+6$ yields $-7+2c=5c+12 \Rightarrow$

$3c=-19 \Rightarrow c=-\frac{19}{3}$.

[53] (a) $\dfrac{7x}{x-5}=\dfrac{42}{x-5} \Rightarrow 7x=42 \Rightarrow x=6$,

and the two equations are equivalent since they have the same solution.

(b) No, 5 is not a solution of the first equation since it is undefined if $x=5$.

[55] Substituting $\frac{5}{3}$ for $x$ yields $\frac{5}{3}a+b=0$, or, equivalently, $b=-\frac{5}{3}a$.

Choose any $a$ and $b$ such that $b=-\frac{5}{3}a$. For example, let $a=3$ and $b=-5$.

[57] Going from the second line to the third line, we must remember that division by the variable expression $x-2$ is not allowed. ★ $x+1=x+2$

[59] $EK+L=D-TK \Rightarrow EK+TK=D-L \Rightarrow K(E+T)=D-L \Rightarrow$

$K=\dfrac{D-L}{E+T}$

[61] $M=\dfrac{Q+1}{Q} \Rightarrow MQ=Q+1 \Rightarrow MQ-Q=1 \Rightarrow Q(M-1)=1 \Rightarrow Q=\dfrac{1}{M-1}$

[63] $I=Prt \Rightarrow \dfrac{I}{rt}=\dfrac{Prt}{rt} \Rightarrow P=\dfrac{I}{rt}$

[65] $A=\frac{1}{2}bh \Rightarrow 2A=bh \Rightarrow h=\dfrac{2A}{b}$

[67] $F=g\dfrac{mM}{d^2} \Rightarrow Fd^2=gmM \Rightarrow m=\dfrac{Fd^2}{gM}$

[69] Since we are solving for $w$, we will get all terms containing $w$ on one side.

$$P=2l+2w \Rightarrow P-2l=2w \Rightarrow w=\frac{P-2l}{2}$$

[71] $A=\frac{1}{2}(b_1+b_2)h \Rightarrow \dfrac{2A}{h}=b_1+b_2 \Rightarrow b_1=\dfrac{2A}{h}-b_2$, or $b_1=\dfrac{2A-hb_2}{h}$

[73]

| | |
|---|---|
| $S = \dfrac{p}{q + p(1-q)}$ | given |
| $Sq + Sp(1-q) = p$ | eliminate the fraction |
| $Sq + Sp - Spq = p$ | multiply terms |
| $Sq - Spq = p - Sp$ | isolate terms containing $q$ |
| $Sq(1-p) = p(1-S)$ | factor |
| $q = \dfrac{p(1-S)}{S(1-p)}$ | solve for $q$ |

[75] $\frac{1}{f} = \frac{1}{p} + \frac{1}{q}$ {multiply by the lcd, $fpq$} $\Rightarrow$

$$pq = fq + fp \;\Rightarrow\; pq - fq = fp \;\Rightarrow\; q(p-f) = fp \;\Rightarrow\; q = \frac{fp}{p-f}$$

[77] The $y$-value decreases 1.2 units for each 1 unit increase in the $x$–value. The data is best described by equation (1), $y = -1.2x + 2$.

## *2.2 Exercises*

[1] Let $x$ denote the score on the next test.

$$\frac{75+82+71+84+x}{5} = 80 \;\Rightarrow\; 312 + x = 5(80) \;\Rightarrow\; x = 400 - 312 = 88$$

[3] Let $x$ denote the gross pay. Gross pay − deductions = Net (take-home) pay $\Rightarrow$

$$x - 0.40x = 492 \;\Rightarrow\; 0.60x = 492 \;\Rightarrow\; x = \frac{492}{0.60} \;\Rightarrow\; x = 820.$$

[5] (a) $\text{IQ} = \dfrac{\text{mental age (MA)}}{\text{chronological age (CA)}} \times 100 = \dfrac{15}{12} \times 100 = 125.$

(b) CA = 15 and IQ = 140 $\Rightarrow$

$$140 = \frac{\text{MA}}{15} \times 100 \;\Rightarrow\; \frac{140}{100} = \frac{\text{MA}}{15} \;\Rightarrow\; \text{MA} = \frac{15(140)}{100} = 21.$$

[7] Let $x$ denote the number of months needed to recover the cost of the insulation. The savings in one month is 10% of \$60 = \$6. $6x = 1080 \;\Rightarrow\; x = 180$ months (or 15 yr).

[9] Let $x$ denote the amount invested in the 8% account and $100{,}000 - x$ the amount invested in the 6.4% account. Use the formula $I = Prt$ (interest = principal × rate × time) with $t = 1$.

$\text{Interest}_{\text{first account}} + \text{Interest}_{\text{second account}} = \text{Interest}_{\text{total}} \;\Rightarrow$

$x(0.08) + (100{,}000 - x)(0.064) = 7500 \;\Rightarrow$

$0.08x + 6400 - 0.064x = 7500 \;\Rightarrow$

$0.016x = 1100 \;\Rightarrow\; x = 68{,}750$. Since only \$50,000 can be insured in the 8% account, we *cannot* fully insure the money and earn annual interest of \$7,500.

[11] Let $x$ denote the number of children and $600 - x$ the number of adults.

$\text{Receipts}_{\text{children}} + \text{Receipts}_{\text{adults}} = \text{Receipts}_{\text{total}} \Rightarrow$

$x(2) + (600 - x)(5) = 2400 \Rightarrow -3x = -600 \Rightarrow x = 200$ children.

[13] Let $x$ denote the number of ounces of glucose solution and $7 - x$ the number of ounces of water.

$\text{Glucose}_{30\%} + \text{Water}_{0\%} = \text{Glucose}_{20\%} \Rightarrow$

$x(0.30) + (7 - x)(0) = 7(0.20) \Rightarrow$

$0.3x = 1.4 \Rightarrow x = \frac{14}{3}$.

Use $\frac{14}{3}$ oz of the 30% glucose solution and $7 - \frac{14}{3} = \frac{7}{3}$ oz of water.

[15] Let $x$ denote the number of grams of British sterling silver and $200 - x$ the number of grams of pure copper. We will compare the amounts of pure copper.

{ The percentages are 7.5%, 100%, and 10% or, equivalently, 0.075, 1, and 0.10. }

$\text{Copper}_{\text{British sterling silver}} + \text{Copper}_{\text{pure}} = \text{Copper}_{\text{alloy}} \Rightarrow$

$(0.075)x + 1(200 - x) = (0.10)(200) \Rightarrow 180 = 0.925x \Rightarrow$

$x = 194.6$. Use 194.6 g of British sterling silver and 5.4 g of copper.

[17] (a) They will meet when the sum of their distances is 224. Let $t$ denote the desired number of seconds. Using distance = rate × time, we have

$1.5t + 2t = 224 \Rightarrow 3.5t = 224 \Rightarrow t = 64$ sec

(b) $64(1.5) = 96$ m and $64(2) = 128$ m, respectively

[19] Let $r$ denote the rate of the snowplow. At 8:30 A.M., the car has traveled 15 miles and the snowplow has been traveling for $2\frac{1}{2} = \frac{5}{2}$ hours.

Using the relationship time × rate = distance, $\frac{5}{2}r = 15 \Rightarrow r = 6$ mi/hr.

[21] (a) Let $r$ denote the rate of the river's current.

The rates of the boat upstream and downstream are $5 - r$ and $5 + r$, respectively.

$\text{Distance}_{\text{upstream}} = \text{Distance}_{\text{downstream}} \Rightarrow$

$(5 - r)\frac{15}{60} = (5 + r)\frac{12}{60} \Rightarrow (5 - r)\frac{1}{4} = (5 + r)\frac{1}{5} \Rightarrow$

$5(5 - r) = 4(5 + r) \Rightarrow 25 - 5r = 20 + 4r \Rightarrow 5 = 9r \Rightarrow r = \frac{5}{9}$ mi/hr.

(b) The distance upstream is $(5 - \frac{5}{9})\frac{1}{4} = \frac{10}{9}$. The total distance is $2 \cdot \frac{10}{9} = \frac{20}{9}$, or $2\frac{2}{9}$ mi.

[23] Let $x$ denote the distance to the target. We know the total time involved and need a formula for time. Solving $d = rt$ for $t$ gives us $t = d/r$.

$\text{Time}_{\text{to target}} + \text{Time}_{\text{from target}} = \text{Time}_{\text{total}} \Rightarrow$

$\frac{x}{3300} + \frac{x}{1100} = 1.5$ { multiply by the lcd, 3300 } $\Rightarrow$

$x + 3x = 1.5(3300) \Rightarrow 4x = 4950 \Rightarrow x = 1237.5$ ft.

[25] Let $l$ denote the length of the side parallel to the river bank. $P = 2w + l$

(a) $P = 2w + 2w = 4w$; $4w = 180 \Rightarrow w = 45$ ft and $A = (45)(90) = 4050$ ft$^2$.

(b) $P = 2w + \frac{1}{2}w = \frac{5}{2}w$; $\frac{5}{2}w = 180 \Rightarrow w = 72$ ft and $A = (72)(36) = 2592$ ft$^2$.

(c) $P = 2w + w = 3w$; $3w = 180 \Rightarrow w = 60$ ft and $A = (60)(60) = 3600$ ft$^2$.

[27] $\text{Area}_{\text{semicircle}} + \text{Area}_{\text{rectangle}} = \text{Area}_{\text{total}} \Rightarrow$

$\frac{1}{2}\pi r^2 + lw = 24 \Rightarrow \frac{1}{2}\pi(\frac{3}{2})^2 + (h - \frac{3}{2})3 = 24 \Rightarrow$

$(h - \frac{3}{2})3 = 24 - \frac{9\pi}{8} \Rightarrow h - \frac{3}{2} = 8 - \frac{3\pi}{8} \Rightarrow h = \frac{19}{2} - \frac{3\pi}{8} \approx 8.32$ ft.

[29] Let $h_1$ denote the height of the cylinder.

$V = \frac{2}{3}\pi r^3 + \pi r^2 h_1 = 11{,}250\pi$ and $r = 15 \Rightarrow$

$2250\pi + 225\pi h_1 = 11{,}250\pi \Rightarrow h_1 = 40$. The total height is 40 ft + 15 ft = 55 ft.

[31] Let $x$ denote the desired time. Using the rates (in minutes), $\frac{1}{90} + \frac{1}{60} = \frac{1}{x} \Rightarrow$

{multiply by the lcd, $180x$} $2x + 3x = 180 \Rightarrow 5x = 180 \Rightarrow x = 36$ min.

[33] Let $x$ denote the desired time. Using the rates (in minutes), $\frac{1}{45} + \frac{1}{x} = \frac{1}{20} \Rightarrow$

{multiply by the lcd, $180x$} $4x + 180 = 9x \Rightarrow 180 = 5x \Rightarrow x = 36$ min.

[35] First, a simple example of calculating a GPA. Suppose a student gets a 3-credit A (worth 4 honor points) and a 4-credit C (worth 2 honor points). Then,

$$\text{GPA} = \frac{\text{total weighted honor points}}{\text{total credit hours}} = \frac{3(4.0) + 4(2.0)}{3 + 4} = \frac{12 + 8}{7} = \frac{20}{7} \approx 2.86.$$

Let $x$ denote the number of additional credit hours.

$\text{GPA} = 3.2 \Rightarrow \dfrac{48(2.75) + x(4.0)}{48 + x} = 3.2 \Rightarrow 132 + 4x = 3.2(48 + x) \Rightarrow$

$132 + 4x = 153.6 + 3.2x \Rightarrow 21.6 = 0.8x \Rightarrow x = \frac{21.6}{0.8} = 27.$

[37] (a) $h = 5280$ and $T_0 = 70 \Rightarrow T = 70 - \left(\frac{5.5}{1000}\right)5280 = 40.96^\circ$F.

(b) $T = 32 \Rightarrow 32 = 70 - \left(\frac{5.5}{1000}\right)h \Rightarrow h = (70 - 32)\left(\frac{1000}{5.5}\right) \approx 6909$ ft.

[39] $B = 55$ and $h = 10{,}000 - 4000 = 6000 \Rightarrow T = 55 - \left(\frac{3}{1000}\right)(6000) = 37^\circ$F.

## 2.3 Exercises

[1] $6x^2 + x - 12 = 0 \Rightarrow (2x + 3)(3x - 4) = 0 \Rightarrow$

$2x + 3 = 0$ or $3x - 4 = 0 \Rightarrow x = -\frac{3}{2}, \frac{4}{3}$

[3] $15x^2 - 12 = -8x$ {get all terms on one side of the equals sign, zero on the other side}

$\Rightarrow 15x^2 + 8x - 12 = 0$ {factor} $\Rightarrow (5x + 6)(3x - 2) = 0 \Rightarrow x = -\frac{6}{5}, \frac{2}{3}$

5 A common mistake for this exercise is to write $2x = 27$ or $4x + 15 = 27$.

However, remember that you want to get 0 on one side of the equals sign.

$2x(4x+15) = 27 \Rightarrow 8x^2 + 30x - 27 = 0 \Rightarrow (2x+9)(4x-3) = 0 \Rightarrow x = -\frac{9}{2}, \frac{3}{4}$

7 Divide both sides by a nonzero constant whenever possible.

In this case, 5 divides evenly into both sides.

$75x^2 + 35x - 10 = 0$ {divide by 5} $\Rightarrow 15x^2 + 7x - 2 = 0$ {factor} $\Rightarrow$

$(3x+2)(5x-1) = 0$ {zero factor theorem} $\Rightarrow x = -\frac{2}{3}, \frac{1}{5}$

9 $12x^2 + 60x + 75 = 0$ {factor out the gcf, 3} $\Rightarrow$

$3(4x^2 + 20x + 25) = 0$ {divide by 3} $\Rightarrow$

$4x^2 + 20x + 25 = 0 \Rightarrow (2x+5)^2 = 0 \Rightarrow x = -\frac{5}{2}$

11 We will use the same process for solving rational equations as outlined in §2.1.

Here, the lcd is $x(x+3)$ and we need to remember that $x \neq 0, -3$.

$\left[\frac{2x}{x+3} + \frac{5}{x} - 4 = \frac{18}{x^2+3x}\right] \cdot x(x+3) \Rightarrow 2x(x) + 5(x+3) - 4(x^2+3x) = 18 \Rightarrow$

$0 = 2x^2 + 7x + 3 \Rightarrow (2x+1)(x+3) = 0 \Rightarrow$

$x = -\frac{1}{2}$ {$-3$ is not in the domain of the given expressions}

13 $\left[\frac{5x}{x-3} + \frac{4}{x+3} = \frac{90}{x^2-9}\right] \cdot (x+3)(x-3) \Rightarrow 5x(x+3) + 4(x-3) = 90 \Rightarrow$

$5x^2 + 19x - 102 = 0 \Rightarrow (5x+34)(x-3) = 0 \Rightarrow$

$x = -\frac{34}{5}$ {3 is not in the domain of the given expressions}

15 (a) The first equation, $x^2 = 16$, has solutions $x = \pm 4$. The equations are not equivalent since $-4$ is not a solution of the second equation, $x = 4$.

(b) First note that $x = \sqrt{9} = 3$.

Thus, the equations are equivalent since they have exactly the same solutions.

17 Using the special quadratic equation in this section, $x^2 = 169 \Rightarrow$

$x = \pm\sqrt{169} = \pm 13$. Note that this is *not* the same as saying $\sqrt{169} = \pm 13$.

19 $25x^2 = 9 \Rightarrow x^2 = \frac{9}{25} \Rightarrow x = \pm\sqrt{\frac{9}{25}} = \pm\frac{3}{5}$

21 $(x-3)^2 = 17 \Rightarrow x - 3 = \pm\sqrt{17} \Rightarrow x = 3 \pm \sqrt{17}$

23 $4(x+2)^2 = 11 \Rightarrow (x+2)^2 = \frac{11}{4} \Rightarrow x + 2 = \pm\sqrt{\frac{11}{4}} \Rightarrow x = -2 \pm \frac{1}{2}\sqrt{11}$

25 For this exercise, consider the general expression $x^2 + bx + c$.

(a) In general, $d = (\frac{1}{2}b)^2$. In this case, $d = \left[\frac{1}{2}(9)\right]^2 = \frac{81}{4}$.

(b) As in part (a), $d = (\frac{1}{2}b)^2 = \left[\frac{1}{2}(-8)\right]^2 = 16$. *Note:* It is appropriate to use 8 or $-8$.

(c) In general, $d = 2(\pm\sqrt{c})$ for $c > 0$.

In this case, $c = 36 \Rightarrow \sqrt{c} = 6$, and $d = 2(\pm 6) = \pm 12$.

(d) $c = \frac{49}{4} \Rightarrow \sqrt{c} = \frac{7}{2}$, and $d = 2(\pm\frac{7}{2}) = \pm 7$.

**27** $x^2+6x+7=0$ {add 9 to both sides to complete the square with $x^2+6x$} $\Rightarrow$

$$x^2+6x+\underline{\ 9\ }=-7+\underline{\ 9\ } \Rightarrow (x+3)^2=2 \Rightarrow x+3=\pm\sqrt{2} \Rightarrow x=-3\pm\sqrt{2}$$

**29**

| | |
|---|---|
| $4x^2-12x-11=0$ | given |
| $x^2-3x-\frac{11}{4}=0$ | divide by 4 |
| $x^2-3x=\frac{11}{4}$ | isolate $x^2$ and $x$ terms |
| $x^2-3x+\frac{9}{4}=\frac{11}{4}+\frac{9}{4}$ | add $(\frac{1}{2}\cdot 3)^2=\frac{9}{4}$ |
| $(x-\frac{3}{2})^2=5$ | factor and simplify |
| $x-\frac{3}{2}=\pm\sqrt{5}$ | take the square root |
| $x=\frac{3}{2}\pm\sqrt{5}$ | solve for $x$ |

**31** $6x^2-x=2 \Rightarrow 6x^2-x-2=0.$

Use the quadratic formula, $x=\dfrac{-b\pm\sqrt{b^2-4ac}}{2a}$, with $a=6$, $b=-1$, and $c=-2$.

$$x=\frac{-(-1)\pm\sqrt{(-1)^2-4(6)(-2)}}{2(6)}=\frac{1\pm\sqrt{1+48}}{12}=\frac{1\pm 7}{12}=-\frac{1}{2},\frac{2}{3}$$

**33** $x^2+4x+2=0 \Rightarrow x=\dfrac{-4\pm\sqrt{16-8}}{2(1)}=\dfrac{-4\pm 2\sqrt{2}}{2}=-2\pm\sqrt{2}$

**35** $2x^2-3x-4=0 \Rightarrow x=\dfrac{-(-3)\pm\sqrt{(-3)^2-4(2)(-4)}}{2(2)}=\dfrac{3\pm\sqrt{9+32}}{4}=\dfrac{3}{4}\pm\dfrac{1}{4}\sqrt{41}$

*Note:* A common mistake is to not divide 4 into *both* terms of the numerator.

**37** $\frac{3}{2}z^2-4z-1=0$ {multiply by 2} $\Rightarrow 3z^2-8z-2=0 \Rightarrow$

$$z=\frac{8\pm\sqrt{64+24}}{2\cdot 3}=\frac{8\pm\sqrt{88}}{2\cdot 3}=\frac{2(4\pm\sqrt{22})}{2\cdot 3}=\frac{4}{3}\pm\frac{1}{3}\sqrt{22}$$

**39** Multiply by the lcd, $w^2$. $\left[\dfrac{5}{w^2}-\dfrac{10}{w}+2=0\right]\cdot w^2 \Rightarrow 5-10w+2w^2=0 \Rightarrow$

$$w=\frac{10\pm\sqrt{100-40}}{4}=\frac{10\pm 2\sqrt{15}}{4}=\frac{5}{2}\pm\frac{1}{2}\sqrt{15}$$

**41** $4x^2+81=36x \Rightarrow 4x^2-36x+81=0 \Rightarrow x=\dfrac{36\pm\sqrt{1296-1296}}{8}=\dfrac{36}{8}=\dfrac{9}{2}$

**43** $\dfrac{5x}{x^2+9}=-1 \Rightarrow 5x=-x^2-9 \Rightarrow x^2+5x+9=0 \Rightarrow x=\dfrac{-5\pm\sqrt{25-36}}{2}.$

Since the discriminant is negative, there are no real solutions.

**45** The *expression* is $x^2 + x - 30$. The associated quadratic *equation* is $x^2 + x - 30 = 0$.

Using the quadratic formula, $x = \dfrac{-b \pm \sqrt{b^2 - 4ac}}{2a}$ to solve for $x$ with $a = 1$, $b = 1$,

and $c = -30$ gives us: $$x = \frac{-1 \pm \sqrt{1^2 - 4(1)(-30)}}{2(1)} = \frac{-1 \pm \sqrt{1 + 120}}{2}$$
$$= \frac{-1 \pm \sqrt{121}}{2} = \frac{-1 \pm 11}{2} = \frac{10}{2}, \frac{-12}{2} = 5, -6$$

Write the equation as a product of linear factors: $[x - (5)]\,[x - (-6)] = 0$

Now simplify: $(x - 5)(x + 6) = 0$

So the final factored form of $x^2 + x - 30$ is $(x - 5)(x + 6)$.

**47** $12x^2 - 16x - 3 = 0$ $\{a = 12,\ b = -16,\ c = -3\}$,

so $x = \dfrac{16 \pm \sqrt{256 + 144}}{24} = \dfrac{16 \pm 20}{24} = \dfrac{3}{2}, -\dfrac{1}{6}$.

Write the equation as a product of linear factors: $\left[x - \frac{3}{2}\right]\left[x - (-\frac{1}{6})\right] = 0$

Now multiply the first factor by 2 and the second factor by 6. $(2x - 3)(6x + 1) = 0$

So the final factored form of $12x^2 - 16x - 3$ is $(2x - 3)(6x + 1)$.

**49** (a) For this exercise, we must recognize the equation as a quadratic in $x$, that is,
$$(\mathrm{A})x^2 + (\mathrm{B})x + (\mathrm{C}) = 0,$$
where A is the coefficient of $x^2$, B is the coefficient of $x$, and C is the collection of all terms that do not contain $x^2$ or $x$.

$4x^2 - 4xy + 1 - y^2 = 0 \;\Rightarrow\; (4)x^2 + (-4y)x + (1 - y^2) = 0 \;\Rightarrow$

$$x = \frac{4y \pm \sqrt{16y^2 - 16(1 - y^2)}}{2(4)} = \frac{4y \pm \sqrt{16[y^2 - (1 - y^2)]}}{2(4)} =$$
$$\frac{4y \pm 4\sqrt{2y^2 - 1}}{2(4)} = \frac{y \pm \sqrt{2y^2 - 1}}{2}$$

(b) Similar to part (a), we must now recognize the equation as a quadratic equation in $y$. $4x^2 - 4xy + 1 - y^2 = 0 \;\Rightarrow\; (-1)y^2 + (-4x)y + (4x^2 + 1) = 0 \;\Rightarrow$

$$y = \frac{4x \pm \sqrt{16x^2 + 4(4x^2 + 1)}}{-2} = \frac{4x \pm 2\sqrt{8x^2 + 1}}{-2} = -2x \pm \sqrt{8x^2 + 1}$$

**51** $K = \frac{1}{2}mv^2 \;\Rightarrow\; v^2 = \frac{2K}{m} \;\Rightarrow\; v = \pm\sqrt{\frac{2K}{m}} \;\Rightarrow\; v = \sqrt{\frac{2K}{m}}$ since $v > 0$.

53 $A = 2\pi r(r+h) \Rightarrow A = 2\pi r^2 + 2\pi rh \Rightarrow$

$(2\pi)r^2 + (2\pi h)r - A = 0$ {a quadratic equation in $r$} $\Rightarrow$

$$r = \frac{-(2\pi h) \pm \sqrt{(2\pi h)^2 - 4(2\pi)(-A)}}{2(2\pi)} = \frac{-2\pi h \pm \sqrt{4\pi^2h^2 + 8\pi A}}{2(2\pi)} =$$

$$\frac{-2\pi h \pm 2\sqrt{\pi^2h^2 + 2\pi A}}{2(2\pi)} = \frac{-\pi h \pm \sqrt{\pi^2h^2 + 2\pi A}}{2\pi}.$$

Since $r > 0$, we must use the plus sign, and $r = \dfrac{-\pi h + \sqrt{\pi^2h^2 + 2\pi A}}{2\pi}$.

55 $V = V_{\text{max}}\left[1 - \left(\frac{r}{r_0}\right)^2\right] \Rightarrow \frac{V}{V_{\text{max}}} = 1 - \left(\frac{r}{r_0}\right)^2 \Rightarrow \left(\frac{r}{r_0}\right)^2 = 1 - (V/V_{\text{max}}) \Rightarrow$

$$r^2 = r_0^2[1 - (V/V_{\text{max}})] \; \{r > 0\} \Rightarrow r = r_0\sqrt{1 - (V/V_{\text{max}})}$$

57 Using $V = \pi r^2 h$ with $V = 3000$ and $h = 20$ gives us:

$$3000 = \pi r^2(20) \Rightarrow r^2 = 150/\pi \Rightarrow r = \sqrt{150/\pi} \approx 6.9 \text{ cm}$$

59 (a) $s = 48 \Rightarrow -16t^2 + 64t = 48 \Rightarrow t^2 - 4t + 3 = 0 \Rightarrow (t-1)(t-3) = 0 \Rightarrow$

$t = 1, 3$. After 1 sec and after 3 sec

(b) It will hit the ground when $s = 0$.

$s = 0 \Rightarrow -16t^2 + 64t = 0 \Rightarrow t(t-4) = 0 \Rightarrow t = 0, 4$. After 4 seconds

61 (a) $T = 98 \Rightarrow h = 1000(100 - T) + 580(100 - T)^2 = 1000(2) + 580(2)^2 = 4320$ m.

(b) If $x = 100 - T$ and $h = 8840$, then $8840 = 1000x + 580x^2 \Rightarrow$

$29x^2 + 50x - 442 = 0 \Rightarrow x = \dfrac{-25 \pm \sqrt{13{,}443}}{29} \approx -4.86, 3.14$.

$x = -4.86 \Rightarrow T = 100 - x = 104.86\,^\circ\text{C}$,

which is outside the allowable range of $T$.

$x = 3.14 \Rightarrow T = 100 - x = 96.86\,^\circ\text{C}$ for $95 \le T \le 100$.

63 Let $x$ denote the width of the walk. The area (including the walk) has dimensions $(26 + x + x)$ by $(30 + x + x)$ or, equivalently, $(26 + 2x)$ by $(30 + 2x)$.

$\text{Area}_{\text{plot}} + \text{Area}_{\text{walk}} = \text{Area}_{\text{total}} \Rightarrow 26 \cdot 30 + 240 = (26 + 2x)(30 + 2x) \Rightarrow$

$26 \cdot 30 + 240 = 26 \cdot 30 + 52x + 60x + 4x^2 \Rightarrow 240 = 4x^2 + 112x \Rightarrow$

$x^2 + 28x - 60 = 0 \Rightarrow (x + 30)(x - 2) = 0 \Rightarrow x = 2$ ft, since $x$ is positive.

65 Let $x$ denote the length of one side.

$\text{Cost}_{\text{preparation}} + \text{Cost}_{\text{fence}} = \text{Cost}_{\text{total}} \Rightarrow$

$x^2(\$0.50) + 4x(\$1) = \$120 \Rightarrow$

$x^2 + 8x - 240 = 0 \Rightarrow (x + 20)(x - 12) = 0 \Rightarrow x = 12$.

The size of the garden should be 12 ft by 12 ft.

[67] Let $d(A, P) = x$ and $d(P, B) = 6 - x$.

$x^2 + (6 - x)^2 = 5^2 \Rightarrow 2x^2 - 12x + 11 = 0 \Rightarrow x = 3 \pm \frac{1}{2}\sqrt{14} \approx 4.9,\ 1.1$ mi.

There are 4 possible roads since $P$ could be on either side of segment $AB$.

[69] (a) The distances of the northbound and eastbound planes are $100 + 200t$ and $400t$, respectively. Using the Pythagorean theorem,

$$d = \sqrt{(100 + 200t)^2 + (400t)^2} = \sqrt{100^2(1 + 2t)^2 + 100^2(4t)^2} = 100\sqrt{20t^2 + 4t + 1}.$$

(b) $d = 500 \Rightarrow 500 = 100\sqrt{20t^2 + 4t + 1} \Rightarrow 5 = \sqrt{20t^2 + 4t + 1} \Rightarrow$

$5^2 = 20t^2 + 4t + 1 \Rightarrow 5t^2 + t - 6 = 0 \Rightarrow$

$(5t + 6)(t - 1) = 0 \Rightarrow t = 1$ hour after 2:30 P.M., or 3:30 P.M.

[71] Let $x$ denote the outer width of the box, $x - 2$ the inner width.

Since the base is square, $(x - 2)^2 = 144 \Rightarrow x = 14$.

The length is $3(1) + 2(x - 2) = 2x - 1$. Thus, the size is 14 in. by 27 in.

[73] Let $x$ denote the rate of the canoeist in still water.

$x - 5$ is the rate upstream and $x + 5$ is the rate downstream.

$\text{Time}_{\text{up}} = \text{Time}_{\text{down}} + \frac{1}{2} \Rightarrow \left\{t = \frac{d}{r}\right\} \frac{1.2}{x - 5} = \frac{1.2}{x + 5} + \frac{1}{2} \Rightarrow$

$2.4(x + 5) = 2.4(x - 5) + x^2 - 25 \Rightarrow x^2 = 49 \Rightarrow x = 7$ mi/hr.

[75] Let $x$ denote the number of pairs ordered. The price per pair is the discount subtracted from \$40. Since the discount is \$0.04 times the number ordered, $x$, the cost per pair is $40 - 0.04x$. Cost = (# of pairs)(cost per pair) $\Rightarrow$

$8400 = x(40 - 0.04x) \Rightarrow \frac{1}{25}x^2 - 40x + 8400 = 0$ {multiply by 25} $\Rightarrow$

$x^2 - 1000x + 210{,}000 = 0 \Rightarrow (x - 300)(x - 700) = 0 \Rightarrow x = 300$ for $0 \le x \le 600$.

[77] The total surface area is the sum of the surface area of the cylinder and that of the top and bottom. $S = 2\pi rh + 2\pi r^2 \Rightarrow 10\pi = 8\pi r + 2\pi r^2$ {divide by $2\pi$} $\Rightarrow$

$r^2 + 4r - 5 = 0 \Rightarrow (r + 5)(r - 1) = 0 \Rightarrow r = 1$, and the diameter is 2 ft.

[79] $V$ is 95% of $V_0 \Rightarrow V = 0.95V_0 \Rightarrow \dfrac{V}{V_0} = 0.95$.

$0.95 = 0.8197 + 0.007752t + 0.0000281t^2 \Rightarrow 0.281t^2 + 77.52t - 1303 = 0 \Rightarrow$

$$t = \frac{-77.52 \pm \sqrt{(77.52)^2 - 4(0.281)(-1303)}}{2(0.281)} \approx -291.76,\ 15.89.$$

Thus, the volume of the fireball will be 95% of the maximum volume approximately 15.89 seconds after the explosion.

[81] (a) $x = \dfrac{-4{,}500{,}000 \pm \sqrt{4{,}500{,}000^2 - 4(1)(-0.96)}}{2} \approx 0$ and $-4{,}500{,}000$

(b) $x = \dfrac{-b \pm \sqrt{b^2-4ac}}{2a} \cdot \dfrac{-b \mp \sqrt{b^2-4ac}}{-b \mp \sqrt{b^2-4ac}} = \dfrac{b^2-(b^2-4ac)}{2a(-b \mp \sqrt{b^2-4ac})} =$

$\dfrac{4ac}{2a(-b \mp \sqrt{b^2-4ac})} = \dfrac{2c}{-b \mp \sqrt{b^2-4ac}}$. The root near zero was obtained in part (a) using the plus sign. In the second formula, it corresponds to the minus sign.

$$x = \frac{2(-0.96)}{-4{,}500{,}000 - \sqrt{4{,}500{,}000^2 - 4(1)(-0.96)}} \approx 2.13 \times 10^{-7}$$

[83] (a) Let $Y_1 = T_1 = -1.09L + 96.01$ and $Y_2 = T_2 = -0.011L^2 - 0.126L + 81.45$.

Table each equation and compare them to the actual temperatures.

| $x$ ($L$) | $Y_1$ | $Y_2$ | S. Hem. |
|---|---|---|---|
| 85 | 3.36 | −8.74 | −5 |
| 75 | 14.26 | 10.13 | 10 |
| 65 | 25.16 | 26.79 | 27 |
| 55 | 36.06 | 41.25 | 42 |
| 45 | 46.96 | 53.51 | 53 |
| 35 | 57.86 | 63.57 | 65 |
| 25 | 68.76 | 71.43 | 75 |
| 15 | 79.66 | 77.09 | 78 |
| 5 | 90.56 | 80.55 | 79 |

Comparing $Y_1$ ($T_1$) with $Y_2$ ($T_2$), we can see that the linear equation $T_1$ is not as accurate as the quadratic equation $T_2$.

(b) $L = 50 \Rightarrow T_2 = -0.011(50)^2 - 0.126(50) + 81.45 = 47.65^\circ$F.

## 2.4 Exercises

[1] $(5-2i)+(-3+6i) = [5+(-3)]+(-2+6)i = 2+4i$

[3] $(7-6i)-(-11-3i) = (7+11)+(-6+3)i = 18-3i$

[5] $(3+5i)(2-7i) = (3+5i)2+(3+5i)(-7i)$ {distributive property}

$= 6+10i-21i-35i^2$ {multiply terms}

$= (6-35i^2)+(10-21)i$ {group the real parts and the imaginary parts}

$= (6+35)-11i$ {$i^2=-1$}

$= 41-11i$

[7] $(1-3i)(2+5i)=(2-15i^2)+(5-6)i=(2+15)-i=17-i$

[9] Use the special product formula for $(x-y)^2$ on the inside front cover of the text.

$$(5-2i)^2=5^2-2(5)(2i)+(2i)^2=25-20i+4i^2=(25-4)-20i=21-20i$$

[11] $i(3+4i)^2=i[3^2+2(3)(4i)+(4i)^2]=i[(9-16)+24i]=i(-7+24i)=-24-7i$

[13] $(3+4i)(3-4i)$ { note that this difference of squares ... } =

$3^2-(4i)^2=9-(-16)=$ {... becomes a "sum of squares" } $9+16=25$

[15] Since $i^k=1$ if $k$ is a multiple of 4, we will write $i^{43}$ as $i^{40}i^3$, knowing that $i^{40}$ will reduce to 1. $\quad i^{43}=i^{40}i^3=(i^4)^{10}(-i)=1^{10}(-i)=-i$

[17] Since $i^k=1$ if $k$ is a multiple of 4, we will write $i^{73}$ as $i^{72}i^1$, knowing that $i^{72}$ will reduce to 1. $\quad i^{73}=i^{72}i=(i^4)^{18}i=1^{18}i=i$

[19] Multiply both the numerator and the denominator by the conjugate of the denominator to eliminate all $i$'s in the denominator. The new denominator is the sum of the squares of the coefficients—in this case, $2^2$ and $4^2$.

$$\frac{3}{2+4i}\cdot\frac{2-4i}{2-4i}=\frac{6-12i}{4-(-16)}=\frac{6-12i}{20}=\frac{3}{10}-\frac{3}{5}i$$

[21] Multiply both the numerator and the denominator by the conjugate of the denominator to eliminate all $i$'s in the denominator. The new denominator is the sum of the squares of the coefficients—in this case, $6^2$ and $2^2$.

$$\frac{1-7i}{6-2i}\cdot\frac{6+2i}{6+2i}=\frac{(6+14)+(2-42)i}{36-(-4)}=\frac{20-40i}{40}=\frac{1}{2}-i$$

[23] $\dfrac{-4+6i}{2+7i}\cdot\dfrac{2-7i}{2-7i}=\dfrac{(-8+42)+(28+12)i}{4-(-49)}=\dfrac{34+40i}{53}=\dfrac{34}{53}+\dfrac{40}{53}i$

[25] Multiplying the denominator by $i$ will eliminate the $i$'s in the denominator.

$$\frac{4-2i}{-5i}=\frac{4-2i}{-5i}\cdot\frac{i}{i}=\frac{4i-2i^2}{-5i^2}=\frac{2+4i}{5}=\frac{2}{5}+\frac{4}{5}i$$

[27] Use the special product formula for $(x+y)^3$ on the inside front cover of the text.

$$\begin{aligned}(2+5i)^3 &= (2)^3+3(2)^2(5i)+3(2)(5i)^2+(5i)^3\\ &= (8+150i^2)+(60i+125i^3)=(8-150)+(60-125)i=-142-65i\end{aligned}$$

[29] A common mistake is to multiply $\sqrt{-4}\sqrt{-16}$ and obtain $\sqrt{64}$, or 8.

The correct procedure is $\sqrt{-4}\sqrt{-16}=\sqrt{4}\,i\cdot\sqrt{16}\,i=(2i)(4i)=8i^2=-8$.

$$(2-\sqrt{-4})(3-\sqrt{-16})=(2-2i)(3-4i)=(6-8)+(-6i-8i)=-2-14i$$

[31] $\dfrac{4+\sqrt{-81}}{7-\sqrt{-64}}=\dfrac{4+9i}{7-8i}\cdot\dfrac{7+8i}{7+8i}=\dfrac{(28-72)+(32+63)i}{49-(-64)}=\dfrac{-44+95i}{113}=-\dfrac{44}{113}+\dfrac{95}{113}i$

[33] $\dfrac{\sqrt{-36}\sqrt{-49}}{\sqrt{-16}}=\dfrac{(6i)(7i)}{4i}\cdot\dfrac{-i}{-i}=\dfrac{(-42)(-i)}{-4i^2}=\dfrac{42i}{4}=\dfrac{21}{2}i$

[35] We need to equate the real parts and the imaginary parts on each side of " = ".

$4+(x+2y)i = x+2i \Rightarrow 4 = x$ and $x+2y=2 \Rightarrow$

$x=4$ and $4+2y=2 \Rightarrow 2y=-2 \Rightarrow y=-1$, so $x=4$ and $y=-1$.

[37] $(2x-y)-16i = 10+4yi \Rightarrow 2x-y=10$ and $-16=4y \Rightarrow$

$y=-4$ and $2x-(-4)=10 \Rightarrow 2x+4=10 \Rightarrow 2x=6 \Rightarrow x=3$,

so $x=3$ and $y=-4$.

[39] $x^2-6x+13=0 \Rightarrow$

$$x = \frac{-(-6)\pm\sqrt{(-6)^2-4(1)(13)}}{2(1)} = \frac{6\pm\sqrt{36-52}}{2} = \frac{6\pm\sqrt{-16}}{2} = \frac{6\pm 4i}{2} = 3\pm 2i$$

[41] $x^2+4x+13=0 \Rightarrow$

$$x = \frac{-4\pm\sqrt{4^2-4(1)(13)}}{2(1)} = \frac{-4\pm\sqrt{16-52}}{2} = \frac{-4\pm\sqrt{-36}}{2} = \frac{-4\pm 6i}{2} = -2\pm 3i$$

[43] $x^2-5x+20=0 \Rightarrow$

$$x = \frac{-(-5)\pm\sqrt{(-5)^2-4(1)(20)}}{2(1)} = \frac{5\pm\sqrt{25-80}}{2} = \frac{5\pm\sqrt{-55}}{2} = \frac{5}{2}\pm\frac{1}{2}\sqrt{55}\,i$$

[45] $4x^2+x+3=0 \Rightarrow$

$$x = \frac{-1\pm\sqrt{1^2-4(4)(3)}}{2(4)} = \frac{-1\pm\sqrt{1-48}}{8} = \frac{-1\pm\sqrt{-47}}{8} = -\frac{1}{8}\pm\frac{1}{8}\sqrt{47}\,i$$

[47] Solving $x^3=-125$ would only give us the solution $x=-5$. So first we need to factor $x^3+125$ as the sum of cubes. $x^3+125=0 \Rightarrow (x+5)(x^2-5x+25)=0 \Rightarrow$

$$x=-5 \text{ or } x = \frac{5\pm\sqrt{25-100}}{2} = \frac{5\pm 5\sqrt{3}\,i}{2}.$$ The three solutions are $-5$, $\frac{5}{2}\pm\frac{5}{2}\sqrt{3}\,i$.

[49] $x^4=256 \Rightarrow x^4-256=0 \Rightarrow (x^2-16)(x^2+16)=0 \Rightarrow$

$x^2=16, -16 \Rightarrow x=\pm\sqrt{16}, \pm\sqrt{-16} \Rightarrow x=\pm 4, \pm 4i$

[51] $4x^4+25x^2+36=0 \Rightarrow (x^2+4)(4x^2+9)=0 \Rightarrow$

$x^2=-4, -\frac{9}{4} \Rightarrow x=\pm\sqrt{-4}, \pm\sqrt{-\frac{9}{4}} \Rightarrow x=\pm 2i, \pm\frac{3}{2}i$

[53] $x^3+3x^2+4x=0 \Rightarrow x(x^2+3x+4)=0 \Rightarrow$

$$x=0 \text{ or } x = \frac{-3\pm\sqrt{9-16}}{2} = \frac{-3\pm\sqrt{7}\,i}{2}.$$ The three solutions are $0$, $-\frac{3}{2}\pm\frac{1}{2}\sqrt{7}\,i$.

*Note:* Exer. 55–60: Let $z = a + bi$ and $w = c + di$.

[55] If $z = a + bi$ and $w = c + di$, then

| | |
|---|---|
| $\overline{z+w} = \overline{(a+bi)+(c+di)}$ | definition of $z$ and $w$ |
| $= \overline{(a+c)+(b+d)i}$ | write in complex number form |
| $= (a+c)-(b+d)i$ | definition of conjugate |
| $= (a-bi)+(c-di)$ | rearrange terms |
| $= \overline{z}+\overline{w}$. | definition of conjugates of $z$ and $w$ |

In the step described by "rearrange terms,"

we are really looking ahead to the terms we want to obtain, $\overline{z}$ and $\overline{w}$.

[57] $\overline{z \cdot w} = \overline{(a+bi)\cdot(c+di)} = \overline{(ac-bd)+(ad+bc)i} =$

$(ac-bd)-(ad+bc)i = ac - adi - bd - bci = a(c-di) - bi(c-di) =$

$(a-bi)\cdot(c-di) = \overline{z}\cdot\overline{w}$

[59] If $\overline{z} = z$, then $a - bi = a + bi$ and hence $-bi = bi$, or $2bi = 0$.

Thus, $b = 0$ and $z = a$ is real. Conversely, if $z$ is real, then $b = 0$ and hence

$\overline{z} = \overline{a+0i} = a - 0i = a + 0i = z$.

## 2.5 Exercises

[1] $|x+4| = 11 \Rightarrow x+4 = 11$ or $x+4 = -11 \Rightarrow x = 7$ or $x = -15$

[3] We must first isolate the absolute value term before proceeding.

$|3x-2|+3 = 7 \Rightarrow |3x-2| = 4 \Rightarrow 3x-2 = 4$ or $3x-2 = -4 \Rightarrow$

$3x = 6$ or $3x = -2 \Rightarrow x = 2$ or $x = -\frac{2}{3}$

[5] $3|x+1|-2 = -11 \Rightarrow 3|x+1| = -9 \Rightarrow |x+1| = -3$.

Since the absolute value of an expression is nonnegative, $|x+1| = -3$ has no solution.

[7] Since there are four terms, we first try factoring by grouping.

$9x^3 - 18x^2 - 4x + 8 = 0 \Rightarrow 9x^2(x-2) - 4(x-2) = 0 \Rightarrow$

$(9x^2-4)(x-2) = 0 \Rightarrow x^2 = \frac{4}{9}$ or $x = 0 \Rightarrow x = \pm\frac{2}{3}, 2$

[9] Notice that we can factor an $x$ out of each term, and then factor by grouping.

$4x^4 + 10x^3 = 6x^2 + 15x \Rightarrow x(4x^3 + 10x^2 - 6x - 15) = 0 \Rightarrow$

$x\left[2x^2(2x+5) - 3(2x+5)\right] = 0 \Rightarrow x(2x^2-3)(2x+5) = 0 \Rightarrow$

$x = 0$ or $x^2 = \frac{3}{2}$ or $x = -\frac{5}{2} \Rightarrow x = 0, \pm\frac{1}{2}\sqrt{6}, -\frac{5}{2}$

*Note:* $x^2 = \frac{3}{2} \Rightarrow x = \pm\sqrt{\frac{3}{2}} = \pm\frac{\sqrt{3}}{\sqrt{2}}\cdot\frac{\sqrt{2}}{\sqrt{2}} = \pm\frac{\sqrt{6}}{2} = \pm\frac{1}{2}\sqrt{6}$.

There are several ways to write this answer—your professor may have a preference.

[11] $y^{3/2} = 5y \Rightarrow y^{3/2} - 5y = 0 \Rightarrow y(y^{1/2} - 5) = 0 \Rightarrow y = 0$ or $y^{1/2} = 5$.

$y^{1/2} = 5 \Rightarrow (y^{1/2})^2 = 5^2 \Rightarrow y = 25$. The solutions are $y = 0$ and $y = 25$.

*Note:* The following guidelines may be helpful when solving radical equations.

**Guidelines for Solving a Radical Equation**

(1) Isolate the radical. If we cannot get the radical isolated on one side of the equals sign because there is more than one radical, then we will split up the radical terms as evenly as possible on each side of the equals sign. For example, if there are 2 radicals, we put one on each side; if there are 3 radicals, we put 2 on one side and 1 on the the other.

(2) Raise both sides to the same power as the root index. *Note:* Remember here that

$$\boxed{(a + b\sqrt{n})^2 = a^2 + 2ab\sqrt{n} + b^2 n} \quad \text{and that} \quad (a + b\sqrt{n})^2 \neq a^2 + b^2 n.$$

(3) If your equation contains no radicals, proceed to part (4). If there are still radicals in the equation, go back to part (1).

(4) Solve the resulting equation.

(5) Check the answers found in part (4) in the original equation to determine the valid solutions. *Note:* You may check the solutions in any equivalent equation of the original equation, that is, an equation which occurs prior to raising both sides to a power. Also, extraneous real number solutions are introduced when raising both sides to an even power. Hence, all solutions *must* be checked in this case. Checking solutions when raising each side to an odd power is up to the individual professor.

[13] $\sqrt{7 - 5x} = 8 \Rightarrow (\sqrt{7 - 5x})^2 = 8^2 \Rightarrow 7 - 5x = 64 \Rightarrow x = -\frac{57}{5}$

[15] $2 + \sqrt[3]{1 - 5t} = 0 \Rightarrow \sqrt[3]{1 - 5t} = -2 \Rightarrow (\sqrt[3]{1 - 5t})^3 = (-2)^3 \Rightarrow$

$1 - 5t = -8 \Rightarrow -5t = -9 \Rightarrow t = \frac{9}{5}$

[17] $\sqrt[5]{2x^2 + 1} - 2 = 0 \Rightarrow \sqrt[5]{2x^2 + 1} = 2 \Rightarrow (\sqrt[5]{2x^2 + 1})^5 = 2^5 \Rightarrow$

$2x^2 + 1 = 32 \Rightarrow x^2 = \frac{31}{2} \Rightarrow x = \pm\sqrt{\frac{31}{2}} \Rightarrow x = \pm\frac{1}{2}\sqrt{62}$

[19] $\sqrt{7 - x} = x - 5$ {square both sides} $\Rightarrow$

$7 - x = x^2 - 10x + 25$ {set equal to zero} $\Rightarrow$

$x^2 - 9x + 18 = 0$ {factor} $\Rightarrow (x - 3)(x - 6) = 0 \Rightarrow x = 3, 6$.

Check $x = 6$: LS $= \sqrt{7 - 6} = 1$; RS $= 6 - 5 = 1$.

Since both sides have the same value, $x = 6$ is a valid solution.

Check $x = 3$: LS $= \sqrt{7 - 3} = 2$; RS $= 3 - 5 = -2$.

Since both sides do not have the same value, $x = 3$ is an extraneous solution.

**21** $3\sqrt{2x-3}+2\sqrt{7-x}=11$ given

$3\sqrt{2x-3}=11-2\sqrt{7-x}$ split radicals evenly

$9(2x-3)=121-44\sqrt{7-x}+4(7-x)$ square both sides

$44\sqrt{7-x}=-22x+176$ isolate radical again, simplify

$2\sqrt{7-x}=8-x$ divide by gcf, 22

$4(7-x)=64-16x+x^2$ square both sides

$x^2-12x+36=0$ collect terms on one side

$(x-6)^2=0$ factor

$x-6=0$ take the square root

$x=6$ solve for $x$

Check $x=6$: LS $=3(3)+2(1)=11=$ RS, $\Rightarrow$ $x=6$ is the solution.

**23** Remember to isolate the radical term first.

$x=4+\sqrt{4x-19} \Rightarrow x-4=\sqrt{4x-19} \Rightarrow x^2-8x+16=4x-19 \Rightarrow$

$x^2-12x+35=0 \Rightarrow (x-5)(x-7)=0 \Rightarrow x=5,\ 7$

**25** $x+\sqrt{5x+19}=-1 \Rightarrow \sqrt{5x+19}=-x-1 \Rightarrow 5x+19=x^2+2x+1 \Rightarrow$

$x^2-3x-18=0 \Rightarrow (x-6)(x+3)=0 \Rightarrow x=-3,\ 6.$

Check $x=-3$: LS $=-3+2=-1=$ RS $\Rightarrow$ $x=-3$ is a solution.

Check $x=6$: LS $=6+7=13\neq$ RS $\Rightarrow$ $x=6$ is an extraneous solution.

**27** $\sqrt{7-2x}-\sqrt{5+x}=\sqrt{4+3x}$ {square both sides} $\Rightarrow$

$(7-2x)-2\sqrt{(7-2x)(5+x)}+(5+x)=4+3x$ {isolate the radical} $\Rightarrow$

$-4x+8=2\sqrt{-2x^2-3x+35}$ {divide by 2} $\Rightarrow$

$-2x+4=\sqrt{-2x^2-3x+35}$ {square both sides} $\Rightarrow$

$4x^2-16x+16=-2x^2-3x+35$ {simplify} $\Rightarrow$

$6x^2-13x-19=0 \Rightarrow (x+1)(6x-19)=0 \Rightarrow x=-1,\ \frac{19}{6}.$

Check $x=-1$: LS $=3-2=1=$ RS $\Rightarrow$ $x=-1$ is a solution.

Check $x=\frac{19}{6}$: LS $=\sqrt{\frac{2}{3}}-\sqrt{\frac{49}{6}}$ {note that this is negative} $\neq\sqrt{\frac{27}{2}}=$ RS $\Rightarrow$

$x=\frac{19}{6}$ is an extraneous solution.

**29** $\sqrt{11+8x}+1=\sqrt{9+4x} \Rightarrow (11+8x)+2\sqrt{11+8x}+1=9+4x \Rightarrow$

$2\sqrt{8x+11}=-4x-3 \Rightarrow 4(8x+11)=16x^2+24x+9 \Rightarrow$

$16x^2-8x-35=0 \Rightarrow (4x-7)(4x+5)=0 \Rightarrow$

$x=-\frac{5}{4}$ and $\frac{7}{4}$ is an extraneous solution.

[31] $\sqrt{2\sqrt{x+1}} = \sqrt{3x-5} \Rightarrow 2\sqrt{x+1} = 3x-5 \Rightarrow 4(x+1) = 9x^2 - 30x + 25 \Rightarrow$

$9x^2 - 34x + 21 = 0 \Rightarrow (x-3)(9x-7) = 0 \Rightarrow x = 3, \frac{7}{9}$.

Check $x = 3$: LS $= 2 =$ RS $\Rightarrow x = 3$ is a solution.

Check $x = \frac{7}{9}$: LS $= \sqrt{2 \cdot \frac{4}{3}} = \sqrt{\frac{8}{3}} \neq \sqrt{-\frac{8}{3}} =$ RS $\Rightarrow x = \frac{7}{9}$ is an extraneous solution.

[33] $\sqrt{1+4\sqrt{x}} = \sqrt{x}+1 \Rightarrow 1+4\sqrt{x} = x+2\sqrt{x}+1 \Rightarrow 2\sqrt{x} = x \Rightarrow$

$4x = x^2 \Rightarrow x(4-x) = 0 \Rightarrow x = 0, 4$

[35] $x^4 - 25x^2 + 144 = 0 \Rightarrow (x^2-9)(x^2-16) = 0 \Rightarrow$

$x^2 = 9, 16 \Rightarrow x = \pm\sqrt{9}, \pm\sqrt{16} \Rightarrow x = \pm 3, \pm 4$

[37] We recognize this equation as a quadratic equation in $y^2$ and apply the quadratic formula, solving for $y^2$, *not* $y$.

$$5y^4 - 7y^2 + 1 = 0 \Rightarrow y^2 = \frac{7 \pm \sqrt{29}}{10} \cdot \frac{10}{10} = \frac{70 \pm 10\sqrt{29}}{100} \Rightarrow y = \pm\frac{1}{10}\sqrt{70 \pm 10\sqrt{29}}$$

Alternatively, let $u = y^2$ and solve $5u^2 - 7u + 1 = 0$.

[39] $36x^{-4} - 13x^{-2} + 1 = 0 \Rightarrow (4x^{-2}-1)(9x^{-2}-1) = 0 \Rightarrow$

$x^{-2} = \frac{1}{4}, \frac{1}{9} \Rightarrow x^2 = 4, 9 \Rightarrow x = \pm 2, \pm 3$

Alternatively, let $u = x^{-2}$ and solve $36u^2 - 13u + 1 = 0$.

[41] $3x^{2/3} + 4x^{1/3} - 4 = 0 \Rightarrow (3x^{1/3}-2)(x^{1/3}+2) = 0 \Rightarrow$

$\sqrt[3]{x} = \frac{2}{3}, -2 \Rightarrow x = \frac{8}{27}, -8$

Alternatively, let $u = x^{1/3}$ and solve $3u^2 + 4u - 4 = 0$.

[43] $6w - 23w^{1/2} + 20 = 0 \Rightarrow (2w^{1/2}-5)(3w^{1/2}-4) = 0 \Rightarrow$

$\sqrt{w} = \frac{5}{2}, \frac{4}{3} \Rightarrow w = \frac{25}{4}, \frac{16}{9}$

Alternatively, let $u = w^{1/2}$ and solve $6u^2 - 23u + 20 = 0$.

[45] $\left(\frac{t}{t+1}\right)^2 - \frac{2t}{t+1} - 8 = 0 \Rightarrow \left(\frac{t}{t+1} - 4\right)\left(\frac{t}{t+1} + 2\right) = 0 \Rightarrow \frac{t}{t+1} = 4$ or $-2 \Rightarrow$

$t = 4(t+1)$ or $-2(t+1) \Rightarrow t = 4t+4$ or $t = -2t-2 \Rightarrow$

$-3t = 4$ or $3t = -2 \Rightarrow t = -\frac{4}{3}, -\frac{2}{3}$

Alternatively, let $u = \frac{t}{t+1}$ and solve $u^2 - 2u - 8 = 0$.

[47] For this exercise, we must note that the variable terms are both cubed, and then combine them into one term.

$27x^3 = (x+5)^3 \Rightarrow 27 = \frac{(x+5)^3}{x^3} \Rightarrow \left(\frac{x+5}{x}\right)^3 = 3^3 \Rightarrow \frac{x+5}{x} = 3 \Rightarrow$

$x + 5 = 3x \Rightarrow 5 = 2x \Rightarrow x = \frac{5}{2}$

49 The least common multiple of 3 and 4 is 12—so by raising both sides to the 12th power we will eliminate the radicals. $\sqrt[3]{x} = 2\sqrt[4]{x} \Rightarrow (\sqrt[3]{x})^{12} = (2\sqrt[4]{x})^{12} \Rightarrow$

$$x^4 = 2^{12}x^3 \Rightarrow x^4 - 4096x^3 = 0 \Rightarrow x^3(x - 4096) = 0 \Rightarrow x = 0,\ 4096.$$

Check $x = 0$: LS $= 0 =$ RS $\Rightarrow$ $x = 0$ is a solution.

Check $x = 4096 = 2^{12}$: LS $= \sqrt[3]{2^{12}} = 2^4$; RS $= 2\sqrt[4]{2^{12}} = 2 \cdot 2^3 = 2^4 \Rightarrow$

$x = 4096$ is a solution.

51 See the illustrations and discussion on page 105 in the text for help on solving equations by raising both sides to a reciprocal power. Note that if $x^{m/n}$ is in the given equation and $m$ is even, we have to use the $\pm$ symbol for the solutions. Here are a few more examples:

| Equation | Solution |
|---|---|
| $x^{1/2} = 4$ | $(x^{1/2})^{2/1} = 4^{2/1} \Rightarrow x = 16$ |
| $x^{-1/2} = 5$ | $(x^{-1/2})^{-2/1} = 5^{-2/1} \Rightarrow x = \frac{1}{25}$ |
| $x^{3/4} = 8$ | $(x^{3/4})^{4/3} = 8^{4/3} \Rightarrow x = 16$ |
| $x^{4/3} = 81$ | $(x^{4/3})^{3/4} = 81^{3/4} \Rightarrow x = \pm 27$ |

(a) $x^{5/3} = 32 \Rightarrow (x^{5/3})^{3/5} = (32)^{3/5} \Rightarrow x = (\sqrt[5]{32})^3 = 2^3 = 8$

(b) $x^{4/3} = 16 \Rightarrow (x^{4/3})^{3/4} = \pm(16)^{3/4} \Rightarrow x = \pm(\sqrt[4]{16})^3 = \pm 2^3 = \pm 8$

(c) $x^{2/3} = -36 \Rightarrow (x^{2/3})^{3/2} = \pm(-36)^{3/2} \Rightarrow x = \pm(\sqrt{-36})^3$,

which are not real numbers. No real solutions

(d) $x^{3/4} = 125 \Rightarrow (x^{3/4})^{4/3} = (125)^{4/3} \Rightarrow x = (\sqrt[3]{125})^4 = 5^4 = 625$

(e) $x^{3/2} = -27 \Rightarrow (x^{3/2})^{2/3} = (-27)^{2/3} \Rightarrow x = (\sqrt[3]{-27})^2 = (-3)^2 = 9$,

which is an extraneous solution. No real solutions

53 $T = 2\pi\sqrt{\frac{l}{g}} \Rightarrow \frac{T}{2\pi} = \sqrt{\frac{l}{g}} \Rightarrow \frac{T^2}{4\pi^2} = \frac{l}{g} \Rightarrow l = \frac{gT^2}{4\pi^2}$

55 $S = \pi r\sqrt{r^2 + h^2} \Rightarrow \frac{S}{\pi r} = \sqrt{r^2 + h^2} \Rightarrow \frac{S^2}{\pi^2 r^2} = h^2 + r^2 \Rightarrow$

$$\frac{S^2}{\pi^2 r^2} - r^2 = h^2 \Rightarrow \frac{S^2}{\pi^2 r^2} - \frac{\pi^2 r^4}{\pi^2 r^2} = h^2 \Rightarrow h^2 = \frac{1}{\pi^2 r^2}(S^2 - \pi^2 r^4) \Rightarrow$$

$$h = \pm\frac{1}{\pi r}\sqrt{S^2 - \pi^2 r^4} \Rightarrow h = \frac{1}{\pi r}\sqrt{S^2 - \pi^2 r^4} \text{ since } h > 0$$

57 From the Pythagorean theorem, $d^2 + h^2 = L^2$. Since $d$ is to be 25% of $L$, we have $d = \frac{1}{4}L$, so $(\frac{1}{4}L)^2 + h^2 = L^2 \Rightarrow h^2 = L^2 - (\frac{1}{4}L)^2 \Rightarrow h^2 = 1L^2 - \frac{1}{16}L^2 \Rightarrow$

$$h^2 = \tfrac{15}{16}L^2 \Rightarrow h = \sqrt{\tfrac{15}{16}L^2} \text{ (since } h > 0) = \frac{\sqrt{15}}{4}L \approx 0.97L. \text{ Thus, } h \approx 97\%L.$$

[59] $P = 0.31ED^2V^3 \Rightarrow V = \left(\frac{P}{0.31ED^2}\right)^{1/3} = \left(\frac{10{,}000}{(0.31)(0.42)10^2}\right)^{1/3} \approx 9.16$ ft/sec.

Multiplying by $\frac{60}{88}$ (or $\frac{15}{22}$) to convert to mi/hr gives us approximately 6.24 mi/hr.

[61] $k = 10^5$ and $c = \frac{1}{2} \Rightarrow Q = kP^{-c} = 10^5P^{-1/2} \Rightarrow Q = \frac{10^5}{\sqrt{P}} \Rightarrow \sqrt{P} = \frac{10^5}{Q} \Rightarrow$

$$P = \left(\frac{10^5}{Q}\right)^2 = \left(\frac{100{,}000}{5000}\right)^2 = (20)^2 = 400 \text{ cents, or, \$4.00.}$$

[63] $V = 144$ and $V = \frac{1}{3}\pi r^2 h \Rightarrow 144 = \frac{1}{3}\pi r^3$ {since $r = h$} $\Rightarrow$

$$r^3 = \frac{3 \cdot 144}{\pi} \Rightarrow r = \sqrt[3]{\frac{432}{\pi}}, \text{ and the diameter is } 2 \cdot \sqrt[3]{\frac{432}{\pi}} \approx 10.3 \text{ cm.}$$

[65] $y = 60\%$ { $= \frac{3}{5}$} $\Rightarrow \frac{x^3}{x^3 + (1-x)^3} = \frac{3}{5} \Rightarrow 5x^3 = 3x^3 + 3(1-x)^3 \Rightarrow$

$$2x^3 = 3(1-x)^3 \Rightarrow \left(\frac{x}{1-x}\right)^3 = \frac{3}{2} \Rightarrow \frac{x}{1-x} = \sqrt[3]{1.5} \Rightarrow x = \sqrt[3]{1.5} - \sqrt[3]{1.5}\,x \Rightarrow$$

$$x + \sqrt[3]{1.5}\,x = \sqrt[3]{1.5} \Rightarrow (1 + \sqrt[3]{1.5})x = \sqrt[3]{1.5} \Rightarrow x = \frac{\sqrt[3]{1.5}}{1 + \sqrt[3]{1.5}} \approx 0.534, \text{ or } 53.4\%.$$

[67] $\text{Cost}_{\text{underwater}} + \text{Cost}_{\text{overland}} = \text{Cost}_{\text{total}} \Rightarrow$

$7500 \cdot (\text{underwater miles}) + 6000 \cdot (\text{overland miles}) = 35{,}000 \Rightarrow$

$7500\sqrt{x^2+1} + 6000(5-x) = 35{,}000 \Rightarrow$

$7500\sqrt{x^2+1} + 30{,}000 - 6000x = 35{,}000 \Rightarrow 7500\sqrt{x^2+1} = 6000x + 5{,}000 \Rightarrow$

$15\sqrt{x^2+1} = 12x + 10$ {divide by 500} $\Rightarrow 225(x^2+1) = 144x^2 + 240x + 100 \Rightarrow$

$225x^2 + 225 = 144x^2 + 240x + 100 \Rightarrow 81x^2 - 240x + 125 = 0 \Rightarrow$

$$x = \frac{240 \pm \sqrt{17{,}100}}{162} = \frac{6 \cdot 40 \pm \sqrt{900 \cdot 19}}{6 \cdot 27} = \frac{6 \cdot 40 \pm 30\sqrt{19}}{6 \cdot 27} = \frac{40 \pm 5\sqrt{19}}{27} \approx$$

2.2887, 0.6743 mi. There are two possible routes.

[69] (a) Let $Y_1 = D_1 = 6.096L + 685.7$ and

$Y_2 = D_2 = 0.00178L^3 - 0.072L^2 + 4.37L + 719.$

Table each equation and compare them to the actual values.

| $x$ ($L$) | $Y_1$ | $Y_2$ | Summer |
|---|---|---|---|
| 0 | 686 | 719 | 720 |
| 10 | 747 | 757 | 755 |
| 20 | 808 | 792 | 792 |
| 30 | 869 | 833 | 836 |
| 40 | 930 | 893 | 892 |
| 50 | 991 | 980 | 978 |
| 60 | 1051 | 1106 | 1107 |

(continued)

Comparing $Y_1$ $(D_1)$ with $Y_2$ $(D_2)$ we can see that the linear equation $D_1$ is not as accurate as the cubic equation $D_2$.

(b) $L = 35 \Rightarrow D_2 = 0.00178(35)^3 - 0.072(35)^2 + 4.37(35) + 719 \approx 860$ min.

[71] The volume of the box is $V = hw^2 = 25$, where $h$ is the height and $w$ is the length of a side of the square base. The amount of cardboard will be minimized when the surface area of the box is a minimum. The surface area is given by $S = w^2 + 4wh$. Since $h = 25/w^2$, we have $S = w^2 + 100/w$. Form a table for $w$ and $S$.

| $w$ | $S$ | $w$ | $S$ |
|---|---|---|---|
| 3.4 | 40.972 | 3.7 | 40.717 |
| 3.5 | 40.821 | 3.8 | 40.756 |
| 3.6 | 40.738 | 3.9 | 40.851 |

The minimum surface area is $S \approx 40.717$ when $w \approx 3.7$ and $h = 25/w^2 \approx 1.8$.

## 2.6 Exercises

[1] (a) 5 is added to both sides: $-7 + 5 < -3 + 5 \Rightarrow -2 < 2$

(b) 4 is subtracted from both sides: $-7 - 4 < -3 - 4 \Rightarrow -11 < -7$

(c) both sides are multiplied by $\frac{1}{3}$: $-7 \cdot \frac{1}{3} < -3 \cdot \frac{1}{3} \Rightarrow -\frac{7}{3} < -1$

(d) both sides are multiplied by $-\frac{1}{3}$: $-7 \cdot -\frac{1}{3} > -3 \cdot -\frac{1}{3} \Rightarrow \frac{7}{3} > 1 \Rightarrow 1 < \frac{7}{3}$

*Note:* The bracket symbols "[" and "]", are used with $\leq$ or $\geq$ to denote that the end point of the interval is part of the solution. Parentheses, "(" and ")" are used with $<$ or $>$ and denote that the end point is *not* part of the solution.

[3] $x < -2 \Leftrightarrow (-\infty, -2)$

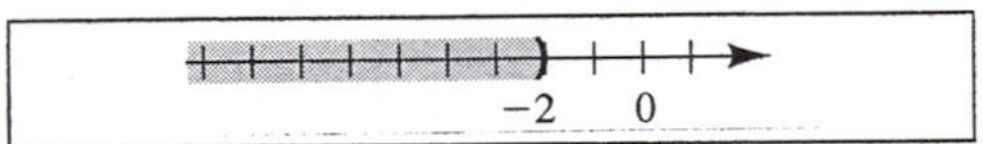


*Figure 3*

[5] $x \geq 4 \Leftrightarrow [4, \infty)$

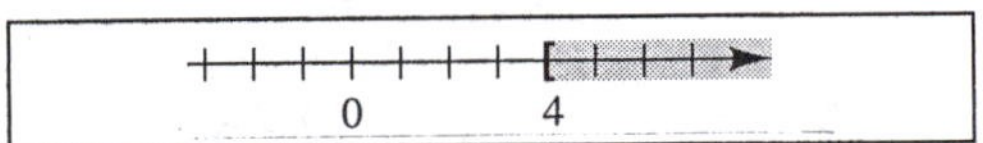


*Figure 5*

[7] $-2 < x \leq 4 \Leftrightarrow (-2, 4]$

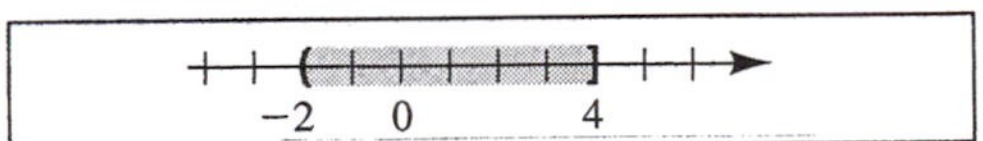


*Figure 7*

[9] $3 \leq x \leq 7 \Leftrightarrow [3, 7]$

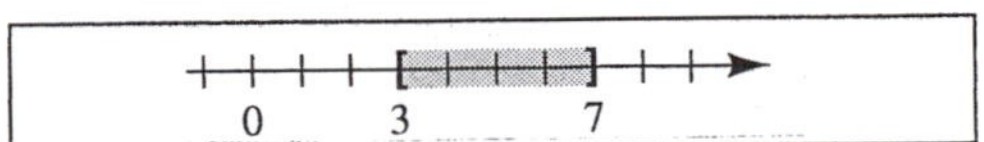


*Figure 9*

[11] $5 > x \geq -2 \Leftrightarrow -2 \leq x < 5 \Leftrightarrow [-2, 5)$

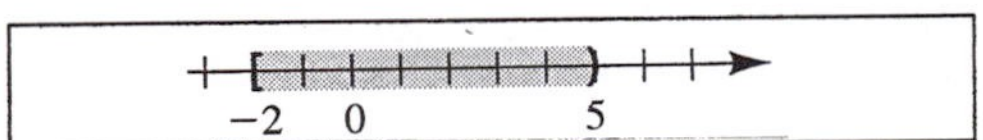


*Figure 11*

[13] $(-5, 8] \Leftrightarrow -5 < x \leq 8$

[15] $[-4, -1] \Leftrightarrow -4 \leq x \leq -1$

[17] $[4, \infty) \Leftrightarrow x \geq 4$

[19] $(-\infty, -5) \Leftrightarrow x < -5$

[21] $3x - 2 > 14 \Rightarrow 3x > 16 \Rightarrow x > \frac{16}{3} \Leftrightarrow (\frac{16}{3}, \infty)$

[23] $-2 - 3x \geq 2 \Rightarrow -3x \geq 4$ { Remember to change the direction of the inequality when multiplying or dividing by a negative value. } $\Rightarrow x \leq -\frac{4}{3} \Leftrightarrow (-\infty, -\frac{4}{3}]$

[25] $2x + 5 < 3x - 7 \Rightarrow -x < -12 \Rightarrow x > 12 \Leftrightarrow (12, \infty)$

[27] $\left[9 + \frac{1}{3}x \geq 4 - \frac{1}{2}x\right] \cdot 6$ { multiply by the lcd, 6 } $\Rightarrow 54 + 2x \geq 24 - 3x \Rightarrow$
$5x \geq -30 \Rightarrow x \geq -6 \Leftrightarrow [-6, \infty)$

[29] $-3 < 2x - 5 < 7$ { add 5 to all three expressions } $\Rightarrow$
$2 < 2x < 12$ { divide all three expressions by 2 } $\Rightarrow 1 < x < 6 \Leftrightarrow (1, 6)$

[31] $\left[3 \leq \frac{2x-3}{5} < 7\right] \cdot 5$ { given inequality } $\Rightarrow$

$15 \leq 2x - 3 < 35$ { multiply by the lcd, 5 } $\Rightarrow$

$18 \leq 2x < 38$ { add 3 to all three parts } $\Rightarrow$

$9 \leq x < 19$ { divide all three parts by 2 } $\Leftrightarrow [9, 19)$ { equivalent interval notation }

[33] $4 > \frac{2-3x}{7} \geq -2$ { multiply all three expressions by 7 } $\Rightarrow 28 > 2 - 3x \geq -14 \Rightarrow$

$26 > -3x \geq -16$ { divide by $-3$ and change directions of *both* inequality signs } $\Rightarrow$
$-\frac{26}{3} < x \leq \frac{16}{3} \Leftrightarrow (-\frac{26}{3}, \frac{16}{3}]$

[35] $0 \leq 4 - \frac{1}{3}x < 2 \Rightarrow -4 \leq -\frac{1}{3}x < -2 \Rightarrow 12 \geq x > 6 \Rightarrow 6 < x \leq 12 \Leftrightarrow (6, 12]$

[37] $(2x - 3)(4x + 5) \leq (8x + 1)(x - 7) \Rightarrow 8x^2 - 2x - 15 \leq 8x^2 - 55x - 7 \Rightarrow$
$53x \leq 8 \Rightarrow x \leq \frac{8}{53} \Leftrightarrow (-\infty, \frac{8}{53}]$

[39] $(x - 4)^2 > x(x + 12) \Rightarrow x^2 - 8x + 16 > x^2 + 12x \Rightarrow -20x > -16 \Rightarrow$
$x < \frac{4}{5} \Leftrightarrow (-\infty, \frac{4}{5})$

[41] By the law of signs, a quotient is positive if the sign of the numerator and the sign of the denominator are the same. Since the numerator is positive, $\frac{4}{3x+2} > 0 \Rightarrow$ $3x + 2 > 0 \Rightarrow x > -\frac{2}{3} \Leftrightarrow (-\frac{2}{3}, \infty)$. The expression is never equal to 0 since the numerator is never 0. Thus, the solution of $\frac{4}{3x+2} \geq 0$ is $(-\frac{2}{3}, \infty)$.

[43] $\frac{-2}{4-3x} > 0 \Rightarrow 4 - 3x < 0$ { denominator must also be negative } $\Rightarrow$
$4 < 3x \Rightarrow 3x > 4 \Rightarrow x > \frac{4}{3} \Leftrightarrow (\frac{4}{3}, \infty)$

[45] $(1 - x)^2 > 0\ \forall x$ except 1. Thus, $\frac{2}{(1-x)^2} > 0$ has solution $\mathbb{R} - \{1\}$.

[47] When reading this inequality, think of the concept "I want all the numbers that lie less than 3 units from the origin," and your answer should make some common sense.
$|x| < 3 \Rightarrow -3 < x < 3 \Leftrightarrow (-3, 3)$

**49** When reading this inequality, think of the concept "I want all the numbers that lie at least 5 units from the origin," and your answer should make some common sense.

$$|x| \geq 5 \;\Rightarrow\; x \geq 5 \text{ or } x \leq -5 \;\Leftrightarrow\; (-\infty, -5] \cup [5, \infty)$$

**51** $|x+3| < 0.01 \;\Rightarrow\; -0.01 < x+3 < 0.01 \;\Rightarrow$

$$-3.01 < x < -2.99 \;\Leftrightarrow\; (-3.01, -2.99)$$

**53** $|x+2| + 0.1 \geq 0.2$ {isolate the absolute value expression} $\Rightarrow$

$|x+2| \geq 0.1 \;\Rightarrow\; x+2 \geq 0.1$ or $x+2 \leq -0.1 \;\Rightarrow$

$$x \geq -1.9 \text{ or } x \leq -2.1 \;\Leftrightarrow\; (-\infty, -2.1] \cup [-1.9, \infty)$$

**55** $|2x+5| < 4 \;\Rightarrow\; -4 < 2x+5 < 4 \;\Rightarrow\; -9 < 2x < -1 \;\Rightarrow$

$$-\tfrac{9}{2} < x < -\tfrac{1}{2} \;\Leftrightarrow\; (-\tfrac{9}{2}, -\tfrac{1}{2})$$

**57** $-\frac{1}{3}|6-5x| + 2 \geq 1 \;\Rightarrow\; -\frac{1}{3}|6-5x| \geq -1 \;\Rightarrow\; |6-5x| \leq 3 \;\Rightarrow$

$$-3 \leq 6-5x \leq 3 \;\Rightarrow\; -9 \leq -5x \leq -3 \;\Rightarrow\; \tfrac{9}{5} \geq x \geq \tfrac{3}{5} \;\Leftrightarrow\; [\tfrac{3}{5}, \tfrac{9}{5}]$$

**59** Since $|7x+2| \geq 0 \;\forall x$, $|7x+2| > -2$ has solution $(-\infty, \infty)$.

**61** $|3x-9| > 0 \;\forall x$ except when $3x-9=0$, or $x=3$. The solution is $(-\infty, 3) \cup (3, \infty)$.

**63** $\left|\dfrac{2-3x}{5}\right| \geq 2 \;\Rightarrow\; \dfrac{|2-3x|}{|5|} \geq 2 \;\Rightarrow\; |2-3x| \geq 10 \;\Rightarrow$

$2-3x \geq 10$ or $2-3x \leq -10 \;\Rightarrow\; -3x \geq 8$ or $-3x \leq -12 \;\Rightarrow$

$$x \leq -\tfrac{8}{3} \text{ or } x \geq 4 \;\Leftrightarrow\; (-\infty, -\tfrac{8}{3}] \cup [4, \infty)$$

**65** Since $|5-2x| \geq 0 \;\forall x$, we can multiply the inequality by $|5-2x|$ without changing the direction of the inequality sign. We must exclude $x = \frac{5}{2}$ from the solution since it makes the original inequality undefined.

$\dfrac{3}{|5-2x|} < 2 \;\Rightarrow\; |5-2x| > \frac{3}{2} \;\Rightarrow\; 5-2x > \frac{3}{2}$ or $5-2x < -\frac{3}{2} \;\Rightarrow$

$-2x > -\frac{7}{2}$ or $-2x < -\frac{13}{2} \;\Rightarrow\; x < \frac{7}{4}$ or $x > \frac{13}{4}$ {$\frac{5}{2}$ doesn't fall in this region} $\Leftrightarrow$

$$(-\infty, \tfrac{7}{4}) \cup (\tfrac{13}{4}, \infty)$$

**67** $-2 < |x| < 4 \;\Rightarrow\; -2 < |x|$ *and* $|x| < 4$.

Since $-2$ is always less than $|x|$ {because $|x| \geq 0$}, we only need to consider

$$|x| < 4. \quad |x| < 4 \;\Rightarrow\; -4 < x < 4 \;\Leftrightarrow\; (-4, 4)$$

**69** From the definition of absolute value, $|x-2|$ equals either $x-2$ or $-(x-2)$.

Thus, $1 < |x-2| < 4 \;\Rightarrow\; 1 < x-2 < 4$ or $1 < -(x-2) < 4 \;\Rightarrow$

$1 < x-2 < 4$ or $-1 > x-2 > -4 \;\Rightarrow\; 3 < x < 6$ or $1 > x > -2 \;\Leftrightarrow\; (-2, 1) \cup (3, 6)$.

An alternative method is to rewrite the inequality as $|x-2| > 1$ *and* $|x-2| < 4$.

Solving independently gives us

$x-2 > 1$ or $x-2 < -1 \;\Rightarrow\; x > 3$ or $x < 1$ *and* $-4 < x-2 < 4 \;\Rightarrow\; -2 < x < 6$.

Taking the *intersection* of these intervals gives $(-2, 1) \cup (3, 6)$.

**71** (a) $|x+5| = 3 \Rightarrow x+5=3$ or $x+5=-3 \Rightarrow x=-2$ or $x=-8$.

(b) $|x+5| < 3$ has solutions between the values found in part (a), that is, $(-8, -2)$.

(c) The solutions of $|x+5| > 3$ are the portions of the real line that are not in parts (a) and (b), that is, $(-\infty, -8) \cup (-2, \infty)$.

**73** We could think of this statement as "the difference between $w$ and 148 is at most 2." In symbols, we have $|w - 148| \leq 2$. Intuitively, we know that this inequality must describe the weights from 146 to 150.

**75** The difference of two temperatures $T_1$ and $T_2$ can be represented by $T_1 - T_2$. Since there is no indication as to whether $T_1$ is larger than $T_2$, or vice versa, we will use $|T_1 - T_2|$. Thus, the inequality is $5 < |T_1 - T_2| < 10$.

**77** $30 \leq C \leq 40 \Rightarrow 30 \leq \frac{5}{9}(F-32) \leq 40 \Rightarrow 54 \leq F - 32 \leq 72 \Rightarrow 86 \leq F \leq 104$

**79** Since $V = 110$, $R = \dfrac{110}{I}$, or equivalently, $I = \dfrac{110}{R}$. If the current is not to exceed 10, we want to solve the inequality $I \leq 10$. $I \leq 10 \Rightarrow \dfrac{110}{R} \leq 10 \Rightarrow 110 \leq 10R \Rightarrow$ $\{R > 0$, so we may multiply by $R$ without changing the direction of the inequality$\}$ $10R \geq 110 \Rightarrow R \geq 11$

**81** We want to know what condition will assure us that an object's image is at least 3 times as large as the object, or, equivalently, when $M \geq 3$.

$M \geq 3 \{f = 6\} \Rightarrow \dfrac{6}{6-p} \geq 3 \Rightarrow 6 \geq 18 - 3p$ $\{$since $6 - p > 0$, we can multiply by $6 - p$ and not change the direction of the inequality$\}$ $\Rightarrow 3p \geq 12 \Rightarrow$ $p \geq 4$, but $p < 6$ since $p < f$. Thus, $4 \leq p < 6$.

**83** Let $x$ denote the number of years before $A$ becomes more economical than $B$. The costs are the initial costs plus the yearly costs times the number of years.

$\text{Cost}_A < \text{Cost}_B \Rightarrow 50{,}000 + 4000x < 40{,}000 + 5500x \Rightarrow 10{,}000 < 1500x \Rightarrow$ $x > \frac{20}{3}$, or $6\frac{2}{3}$ yr.

**85** (a) 5 ft 9 in $= 69$ in. In a 40 year period, a person's height will decrease by $40 \times 0.024 = 0.96$ in $\approx 1$ in. The person will be approximately one inch shorter, or 5 ft 8 in. at age 70.

(b) 5 ft 6 in $= 66$ in. In 20 years, a person's height $(h = 66)$ will change by $0.024 \times 20 = 0.48$ in. Thus, $66 - 0.48 \leq h \leq 66 + 0.48 \Rightarrow 65.52 \leq h \leq 66.48$.

## 2.7 Exercises

*Note:* Many solutions for exercises involving inequalities contain a sign diagram. You may want to read Example 3 again if you have trouble interpreting the sign diagrams.

[1] $(3x+1)(5-10x)>0$ has solutions in the interval $(-\frac{1}{3}, \frac{1}{2})$. See *Diagram 1* for details concerning the signs of the individual factors and the resulting sign. We'll use the $\oplus$ symbol to denote that the expression is positive and the $\ominus$ symbol to denote that the expression is negative.

| | | | |
|---|---|---|---|
| Resulting sign: | $\ominus$ | $\oplus$ | $\ominus$ |
| Sign of $5-10x$: | $+$ | $+$ | $-$ |
| Sign of $3x+1$: | $-$ | $+$ | $+$ |
| $x$ values: | $-1/3$ | $1/2$ | |

*Diagram 1*

[3] $(x+2)(x-1)(4-x)\le 0$. From *Diagram 3*, we see the product is negative for $x\in(-2, 1)\cup(4, \infty)$. Since we want to also include the values that make the product equal to zero $\{-2, 1, \text{and } 4\}$, the solution is $[-2, 1]\cup[4, \infty)$.

| | | | | |
|---|---|---|---|---|
| Resulting sign: | $\oplus$ | $\ominus$ | $\oplus$ | $\ominus$ |
| Sign of $4-x$: | $+$ | $+$ | $+$ | $-$ |
| Sign of $x-1$: | $-$ | $-$ | $+$ | $+$ |
| Sign of $x+2$: | $-$ | $+$ | $+$ | $+$ |
| $x$ values: | $-2$ | $1$ | $4$ | |

*Diagram 3*

[5] $x^2-x-6<0 \;\Rightarrow\; (x-3)(x+2)<0$; $(-2, 3)$

| | | | |
|---|---|---|---|
| Resulting sign: | $\oplus$ | $\ominus$ | $\oplus$ |
| Sign of $x-3$: | $-$ | $-$ | $+$ |
| Sign of $x+2$: | $-$ | $+$ | $+$ |
| $x$ values: | $-2$ | $3$ | |

*Diagram 5*

[7] $x^2-2x-5>3 \;\Rightarrow\; x^2-2x-8>0 \;\Rightarrow\; (x-4)(x+2)>0$; $(-\infty, -2)\cup(4, \infty)$

| | | | |
|---|---|---|---|
| Resulting sign: | $\oplus$ | $\ominus$ | $\oplus$ |
| Sign of $x-4$: | $-$ | $-$ | $+$ |
| Sign of $x+2$: | $-$ | $+$ | $+$ |
| $x$ values: | $-2$ | $4$ | |

*Diagram 7*

[9] $x(2x+3)\ge 5 \;\Rightarrow\; 2x^2+3x-5\ge 0 \;\Rightarrow\; (2x+5)(x-1)\ge 0$; $(-\infty, -\frac{5}{2}]\cup[1, \infty)$

| | | | |
|---|---|---|---|
| Resulting sign: | $\oplus$ | $\ominus$ | $\oplus$ |
| Sign of $x-1$: | $-$ | $-$ | $+$ |
| Sign of $2x+5$: | $-$ | $+$ | $+$ |
| $x$ values: | $-5/2$ | $1$ | |

*Diagram 9*

[11] $6x - 8 > x^2 \Rightarrow x^2 - 6x + 8 < 0 \Rightarrow (x-2)(x-4) < 0$; $(2, 4)$

| Resulting sign: | $\oplus$ | $\ominus$ | $\oplus$ |
|---|---|---|---|
| Sign of $x - 4$: | $-$ | $-$ | $+$ |
| Sign of $x - 2$: | $-$ | $+$ | $+$ |
| $x$ values: | 2 | 4 | |

Diagram 11

*Note:* Solving $x^2 <$ (or $>$) $a^2$ for $a > 0$ may be solved using factoring, that is, $x^2 - a^2 < 0 \Rightarrow (x+a)(x-a) < 0 \Rightarrow -a < x < a$; or by taking the square root of each side, that is, $\sqrt{x^2} < \sqrt{a^2} \Rightarrow |x| < a \Rightarrow -a < x < a$.

[13] *Note:* The most common mistake is not remembering that $\sqrt{x^2} = |x|$.

$$x^2 < 16 \Rightarrow |x| < 4 \Rightarrow -4 < x < 4 \Leftrightarrow (-4, 4).$$

[15] $25x^2 - 9 < 0 \Rightarrow x^2 < \frac{9}{25} \Rightarrow |x| < \frac{3}{5} \Rightarrow -\frac{3}{5} < x < \frac{3}{5} \Leftrightarrow (-\frac{3}{5}, \frac{3}{5})$

[17] $16x^2 \geq 9x \Rightarrow x(16x - 9) \geq 0$; $(-\infty, 0] \cup [\frac{9}{16}, \infty)$

| Resulting sign: | $\oplus$ | $\ominus$ | $\oplus$ |
|---|---|---|---|
| Sign of $16x - 9$: | $-$ | $-$ | $+$ |
| Sign of $x$: | $-$ | $+$ | $+$ |
| $x$ values: | 0 | 9/16 | |

Diagram 17

[19] $x^4 + 5x^2 \geq 36 \Rightarrow x^4 + 5x^2 - 36 \geq 0 \Rightarrow (x^2+9)(x^2-4) \geq 0 \Rightarrow$
$x^2 - 4 \geq 0$ { $x^2 + 9 > 0$, so we don't have to consider its sign } $\Rightarrow$

$$x^2 \geq 4 \Rightarrow |x| \geq 2 \Rightarrow x \geq 2 \text{ or } x \leq -2 \Leftrightarrow (-\infty, -2] \cup [2, \infty)$$

[21] $x^3 + 2x^2 - 4x - 8 \geq 0 \Rightarrow x^2(x+2) - 4(x+2) \geq 0 \Rightarrow (x^2-4)(x+2) \geq 0 \Rightarrow$ $(x-2)(x+2)^2 \geq 0$. The expression $(x+2)^2$ is positive except when $x = -2$. The sign is determined by the sign of $x - 2$, which is positive if $x > 2$. Since $x = \pm 2$ make the expression zero, the solution is $x \geq 2$ or $x = -2 \Leftrightarrow \{-2\} \cup [2, \infty)$.

[23] $\dfrac{x^2(x+2)}{(x+2)(x+1)} \leq 0 \Rightarrow \dfrac{x^2}{x+1} \leq 0$ { we will exclude $x = -2$ since it makes the original expression undefined } $\Rightarrow \dfrac{1}{x+1} \leq 0$ { we can divide by $x^2$ since $x^2 \geq 0$ and we will include $x = 0$ since it makes $x^2$ equal to zero and we want all solutions less than *or equal to* zero } $\Rightarrow x + 1 < 0$ { the fraction cannot equal zero and $x + 1$ must be negative so that the fraction is negative } $\Rightarrow x < -1$; $(-\infty, -2) \cup (-2, -1) \cup \{0\}$

[25] $\dfrac{x^2 - x}{x^2 + 2x} \leq 0 \Rightarrow \dfrac{x(x-1)}{x(x+2)} \leq 0 \Rightarrow \dfrac{x-1}{x+2} \leq 0$ { we will exclude $x = 0$ from the solution }; $(-2, 0) \cup (0, 1]$

| Resulting sign: | $\oplus$ | $\ominus$ | $\oplus$ |
|---|---|---|---|
| Sign of $x - 1$: | $-$ | $-$ | $+$ |
| Sign of $x + 2$: | $-$ | $+$ | $+$ |
| $x$ values: | $-2$ | 1 | |

Diagram 25

[27] $\dfrac{x-2}{x^2-3x-10} \geq 0 \;\Rightarrow\; \dfrac{x-2}{(x-5)(x+2)} \geq 0$ {$x=2$ is a solution since it makes the fraction equal to zero, $x=5$ and $x=-2$ are excluded since these values make the fraction undefined}; $(-2,\, 2] \cup (5,\, \infty)$

| Resulting sign: | ⊖ | ⊕ | ⊖ | ⊕ |
|---|---|---|---|---|
| Sign of $x-5$: | $-$ | $-$ | $-$ | $+$ |
| Sign of $x-2$: | $-$ | $-$ | $+$ | $+$ |
| Sign of $x+2$: | $-$ | $+$ | $+$ | $+$ |
| $x$ values: | $-2$ | 2 | 5 | |

*Diagram 27*

[29] $\dfrac{-3x}{x^2-9} > 0 \;\Rightarrow\; \dfrac{x}{(x+3)(x-3)} < 0$ {divide by $-3$}; $(-\infty,\, -3) \cup (0,\, 3)$

| Resulting sign: | ⊖ | ⊕ | ⊖ | ⊕ |
|---|---|---|---|---|
| Sign of $x-3$: | $-$ | $-$ | $-$ | $+$ |
| Sign of $x$: | $-$ | $-$ | $+$ | $+$ |
| Sign of $x+3$: | $-$ | $+$ | $+$ | $+$ |
| $x$ values: | $-3$ | 0 | 3 | |

*Diagram 29*

[31] $\dfrac{x+1}{2x-3} > 2 \;\Rightarrow\; \dfrac{x+1-2(2x-3)}{2x-3} > 0 \;\Rightarrow\; \dfrac{-3x+7}{2x-3} > 0.$

From *Diagram 31*, the solution is $(\frac{3}{2},\, \frac{7}{3})$. Note that you should *not* multiply by the factor $2x-3$ as we did with rational *equations* because $2x-3$ may be positive or negative, and multiplying by it would require solving two inequalities. This method of solution tends to be more difficult than the sign diagram method.

| Resulting sign: | ⊖ | ⊕ | ⊖ |
|---|---|---|---|
| Sign of $-3x+7$: | $+$ | $+$ | $-$ |
| Sign of $2x-3$: | $-$ | $+$ | $+$ |
| $x$ values: | $3/2$ | $7/3$ | |

*Diagram 31*

[33] $\dfrac{1}{x-2} \geq \dfrac{3}{x+1} \;\Rightarrow\; \dfrac{1(x+1)-3(x-2)}{(x-2)(x+1)} \geq 0 \;\Rightarrow\; \dfrac{-2x+7}{(x-2)(x+1)} \geq 0$; $(-\infty,\, -1) \cup (2,\, \frac{7}{2}]$

| Resulting sign: | ⊕ | ⊖ | ⊕ | ⊖ |
|---|---|---|---|---|
| Sign of $-2x+7$: | $+$ | $+$ | $+$ | $-$ |
| Sign of $x-2$: | $-$ | $-$ | $+$ | $+$ |
| Sign of $x+1$: | $-$ | $+$ | $+$ | $+$ |
| $x$ values: | $-1$ | 2 | $7/2$ | |

*Diagram 33*

**35** $\dfrac{4}{3x-2} \le \dfrac{2}{x+1} \;\Rightarrow\; \dfrac{4(x+1)-2(3x-2)}{(3x-2)(x+1)} \le 0 \;\Rightarrow\; \dfrac{-2x+8}{(3x-2)(x+1)} \le 0$; $(-1, \frac{2}{3}) \cup [4, \infty)$

| Resulting sign: | ⊕ | ⊖ | ⊕ | ⊖ |
|---|---|---|---|---|
| Sign of $-2x+8$: | + | + | + | − |
| Sign of $3x-2$: | − | − | + | + |
| Sign of $x+1$: | − | + | + | + |
| $x$ values: | | −1 | 2/3 | 4 |

Diagram 35

**37** $\dfrac{x}{3x-5} \le \dfrac{2}{x-1} \;\Rightarrow\; \dfrac{x(x-1)-2(3x-5)}{(3x-5)(x-1)} \le 0 \;\Rightarrow\; \dfrac{(x-2)(x-5)}{(3x-5)(x-1)} \le 0$; $(1, \frac{5}{3}) \cup [2, 5]$

| Res. sign: | ⊕ | ⊖ | ⊕ | ⊖ | ⊕ |
|---|---|---|---|---|---|
| $x-5$: | − | − | − | − | + |
| $x-2$: | − | − | − | + | + |
| $3x-5$: | − | − | + | + | + |
| $x-1$: | − | + | + | + | + |
| $x$ values: | | 1 | 5/3 | 2 | 5 |

Diagram 37

**39** $x^3 > x \;\Rightarrow\; x^3 - x > 0 \;\Rightarrow\; x(x^2-1) > 0 \;\Rightarrow\; x(x+1)(x-1) > 0$; $(-1, 0) \cup (1, \infty)$

| Resulting sign: | ⊖ | ⊕ | ⊖ | ⊕ |
|---|---|---|---|---|
| Sign of $x-1$: | − | − | − | + |
| Sign of $x$: | − | − | + | + |
| Sign of $x+1$: | − | + | + | + |
| $x$ values: | | −1 | 0 | 1 |

Diagram 39

**41** $v \ge k \;\Rightarrow\; t^3 - 3t^2 - 4t + 20 \ge 8 \;\Rightarrow\; t^3 - 3t^2 - 4t + 12 \ge 0 \;\Rightarrow$

$t^2(t-3) - 4(t-3) \ge 0 \;\Rightarrow\; (t^2-4)(t-3) \ge 0 \;\Rightarrow\; (t+2)(t-2)(t-3) \ge 0.$

For $[0, 5]$, we have the solution $[0, 2] \cup [3, 5]$.

| Resulting sign: | ⊖ | ⊕ | ⊖ | ⊕ |
|---|---|---|---|---|
| Sign of $t-3$: | − | − | − | + |
| Sign of $t-2$: | − | − | + | + |
| Sign of $t+2$: | − | + | + | + |
| $t$ values: | | −2 | 2 | 3 |

Diagram 41

**43** $s > 9 \;\Rightarrow\; -16t^2 + 24t + 1 > 9 \;\Rightarrow\; -16t^2 + 24t - 8 > 0 \;\Rightarrow$

$2t^2 - 3t + 1 < 0$ {divide by $-8$} $\Rightarrow$

$(2t-1)(t-1) < 0$ {use a sign diagram} $\Rightarrow\; \frac{1}{2} < t < 1.$

The dog is more than 9 ft off the ground for $1 - \frac{1}{2} = \frac{1}{2}$ sec.

**45** $d < 75 \;\Rightarrow\; v + \frac{1}{20}v^2 < 75$ {multiply by 20} $\Rightarrow\; 20v + v^2 < 1500 \;\Rightarrow$

$v^2 + 20v - 1500 < 0 \;\Rightarrow\; (v+50)(v-30) < 0$ {use a sign diagram} $\Rightarrow$

$-50 < v < 30 \;\Rightarrow\; 0 \le v < 30$ {since $v \ge 0$}

47 $R > S \Rightarrow \dfrac{4500\,S}{S+500} > S \Rightarrow \dfrac{S(S-4000)}{S+500} < 0$ { use a sign diagram } $\Rightarrow$

$S < -500$ or $0 < S < 4000 \Rightarrow 0 < S < 4000$ {since $S > 0$}

49 $W < 5 \Rightarrow 125\left(\dfrac{6400}{6400+x}\right)^2 < 5 \Rightarrow \left(\dfrac{6400}{6400+x}\right)^2 < \frac{1}{25} \Rightarrow$

$\left(\dfrac{6400}{6400+x}\right)^2 < \left(\frac{1}{5}\right)^2$ { take the square root } $\Rightarrow$

$\dfrac{6400}{6400+x} < \frac{1}{5}$ $\left\{\text{since } \dfrac{6400}{6400+x} > 0\right\}$ $\Rightarrow 32{,}000 < x + 6400 \Rightarrow x > 25{,}600$ km.

51 $7500 \le 0.00334V^2S \le 10{,}000 \Rightarrow 7500 \le 0.00334(210)V^2 \le 10{,}000 \Rightarrow$

$\dfrac{7500}{0.7014} \le V^2 \le \dfrac{10{,}000}{0.7014} \Rightarrow \sqrt{\dfrac{7500}{0.7014}} \le V \le \sqrt{\dfrac{10{,}000}{0.7014}} \Rightarrow$

$103.4 \le V \le 119.4$ ft/sec $\Rightarrow 70.5 \le V \le 81.4$ mi/hr. To convert ft/sec to mi/hr, multiply by $\frac{60}{88}$ or $\frac{15}{22}$, which are reduced forms of $\dfrac{1\text{ foot}}{1\text{ second}} \times \dfrac{3600\text{ seconds}}{1\text{ hour}} \times \dfrac{1\text{ mile}}{5280\text{ feet}}$.

53 By using a table it can be shown that the expression is equal to zero when $x = -3$, $-2$, 2, 4. The expression $Y_1 = x^4 - x^3 - 16x^2 + 4x + 48$ is negative when $x \in (-3, -2) \cup (2, 4)$.

| $x$ | $Y_1$ | $x$ | $Y_1$ |
|---|---|---|---|
| −3.5 | 30.938 | 1.0 | 36 |
| −3.0 | 0 | 1.5 | 19.688 |
| −2.5 | −7.313 | 2.0 | 0 |
| −2.0 | 0 | 2.5 | −18.56 |
| −1.5 | 14.438 | 3.0 | −30 |
| −1.0 | 30 | 3.5 | −26.81 |
| −0.5 | 42.188 | 4.0 | 0 |
| 0.0 | 48 | 4.5 | 60.938 |
| 0.5 | 45.938 | 5.0 | 168 |

## *Chapter 2 Review Exercises*

1 $\left[\dfrac{3x+1}{5x+7} = \dfrac{6x+11}{10x-3}\right] \cdot (5x+7)(10x-3) \Rightarrow (3x+1)(10x-3) =$

$(6x+11)(5x+7) \Rightarrow 30x^2 + x - 3 = 30x^2 + 97x + 77 \Rightarrow 96x = -80 \Rightarrow x = -\frac{5}{6}$

3 $\left[\dfrac{2}{x+5} - \dfrac{3}{2x+1} = \dfrac{5}{6x+3}\right] \cdot 3(x+5)(2x+1) \Rightarrow 6(2x+1) - 9(x+5) = 5(x+5) \Rightarrow$

$3x - 39 = 5x + 25 \Rightarrow -2x = 64 \Rightarrow x = -32$

[4] $\left[\frac{7}{x-2}-\frac{6}{x^2-4}=\frac{3}{2x+4}\right]\cdot 2(x+2)(x-2) \Rightarrow 14(x+2)-12=3(x-2) \Rightarrow$

$11x=-22 \Rightarrow x=-2$, which is not in the domain of the given expressions.

No solution

[5] $\text{LS}=\frac{1}{\sqrt{x}}-2=\frac{1-2\sqrt{x}}{\sqrt{x}}=\text{RS}$, an identity.

The given equation is true for every $x>0$.

[7] $x(3x+4)=5 \Rightarrow 3x^2+4x-5=0 \Rightarrow$

$$x=\frac{-4\pm\sqrt{16+60}}{6}=\frac{-4\pm 2\sqrt{19}}{6}=-\tfrac{2}{3}\pm\tfrac{1}{3}\sqrt{19}$$

[9] $(x-2)(x+1)=3 \Rightarrow x^2-x-2=3 \Rightarrow x^2-x-5=0 \Rightarrow$

$$x=\frac{1\pm\sqrt{1+20}}{2}=\tfrac{1}{2}\pm\tfrac{1}{2}\sqrt{21}$$

[11] $x^{2/3}-2x^{1/3}-15=0 \Rightarrow (x^{1/3}+3)(x^{1/3}-5)=0 \Rightarrow \sqrt[3]{x}=-3,\ 5 \Rightarrow$

$$x=(-3)^3,\ 5^3 \Rightarrow x=-27,\ 125$$

[13] $5x^2=2x-3 \Rightarrow 5x^2-2x+3=0 \Rightarrow x=\frac{2\pm\sqrt{4-60}}{10}=\frac{2\pm 2\sqrt{14}i}{10}=\tfrac{1}{5}\pm\tfrac{1}{5}\sqrt{14}i$

[15] $6x^4+29x^2+28=0 \Rightarrow (2x^2+7)(3x^2+4)=0 \Rightarrow x^2=-\tfrac{7}{2},\ -\tfrac{4}{3} \Rightarrow$

$$x=\pm\tfrac{1}{2}\sqrt{14}i,\ \pm\tfrac{2}{3}\sqrt{3}i$$

[16] $x^4-3x^2+1=0 \Rightarrow x^2=\frac{3\pm\sqrt{5}}{2}\cdot\frac{2}{2}=\frac{6\pm 2\sqrt{5}}{4} \Rightarrow x=\pm\sqrt{\frac{6\pm 2\sqrt{5}}{4}} \Rightarrow$

$$x=\pm\tfrac{1}{2}\sqrt{6\pm 2\sqrt{5}}\approx \pm 1.62,\ \pm 0.62$$

[17] $|4x-1|=7 \Rightarrow 4x-1=7$ or $4x-1=-7 \Rightarrow$

$4x=8$ or $4x=-6 \Rightarrow x=2$ or $x=-\tfrac{3}{2}$

[18] $2|2x+1|+1=19 \Rightarrow 2|2x+1|=18 \Rightarrow |2x+1|=9 \Rightarrow$

$2x+1=9$ or $2x+1=-9 \Rightarrow 2x=8$ or $2x=-10 \Rightarrow x=4$ or $x=-5$

[19] $\left[\frac{1}{x}+6=\frac{5}{\sqrt{x}}\right]\cdot x \Rightarrow 1+6x=5\sqrt{x} \Rightarrow$

$6x-5\sqrt{x}+1=0$ {factoring or substituting would be appropriate} $\Rightarrow$

$$(2\sqrt{x}-1)(3\sqrt{x}-1)=0 \Rightarrow \sqrt{x}=\tfrac{1}{2},\tfrac{1}{3} \Rightarrow x=\tfrac{1}{4},\tfrac{1}{9}$$

Check $x=\tfrac{1}{4}$: LS $=4+6=10$; RS $=5/\tfrac{1}{2}=10 \Rightarrow x=\tfrac{1}{4}$ is a solution.

Check $x=\tfrac{1}{9}$: LS $=9+6=15$; RS $=5/\tfrac{1}{3}=15 \Rightarrow x=\tfrac{1}{9}$ is a solution.

[21] $\sqrt{7x+2}+x=6 \Rightarrow (\sqrt{7x+2})^2=(6-x)^2 \Rightarrow 7x+2=36-12x+x^2 \Rightarrow$

$x^2-19x+34=0 \Rightarrow (x-2)(x-17)=0 \Rightarrow$

$x=2$ and 17 is an extraneous solution.

[23] $\sqrt{3x+1} - \sqrt{x+4} = 1 \Rightarrow \sqrt{3x+1} = 1 + \sqrt{x+4} \Rightarrow$

$3x+1 = 1 + 2\sqrt{x+4} + x + 4 \Rightarrow 2\sqrt{x+4} = 2x - 4 \Rightarrow$

$\sqrt{x+4} = x - 2 \Rightarrow (\sqrt{x+4})^2 = (x-2)^2 \Rightarrow$

$x + 4 = x^2 - 4x + 4 \Rightarrow x^2 - 5x = 0 \Rightarrow x(x-5) = 0 \Rightarrow x = 0, 5.$

Check $x = 0$: LS $= 1 - 2 = -1 \neq$ RS $\Rightarrow x = 0$ is an extraneous solution.

Check $x = 5$: LS $= 4 - 3 = 1 =$ RS $\Rightarrow x = 5$ is a solution.

[25] $3x^2 - 12x + 3 = 0 \Rightarrow x^2 - 4x + 1 = 0 \Rightarrow x^2 - 4x + 4 = -1 + 4 \Rightarrow$

$(x-2)^2 = 3 \Rightarrow x - 2 = \pm\sqrt{3} \Rightarrow x = 2 \pm \sqrt{3}$

[27] The expression $(x-3)^2$ is never less than 0, but it is equal to 0 when $x = 3$.

Thus, $(x-3)^2 \leq 0$ has solution $x = 3$.

[29] $\left[-\frac{1}{2} < \frac{2x+3}{5} < \frac{3}{2}\right] \cdot 10 \Rightarrow -5 < 4x + 6 < 15 \Rightarrow -11 < 4x < 9 \Rightarrow$

$-\frac{11}{4} < x < \frac{9}{4} \Leftrightarrow (-\frac{11}{4}, \frac{9}{4})$

[31] $\frac{6}{10x+3} < 0 \Rightarrow 10x + 3 < 0 \{\text{since } 6 > 0\} \Rightarrow x < -\frac{3}{10} \Leftrightarrow (-\infty, -\frac{3}{10})$

[33] $2\,|3 - x| + 1 > 5 \Rightarrow 2\,|3 - x| > 4 \Rightarrow |3 - x| > 2 \Rightarrow$

$3 - x > 2$ or $3 - x < -2 \Rightarrow 1 > x$ or $5 < x \Rightarrow x < 1$ or $x > 5 \Leftrightarrow$

$(-\infty, 1) \cup (5, \infty)$

[35] $|16 - 3x| \geq 5 \Rightarrow 16 - 3x \geq 5$ or $16 - 3x \leq -5 \Rightarrow$

$-3x \geq -11$ or $-3x \leq -21 \Rightarrow x \leq \frac{11}{3}$ or $x \geq 7 \Leftrightarrow (-\infty, \frac{11}{3}] \cup [7, \infty)$

[37] $10x^2 + 11x > 6 \Rightarrow 10x^2 + 11x - 6 > 0 \Rightarrow (2x+3)(5x-2) > 0$; $(-\infty, -\frac{3}{2}) \cup (\frac{2}{5}, \infty)$

| Resulting sign: | ⊕ | ⊖ | ⊕ |
|---|---|---|---|
| Sign of $5x-2$: | − | − | + |
| Sign of $2x+3$: | − | + | + |
| $x$ values: | | −3/2 | 2/5 |

Diagram 37

[38] $x(x-3) \leq 10 \Rightarrow x^2 - 3x - 10 \leq 0 \Rightarrow (x-5)(x+2) \leq 0$; $[-2, 5]$

| Resulting sign: | ⊕ | ⊖ | ⊕ |
|---|---|---|---|
| Sign of $x - 5$: | − | − | + |
| Sign of $x+2$: | − | + | + |
| $x$ values: | | −2 | 5 |

Diagram 38

[39] $\frac{x^2(3-x)}{x+2} \leq 0 \Rightarrow \frac{3-x}{x+2} \leq 0$ {include 0}; $(-\infty, -2) \cup \{0\} \cup [3, \infty)$

| Resulting sign: | ⊖ | ⊕ | ⊖ |
|---|---|---|---|
| Sign of $3-x$: | + | + | − |
| Sign of $x+2$: | − | + | + |
| $x$ values: | | −2 | 3 |

Diagram 39

[41] $\dfrac{3}{2x+3} < \dfrac{1}{x-2} \Rightarrow \dfrac{3(x-2)-1(2x+3)}{(2x+3)(x-2)} < 0 \Rightarrow \dfrac{x-9}{(2x+3)(x-2)} < 0;$

$(-\infty, -\frac{3}{2}) \cup (2, 9)$

| Resulting sign: | ⊖ | ⊕ | ⊖ | ⊕ |
|---|---|---|---|---|
| Sign of $x-9$: | − | − | − | + |
| Sign of $x-2$: | − | − | + | + |
| Sign of $2x+3$: | − | + | + | + |
| $x$ values: | −3/2 | 2 | 9 | |

Diagram 41

[43] $x^3 > x^2 \Rightarrow x^2(x-1) > 0 \; \{x^2 \geq 0\} \Rightarrow x - 1 > 0 \Rightarrow x > 1 \Leftrightarrow (1, \infty)$

[44] $(x^2 - x)(x^2 - 5x + 6) < 0 \Rightarrow$

$x(x-1)(x-2)(x-3) < 0$; $(0, 1) \cup (2, 3)$

| Res. sign: | ⊕ | ⊖ | ⊕ | ⊖ | ⊕ |
|---|---|---|---|---|---|
| $x-3$: | − | − | − | − | + |
| $x-2$: | − | − | − | + | + |
| $x-1$: | − | − | + | + | + |
| $x$: | − | + | + | + | + |
| $x$ values: | 0 | 1 | 2 | 3 | |

Diagram 44

[45] $P + N = \dfrac{C+2}{C} \Rightarrow C(P+N) = C+2 \Rightarrow CP + CN = C + 2 \Rightarrow$

$CP + CN - C = 2 \Rightarrow C(P + N - 1) = 2 \Rightarrow C = \dfrac{2}{P+N-1}$

[47] $V = \frac{4}{3}\pi r^3 \Rightarrow r^3 = \dfrac{3V}{4\pi} \Rightarrow r = \sqrt[3]{\dfrac{3V}{4\pi}}$

[49] $c = \sqrt{4h(2R-h)} \Rightarrow c^2 = 8Rh - 4h^2 \Rightarrow 4h^2 - 8Rh + c^2 = 0 \Rightarrow$

$$h = \frac{8R \pm \sqrt{64R^2 - 16c^2}}{8} = \frac{8R \pm 4\sqrt{4R^2 - c^2}}{8} = R \pm \tfrac{1}{2}\sqrt{4R^2 - c^2}$$

[50] $V = \frac{1}{3}\pi h(r^2 + R^2 + rR) \Rightarrow r^2 + Rr + R^2 - \dfrac{3V}{\pi h} = 0 \Rightarrow$

$(\pi h)r^2 + (\pi hR)r + (\pi hR^2 - 3V) = 0 \Rightarrow$

$$r = \frac{-\pi hR \pm \sqrt{\pi^2 h^2 R^2 - 4\pi h(\pi hR^2 - 3V)}}{2\pi h} = \frac{-\pi hR \pm \sqrt{12\pi hV - 3\pi^2 h^2 R^2}}{2\pi h}.$$

Since $r > 0$, we must use the plus sign, and $r = \dfrac{-\pi hR + \sqrt{12\pi hV - 3\pi^2 h^2 R^2}}{2\pi h}$.

[51] $(7 + 5i) - (-8 + 3i) = (7 + 8) + (5 - 3)i = 15 + 2i$

[53] $(3 + 8i)^2 = 3^2 + 2(3)(8i) + (8i)^2 = (9 - 64) + 48i = -55 + 48i$

[55] $\dfrac{6-3i}{2+7i} = \dfrac{6-3i}{2+7i} \cdot \dfrac{2-7i}{2-7i} = \dfrac{(12-21) + (-42-6)i}{53} = -\dfrac{9}{53} - \dfrac{48}{53}i$

[58] $R = 2$ and $R_1 = 5 \Rightarrow \left[\frac{1}{2} = \frac{1}{5} + \frac{1}{R_2}\right] \cdot 10R_2 \Rightarrow 5R_2 = 2R_2 + 10 \Rightarrow R_2 = \frac{10}{3}$ ohms

[59] Let $P$ denote the principal that will be invested, and $r$ the yield rate of the stock fund. $\text{Income}_{\text{stocks}} - 28\%$ federal tax $- 7\%$ state tax $= \text{Income}_{\text{bonds}}$ $\Rightarrow$

$(Pr) - 0.28(Pr) - 0.07(Pr) = 0.07186P$ {divide by $P$} $\Rightarrow$

$1r - 0.28r - 0.07r = 0.07186 \Rightarrow 0.65r = 0.07186 \Rightarrow r \approx 0.11055$, or, 11.055%.

[60] Let $x$ denote the number of $\text{cm}^3$ of gold. $\text{Grams}_{\text{gold}} + \text{Grams}_{\text{silver}} = \text{Grams}_{\text{total}}$ $\Rightarrow$

$x(19.3) + (5 - x)(10.5) = 80 \Rightarrow 8.8x = 27.5 \Rightarrow x = 3.125$.

The number of grams of gold is $19.3x = 60.3125 \approx 60.3$.

[62] Let $x$ denote the number of grams of 95% ethyl alcohol solution used,

$400 - x$ the number of grams of water.

$95(x) + 0(400 - x) = 75(400)$ {all in %} $\Rightarrow 95x = 75(400) \Rightarrow$

$x = \frac{6000}{19} \approx 315.8$. Use 315.8 g of ethyl alcohol and 84.2 g of water.

[64] Let $x$ denote the distance upstream. 10 gallons of gas @ 16 mi/gal = 160 miles.

At 20 mi/hr, there is enough fuel for 8 hours of travel.

The rate of the boat upstream is 15 mi/hr and the rate downstream is 25 mi/hr.

$\text{Time}_{\text{up}} + \text{Time}_{\text{down}} = \text{Time}_{\text{total}}$ $\Rightarrow$

$$\left[\frac{x}{15} + \frac{x}{25} = 8\right] \cdot 75 \Rightarrow 5x + 3x = 600 \Rightarrow 8x = 600 \Rightarrow x = 75 \text{ mi.}$$

[66] Let $x$ denote the speed of the wind.

$\text{Distance}_{\text{with wind}} = \text{Distance}_{\text{against wind}} \Rightarrow (320 + x)\frac{1}{2} = (320 - x)\frac{3}{4}$ $\{d = rt\}$ $\Rightarrow$

$640 + 2x = 960 - 3x \Rightarrow 5x = 320 \Rightarrow x = 64$ mi/hr

[67] Let $50 + r$ denote the rate the automobile, that is, $r$ is the rate over 50 mi/hr.

The automobile must travel $40 + 20 = 60$ ft more than the truck (traveling at 50 mi/hr) in 5 seconds. Since 1 mi/hr $= \frac{5280}{3600} = \frac{22}{15}$ ft/sec, the automobile's rate in *excess* of 50 mi/hr is $\frac{22}{15}r$. Thus, $d = rt \Rightarrow 60 = (\frac{22}{15}r)(5) \Rightarrow r = \frac{90}{11}$.

The rate is $50 + \frac{90}{11} = \frac{640}{11} \approx 58.2$ mi/hr.

[68] Let $x$ denote the number of hours needed to fill an empty bin.

Using the hourly rates, $\left[\frac{1}{2} - \frac{1}{5} = \frac{1}{x}\right] \cdot 10x \Rightarrow 5x - 2x = 10 \Rightarrow 3x = 10 \Rightarrow$

$x = \frac{10}{3}$ hr. Since the bin was half-full at the start, $\frac{1}{2}x = \frac{1}{2} \cdot \frac{10}{3} = \frac{5}{3}$ hr, or, 1 hr 40 min.

[70] Let $d$ denote the distance from the center of the city to a corner and $2x$ denote the length of one side of the city. $x^2 + x^2 = d^2 \Rightarrow d = \sqrt{2}x$.

$A$ = area of the city $= (2x)^2 = 4x^2$, or $2d^2$. Currently: $d = 10 \Rightarrow A = 200$. One decade ago: $A = 150 \Rightarrow d = \sqrt{75} = 5\sqrt{3}$. The change in $d$ is $10 - 5\sqrt{3} \approx 1.34$ mi.

[72] (a) The eastbound car has distance $20t$ and the southbound car has distance

$(-2+50t)$. $d^2 = (20t)^2 + (-2+50t)^2 \Rightarrow d = \sqrt{2900t^2 - 200t + 4}$

(b) $104 = \sqrt{2900t^2 - 200t + 4} \Rightarrow 2900t^2 - 200t - 10{,}812 = 0 \Rightarrow$

$$725t^2 - 50t - 2703 = 0 \Rightarrow t = \frac{50 \pm \sqrt{7{,}841{,}200}}{1450}\ \{t > 0\} = \frac{5 + 2\sqrt{19{,}603}}{145} \approx$$

1.97, or approximately 11:58 A.M.

[74] Let $x$ denote the length of one side of an end.

(a) $V = lwh \Rightarrow 48 = 6 \cdot x \cdot x \Rightarrow x^2 = 8 \Rightarrow x = 2\sqrt{2}$ ft

(b) $S = lw + 2wh + 2lh \Rightarrow 44 = 6x + 2(x^2) + 2(6x) \Rightarrow 44 = 2x^2 + 18x \Rightarrow$

$x^2 + 9x - 22 = 0 \Rightarrow (x+11)(x-2) = 0 \Rightarrow x = 2$ ft

[76] Let $x$ denote the width of the tiled area, $2x$ the length.

The bathing area has measurements $x - 2$ and $2x - 2$. For the bathing area,

width · length = area $\Rightarrow (x-2)(2x-2) = 40 \Rightarrow 2x^2 - 6x + 4 = 40 \Rightarrow$

$x^2 - 3x + 2 = 20 \Rightarrow x^2 - 3x - 18 = 0 \Rightarrow (x-6)(x+3) = 0 \Rightarrow x = 6\ \{x > 0\}$.

The tiled area is 12 ft by 6 ft and the bathing area is 10 ft by 4 ft.

[78] $pv = 200 \Rightarrow v = \frac{200}{p}$. $25 \le v \le 50 \Rightarrow 25 \le \frac{200}{p} \le 50 \Rightarrow \frac{1}{25} \ge \frac{p}{200} \ge \frac{1}{50} \Rightarrow$

$8 \ge p \ge 4 \Rightarrow 4 \le p \le 8$

[80] $v > 1100 \Rightarrow 1087\sqrt{\frac{T}{273}} > 1100 \Rightarrow$

$$\sqrt{\frac{T}{273}} > \frac{1100}{1087} \Rightarrow \frac{T}{273} > \frac{(1100)^2}{(1087)^2} \Rightarrow T > \frac{273(1100)^2}{(1087)^2} \Rightarrow T > 279.57 \text{ K}$$

[81] $T = 2\pi\sqrt{\frac{l}{980}} \Rightarrow l = \frac{980\,T^2}{4\pi^2}$. $98 \le l \le 100 \Rightarrow 98 \le \frac{980\,T^2}{4\pi^2} \le 100 \Rightarrow$

$$\frac{2\pi^2}{5} \le T^2 \le \frac{20\pi^2}{49} \Rightarrow \frac{10\pi^2}{25} \le T^2 \le \frac{20\pi^2}{49} \Rightarrow \frac{\pi}{5}\sqrt{10} \le T \le \frac{2\pi}{7}\sqrt{5}\ \{T \ge 0\},$$

or, approximately, $1.987 \le T \le 2.007$ sec.

[82] $v = \dfrac{626.4}{\sqrt{h + 6372}} \Rightarrow h = \dfrac{(626.4)^2}{v^2} - 6372$ and $h > 100 \Rightarrow$

$$\frac{(626.4)^2}{v^2} - 6372 > 100 \Rightarrow \frac{(626.4)^2}{v^2} > 6472 \Rightarrow$$

$$v^2 < \frac{(626.4)^2}{6472}\ \{v > 0\} \Rightarrow 0 < v < \frac{626.4}{\sqrt{6472}} \approx 7.786 \text{ km/sec}$$

84 Let $x$ denote the number of trees over 24. Then $24 + x$ represents the total number of trees planted per acre, and $600 - 12x$ represents the number of apples per tree.

Total apples = (# of trees)(# of apples per tree)

$= (24 + x)(600 - 12x) = -12x^2 + 312x + 14{,}400.$

Apples $\geq 16{,}416 \Rightarrow -12x^2 + 312x + 14{,}400 \geq 16{,}416 \Rightarrow$

$-12x^2 + 312x - 2016 \geq 0 \Rightarrow x^2 - 26x + 168 \leq 0 \Rightarrow (x - 12)(x - 14) \leq 0 \Rightarrow$

$12 \leq x \leq 14$. Hence, 36 to 38 trees per acre should be planted.

86 The $y$-values are increasing slowly and can best be described by equation (3), $y = 3\sqrt{x - 0.5}$.

## *Chapter 2 Discussion Exercises*

1 We need to solve the equation $x^2 - xy + y^2 = 0$ for $x$.

Use the quadratic formula with $a = 1$, $b = -y$, and $c = y^2$.

$$x = \frac{-(-y) \pm \sqrt{(-y)^2 - 4(1)(y^2)}}{2(1)} = \frac{y \pm \sqrt{y^2 - 4y^2}}{2} = \frac{y \pm \sqrt{-3y^2}}{2} = \frac{y \pm |y|\sqrt{3}\,i}{2}.$$

Since this equation has imaginary solutions, $x^2 - xy + y^2$ is not factorable over the reals. A similar argument holds for $x^2 + xy + y^2$.

3 (a) $$\frac{1}{\dfrac{a+bi}{c+di}} = \frac{c+di}{a+bi} \cdot \frac{a-bi}{a-bi} = \frac{ac + bd + (ad - bc)i}{a^2 + b^2} = \frac{ac + bd}{a^2 + b^2} + \frac{ad - bc}{a^2 + b^2}i = p + qi$$

(b) Yes, try an example such as 3/4. Let $a = 3$, $b = 0$, $c = 4$, and $d = 0$. Then, from part (a), $p + qi = \frac{12}{9} + \frac{0}{9}i = \frac{12}{9} = \frac{4}{3}$, which is the multiplicative inverse of 3/4.

(c) $a$ and $b$ cannot both be 0 because then the denominator would be 0.

5 *Hint*: Try these examples to help you get to the general solution.

(1) $x^2 + 1 \geq 0$ {In this case, $a > 0$, $D = -4 < 0$, and by examining a sign chart with $x^2 + 1$ as the only factor, we see that the solution is $x \in \mathbb{R}$.}

(2) $x^2 - 2x - 3 \geq 0$

(3) $-x^2 - 4 \geq 0$

(4) $-x^2 - 2x - 1 \geq 0$

(5) $-x^2 + 2x + 3 \geq 0$

General solutions categorized by $a$ and $D$:

(1) $a > 0$, $D \leq 0$: solution is $x \in \mathbb{R}$

(2) $a > 0$, $D > 0$: let $x_1 = (-b - \sqrt{D})/(2a)$ and $x_2 = (-b + \sqrt{D})/(2a)$ $\Rightarrow$ solution is $(-\infty, x_1] \cup [x_2, \infty)$

(3) $a < 0$, $D < 0$: solution is $\{\ \}$

(4) $a < 0$, $D = 0$: solution is $x = -b/(2a)$

(5) $a < 0$, $D > 0$: solution is $[x_1, x_2]$

7 The first equation, $\sqrt{2x-3} + \sqrt{x+5} = 0$, is a sum of square roots that is equal to 0. The only way this could be true is if both radicals are actually equal to 0. It is easy to see that $\sqrt{x+5}$ is equal to 0 only if $x = -5$, but $-5$ will not make $\sqrt{2x-3}$ equal to 0, so there is no reason to try to solve the first equation.

On the other hand, the second equation, $\sqrt[3]{2x-3} + \sqrt[3]{x+5} = 0$, can be written as $\sqrt[3]{2x-3} = -\sqrt[3]{x+5}$. This just says that one cube root is equal to the negative of another cube root, which could happen since a cube root can be negative. Solving this equation gives us $2x - 3 = -(x+5)$ $\Rightarrow$ $3x = -2$ $\Rightarrow$ $x = -\frac{2}{3}$.

# Chapter 3: Functions and Graphs

## 3.1 Exercises

[1] The points $A(5,\ -2)$, $B(-5,\ -2)$, $C(5,\ 2)$, $D(-5,\ 2)$, $E(3,\ 0)$, and $F(0,\ 3)$ are plotted in *Figure 1.*

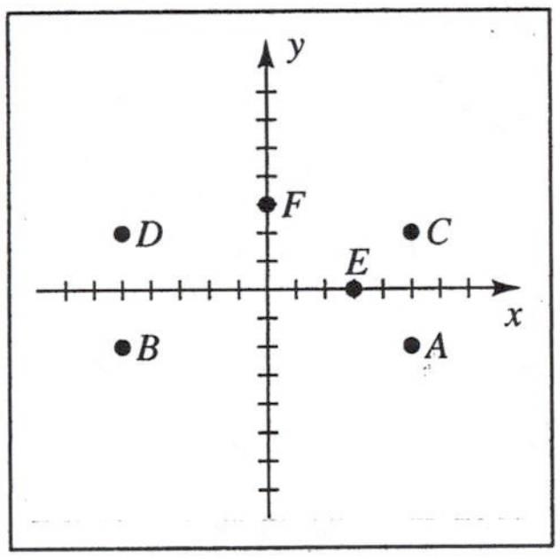


*Figure 1*

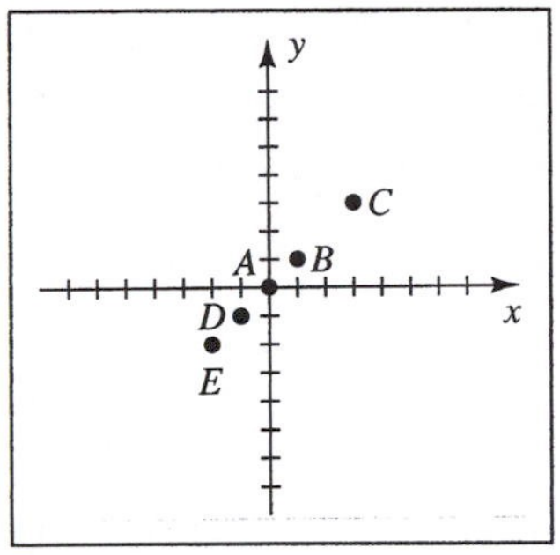


*Figure 3*

[3] The points $A(0,\ 0)$, $B(1,\ 1)$, $C(3,\ 3)$, $D(-1,\ -1)$, and $E(-2,\ -2)$ are plotted in *Figure 3.* The set of all points of the form $(a,\ a)$ is the line bisecting quadrants I and III.

[5] The points are $A(3,\ 3)$, $B(-3,\ 3)$, $C(-3,\ -3)$, $D(3,\ -3)$, $E(3,\ 0)$, and $F(0,\ 3)$.

[7] (a) $x = -2$ is the line parallel to the $y$-axis that intersects the $x$-axis at $(-2,\ 0)$.

(b) $y = 3$ is the line parallel to the $x$-axis that intersects the $y$-axis at $(0,\ 3)$.

(c) $x \geq 0$ { $x$ is zero or positive }
is the set of all points to the right of and on the $y$-axis.

(d) $xy > 0$ { $x$ and $y$ have the same sign, that is, either both are positive or both are negative } is the set of all points in quadrants I and III.

(e) $y < 0$ { $y$ is negative } is the set of all points below the $x$-axis.

(f) $x = 0$ is the set of all points on the $y$-axis.

[9] (a) $A(4,\ -3),\ B(6,\ 2) \Rightarrow d(A,\ B) = \sqrt{(6-4)^2 + [2-(-3)]^2} = \sqrt{4+25} = \sqrt{29}$

(b) $A(4,\ -3),\ B(6,\ 2) \Rightarrow M_{AB} = \left(\frac{4+6}{2},\ \frac{-3+2}{2}\right) = (5,\ -\frac{1}{2})$

[11] (a) $A(-5,\ 0),\ B(-2,\ -2) \Rightarrow$

$$d(A,\ B) = \sqrt{[-2-(-5)]^2 + (-2-0)^2} = \sqrt{9+4} = \sqrt{13}$$

(b) $A(-5,\ 0),\ B(-2,\ -2) \Rightarrow M_{AB} = \left(\frac{-5+(-2)}{2},\ \frac{0+(-2)}{2}\right) = (-\frac{7}{2},\ -1)$

[13] (a) $A(7,\ -3),\ B(3,\ -3) \Rightarrow d(A,\ B) = \sqrt{(3-7)^2 + [-3-(-3)]^2} = \sqrt{16+0} = 4$

(b) $A(7,\ -3),\ B(3,\ -3) \Rightarrow M_{AB} = \left(\frac{7+3}{2},\ \frac{-3+(-3)}{2}\right) = (5,\ -3)$

[15] We need to show that the sides satisfy the Pythagorean theorem. Finding the distances, we have $d(A,\ B) = \sqrt{98}$, $d(B,\ C) = \sqrt{32}$, and $d(A,\ C) = \sqrt{130}$. Since $d(A,\ C)$ is the largest of the three values, it must be the hypotenuse, hence, we need to check if $d(A,\ C)^2 = d(A,\ B)^2 + d(B,\ C)^2$. Since $(\sqrt{130})^2 = (\sqrt{98})^2 + (\sqrt{32})^2$, we know that $\triangle ABC$ is a right triangle. The area of a triangle is given by $A = \frac{1}{2}(\text{base})(\text{height})$. We can use $d(B,\ C)$ for the base and $d(A,\ B)$ for the height. Hence, area $= \frac{1}{2}bh = \frac{1}{2}(\sqrt{32})(\sqrt{98}) = \frac{1}{2}(4\sqrt{2})(7\sqrt{2}) = \frac{1}{2}(28)(2) = 28$.

[17] We need to show that all 4 sides are the same length. Checking, we find that $d(A,\ B) = d(B,\ C) = d(C,\ D) = d(D,\ A) = \sqrt{29}$. This guarantees that we have a rhombus {a parallelogram with 4 equal sides}. Thus, we also need to show that adjacent sides meet at right angles. This can be done by showing that two adjacent sides and a diagonal form a right triangle. Using $\triangle ABC$, we see that $d(A,\ C) = \sqrt{58}$ and hence $d(A,\ C)^2 = d(A,\ B)^2 + d(B,\ C)^2$. We conclude that $ABCD$ is a square.

[19] Let $B = (x,\ y)$. $A(-3,\ 8) \Rightarrow M_{AB} = \left(\frac{-3+x}{2}, \frac{8+y}{2}\right)$.

Since $M_{AB} = C(5,\ -10)$, we must have $\frac{-3+x}{2} = 5$ and $\frac{8+y}{2} = -10 \Rightarrow$

$-3 + x = 2(5)$ and $8 + y = 2(-10) \Rightarrow x = 13$ and $y = -28$. $B = (13,\ -28)$.

[21] The perpendicular bisector of $AB$ is the line that passes through the midpoint of segment $AB$ and intersects segment $AB$ at a right angle. The points on the perpendicular bisector are all equidistant from $A$ and $B$. Thus, we need to show that $d(A,\ C) = d(B,\ C)$. Since each of these is $\sqrt{145}$, we conclude that $C$ is on the perpendicular bisector of $AB$.

[23] We must have $d(A,\ P) = d(B,\ P)$.

$\sqrt{(x+4)^2 + (y+3)^2} = \sqrt{(x-6)^2 + (y-1)^2} \Rightarrow$

$x^2 + 8x + 16 + y^2 + 6y + 9 = x^2 - 12x + 36 + y^2 - 2y + 1$ {square both sides} $\Rightarrow$

$8x + 6y + 25 = -12x - 2y + 37 \Rightarrow 20x + 8y = 12 \Rightarrow 5x + 2y = 3$

[25] Let $O(0,\ 0)$ represent the origin. Applying the distance formula with $O$ and $P(x,\ y)$, we have $d(O,\ P) = 5 \Rightarrow \sqrt{(x-0)^2 + (y-0)^2} = 5 \Rightarrow \sqrt{x^2 + y^2} = 5$.

This is a circle of radius 5 with center at the origin.

[27] Let $Q(0, y)$ be an arbitrary point on the $y$-axis.

Applying the distance formula with $Q$ and $P(5, 3)$, we have $6 = d(P, Q) \Rightarrow$

$6 = \sqrt{(0-5)^2 + (y-3)^2} \Rightarrow 36 = 25 + y^2 - 6y + 9 \Rightarrow$

$y^2 - 6y - 2 = 0 \Rightarrow y = 3 \pm \sqrt{11}$. The points are $(0, 3 + \sqrt{11})$ and $(0, 3 - \sqrt{11})$.

[29] $5 = \sqrt{(2a-1)^2 + (a-3)^2} \Rightarrow 25 = 4a^2 - 4a + 1 + a^2 - 6a + 9 \Rightarrow$

$5a^2 - 10a - 15 = 0 \Rightarrow a^2 - 2a - 3 = 0 \Rightarrow (a-3)(a+1) = 0 \Rightarrow a = 3, -1$.

Since the $y$-coordinate is negative in the third quadrant, $a = -1$,

and $(2a, a) = (-2, -1)$.

[31] With $P(a, 3)$ and $Q(5, 2a)$, $d(P, Q) > \sqrt{26} \Rightarrow \sqrt{(5-a)^2 + (2a-3)^2} > \sqrt{26} \Rightarrow$

$25 - 10a + a^2 + 4a^2 - 12a + 9 > 26 \Rightarrow 5a^2 - 22a + 8 > 0 \Rightarrow (5a-2)(a-4) > 0$.

Using *Diagram 31*, we see that $a < \frac{2}{5}$ or $a > 4$ will assure us that $d(P, Q) > \sqrt{26}$.

| | | | |
|---|---|---|---|
| Resulting sign: | ⊕ | ⊖ | ⊕ |
| Sign of $5a - 2$: | $-$ | $+$ | $+$ |
| Sign of $a - 4$: | $-$ | $-$ | $+$ |
| $a$ values: | 2/5 | 4 | |

*Diagram 31*

[33] Let $M$ be the midpoint of the hypotenuse. Then $M = (\frac{1}{2}a, \frac{1}{2}b)$. Since $A = (a, 0)$, we

have $d(A, M) = \sqrt{(a - \frac{1}{2}a)^2 + (0 - \frac{1}{2}b)^2}$

$= \sqrt{(\frac{1}{2}a)^2 + (-\frac{1}{2}b)^2} = \sqrt{\frac{1}{4}a^2 + \frac{1}{4}b^2} = \sqrt{\frac{1}{4}(a^2 + b^2)} = \frac{1}{2}\sqrt{a^2 + b^2}$.

In a similar fashion, show that $d(B, M) = d(O, M) = \frac{1}{2}\sqrt{a^2 + b^2}$.

[35] Plot the points $A(-5, -3.5)$, $B(-2, 2)$, $C(1, 0.5)$, $D(4, 1)$, and $E(7, 2.5)$.

$[-10, 10]$ by $[-10, 10]$

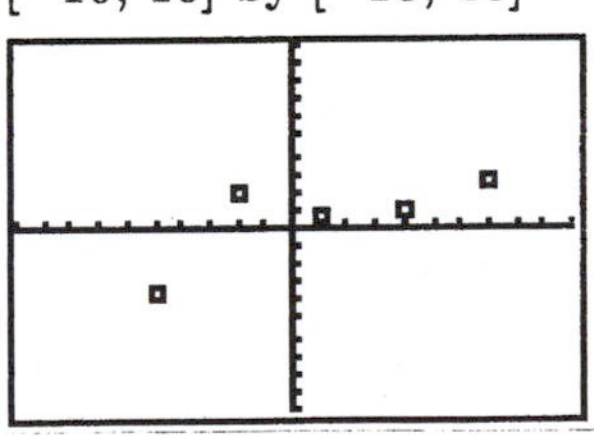

*Figure 35*

$[1988, 1996]$ by $[54 \times 10^6, 61 \times 10^6, 10^6]$

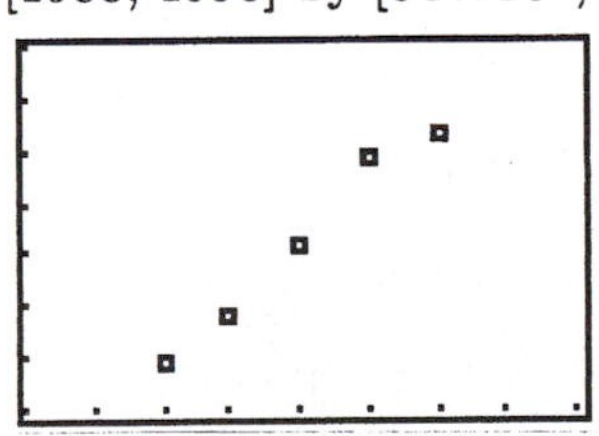

*Figure 37*

[37] (a) Plot the points (1990, 54,871,330), (1991, 55,786,390), (1992, 57,211,600), (1993, 58,834,440), and (1994, 59,332,200).

(b) The number of cable subscribers is increasing each year.

## 3.2 Exercises

[1] As in Example 1, we expect the graph to be a line.

Creating a table of values similar to those in the text, we have:

| $x$ | $-2$ | $-1$ | $0$ | $1$ | $2$ |
|---|---|---|---|---|---|
| $y$ | $-7$ | $-5$ | $-3$ | $-1$ | $1$ |

By plotting these points and connecting them, we obtain *Figure 1.*

To find the $x$-intercept, let $y = 0$ in $y = 2x - 3$, and solve for $x$ to get 1.5.

To find the $y$-intercept, let $x = 0$ in $y = 2x - 3$, and solve for $y$ to get $-3$.

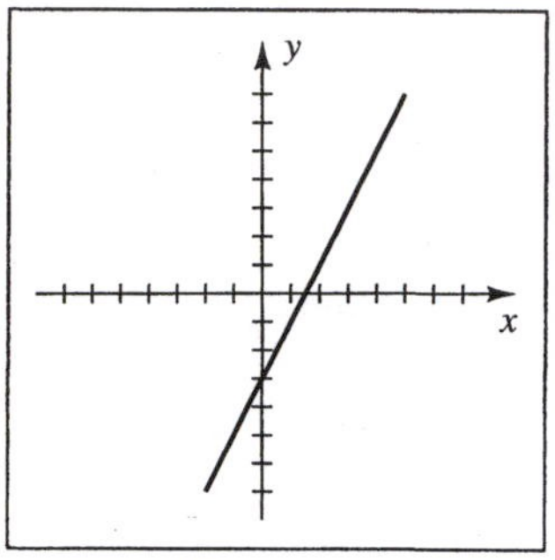


*Figure 1*

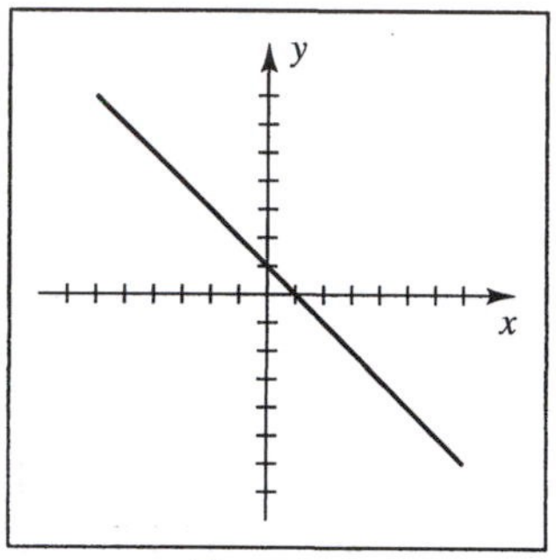


*Figure 3*

[3] $y = -x + 1$ • $x$-intercept: $y = 0 \Rightarrow 0 = -x + 1 \Rightarrow x = 1$

$y$-intercept: $x = 0 \Rightarrow y = -0 + 1 = 1$

[5] $y = -4x^2$ • Multiplying the $y$-values of $y = x^2$ by $-4$ gives us all negative $y$-values for $y = -4x^2$. The vertex of the parabola is at (0, 0), so its $x$-intercept is 0 and its $y$-intercept is 0.

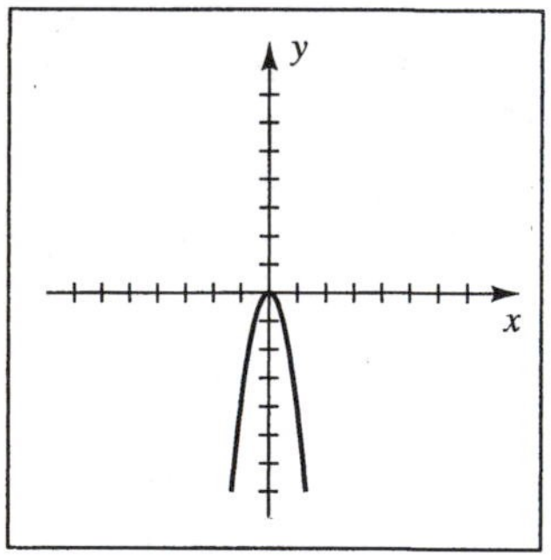


*Figure 5*

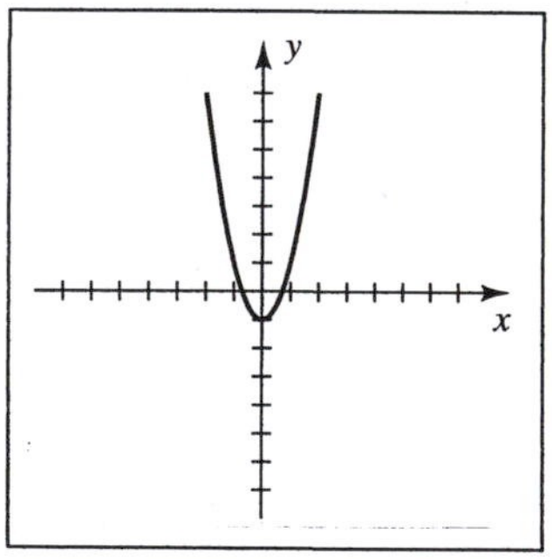


*Figure 7*

[7] $y = 2x^2 - 1$ • $x$-intercepts: $y = 0 \Rightarrow 0 = 2x^2 - 1 \Rightarrow 1 = 2x^2 \Rightarrow$

$x^2 = \frac{1}{2} \Rightarrow x = \pm\sqrt{\frac{1}{2}}$, which can be written as $\pm\frac{1}{\sqrt{2}}$ or $\pm\frac{\sqrt{2}}{2}$ or $\pm\frac{1}{2}\sqrt{2}$

$y$-intercept: $x = 0 \Rightarrow y = -1$ (continued)

Since we can substitute $-x$ for $x$ in the equation and obtain an equivalent equation, we know the graph is symmetric with respect to the $y$-axis. We will make use of this fact when constructing our table. As in Example 2, we obtain a parabola.

| $x$ | $\pm 2$ | $\pm\frac{3}{2}$ | $\pm 1$ | $\pm\frac{1}{2}$ | $0$ |
|---|---|---|---|---|---|
| $y$ | $7$ | $\frac{7}{2}$ | $1$ | $-\frac{1}{2}$ | $-1$ |

[9] $x = \frac{1}{4}y^2$ • This graph is similar to the one in Example 5—multiplying by $\frac{1}{4}$ narrows the parabola $x = y^2$. The $x$-intercept is 0 and the $y$-intercept is 0.

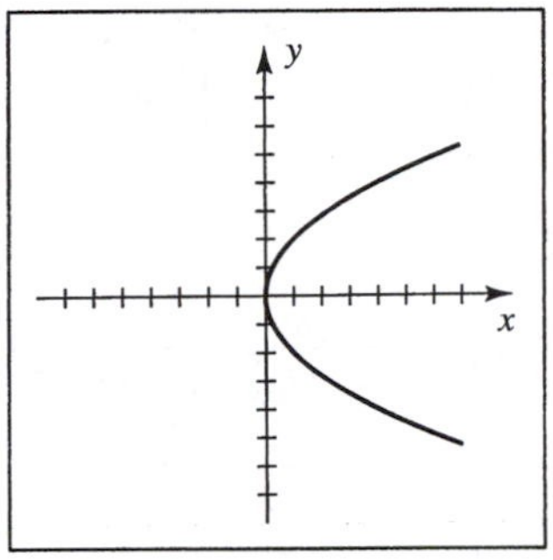

*Figure 9*

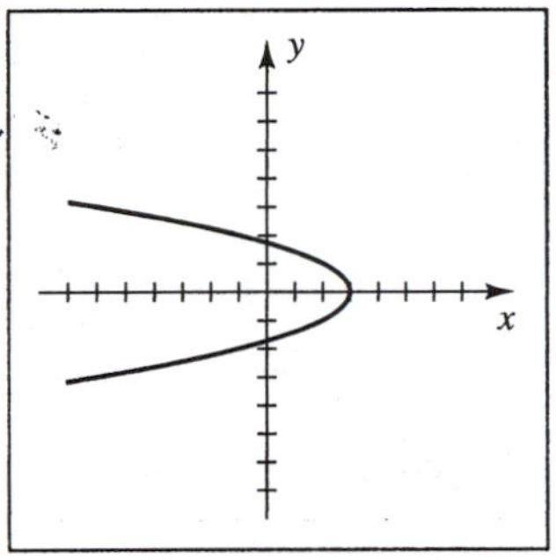

*Figure 11*

[11] $x = -y^2 + 3$ • $x$-intercept: $y = 0 \Rightarrow x = 3$;

$y$-intercepts: $x = 0 \Rightarrow 0 = -y^2 + 3 \Rightarrow y^2 = 3 \Rightarrow y = \pm\sqrt{3}$

Since we can substitute $-y$ for $y$ in the equation and obtain an equivalent equation, we know the graph is symmetric with respect to the $x$-axis. We will make use of this fact when constructing our table. As in Example 5, we obtain a parabola.

| $x$ | $-13$ | $-6$ | $-1$ | $2$ | $3$ |
|---|---|---|---|---|---|
| $y$ | $\pm 4$ | $\pm 3$ | $\pm 2$ | $\pm 1$ | $0$ |

[13] $y = -\frac{1}{2}x^3$ • The graph is similar to the graph of $y = \frac{1}{4}x^3$ in Example 6. The negative has the effect of "flipping" the graph about the $x$-axis. The $x$-intercept is 0 and the $y$-intercept is 0.

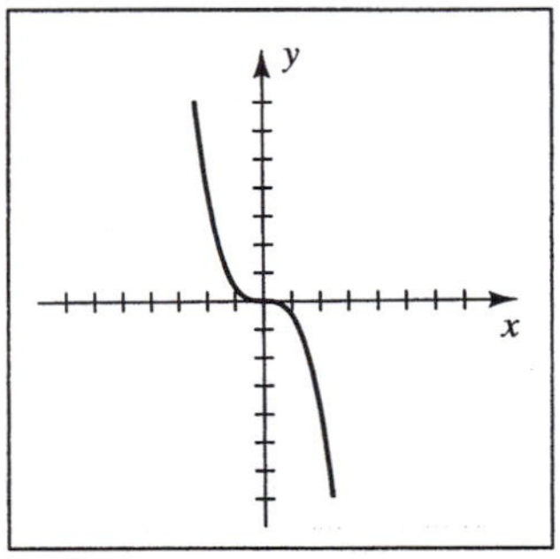

*Figure 13*

[15] $y = x^3 - 8$ • $x$-intercept: $y = 0 \Rightarrow 0 = x^3 - 8 \Rightarrow x^3 = 8 \Rightarrow x = \sqrt[3]{8} = 2$

$y$-intercept: $x = 0 \Rightarrow y = -8$

The effect of the $-8$ is to shift the graph of $y = x^3$ down 8 units.

| $x$ | $-2$ | $-1$ | 0 | 1 | 2 |
|---|---|---|---|---|---|
| $y$ | $-16$ | $-9$ | $-8$ | $-7$ | 0 |

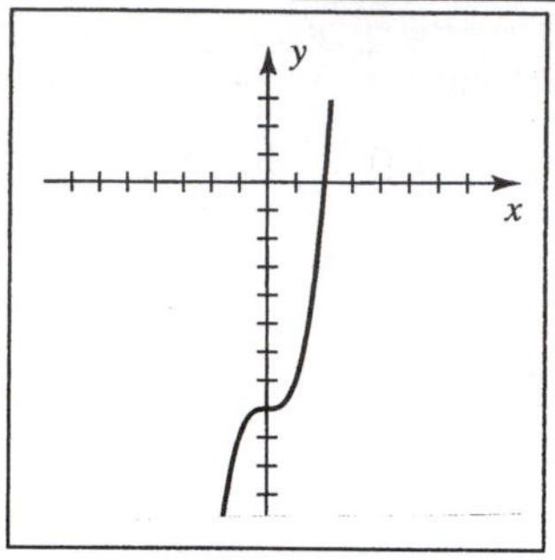


Figure 15

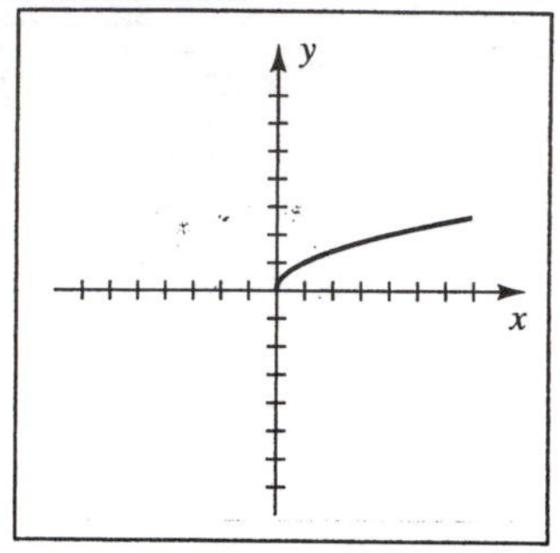


Figure 17

[17] $y = \sqrt{x}$ • $x$-intercept 0; $y$-intercept 0

This is the top half of the parabola $x = y^2$.

An equation of the bottom half is $y = -\sqrt{x}$.

[19] $y = \sqrt{x} - 4$ •

$x$-intercept: $y = 0 \Rightarrow 0 = \sqrt{x} - 4 \Rightarrow$

$4 = \sqrt{x} \Rightarrow x = 16$

$y$-intercept: $x = 0 \Rightarrow y = -4$

| $x$ | 0 | 1 | 4 | 9 | 16 |
|---|---|---|---|---|---|
| $y$ | $-4$ | $-3$ | $-2$ | $-1$ | 0 |

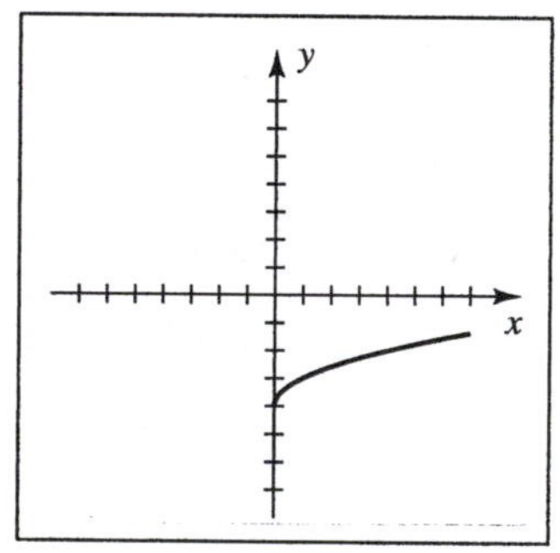


Figure 19

[21] You may be able to do this exercise mentally. For example (using Exercise 1 with $y = 2x - 3$), we see that substituting $-x$ for $x$ gives us $y = -2x - 3$; substituting $-y$ for $y$ gives us $-y = 2x - 3$ or, equivalently, $y = -2x + 3$; and substituting $-x$ for $x$ and $-y$ for $y$ gives us $-y = -2x - 3$ or, equivalently, $y = 2x + 3$. None of the resulting equations are equivalent to the original equation, so there is no symmetry with respect to the $y$-axis, $x$-axis, or the origin.

(a) The graphs of the equations in Exercises 5 and 7 are symmetric with respect to the $y$-axis.

(b) The graphs of the equations in Exercises 9 and 11 are symmetric with respect to the $x$-axis.

(c) The graph of the equation in Exercise 13 is symmetric with respect to the origin.

[23] $x^2 + y^2 = 11$ is a circle of radius $\sqrt{11}$ with center at the origin.

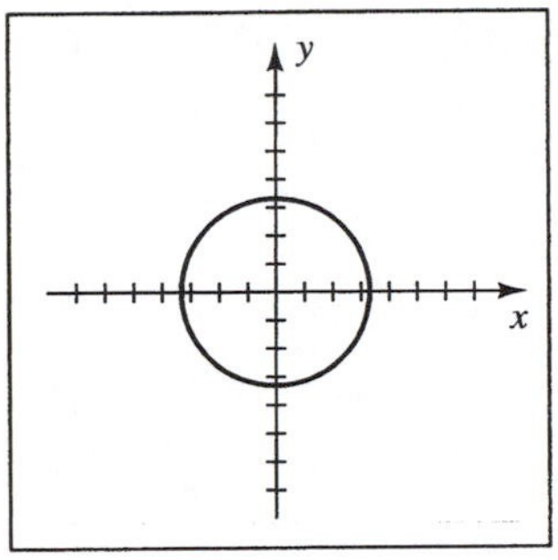


Figure 23

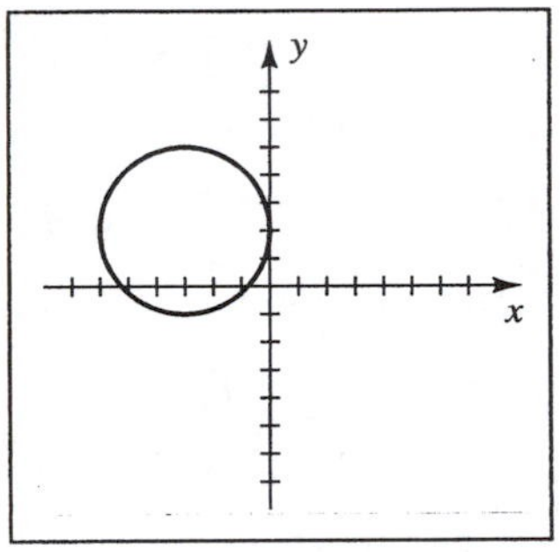


Figure 25

[25] $(x+3)^2 + (y-2)^2 = 9$ is a circle of radius $r = \sqrt{9} = 3$ with center $C(-3,\ 2)$. To determine the center from the given equation, it may help to ask yourself "What values makes the expressions $(x+3)$ and $(y-2)$ equal to zero?" The answers are $-3$ and 2.

[27] $(x+3)^2 + y^2 = 16$ is a circle of radius $r = \sqrt{16} = 4$ with center $C(-3,\ 0)$.

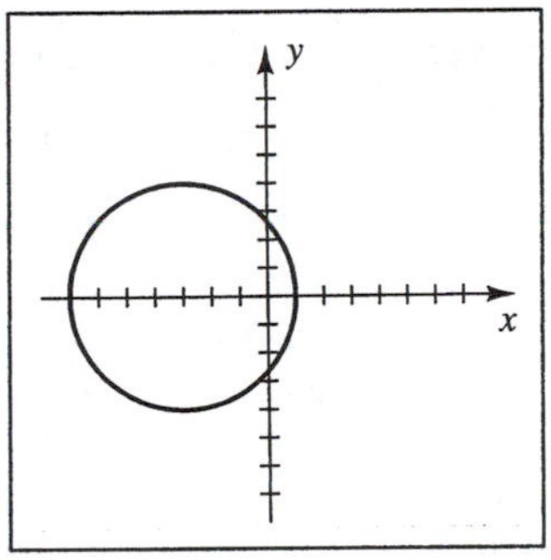


Figure 27

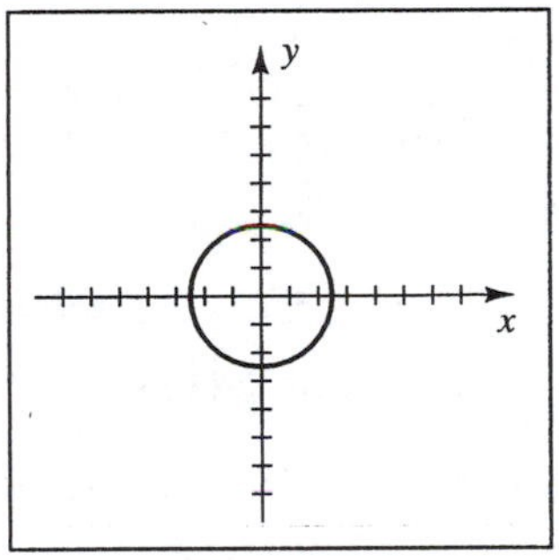


Figure 29

[29] $4x^2 + 4y^2 = 25 \quad \Rightarrow$

$$x^2 + y^2 = \tfrac{25}{4} \text{ is a circle of radius } r = \sqrt{\tfrac{25}{4}} = \tfrac{5}{2} \text{ with center } C(0,\ 0).$$

[31] As in Example 9, $y = -\sqrt{16 - x^2}$ is the lower half of the circle $x^2 + y^2 = 16$.

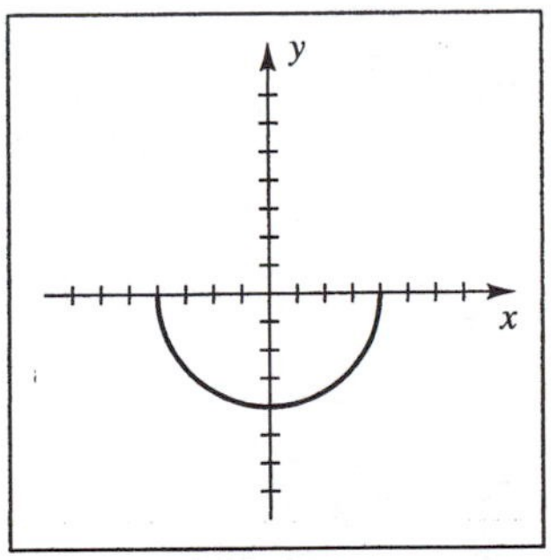


Figure 31

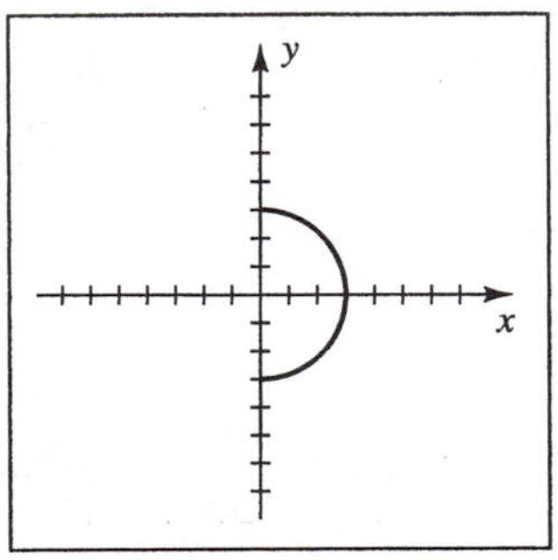


Figure 33

[33] $x = \sqrt{9 - y^2}$ is the right half of the circle $x^2 + y^2 = 9$.

[35] Center $C(2, -3)$, radius 5 • $(x-2)^2+(y+3)^2=5^2=25$

[37] Center $C(\frac{1}{4}, 0)$, radius $\sqrt{5}$ • $(x-\frac{1}{4})^2+y^2=(\sqrt{5})^2=5$

[39] An equation of a circle with center $C(-4, 6)$ is $(x+4)^2+(y-6)^2=r^2$.

Since the circle passes through $P(1, 2)$, we know that $x=1$ and $y=2$ is one solution of the general equation. Letting $x=1$ and $y=2$ yields $5^2+(-4)^2=r^2 \Rightarrow r^2=41$.

An equation is $(x+4)^2+(y-6)^2=41$.

[41] "Tangent to the $y$-axis" means that the circle will intersect the $y$-axis at exactly one point. The distance from the center $C(-3, 6)$ to this point of tangency is 3 units—this is the length of the radius of the circle. An equation is $(x+3)^2+(y-6)^2=9$.

[43] Since the radius is 4 and $C(h, k)$ is in QII, $h=-4$ and $k=4$.

An equation is $(x+4)^2+(y-4)^2=16$.

[45] The center of the circle is the midpoint $M$ of $A(4, -3)$ and $B(-2, 7)$. $M=(1, 2)$.

The radius of the circle is $\frac{1}{2}\cdot d(A, B)=\frac{1}{2}\sqrt{[4-(-2)]^2+(-3-7)^2}=\frac{1}{2}\sqrt{36+100}=$

$\frac{1}{2}\sqrt{136}=\frac{1}{2}\sqrt{4\cdot 34}=\sqrt{34}$. An equation is $(x-1)^2+(y-2)^2=34$.

Alternatively, once we know the center, $(1, 2)$, we also know that the equation has the form $(x-1)^2+(y-2)^2=r^2$. Now substitute 4 for $x$ and $-3$ for $y$ to obtain $3^2+(-5)^2=r^2 \Rightarrow 9+25=r^2$, or $r^2=34$.

[47] $x^2+y^2-4x+6y-36=0$ {complete the square on $x$ and $y$} $\Rightarrow$

$x^2-4x+\underline{4}+y^2+6y+\underline{9}=36+\underline{4}+\underline{9} \Rightarrow$

$(x-2)^2+(y+3)^2=49$. This is a circle with center $C(2, -3)$ and radius $r=7$.

[49] $x^2+y^2+4y-117=0 \Rightarrow x^2+y^2+4y+\underline{4}=117+\underline{4} \Rightarrow x^2+(y+2)^2=121$.

$C(0, -2)$; $r=11$

[51] $2x^2+2y^2-12x+4y-15=0$ {add 15 to both sides and divide by 2} $\Rightarrow$

$x^2+y^2-6x+2y=\frac{15}{2}$ {complete the square on $x$ and $y$} $\Rightarrow$

$x^2-6x+\underline{9}+y^2+2y+\underline{1}=\frac{15}{2}+\underline{9}+\underline{1} \Rightarrow$

$(x-3)^2+(y+1)^2=\frac{35}{2}$. This is a circle with center $C(3, -1)$ and radius $r=\frac{1}{2}\sqrt{70}$.

[53] $x^2+y^2+4x-2y+5=0 \Rightarrow x^2+4x+\underline{4}+y^2-2y+\underline{1}=-5+\underline{4}+\underline{1} \Rightarrow$

$(x+2)^2+(y-1)^2=0$. $C(-2, 1)$; $r=0$ (a point)

[55] $x^2+y^2-2x-8y+19=0 \Rightarrow$

$x^2-2x+\underline{1}+y^2-8y+\underline{16}=-19+\underline{1}+\underline{16} \Rightarrow$

$(x-1)^2+(y-4)^2=-2$. This is not a circle since $r^2$ cannot equal $-2$.

[57] To obtain equations for the upper and lower halves, we solve the given equation for $y$ in terms of $x$. $x^2+y^2=36 \Rightarrow y^2=36-x^2 \Rightarrow y=\pm\sqrt{36-x^2}$.

The upper half is $y=\sqrt{36-x^2}$ and the lower half is $y=-\sqrt{36-x^2}$. (continued)

To obtain equations for the right and left halves, we solve for $x$ in terms of $y$.

$x^2 + y^2 = 36 \;\Rightarrow\; x^2 = 36 - y^2 \;\Rightarrow\; x = \pm\sqrt{36 - y^2}$.

The right half is $x = \sqrt{36 - y^2}$ and the left half is $x = -\sqrt{36 - y^2}$.

**59** To obtain equations for the upper and lower halves, we solve the given equation for $y$ in terms of $x$. $(x-2)^2 + (y+1)^2 = 49 \;\Rightarrow\; (y+1)^2 = 49 - (x-2)^2 \;\Rightarrow$

$y + 1 = \pm\sqrt{49 - (x-2)^2} \;\Rightarrow\; y = -1 \pm \sqrt{49 - (x-2)^2}$.

The upper half is $y = -1 + \sqrt{49 - (x-2)^2}$ and

the lower half is $y = -1 - \sqrt{49 - (x-2)^2}$.

To obtain equations for the right and left halves, we solve for $x$ in terms of $y$.

$(x-2)^2 + (y+1)^2 = 49 \;\Rightarrow\; (x-2)^2 = 49 - (y+1)^2 \;\Rightarrow$

$x - 2 = \pm\sqrt{49 - (y+1)^2} \;\Rightarrow\; x = 2 \pm \sqrt{49 - (y+1)^2}$. The right half is

$x = 2 + \sqrt{49 - (y+1)^2}$ and the left half is $x = 2 - \sqrt{49 - (y+1)^2}$.

**61** From the figure, we see that the diameter of the circle is $1 - (-7) = 8$ units, so the radius is $\frac{1}{2}(8) = 4$. The center $(h, k)$ is at the average of the values shown; that is, $h = \dfrac{-7+1}{2} = -3$ and $k = \dfrac{-2+6}{2} = 2$. Using the standard form of an equation of a circle, $(x-h)^2 + (y-k)^2 = r^2$, we have $[x - (-3)]^2 + (y-2)^2 = 4^2$, or

$(x+3)^2 + (y-2)^2 = 4^2$ {or use 16}.

**63** The figure shows the lower semicircle of a circle centered at the origin having radius 4. The circle has equation $x^2 + y^2 = 4^2$. We want the lower semicircle, so we must have *negative* $y$-values, indicating that we should solve for $y$ and use the negative sign. $x^2 + y^2 = 4^2 \;\Rightarrow\; y^2 = 4^2 - x^2 \;\Rightarrow\; y = \pm\sqrt{4^2 - x^2}$, so $y = -\sqrt{4^2 - x^2}$ is the desired equation.

**65** We need to determine if the distance from $P$ to $C$ is *less than* $r$, *greater than* $r$, or *equal to* $r$ and hence, $P$ will be *inside* the circle, *outside* the circle, or *on* the circle, respectively.

(a) $P(2, 3)$, $C(4, 6) \;\Rightarrow\; d(P, C) = \sqrt{4+9} = \sqrt{13} < r$ $\{r = 4\}$ $\Rightarrow$ $P$ is *inside* $C$.

(b) $P(4, 2)$, $C(1, -2) \;\Rightarrow\; d(P, C) = \sqrt{9+16} = 5 = r$ $\{r = 5\}$ $\Rightarrow$ $P$ is *on* $C$.

(c) $P(-3, 5)$, $C(2, 1) \;\Rightarrow\; d(P, C) = \sqrt{25+16} = \sqrt{41} > r$ $\{r = 6\}$ $\Rightarrow$

$P$ is *outside* $C$.

**67** (a) To find the $x$-intercepts, let $y = 0$ and solve the resulting equation for $x$.

$$x^2 - 4x + 4 = 0 \Rightarrow (x-2)^2 = 0 \Rightarrow x - 2 = 0 \Rightarrow x = 2.$$

(b) To find the $y$-intercepts, let $x = 0$ and solve the resulting equation for $y$.

$$y^2 - 6y + 4 = 0 \Rightarrow y = \frac{6 \pm \sqrt{36-16}}{2} = \frac{6 \pm 2\sqrt{5}}{2} = 3 \pm \sqrt{5}.$$

**69** $x^2 + y^2 + 4x - 6y + 4 = 0 \Leftrightarrow (x+2)^2 + (y-3)^2 = 9$. This is a circle with center $C(-2, 3)$ and radius 3. The circle we want has the same center, $C(-2, 3)$, and radius that is equal to the distance from $C$ to $P(2, 6)$.

$$d(P, C) = \sqrt{16+9} = 5 \text{ and an equation is } (x+2)^2 + (y-3)^2 = 25.$$

**71** The equation of circle $C_2$ is $(x-h)^2 + (y-2)^2 = 2^2$. If we draw a line from the origin to the center of $C_2$, we form a right triangle with hypotenuse $5 - 2$ {$C_2$ radius $-$ $C_1$ radius} $= 3$ and sides of length 2 and $h$. Thus, $h^2 + 2^2 = 3^2 \Rightarrow h = \sqrt{5}$.

**73** The graph of $y_1$ is *below* the graph of $y_2$ to the left of $x = -3$ and to the right of $x = 2$. Writing the <u>$x$-values</u> in interval notation gives us $(-\infty, -3) \cup (2, \infty)$. Note that the $y$-values play no role in writing the answer in interval notation.

**75** $y_1 < y_2$ between $x = -1$ and $x = 1$, excluding $x = 0$ since $y_1 = y_2$ at that value, so the interval is $(-1, 0) \cup (0, 1)$.

**77** The viewing rectangles significantly affect the shape of the circle. Viewing rectangle (2) results in a graph that most looks like a circle.

$[-2, 2]$ by $[-2, 2]$

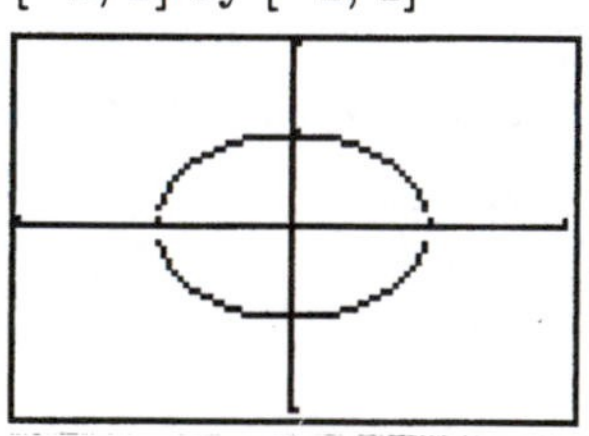

*Figure 77(1)*

$[-3, 3]$ by $[-2, 2]$

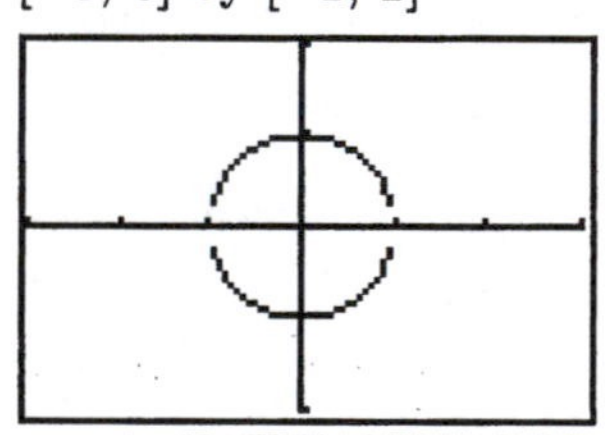

*Figure 77(2)*

$[-2, 2]$ by $[-5, 5]$

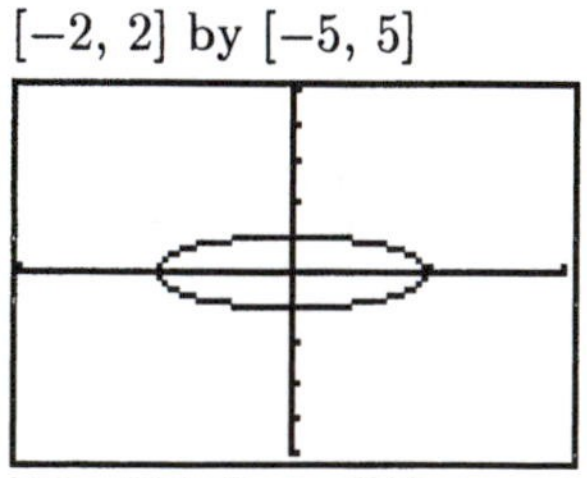

*Figure 77(3)*

$[-5, 5]$ by $[-2, 2]$

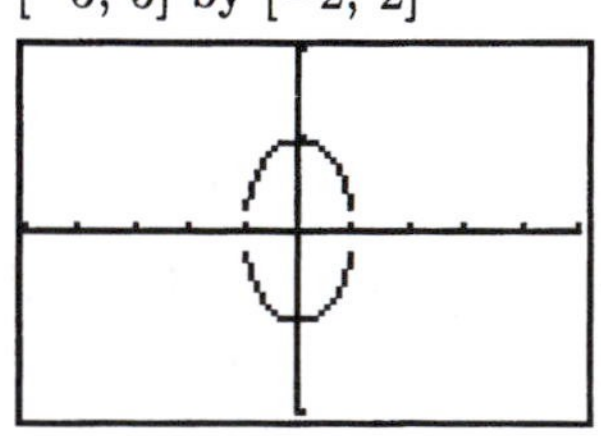

*Figure 77(4)*

[79] Assign $x^3 - \frac{9}{10}x^2 - \frac{43}{25}x + \frac{24}{25}$ to $Y_1$. After trying a standard viewing rectangle, we see that the $x$-intercepts are near the origin and we choose the viewing rectangle [−6, 6] by [−4, 4]. This is simply one choice, not necessarily the best choice. For most graphing calculator exercises, we have selected viewing rectangles that are in a 3 : 2 proportion (horizontal : vertical) to maintain a true proportion. From the graph, there are three $x$-intercepts. Use a root feature to determine that they are −1.2, 0.5, and 1.6.

[−6, 6] by [−4, 4]

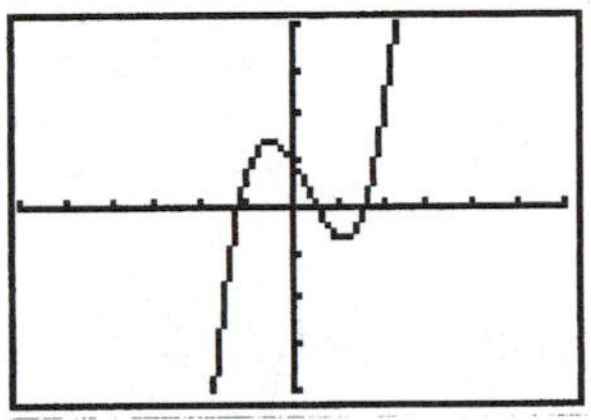

*Figure 79*

[81] Make the assignments $Y_1 = x^3 + x$, $Y_2 = \sqrt{1 - x^2}$, and $Y_3 = -Y_2$.
From the graph, there are two points of intersection.
They are approximately (0.6, 0.8) and (−0.6, −0.8).

[−3, 3] by [−2, 2]

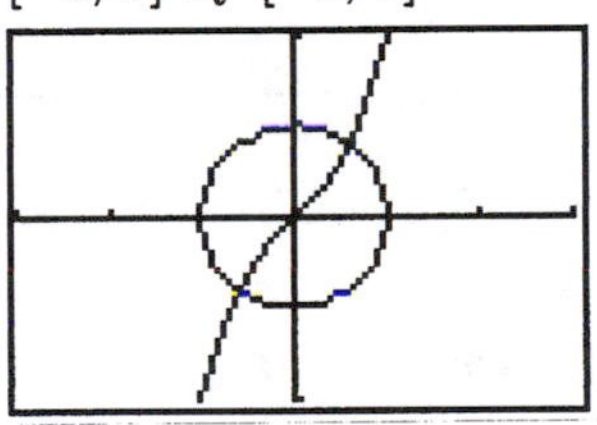

*Figure 81*

[−3, 3] by [−2, 2]

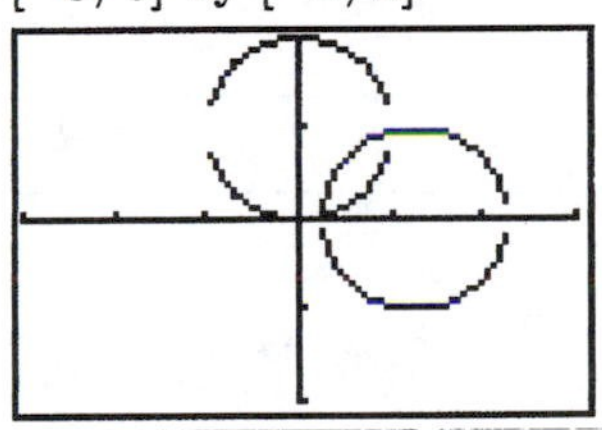

*Figure 83*

[83] Depending on the type of graphing utility used, you may need to solve for $y$ first.

$x^2 + (y-1)^2 = 1 \;\Rightarrow\; y = 1 \pm \sqrt{1 - x^2}$; $(x - \frac{5}{4})^2 + y^2 = 1 \;\Rightarrow\; y = \pm\sqrt{1 - (x - \frac{5}{4})^2}$.

Make the assignments $Y_1 = \sqrt{1 - x^2}$, $Y_2 = 1 + Y_1$, $Y_3 = 1 - Y_1$, $Y_4 = \sqrt{1 - (x - \frac{5}{4})^2}$, and $Y_5 = -Y_4$. Be sure to "turn off" $Y_1$ before graphing. From the graph, there are two points of intersection. They are approximately (0.999, 0.968) and (0.251, 0.032).

**85** The cars are initially 4 miles apart. The distance between them decreases to 0 when they meet on the highway after 2 minutes. Then, the distance between them starts to increase until it is 4 miles after a total of 4 minutes.

[0, 4] by [0, 4]

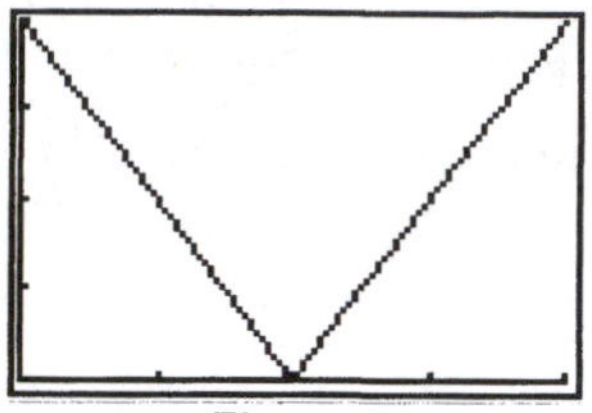

Figure 85

[−50, 50, 10] by [900, 1200, 100]

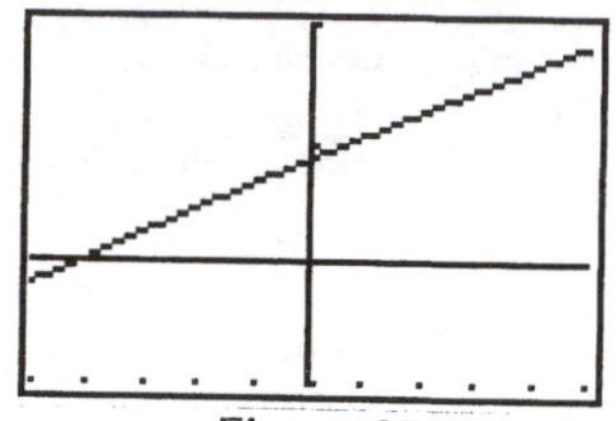

Figure 87

**87** (a) $T = 20°\text{C}$ and $v = 1087\sqrt{\dfrac{T+273}{273}} \Rightarrow v = 1087\sqrt{\dfrac{20+273}{273}} \approx 1126$ ft/sec.

(b) Algebraically: $v = 1000 \Rightarrow 1000 = 1087\sqrt{\dfrac{T+273}{273}} \Rightarrow$

$$\sqrt{\frac{T+273}{273}} = \frac{1000}{1087} \Rightarrow \frac{T+273}{273} = \frac{1000^2}{1087^2} \Rightarrow T + 273 = \frac{273 \cdot 1000^2}{1087^2} \Rightarrow$$

$$T = \frac{273 \cdot 1000^2}{1087^2} - 273 \approx -42°\text{C}.$$

Graphically: Graph $Y_1 = 1087\sqrt{(T+273)/273}$ and $Y_2 = 1000$.

At the point of their intersection, $T \approx -42°\text{C}$.

## 3.3 Exercises

**1** $A(-3, 2)$, $B(5, -4)$ $\Rightarrow$ $m_{AB} = \dfrac{y_2 - y_1}{x_2 - x_1} = \dfrac{(-4) - 2}{5 - (-3)} = \dfrac{-6}{8} = -\dfrac{3}{4}$

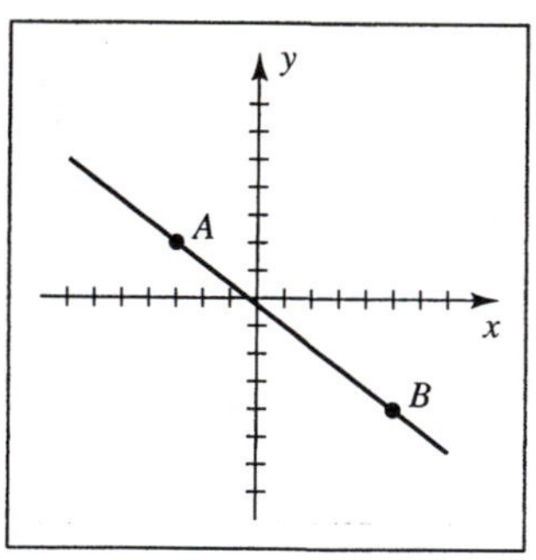


Figure 1

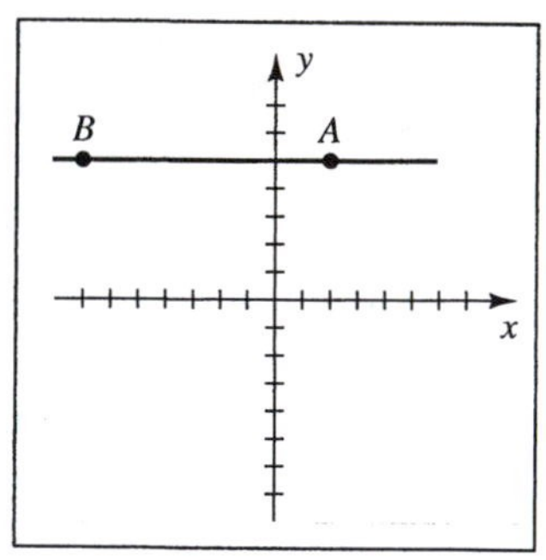


Figure 3

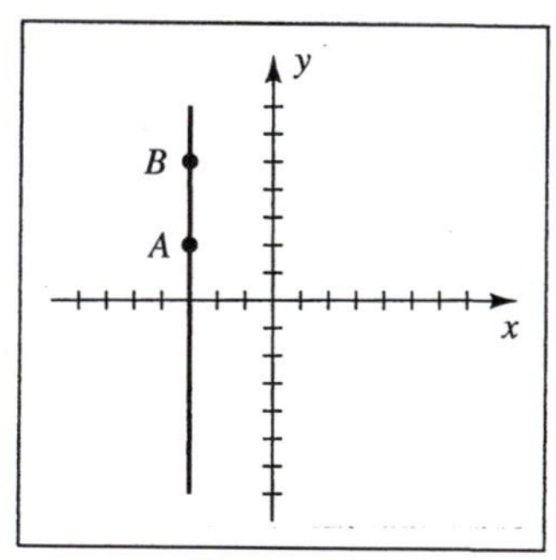


Figure 5

**3** $A(2, 5)$, $B(-7, 5)$ $\Rightarrow$ $m_{AB} = \dfrac{y_2 - y_1}{x_2 - x_1} = \dfrac{5 - 5}{-7 - 2} = \dfrac{0}{-9} = 0$ {horizontal line}

**5** $A(-3, 2)$, $B(-3, 5)$ $\Rightarrow$ $m_{AB} = \dfrac{y_2 - y_1}{x_2 - x_1} = \dfrac{5 - 2}{-3 - (-3)} = \dfrac{3}{0} \Rightarrow$

$m$ is undefined {vertical line}

**7** To show that the polygon is a parallelogram, we must show that the slopes of opposite sides are equal. $A(-3,\ 1)$, $B(5,\ 3)$, $C(3,\ 0)$, $D(-5,\ -2)$ $\Rightarrow$

$$m_{AB} = \tfrac{1}{4} = m_{DC} \quad \text{and} \quad m_{DA} = \tfrac{3}{2} = m_{CB}.$$

**9** To show that the polygon is a rectangle, we must show that the slopes of opposite sides are equal (parallel lines) and the slopes of two adjacent sides are negative reciprocals (perpendicular lines). $A(6,\ 15)$, $B(11,\ 12)$, $C(-1,\ -8)$, $D(-6,\ -5)$ $\Rightarrow$

$$m_{DA} = \tfrac{5}{3} = m_{CB} \quad \text{and} \quad m_{AB} = -\tfrac{3}{5} = m_{DC}.$$

**11** $A(-1,\ -3)$ is 5 units to the left and 5 units down from $B(4,\ 2)$. The fourth vertex $D$ will have the same relative position from $C(-7,\ 5)$, that is, 5 units to the left and 5 units down from $C$. Its coordinates are $(-7-5,\ 5-5) = (-12,\ 0)$.

**13** $m = 3,\ -2,\ \frac{2}{3},\ -\frac{1}{4}$ • Lines with equation $y = mx$ pass through the origin. Draw lines through the origin with slopes 3 {rise $= 3$, run $= 1$}, $-2$ {rise $= -2$, run $= 1$}, $\frac{2}{3}$ {rise $= 2$, run $= 3$}, and $-\frac{1}{4}$ {rise $= -1$, run $= 4$}. A negative "rise" can be thought of as a "drop."

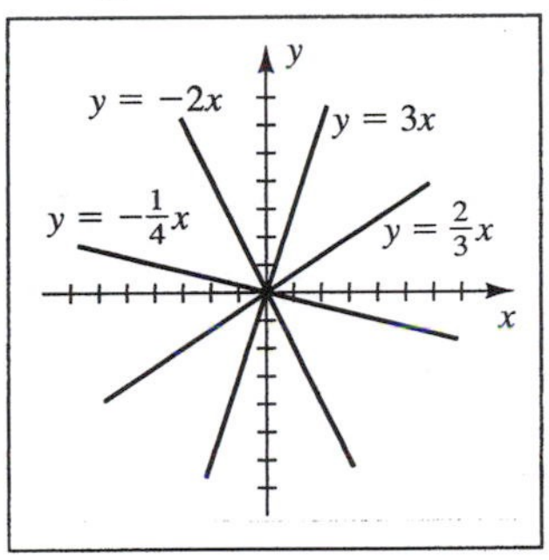


*Figure 13*

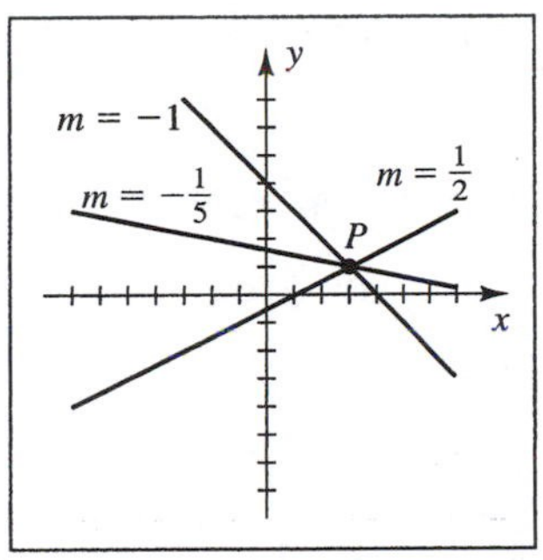


*Figure 15*

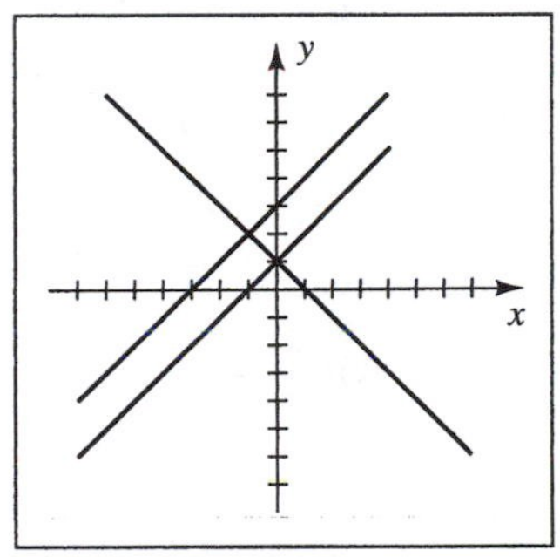


*Figure 19*

**15** Draw lines through the point $P(3,\ 1)$ with slopes $\frac{1}{2}$ {rise $= 1$, run $= 2$}, $-1$ {rise $= -1$, run $= 1$}, and $-\frac{1}{5}$ {rise $= -1$, run $= 5$}.

**17** From the figure, the slope of one of the lines is $\dfrac{\Delta y}{\Delta x} = \dfrac{5}{4}$, so the slopes are $\pm\dfrac{5}{4}$.

Using the point-slope form for the equation of a line with slope $m = \pm\frac{5}{4}$ and point $(x_1,\ y_1) = (2,\ -3)$ gives us $y - (-3) = \pm\frac{5}{4}(x-2)$, or $y + 3 = \pm\frac{5}{4}(x-2)$.

**19** The line $y = 1x + 3$ has slope 1 and $y$-intercept 3.

The line $y = 1x + 1$ has slope 1 and $y$-intercept 1.

The line $y = -1x + 1$ has slope $-1$ and $y$-intercept 1. See *Figure 19.*

**21** (a) "Parallel to the $y$-axis" implies the equation is of the form $x = k$.

The $x$-value of $A(5,\ -2)$ is 5, hence $x = 5$ is the equation.

(b) "Perpendicular to the $y$-axis" implies the equation is of the form $y = k$.

The $y$-value of $A(5, -2)$ is $-2$, hence $y = -2$ is the equation.

[23] Using the point-slope form, the equation of the line through $A(5, -3)$ with slope $-4$ is $y + 3 = -4(x - 5) \Rightarrow y + 3 = -4x + 20 \Rightarrow 4x + y = 17$.

[25] $A(4, 0)$; slope $-3$ {use the point-slope form of a line} $\Rightarrow$

$y - 0 = -3(x - 4) \Rightarrow y = -3x + 12 \Rightarrow 3x + y = 12$.

[27] $A(4, -5)$, $B(-3, 6) \Rightarrow m_{AB} = -\frac{11}{7}$. By the point-slope form, an equation of the line is $y + 5 = -\frac{11}{7}(x - 4) \Rightarrow 7(y + 5) = -11(x - 4) \Rightarrow$

$7y + 35 = -11x + 44 \Rightarrow 11x + 7y = 9$.

[29] $5x - 2y = 4 \Leftrightarrow y = \frac{5}{2}x - 2$. Using the same slope, $\frac{5}{2}$, with $A(2, -4)$, gives us

$y + 4 = \frac{5}{2}(x - 2) \Rightarrow 2(y + 4) = 5(x - 2) \Rightarrow 2y + 8 = 5x - 10 \Rightarrow 5x - 2y = 18$.

[31] $2x - 5y = 8 \Leftrightarrow y = \frac{2}{5}x - \frac{8}{5}$. The slope of this line is $\frac{2}{5}$, so we'll use the negative reciprocal, $-\frac{5}{2}$, for the slope of the new line. $y + 3 = -\frac{5}{2}(x - 7) \Rightarrow$

$2(y + 3) = -5(x - 7) \Rightarrow 2y + 6 = -5x + 35 \Rightarrow 5x + 2y = 29$.

[33] $A(4, 0)$, $B(0, -3) \Rightarrow m = \frac{3}{4}$.

Using the slope-intercept form with $b = -3$ gives us $y = \frac{3}{4}x - 3$.

[35] $A(5, 2)$, $B(-1, 4) \Rightarrow m = -\frac{1}{3}$.

$y - 2 = -\frac{1}{3}(x - 5) \Rightarrow y = -\frac{1}{3}x + \frac{5}{3} + 2 \Rightarrow y = -\frac{1}{3}x + \frac{11}{3}$.

[37] We need the line through the midpoint $M$ of segment $AB$ that is perpendicular to segment $AB$. $A(3, -1)$, $B(-2, 6) \Rightarrow M_{AB} = (\frac{1}{2}, \frac{5}{2})$ and $m_{AB} = -\frac{7}{5}$.

$y - \frac{5}{2} = \frac{5}{7}(x - \frac{1}{2}) \Rightarrow 7(y - \frac{5}{2}) = 5(x - \frac{1}{2}) \Rightarrow 7y - \frac{35}{2} = 5x - \frac{5}{2} \Rightarrow 5x - 7y = -15$.

[39] An equation of the line with slope $-1$ through the origin is

$y - 0 = -1(x - 0)$, or $y = -x$.

[41] We can solve the given equation for $y$ to obtain the slope-intercept form, $y = mx + b$.

$2x = 15 - 3y \Rightarrow 3y = -2x + 15 \Rightarrow y = -\frac{2}{3}x + 5$; $m = -\frac{2}{3}$, $b = 5$

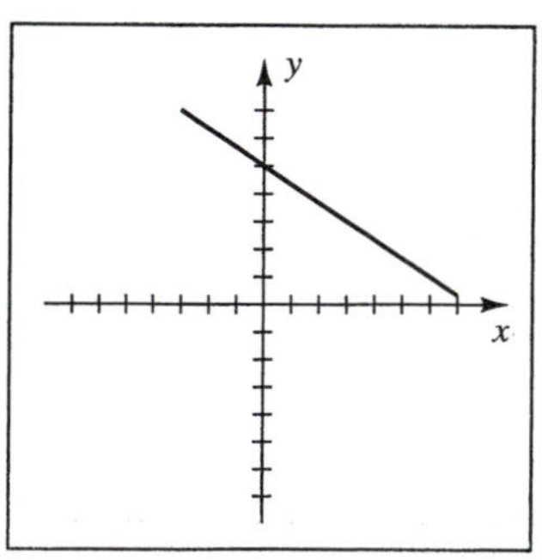

*Figure 41*

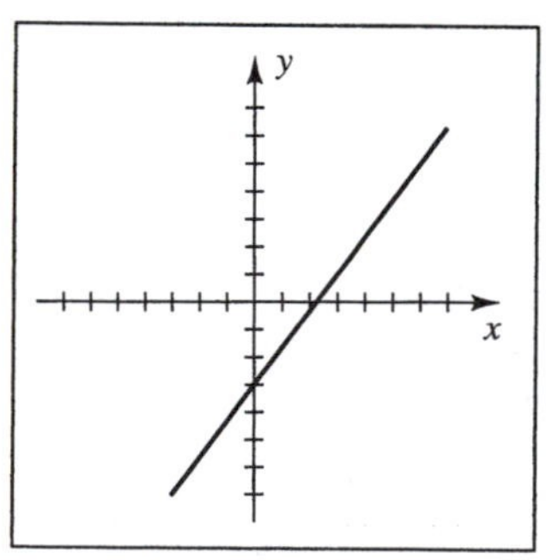

*Figure 43*

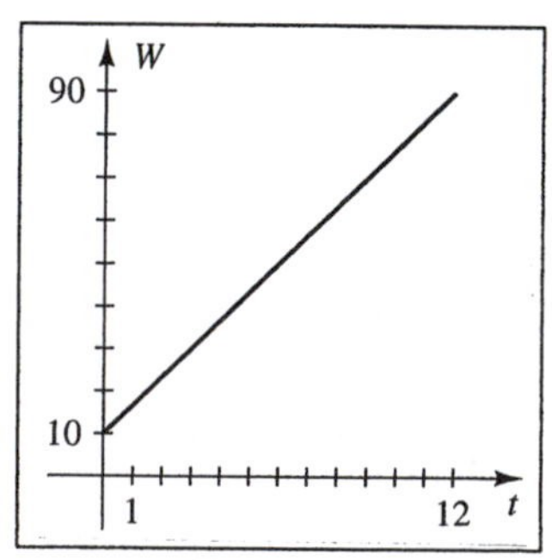

*Figure 57*

[43] $4x - 3y = 9 \Rightarrow -3y = -4x + 9 \Rightarrow y = \frac{4}{3}x - 3$; $m = \frac{4}{3}$, $b = -3$

**45** (a) An equation of the horizontal line with $y$-intercept 3 is $y = 3$.

(b) An equation of the line through the origin with slope $-\frac{1}{2}$ is $y = -\frac{1}{2}x$.

(c) An equation of the line with slope $-\frac{3}{2}$ and $y$-intercept 1 is $y = -\frac{3}{2}x + 1$.

(d) An equation of the line through $(3, -2)$ with slope $-1$ is $y + 2 = -(x - 3)$.

Alternatively, we have a slope of $-1$ and a $y$-intercept of 1, i.e., $y = -x + 1$.

**47** Since we want to obtain a "1" on the right side of the equation, we will divide by 6.

$$[4x - 2y = 6] \cdot \tfrac{1}{6} \Rightarrow \frac{4x}{6} - \frac{2y}{6} = \frac{6}{6} \Rightarrow \frac{2x}{3} - \frac{y}{3} = 1 \Rightarrow \frac{x}{\frac{3}{2}} + \frac{y}{-3} = 1$$

The $x$-intercept is $\frac{3}{2}$ and the $y$-intercept is $-3$.

**49** The radius of the circle is the vertical distance from the center of the circle to the line $y = 5$, that is, $r = 5 - (-2) = 7$. An equation is $(x - 3)^2 + (y + 2)^2 = 49$.

**51** $L = 28 \Rightarrow 1.53t - 6.7 = 28 \Rightarrow t = \dfrac{28 + 6.7}{1.53} \approx 22.68$, or approximately 23 weeks.

**53** (a) $L = 40 \Rightarrow W = 1.70(40) - 42.8 = 25.2$ tons

(b) Error in $L = \pm 2 \Rightarrow$ Error in $W = 1.70(\pm 2) = \pm 3.4$ tons

**55** (a) $y = mx = \dfrac{\text{change in } y \text{ from the beginning of the season}}{\text{change in } x \text{ from the beginning of the season}}(x) = \dfrac{5 - 0}{14 - 0}x = \dfrac{5}{14}x$.

(b) $x = 162 \Rightarrow y = \frac{5}{14}(162) \approx 58$.

**57** (a) Using the slope-intercept form, $W = mt + b = mt + 10$.

$W = 30$ when $t = 3 \Rightarrow 30 = 3m + 10 \Rightarrow m = \frac{20}{3}$ and $W = \frac{20}{3}t + 10$.

(b) $t = 6 \Rightarrow W = \frac{20}{3}(6) + 10 \Rightarrow W = 50$ lb

(c) $W = 70 \Rightarrow 70 = \frac{20}{3}t + 10 \Rightarrow 60 = \frac{20}{3}t \Rightarrow t = 9$ years old

(d) The graph has endpoints at (0, 10) and (12, 90). See *Figure 57* (preceding page).

**59** Using (10, 2480) and (25, 2440) {since an increase of 15°C lowers $H$ by 40}, we have

$$H - 2440 = \frac{2440 - 2480}{25 - 10}(T - 25) \Rightarrow H - 2440 = -\tfrac{8}{3}(T - 25) \Rightarrow$$

$$H - \frac{7320}{3} = -\tfrac{8}{3}T + \frac{200}{3} \Rightarrow H = -\tfrac{8}{3}T + \frac{7520}{3}.$$

**61** (a) Using the slope-intercept form with $m = 0.032$ and $b = 13.5$,

we have $T = 0.032t + 13.5$.

(b) $t = 2000 - 1915 = 85 \Rightarrow T = 0.032(85) + 13.5 = 16.22$°C.

**63** (a) Expenses $(E) = (\$1000) + (5\% \text{ of } R) + (\$2600) + (50\% \text{ of } R) \Rightarrow$

$E = 1000 + 0.05R + 2600 + 0.50R \Rightarrow E = 0.55R + 3600$.

(b) Profit $(P)$ = Revenue $(R)$ – Expenses $(E) \Rightarrow P = R - (0.55R + 3600) \Rightarrow$

$P = R - 0.55R - 3600 \Rightarrow P = 0.45R - 3600$.

(c) *Break even* means $P$ would be 0. $P = 0 \Rightarrow 0 = 0.45R - 3600 \Rightarrow$

$0.45R = 3600 \Rightarrow \frac{45}{100}R = 3600 \Rightarrow R = 3600(\frac{100}{45}) = \$8000$/month

[65] The targets are on the $x$-axis { which is the line $y = 0$ }.

To determine if a target is hit, set $y = 0$ and solve for $x$.

(a) $y - 2 = -1(x - 1) \Rightarrow x + y = 3.$ $y = 0 \Rightarrow x = 3$ and a creature is hit.

(b) $y - \frac{5}{3} = -\frac{4}{9}(x - \frac{3}{2}) \Rightarrow 4x + 9y = 21.$ $y = 0 \Rightarrow x = 5.25$ and no creature is hit.

[67] $s = \dfrac{v_2 - v_1}{h_2 - h_1} \Rightarrow 0.07 = \dfrac{v_2 - 22}{185 - 0} \Rightarrow v_2 = 22 + 0.07(185) = 34.95$ mi/hr.

[69] The slope of $AB$ is $\dfrac{-1.11905 - (-1.3598)}{-0.55 - (-1.3)} = 0.321$. Similarly, the slopes of $BC$ and $CD$ are also 0.321. Therefore, the points all lie on the same line. Since the common slope is 0.321, let $a = 0.321$. $y = 0.321x + b \Rightarrow -1.3598 = 0.321(-1.3) + b \Rightarrow b = -0.9425$. Thus, the points are linearly related by the equation

$$y = 0.321x - 0.9425.$$

[71] $x - 3y = -58 \Leftrightarrow y = (x + 58)/3$ and $3x - y = -70 \Leftrightarrow y = 3x + 70$. Assign $(x + 58)/3$ to $Y_1$ and $3x + 70$ to $Y_2$. Using a standard viewing rectangle, we don't see the lines. Zooming out gives us an indication where the lines intersect and by using an intersect feature, we find that the lines intersect at $(-19, 13)$.

[−30, 3, 2] by [−2, 20, 2]

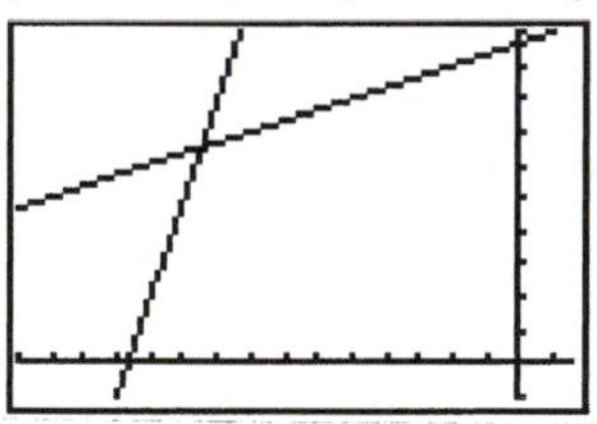

*Figure 71*

[−15, 15] by [−10, 10]

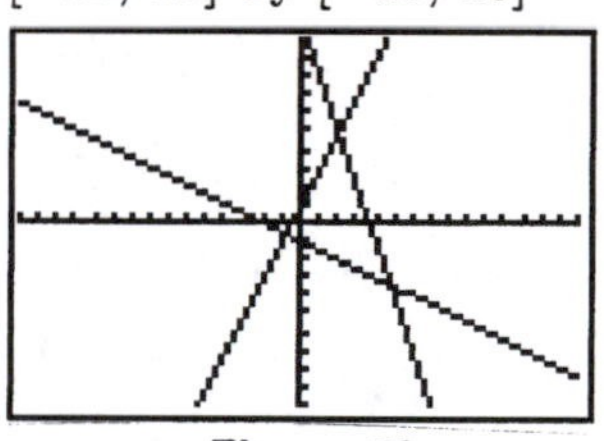

*Figure 73*

[73] From the graph, we can see that the points of intersection are $A(-0.8, -0.6)$, $B(4.8, -3.4)$, and $C(2, 5)$. The lines intersecting at $A$ are perpendicular since they have slopes of 2 and $-\frac{1}{2}$. Since $d(A, B) = \sqrt{39.2}$ and $d(A, C) = \sqrt{39.2}$, the triangle is isosceles. Thus, the polygon is a right isosceles triangle.

[75] The data appear to be linear. Using the two arbitrary points $(-7, -25)$ and $(4.6, 12.2)$, the slope of the line is $\dfrac{12.2 - (-25)}{4.6 - (-7)} \approx 3.2$. An equation of the line is $y + 25 = 3.2(x + 7) \Rightarrow y = 3.2x - 2.6$. If we find the regression line on a calculator, we get the model $y \approx 3.20303x - 2.51553$. See *Figure 75*.

[−8, 5] by [−27, 15, 5]

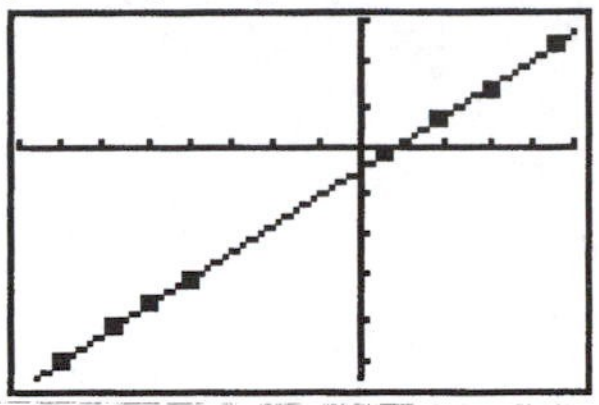

*Figure 75*

[1980, 1988] by [300, 625, 100]

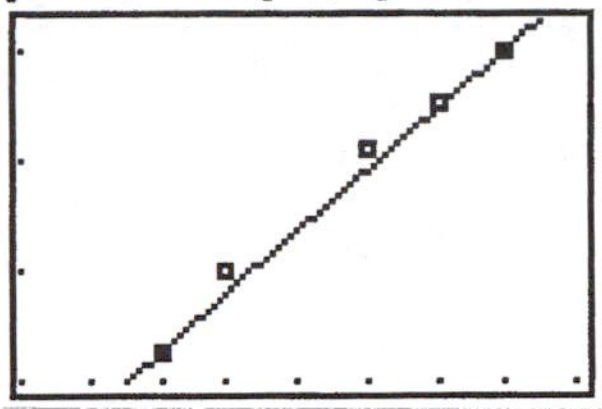

*Figure 77*

77 (a) Plot the points with the form (Year, Cost):

(1982, 325), (1983, 400), (1985, 510), (1986, 550), and (1987, 600)

(b) To find a first approximation for the line use the arbitrary points (1982, 325) and (1987, 600). The resulting line is $y = 55x - 108{,}685$. Adjustments may be made to this equation. If we find the regression line on a calculator, we get the model $y \approx 54.01163x - 106{,}714.477$.

(c) Let $y = 55x - 108{,}685$. When $x = 1984$, $y = 435$ and when $x = 1995$, $y = 1040$.

## *3.4 Exercises*

1 $f(x) = -x^2 - x - 4 \Rightarrow f(-2) = -4 + 2 - 4 = -6$, $f(0) = -4$, and $f(4) = -24$.

3 $f(x) = \sqrt{x-4} - 3x \Rightarrow f(4) = \sqrt{4-4} - 3(4) = \sqrt{0} - 12 = 0 - 12 = -12$.

Similarly, $f(8) = \sqrt{8-4} - 3(8) = \sqrt{4} - 24 = 2 - 24 = -22$

and $f(13) = \sqrt{13-4} - 3(13) = \sqrt{9} - 39 = 3 - 39 = -36$.

Note that $f(a)$, with $a < 4$, would be undefined.

5 (a) $f(x) = 5x - 2 \Rightarrow f(a) = 5(a) - 2 = 5a - 2$

(b) $f(-a) = 5(-a) - 2 = -5a - 2$

(c) $-f(a) = -1 \cdot (5a - 2) = -5a + 2$

(d) $f(a+h) = 5(a+h) - 2 = 5a + 5h - 2$

(e) $f(a) + f(h) = (5a - 2) + (5h - 2) = 5a + 5h - 4$

(f) $\dfrac{f(a+h) - f(a)}{h} = \dfrac{(5a + 5h - 2) - (5a - 2)}{h} = \dfrac{5h}{h} = 5$

7 (a) $f(x) = -x^2 + 4 \Rightarrow f(a) = -(a)^2 + 4 = -a^2 + 4$

(b) $f(-a) = -(-a)^2 + 4 = -a^2 + 4$

(c) $-f(a) = -1 \cdot (-a^2 + 4) = a^2 - 4$

(d) $f(a+h) = -(a+h)^2 + 4 = -(a^2 + 2ah + h^2) + 4 = -a^2 - 2ah - h^2 + 4$

(e) $f(a) + f(h) = (-a^2 + 4) + (-h^2 + 4) = -a^2 - h^2 + 8$

(f) $\dfrac{f(a+h) - f(a)}{h} = \dfrac{(-a^2 - 2ah - h^2 + 4) - (-a^2 + 4)}{h} = \dfrac{-2ah - h^2}{h} = \dfrac{h(-2a - h)}{h}$

$= -2a - h$

[9] (a) $f(x) = x^2 - x + 3 \Rightarrow f(a) = (a)^2 - (a) + 3 = a^2 - a + 3$

(b) $f(-a) = (-a)^2 - (-a) + 3 = a^2 + a + 3$

(c) $-f(a) = -1 \cdot (a^2 - a + 3) = -a^2 + a - 3$

(d) $f(a+h) = (a+h)^2 - (a+h) + 3 = a^2 + 2ah + h^2 - a - h + 3$

(e) $f(a) + f(h) = (a^2 - a + 3) + (h^2 - h + 3) = a^2 + h^2 - a - h + 6$

(f) $\dfrac{f(a+h) - f(a)}{h} = \dfrac{(a^2 + 2ah + h^2 - a - h + 3) - (a^2 - a + 3)}{h} = \dfrac{2ah + h^2 - h}{h} =$

$\dfrac{h(2a + h - 1)}{h} = 2a + h - 1$

[11] (a) $g(x) = 4x^2 \Rightarrow g\left(\frac{1}{a}\right) = 4\left(\frac{1}{a}\right)^2 = 4 \cdot \frac{1}{a^2} = \frac{4}{a^2}$

(b) $g(a) = 4a^2 \Rightarrow \dfrac{1}{g(a)} = \dfrac{1}{4a^2}$

(c) $g(\sqrt{a}) = 4(\sqrt{a})^2 = 4a$

(d) $\sqrt{g(a)} = \sqrt{4a^2} = 2\,|\,a\,| = 2a$ since $a > 0$

[13] (a) $g(x) = \dfrac{2x}{x^2+1} \Rightarrow g\left(\frac{1}{a}\right) = \dfrac{2(1/a)}{(1/a)^2 + 1} = \dfrac{2/a}{1/a^2 + 1} \cdot \dfrac{a^2}{a^2} = \dfrac{2a}{1 + a^2} = \dfrac{2a}{a^2 + 1}$

(b) $\dfrac{1}{g(a)} = \dfrac{1}{\dfrac{2a}{a^2+1}} = \dfrac{a^2+1}{2a}$

(c) $g(\sqrt{a}) = \dfrac{2\sqrt{a}}{(\sqrt{a})^2 + 1} = \dfrac{2\sqrt{a}}{a+1}$

(d) $\sqrt{g(a)} = \sqrt{\dfrac{2a}{a^2+1}} \cdot \dfrac{\sqrt{a^2+1}}{\sqrt{a^2+1}} = \dfrac{\sqrt{2a(a^2+1)}}{a^2+1}$, or, equivalently, $\dfrac{\sqrt{2a^3 + 2a}}{a^2+1}$

[15] All vertical lines intersect the graph in at most one point,

so the graph *is* the graph of a function because it passes the Vertical Line Test.

[17] The domain $D$ is the set of $x$-values; that is, $D = [-4, 1] \cup [2, 4)$. Note that the solid dots on the figure correspond to using brackets {including}, whereas the open dot corresponds to using parentheses {excluding}. The range $R$ is the set of $y$-values; that is, $R = [-3, 3)$.

[19] (a) The domain of a function $f$ is the set of all $x$-values for which the function is defined. In this case, the graph extends from $x = -3$ to $x = 4$. Hence, the domain is $[-3, 4]$.

(b) The range of a function $f$ is the set of all $y$-values that the function takes on. In this case, the graph includes all values from $y = -2$ to $y = 2$. Hence, the range is $[-2, 2]$.

(c) $f(1)$ is the $y$-value of $f$ corresponding to $x = 1$. In this case, $f(1) = 0$.

(d) If we were to draw the horizontal line $y = 1$ on the same coordinate plane, it would intersect the graph at $x = -1, \frac{1}{2}$, and 2. Hence, $f(x) = 1 \Rightarrow x = -1, \frac{1}{2}$, and 2.

(e) The function is above 1 between $x = -1$ and $x = \frac{1}{2}$, and also to the right of $x = 2$. Hence, $f(x) > 1 \Rightarrow x \in (-1, \frac{1}{2}) \cup (2, 4]$.

**21–32** We need to make sure that the radicand { the expression under the radical sign } is greater than or equal to zero and that the denominator is not equal to zero.

**21** $f(x) = \sqrt{2x + 7}$ • $2x + 7 \geq 0 \Rightarrow x \geq -\frac{7}{2} \Leftrightarrow [-\frac{7}{2}, \infty)$

**23** $f(x) = \sqrt{9 - x^2}$ • $9 - x^2 \geq 0 \Rightarrow 9 \geq x^2 \Rightarrow x^2 \leq 9 \Rightarrow |x| \leq 3 \Rightarrow$

$-3 \leq x \leq 3$, or $[-3, 3]$ in interval notation.

**25** $f(x) = \dfrac{x + 1}{x^3 - 4x}$ • For this function we must have the denominator not equal to 0. The denominator is $x^3 - 4x = x(x^2 - 4) = x(x + 2)(x - 2)$, so $x \neq 0, -2, 2$. The solution is then all real numbers *except* 0, −2, 2. In interval notation, we have $(-\infty, -2) \cup (-2, 0) \cup (0, 2) \cup (2, \infty)$. We could also denote this solution as $\mathbb{R} - \{ \pm 2, 0 \}$.

**27** $f(x) = \dfrac{\sqrt{2x - 3}}{x^2 - 5x + 4}$ • For this function we must have the radicand greater than or equal to 0 *and* the denominator not equal to 0. The radicand is greater than or equal to 0 if $2x - 3 \geq 0$, or, equivalently, $x \geq \frac{3}{2}$. The denominator is $(x - 1)(x - 4)$, so $x \neq 1, 4$. The solution is then all real numbers greater than or equal to $\frac{3}{2}$, excluding 4. In interval notation, we have $[\frac{3}{2}, 4) \cup (4, \infty)$.

**29** $f(x) = \dfrac{x - 4}{\sqrt{x - 2}}$ • For this function we must have $x - 2 > 0 \Rightarrow x > 2$.

Note that " > " must be used since the denominator cannot *equal* 0.

In interval notation, we have $(2, \infty)$.

**31** $f(x) = \sqrt{x + 2} + \sqrt{2 - x}$ • We must have $x + 2 \geq 0 \Rightarrow x \geq -2$ and as well as $2 - x \geq 0 \Rightarrow x \leq 2$. The domain is the intersection of $x \geq -2$ and $x \leq 2$, that is,

$[-2, 2]$.

[33] (a) $D = \{x\text{-values}\} = [-5, -3) \cup (-1, 1] \cup (2, 4]$; $R = \{y\text{-values}\} = \{-3\} \cup [-1, 4]$.
Note that the notation for including the single value $-3$ in $R$ uses braces.

(b) ***f* is increasing** on an interval if it goes up as we move from left to right, so $f$ is increasing on $[-4, -3) \cup [3, 4]$.

***f* is decreasing** on an interval if it goes down as we move from left to right, so $f$ is decreasing on $[-5, -4] \cup (2, 3]$. Note that the values $x = -4$ and $x = 3$ are in intervals that are listed as increasing and in intervals that are listed as decreasing.

***f* is constant** on an interval if the $y$-values do not change, so $f$ is constant on $(-1, 1]$.

[35] The graph of the function is increasing on $(-\infty, -3]$ and is decreasing on $[-3, 2]$, so there must be a high point at $x = -3$. Now the graph of the function is increasing on $[2, \infty)$, so there must be a low point at $x = 2$.

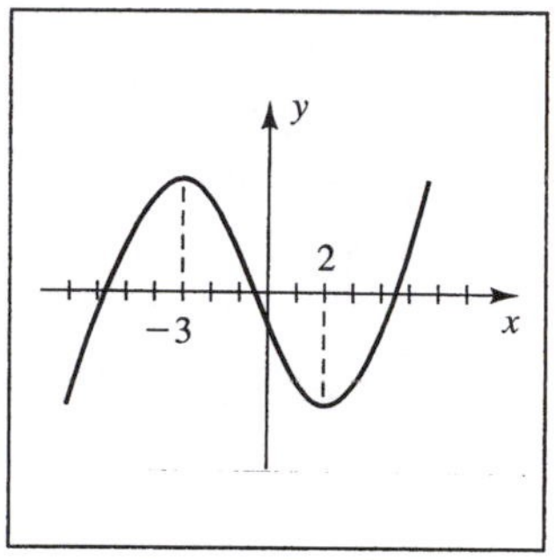

*Figure 35*

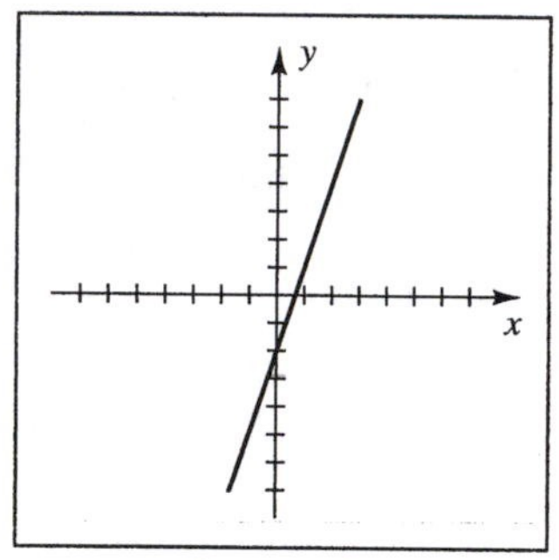

*Figure 37*

[37] (a) $f(x) = 3x - 2$ • This is a line with slope 3 and $y$-intercept 2.

(b) The domain $D$ and the range $R$ are equal to $(-\infty, \infty)$.

(c) $f$ is increasing on its entire domain, that is, $(-\infty, \infty)$.

[39] (a) To sketch the graph of $f(x) = 4 - x^2$, we can make use of the symmetry with respect to the $y$-axis.

| $x$ | $\pm 4$ | $\pm 3$ | $\pm 2$ | $\pm 1$ | 0 |
|---|---|---|---|---|---|
| $y$ | $-12$ | $-5$ | 0 | 3 | 4 |

(b) Since we can substitute any number for $x$, the domain is all real numbers, that is, $D = \mathbb{R}$. By examining *Figure 39*, we see that the values of $y$ are at most 4. Hence, the range of $f$ is all reals less than or equal to 4, that is, $R = (-\infty, 4]$.

(c) A common mistake is to confuse the function values, the $y$'s, with the input values, the $x$'s. We are not interested in the specific $y$-values for determining if the function is increasing, decreasing, or constant. We are only interested if the $y$-values are going up, going down, or staying the same. For the function $f(x) = 4 - x^2$, we say *$f$ is increasing on* $(-\infty, 0]$ since the $y$-values are getting larger as we move from left to right over the $x$-values from $-\infty$ to 0. Also, *$f$ is decreasing on* $[0, \infty)$ since the $y$-values are getting smaller as we move from left to right over the $x$-values from 0 to $\infty$. Note that this answer would have been the same if the function was $f(x) = 500 - x^2$, $f(x) = -300 - x^2$, or any function of the form $f(x) = a - x^2$, where $a$ is any real number.

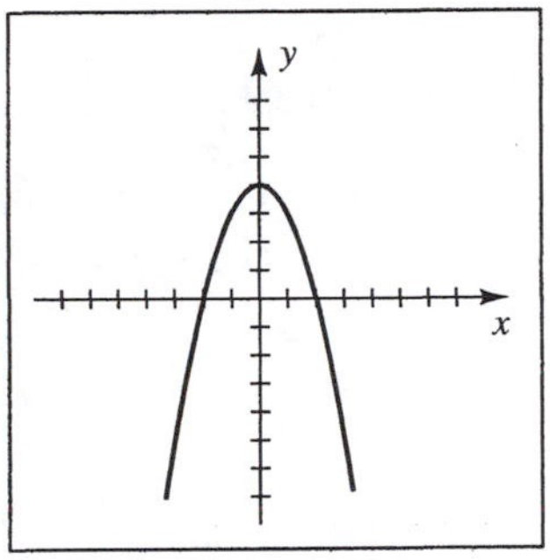

*Figure 39*

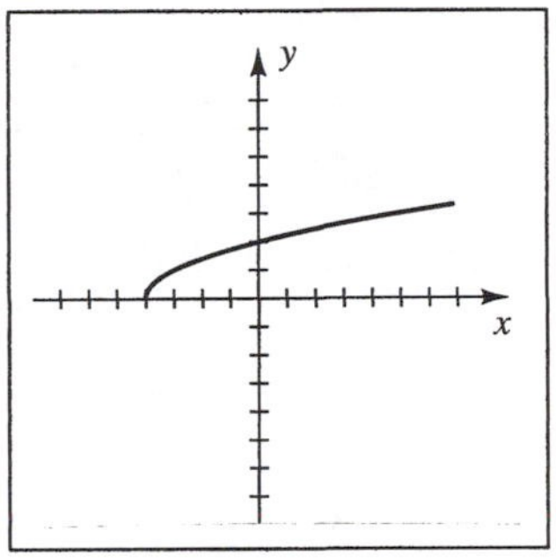

*Figure 41*

**41** (a) $f(x) = \sqrt{x+4}$ • This is half of a parabola opening to the right.

It has an $x$-intercept of $-4$ and a $y$-intercept of 2.

(b) $x + 4 \geq 0 \Rightarrow x \geq -4$, so the domain is $D = [-4, \infty)$.

The $y$-values are all positive or zero, so the range is $R = [0, \infty)$.

(c) $f$ is increasing on its entire domain, that is, on $[-4, \infty)$.

**43** (a) $f(x) = -2$ • This is a horizontal line with $y$-intercept $-2$.

(b) The domain is $D = (-\infty, \infty)$ and the range consists of a single value,

so $R = \{-2\}$.

(c) $f$ is constant on its entire domain, that is, $(-\infty, \infty)$.

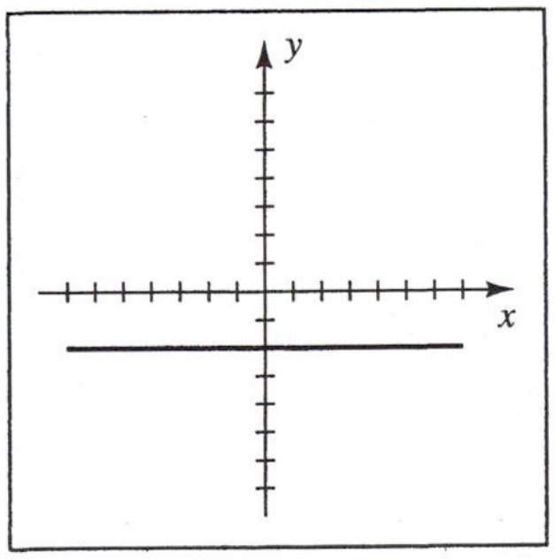

*Figure 43*

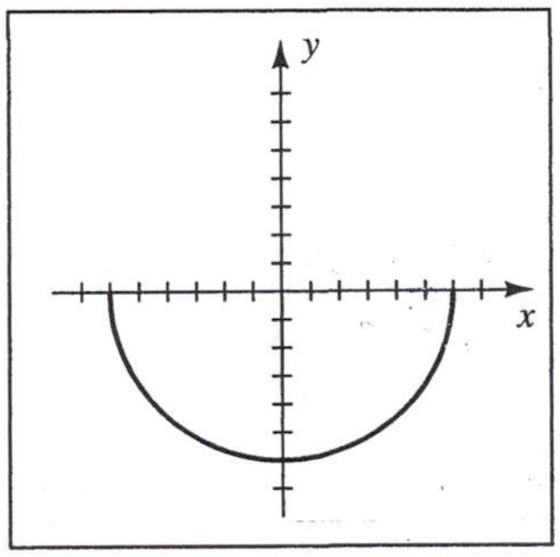

*Figure 45*

[45] (a) We recognize $y = f(x) = -\sqrt{36 - x^2}$ as the lower half of the circle $x^2 + y^2 = 36$.

(b) To find the domain, we solve $36 - x^2 \geq 0$. $36 - x^2 \geq 0 \Rightarrow x^2 \leq 36 \Rightarrow$ $|x| \leq 6 \Rightarrow D = [-6, 6]$. From *Figure 45* on the preceding page, we see that the $y$-values vary from $y = -6$ to $y = 0$. Hence, the range $R$ is $[-6, 0]$.

(c) As we move from left to right, $x = -6$ to $x = 0$, the $y$-values are decreasing. From $x = 0$ to $x = 6$, the $y$-values increase.

Hence, $f$ is decreasing on $[-6, 0]$ and increasing on $[0, 6]$.

[47] $f(x) = x^2 - 3x$, so $f(2) = 2^2 - 3(2) = 4 - 6 = -2$.

$$\frac{f(2+h) - f(2)}{h} = \frac{[(2+h)^2 - 3(2+h)] - (-2)}{h} = \frac{4 + 4h + h^2 - 6 - 3h + 2}{h} =$$

$$\frac{h + h^2}{h} = \frac{h(h+1)}{h} = h + 1$$

[49] $$\frac{f(x+h) - f(x)}{h} = \frac{[(x+h)^2 + 5] - [x^2 + 5]}{h} = \frac{(x^2 + 2xh + h^2 + 5) - (x^2 + 5)}{h} =$$

$$\frac{2xh + h^2}{h} = \frac{h(2x+h)}{h} = 2x + h$$

[51] $$\frac{f(x) - f(a)}{x - a} = \frac{\sqrt{x-3} - \sqrt{a-3}}{x - a} = \frac{\sqrt{x-3} - \sqrt{a-3}}{x - a} \cdot \frac{\sqrt{x-3} + \sqrt{a-3}}{\sqrt{x-3} + \sqrt{a-3}} =$$

$$\frac{(x-3) - (a-3)}{(x-a)(\sqrt{x-3} + \sqrt{a-3})} = \frac{x - a}{(x-a)(\sqrt{x-3} + \sqrt{a-3})} = \frac{1}{\sqrt{x-3} + \sqrt{a-3}}$$

[53] As in Example 7, $a = \frac{2-1}{3-(-3)} = \frac{1}{6}$ and $f$ has the form $f(x) = \frac{1}{6}x + b$.

$f(3) = \frac{1}{6}(3) + b = \frac{1}{2} + b$. But $f(3) = 2$, so $\frac{1}{2} + b = 2 \Rightarrow b = \frac{3}{2}$, and $f(x) = \frac{1}{6}x + \frac{3}{2}$.

[55–64] *Note:* A good question to consider is "Given a particular value of $x$, can a unique value of $y$ be found?" If the answer is yes, the value of $y$ (general formula) is given. If no, two ordered pairs satisfying the relation having $x$ in the first position are given.

[55] $2y = x^2 + 5 \Rightarrow y = \frac{x^2 + 5}{2}$, a function

[57] $x^2 + y^2 = 4 \Rightarrow y^2 = 4 - x^2 \Rightarrow y = \pm\sqrt{4 - x^2}$, not a function, $(0, \pm 2)$

[59] $y = 3$ is a function since for any $x$,

$(x, 3)$ is the only ordered pair in $W$ having $x$ in the first position.

[61] Any ordered pair with $x$-coordinate 0 satisfies $xy = 0$.

Two such ordered pairs are $(0, 0)$ and $(0, 1)$. Not a function

[63] $|y| = |x| \Rightarrow \pm y = \pm x \Rightarrow y = \pm x$, not a function, $(1, \pm 1)$

[65] $V = lwh = (30 - x - x)(20 - x - x)(x)$

$$= (30 - 2x)(20 - 2x)(x) = 2(15 - x) \cdot 2(10 - x)(x) = 4x(15 - x)(10 - x)$$

[67] (a) The formula for the area of a rectangle is $A = lw$ { Area = length × width }.

$$A = 500 \Rightarrow xy = 500 \Rightarrow y = \frac{500}{x}$$

(b) We need to determine the number of linear feet ($P$) first. There are two walls of length $y$, two walls of length $(x - 3)$, and one wall of length $x$.

Thus, $P = \text{Linear feet of wall} = x + 2(y) + 2(x - 3) = 3x + 2\left(\frac{500}{x}\right) - 6$.

The cost $C$ is 100 times $P$, so $C = 100P = 300x + \dfrac{100{,}000}{x} - 600$.

[69] The expression $(h - 25)$ represents the number of feet *above* 25 feet.

$S(h) = 6(h - 25) + 100 = 6h - 150 + 100 = 6h - 50$.

[71] (a) Using (6, 48) and (7, 50.5), we have

$$y - 48 = \frac{50.5 - 48}{7 - 6}(t - 6), \text{ or } y = 2.5t + 33.$$

(b) The slope represents the yearly increase in height, 2.5 in./yr.

(c) $t = 10 \Rightarrow y = 2.5(10) + 33 = 58$ in.

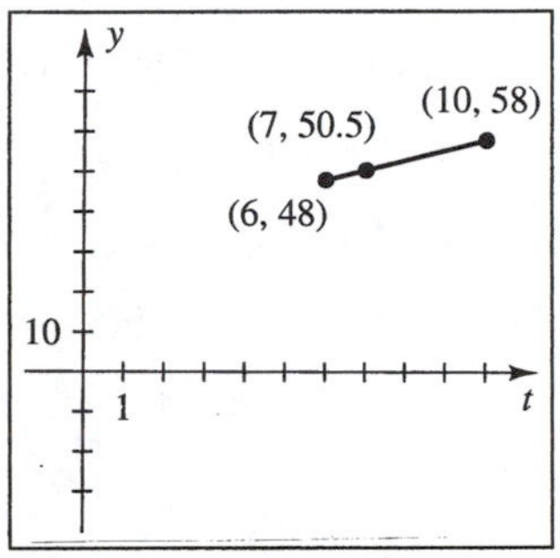


*Figure 71*

[73] The height of the balloon is $2t$. Using the Pythagorean theorem,

$$d^2 = 100^2 + (2t)^2 = 10{,}000 + 4t^2 \Rightarrow d = \sqrt{4(2500 + t^2)} \Rightarrow d = 2\sqrt{t^2 + 2500}.$$

[75] (a) $CTP$ forms a right angle, so the Pythagorean theorem may be applied.

$$(CT)^2 + (PT)^2 = (PC)^2 \Rightarrow r^2 + y^2 = (h + r)^2 \Rightarrow$$

$$r^2 + y^2 = h^2 + 2hr + r^2 \Rightarrow y^2 = h^2 + 2hr \; \{ y > 0 \} \Rightarrow y = \sqrt{h^2 + 2hr}$$

(b) $y = \sqrt{(200)^2 + 2(4000)(200)} = \sqrt{(200)^2(1 + 40)} = 200\sqrt{41} \approx 1280.6$ mi

[77] Form a right triangle with the control booth and the beginning of the runway. Let $y$ denote the distance from the control booth to the beginning of the runway and apply the Pythagorean theorem. $y^2 = 300^2 + 20^2 \Rightarrow y^2 = 90{,}400$. Now form a right triangle, in a different plane, with sides $y$ and $x$ and hypotenuse $d$.

Then $d^2 = y^2 + x^2 \Rightarrow d^2 = 90{,}400 + x^2 \Rightarrow d = \sqrt{90{,}400 + x^2}$.

**79** (b) The maximum $y$-value of 0.75 occurs when $x \approx 0.55$ and
the minimum $y$-value of $-0.75$ occurs when $x \approx -0.55$.

Therefore, the range of $f$ is approximately $[-0.75, 0.75]$.

(c) $f$ is decreasing on $[-2, -0.55]$ and on $[0.55, 2]$. $f$ is increasing on $[-0.55, 0.55]$.

$[-2, 2]$ by $[-2, 2]$

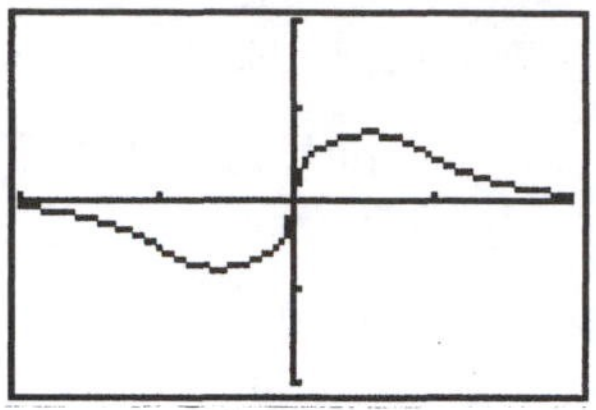

*Figure 79*

$[-0.7, 1.4]$ by $[-1.1, 1]$

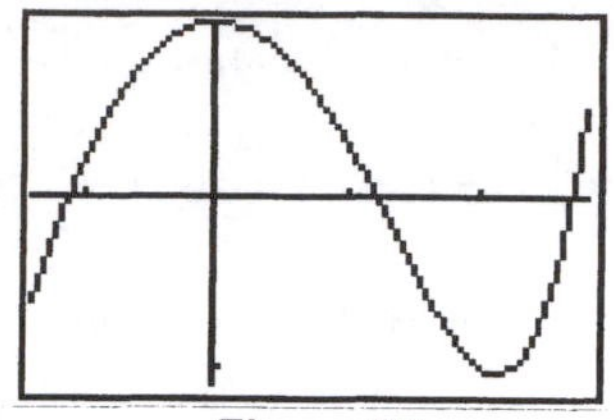

*Figure 81*

**81** (b) The maximum $y$-value of 1 occurs when $x = 0$ and
the minimum $y$-value of $-1.03$ occurs when $x \approx 1.06$.

Therefore, the range of $f$ is approximately $[-1.03, 1]$.

(c) $f$ is decreasing on $[0, 1.06]$. $f$ is increasing on $[-0.7, 0]$ and on $[1.06, 1.4]$.

**83** For each of (a)–(e), an assignment to $Y_1$, an appropriate viewing rectangle, and the solution(s) are listed.

(a) $Y_1 = (x\hat{}5)\hat{}(1/3)$, VR: $[-40, 40, 10]$ by $[-40, 40, 10]$, $x = 8$

(b) $Y_1 = (x\hat{}4)\hat{}(1/3)$, VR: $[-40, 40, 10]$ by $[-40, 40, 10]$, $x = \pm 8$

$[-40, 40, 10]$ by $[-40, 40, 10]$

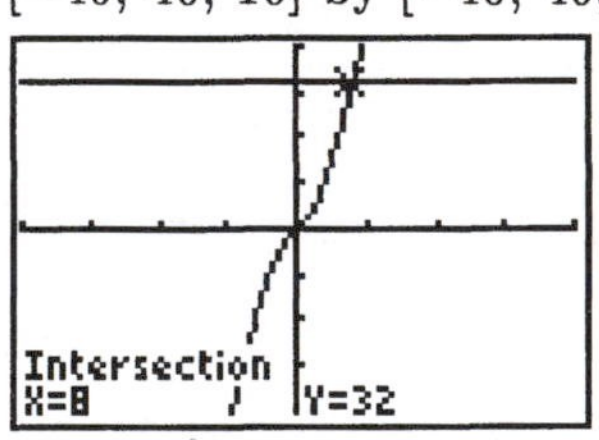


*Figure 83(a)*

$[-40, 40, 10]$ by $[-40, 40, 10]$

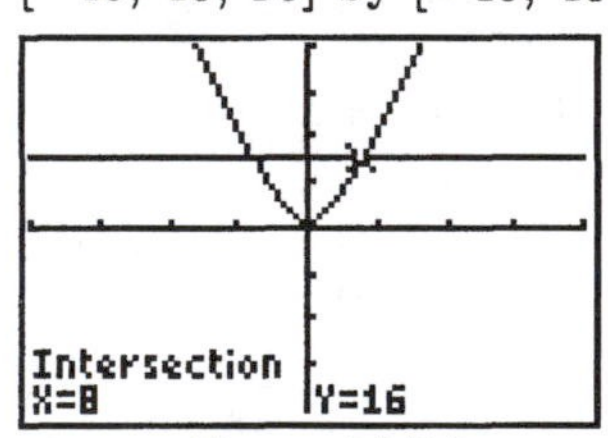


*Figure 83(b)*

(c) $Y_1 = (x\hat{}2)\hat{}(1/3)$, VR: $[-40, 40, 10]$ by $[-40, 40, 10]$, no real solutions

(d) $Y_1 = (x\hat{}3)\hat{}(1/4)$, VR: $[0, 800, 100]$ by $[0, 200, 100]$, $x = 625$

(e) $Y_1 = (x\hat{}3)\hat{}(1/2)$, VR: $[-30, 30, 10]$ by $[-30, 30, 10]$, no real solutions

**85** (a) There are $95 \times 63 = 5985$ total pixels in the screen.

(b) If a function is graphed in dot mode, only one pixel in each column of pixels on the screen can be darkened. Therefore, there are at most 95 pixels darkened. *Note:* In connected mode this may not be true.

[87] (a) First, we must determine an equation of the line that passes through the points (1985, 11,450) and (1994, 20,021).

$y - 11{,}450 = \frac{20{,}021 - 11{,}450}{1994 - 1985}(x - 1985) = \frac{2857}{3}(x - 1985) \Rightarrow$

$y = \frac{2857}{3}x - \frac{5{,}636{,}795}{3}$. Thus, let $f(x) = \frac{2857}{3}x - \frac{5{,}636{,}795}{3}$ and graph $f$.

(b) The average annual increase in the price paid for a new car is equal to the slope:

$\frac{2857}{3} \approx \$952.33$.

(c) Graph $y = \frac{2857}{3}x - \frac{5{,}636{,}795}{3}$ and $y = 25{,}000$ on the same coordinate axes. Their point of intersection is approximately (1999.2, 25,000). Thus, according to this model, in the year 1999 the average price paid for a new car was \$25,000.

[1984, 2005, 5] by [1E4, 3E4, 1E4]

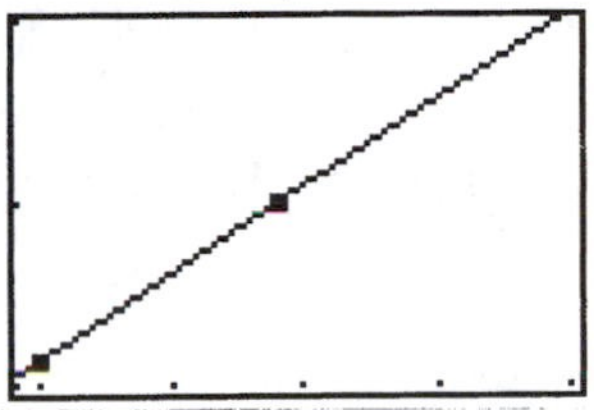

*Figure 87(a)*

[1984, 2005, 5] by [1E4, 3E4, 1E4]

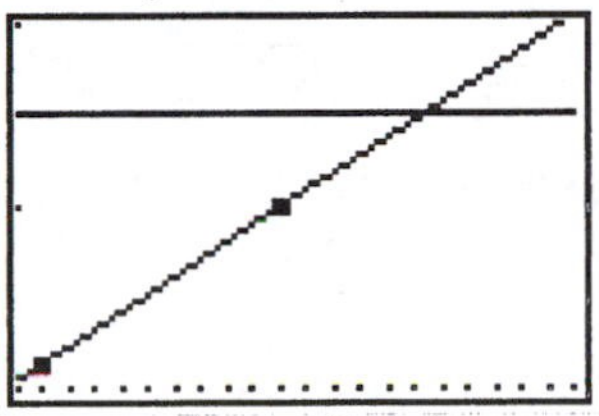

*Figure 87(c)*

## *3.5 Exercises*

[1] Since $f$ is an even function, $f(-x) = f(x)$.

From the table, we see that $f(2) = 7$, so $f(-2) = 7$.

Since $g$ is an odd function, $g(-x) = -g(x)$.

From the table, we see that $g(2) = -6$, so $g(-2) = -(-6) = 6$.

*Note:* Assume that $f$ and $g$ are defined for all real numbers.

If $x = 0$, then the points $(0, b)$ and $(-0, b)$ are on $f$.

But they are the same point and we cannot determine the value of $b$.

If $x = 0$, then the points $(0, b)$ and $(-0, -b)$ are on $g$.

Since the $x$-value is the same, the $y$-values must be equal because a function can't have two $y$-values for the same $x$-value. Thus, $b = -b \Rightarrow 2b = 0 \Rightarrow b = 0$.

[3] $f(x) = 5x^3 + 2x \Rightarrow f(-x) = 5(-x)^3 + 2(-x) = -5x^3 - 2x = -(5x^3 + 2x) = -f(x)$.

Since $f(-x) = -f(x)$, $f$ is odd and its graph is symmetric with respect to the origin.

Note that this means if $(a, b)$ is a point on the graph of $f$,

then the point $(-a, -b)$ is also on the graph.

5. $f(x) = 3x^4 + 2x^2 - 5 \Rightarrow f(-x) = 3(-x)^4 + 2(-x)^2 - 5 = 3x^4 + 2x^2 - 5 = f(x)$.
Since $f(-x) = f(x)$, $f$ is even and its graph is symmetric with respect to the $y$-axis.
Note that this means if $(a, b)$ is a point on the graph of $f$,
then the point $(-a, b)$ is also on the graph.

7. $f(x) = 8x^3 - 3x^2 \Rightarrow f(-x) = 8(-x)^3 - 3(-x)^2 = -8x^3 - 3x^2$
$-f(x) = -1 \cdot f(x) = -1(8x^3 - 3x^2) = -8x^3 + 3x^2$
Since $f(-x) \neq f(x)$ and $f(-x) \neq -f(x)$, $f$ is neither even nor odd.

9. $f(x) = \sqrt{x^2+4} \Rightarrow f(-x) = \sqrt{(-x)^2+4} = \sqrt{x^2+4} = f(x)$.
Since $f(-x) = f(x)$, $f$ is even and its graph is symmetric with respect to the $y$-axis.

11. $f(x) = \sqrt[3]{x^3 - x} \Rightarrow$
$f(-x) = \sqrt[3]{(-x)^3 - (-x)} = \sqrt[3]{-x^3 + x} = \sqrt[3]{-1(x^3 - x)} = \sqrt[3]{-1}\,\sqrt[3]{x^3 - x} = -\sqrt[3]{x^3 - x}$.
$-f(x) = -1 \cdot f(x) = -1 \cdot \sqrt[3]{x^3 - x} = -\sqrt[3]{x^3 - x}$.
Since $f(-x) = -f(x)$, $f$ is odd and its graph is symmetric with respect to the origin.

13. $f(x) = |x| + c$, $c = -3, 1, 3$ •
Shift $g(x) = |x|$ {in Figure 1 in the text} down 3, up 1, and up 3 units, respectively.

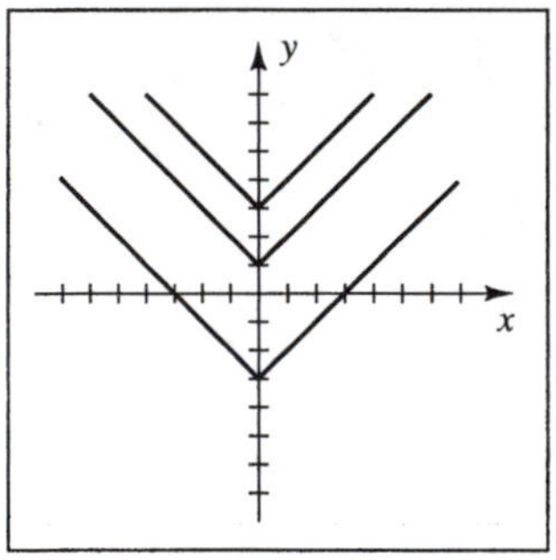

*Figure 13*

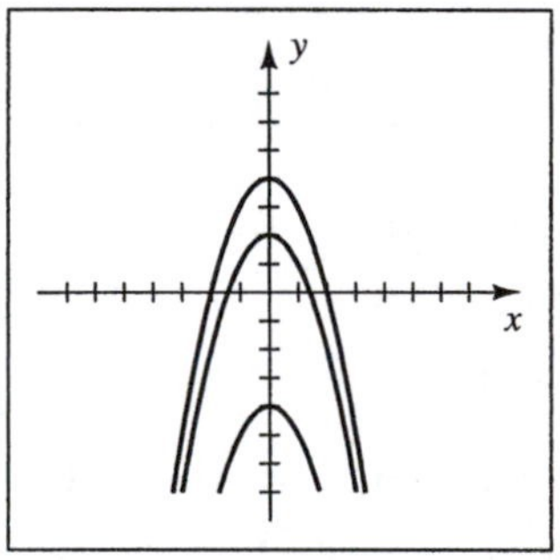

*Figure 15*

15. $f(x) = -x^2 + c$, $c = -4, 2, 4$ •
Shift $g(x) = -x^2$ {in Figure 5 in the text} down 4, up 2, up 4, respectively.

[17] $f(x) = 2\sqrt{x} + c$, $c = -3, 0, 2$ • The graph of $y^2 = x$ is shown in Figure 3 in Section 3.2 in the text. The top half of this graph is the graph of the **square root function**, $h(x) = \sqrt{x}$. The second value of $c$, 0, gives us the graph of $g(x) = 2\sqrt{x}$, which is a vertical stretching of $h$ by a factor of 2. The effect of *adding* $-3$ and 2 is to vertically shift $g$ down 3 units and up 2 units, respectively.

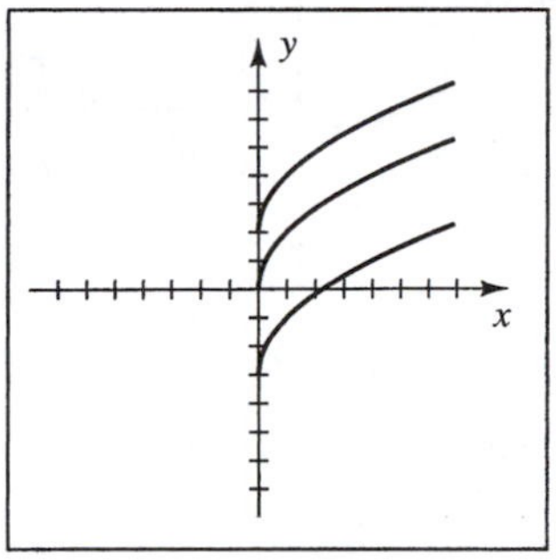


Figure 17

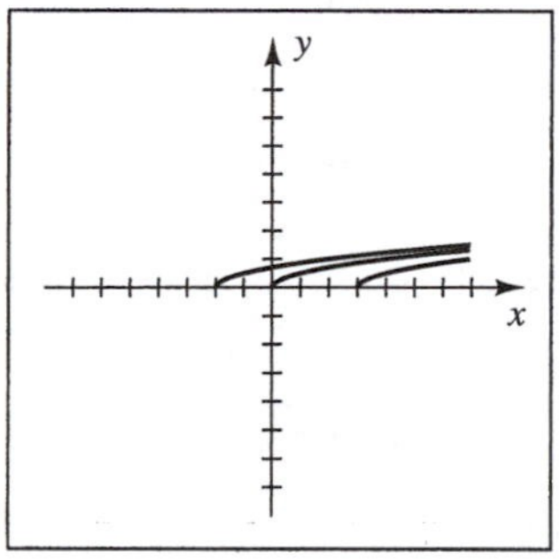


Figure 19

[19] $f(x) = \frac{1}{2}\sqrt{x - c}$, $c = -2, 0, 3$ • The graph of $g(x) = \frac{1}{2}\sqrt{x}$ is a vertical compression of the square root function by a factor of $1/(1/2) = 2$. The effect of *subtracting* $-2$ and 3 from $x$ will be to horizontally shift $g$ left 2 units and right 3 units, respectively. If you forget which way to shift the graph, it is helpful to find the domain of the function. For example, if $h(x) = \sqrt{x - 2}$, then $x - 2$ must be nonnegative. $x - 2 \geq 0 \Rightarrow x \geq 2$, which also indicates a shift of 2 units to the right.

[21] $f(x) = c\sqrt{4 - x^2}$, $c = -2, 1, 3$ • For $c = 1$, the graph of $g(x) = \sqrt{4 - x^2}$ is the upper half of the circle $x^2 + y^2 = 4$. For $c = -2$, reflect $g$ through the $x$-axis and vertically stretch it by a factor of 2. For $c = 3$, vertically stretch $g$ by a factor of 3.

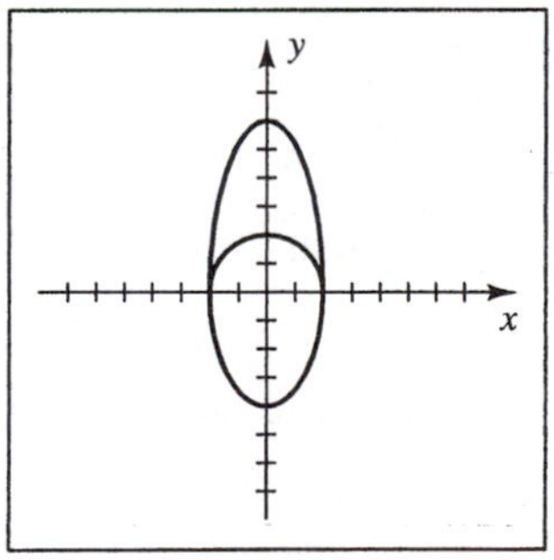


Figure 21

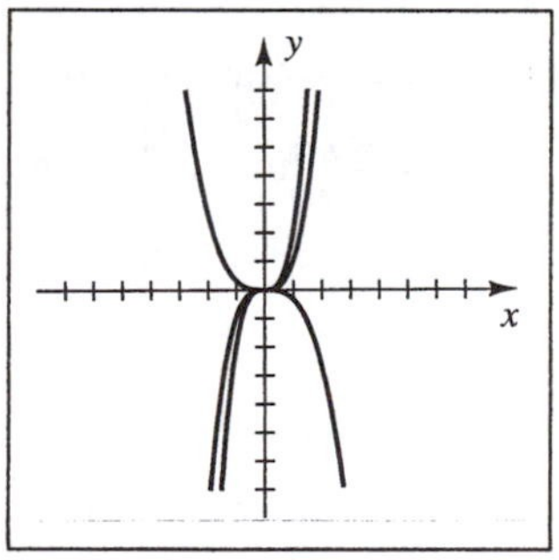


Figure 23

[23] $f(x) = cx^3$, $c = -\frac{1}{3}, 1, 2$ • For $c = 1$, see a graph of the **cubing function** in Appendix I of the text. For $c = -\frac{1}{3}$, reflect $g(x) = x^3$ through the $x$-axis and vertically compress it by a factor of $1/(1/3) = 3$. For $c = 2$, vertically stretch $g$ by a factor of 2.

[25] $f(x) = \sqrt{cx} - 1$, $c = -1, \frac{1}{9}, 4$ • If $c = 1$, then the graph of $g(x) = \sqrt{x} - 1$ is the graph of the square root function vertically shifted down one unit. For $c = -1$, reflect $g$ through the $y$-axis. For $c = \frac{1}{9}$, horizontally stretch $g$ by a factor $1/(1/9) = 9$ { $x$-intercept changes from 1 to 9 }. For $c = 4$, horizontally compress $g$ by a factor 4 { $x$-intercept changes from 1 to $\frac{1}{4}$ }.

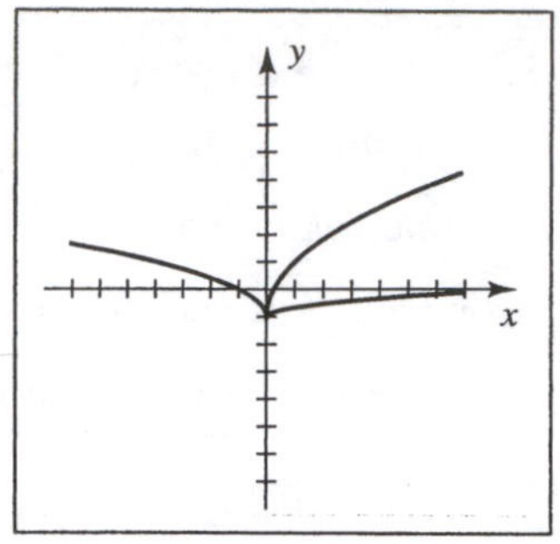

*Figure 25*

[27] You know that $y = f(x+2)$ is the graph of $y = f(x)$ shifted to the left 2 units, so the point $P(0, 5)$ would move to the point $(-2, 5)$. The graph of $y = f(x+2) - 1$ is the graph of $y = f(x+2)$ shifted down 1 unit, so the point $(-2, 5)$ moves to $(-2, 4)$. Summarizing these steps gives us the following:

$P(0, 5)$ { $x + 2$ [subtract 2 from the $x$-coordinate] } $\rightarrow (-2, 5)$
{ $-1$ [subtract 1 from the $y$-coordinate] } $\rightarrow (-2, 4)$

[29] To determine what happens to a point $P$ under this transformation, think of how you would evaluate $y = 2f(x-4) + 1$ for a particular value of $x$. You would first subtract 4 from $x$ and then put that value into the function, obtaining a corresponding $y$-value. Next, you would multiply that $y$-value by 2 and finally, add 1. Summarizing these steps using the given point $P$, we have the following:

$P(3, -2)$ { $x - 4$ [add 4 to the $x$-coordinate] } $\rightarrow (7, -2)$
{ $\times 2$ [multiply the $y$-coordinate by 2] } $\rightarrow (7, -4)$
{ $+1$ [add 1 to the $y$-coordinate] } $\rightarrow (7, -3)$

[31] $P(3, 9)$; $y = \frac{1}{3}f(\frac{1}{2}x) - 1$
{ $\frac{1}{2}x$ [multiply the $x$-coordinate by 2] } $\rightarrow (6, 9)$
{ $\times\frac{1}{3}$ [multiply the $y$-coordinate by $\frac{1}{3}$] } $\rightarrow (6, 3)$
{ $-1$ [subtract 1 from the $y$-coordinate] } $\rightarrow (6, 2)$

[33] For $y = f(x-2) + 3$, the graph of $f$ is shifted 2 units to the right and 3 units up.

[35] For $y = f(-x) - 2$, the graph of $f$ is reflected through the $y$-axis and shifted 2 units down.

[37] For $y = -\frac{1}{2}f(x)$, the graph of $f$ is compressed vertically by a factor of 2 and reflected through the $x$-axis.

[39] For $y = -2f(\frac{1}{3}x)$, the graph of $f$ is stretched horizontally by a factor of 3, stretched vertically by a factor of 2, and reflected through the $x$-axis.

**41** (a) $y = f(x+3)$ • shift $f$ left 3 units

(b) $y = f(x-3)$ • shift $f$ right 3 units

(c) $y = f(x)+3$ • shift $f$ up 3 units

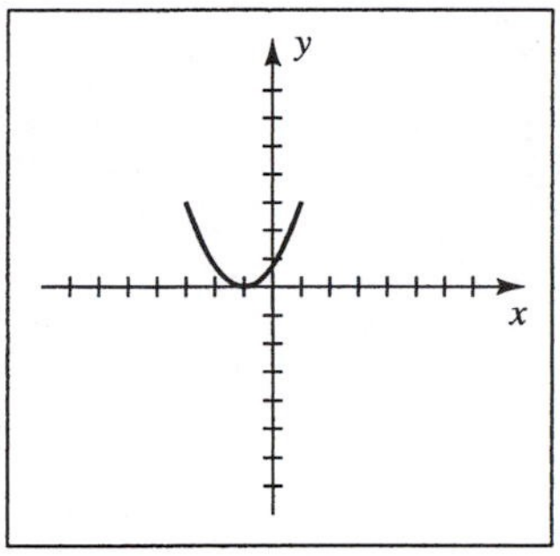


*Figure 41(a)*

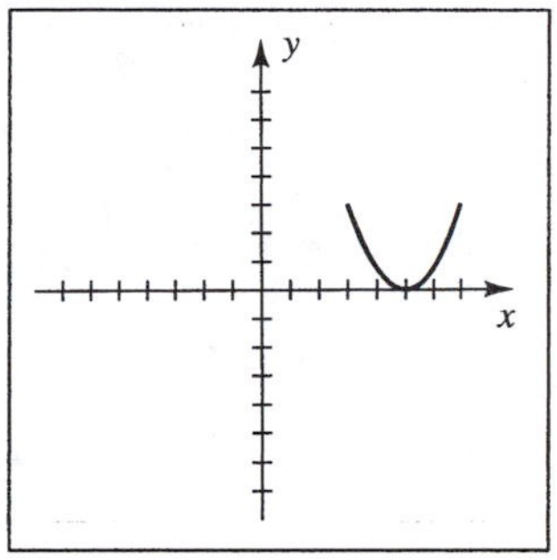


*Figure 41(b)*

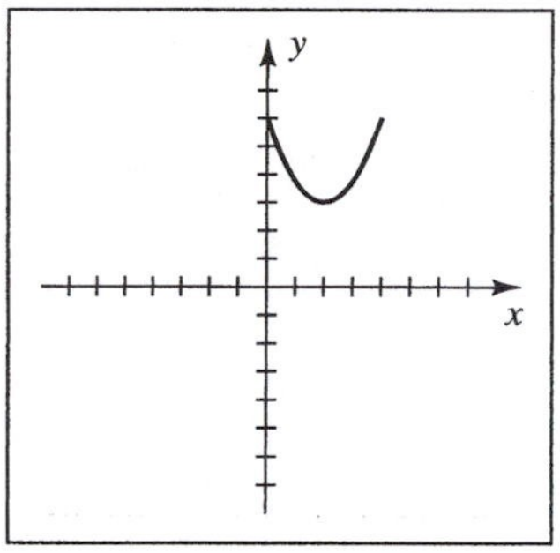


*Figure 41(c)*

(d) $y = f(x)-3$ • shift $f$ down 3 units

(e) $y = -3f(x)$ •

reflect $f$ through the $x$-axis and vertically stretch it by a factor of 3

(f) $y = -\frac{1}{3}f(x)$ • reflect $f$ through the $x$-axis {the effect of the negative sign in front of $\frac{1}{3}$} and vertically compress it by a factor of $1/(1/3) = 3$

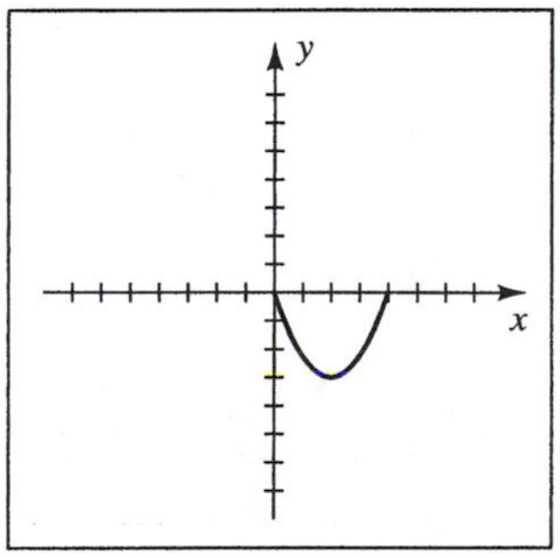


*Figure 41(d)*

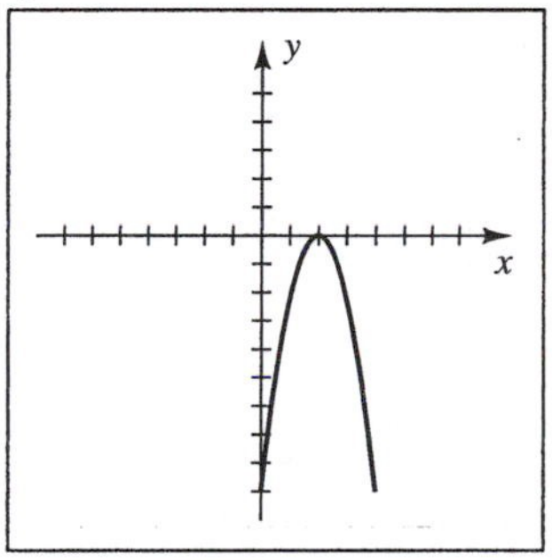


*Figure 41(e)*

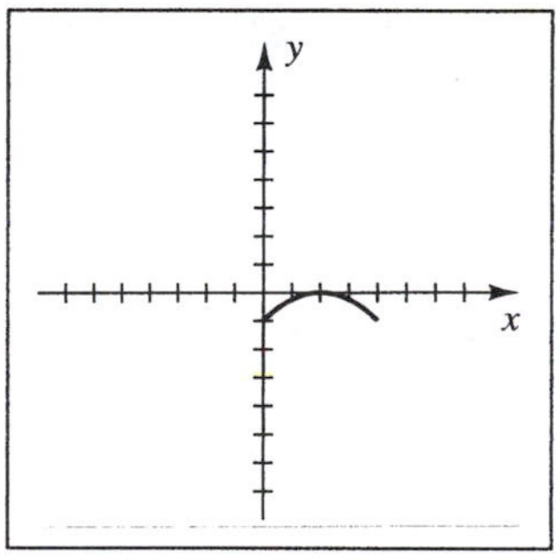


*Figure 41(f)*

(g) $y = f(-\frac{1}{2}x)$ • reflect $f$ through the $y$-axis { the effect of the negative sign inside the parentheses } and horizontally stretch it by a factor of $1/(1/2) = 2$

(h) $y = f(2x)$ • horizontally compress $f$ by a factor of 2

(i) $y = -f(x+2) - 3$ • shift $f$ left 2 units, reflect it through the $x$-axis, and then shift it down 3 units

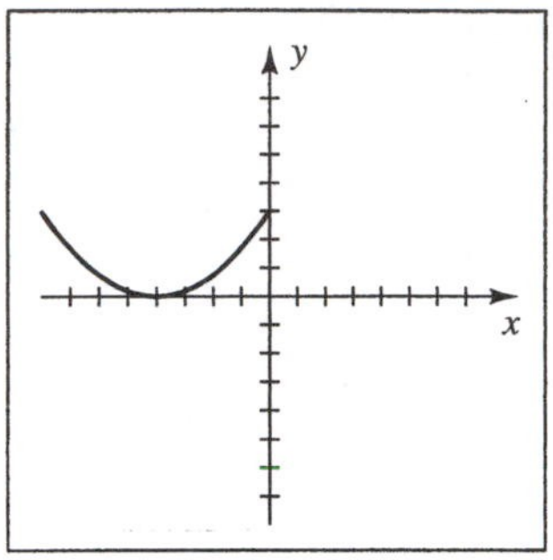

*Figure 41(g)*

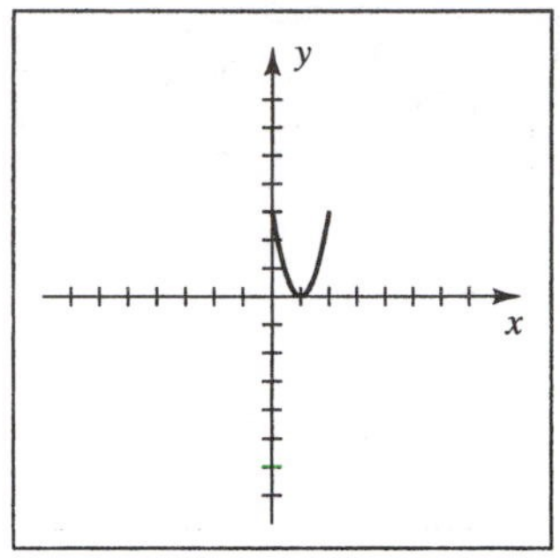

*Figure 41(h)*

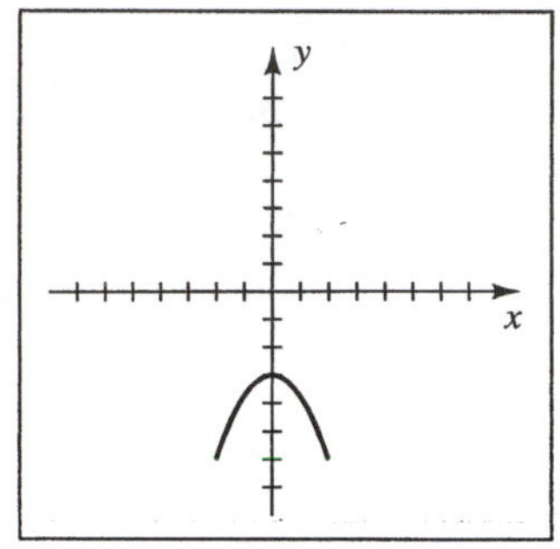

*Figure 41(i)*

(j) $y = f(x-2) + 3$ • shift $f$ right 2 units and up 3

(k) $y = |\, f(x) \,|$ • since no portion of the graph lies below the $x$-axis, the graph is unchanged

(l) $y = f(\,|\,x\,|\,)$ • include the reflection of the given graph through the $y$-axis since all points have positive $x$-coordinates

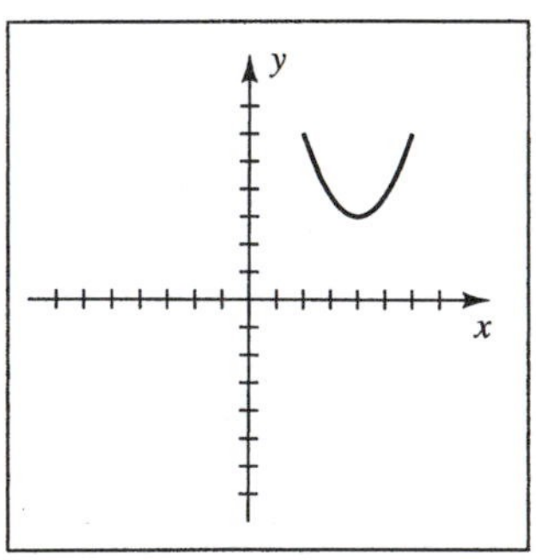

*Figure 41(j)*

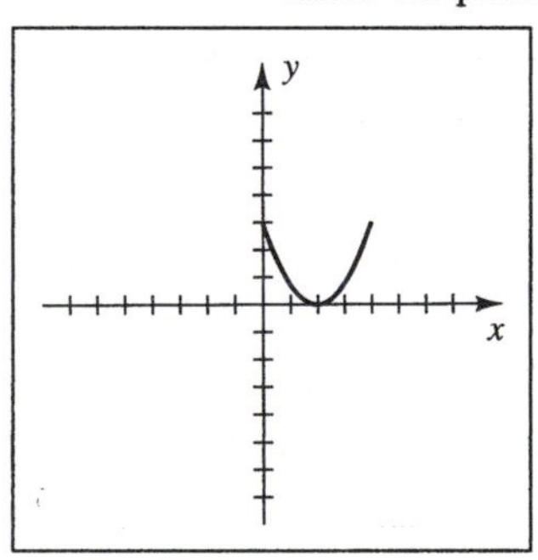

*Figure 41(k)*

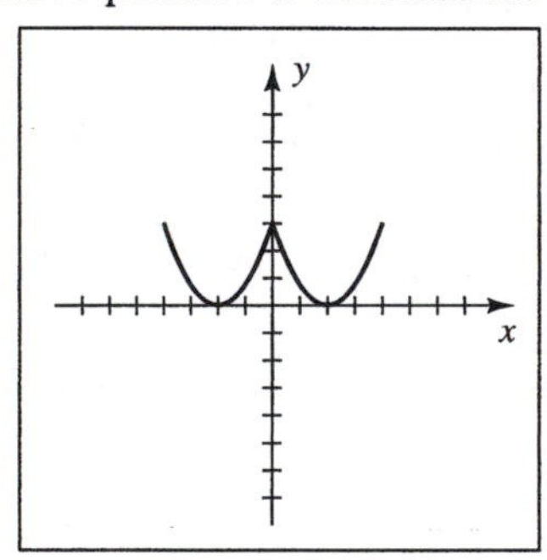

*Figure 41(l)*

**43** (a) The minimum point on $y = f(x)$ is $(2, -1)$.

On the graph labeled (a), the minimum point is $(-7, 0)$.

It has been shifted left 9 units and up 1. Hence, $y = f(x+9) + 1$.

(b) $f$ is reflected through the $x$-axis $\Rightarrow$ $y = -f(x)$

(c) $f$ is reflected through the $x$-axis and shifted left 7 units and down 1 $\Rightarrow$

$y = -f(x+7) - 1$

**45** (a) $f$ is shifted left 4 units $\Rightarrow$ $y = f(x+4)$

(b) $f$ is shifted up 1 unit $\Rightarrow$ $y = f(x)+1$

(c) $f$ is reflected through the $y$-axis $\Rightarrow$ $y = f(-x)$

**47** $f(x) = \begin{cases} 3 & \text{if } x \leq -1 \\ -2 & \text{if } x > -1 \end{cases}$

We can think of $f$ as 2 functions: If $x \leq -1$, then $y = 3$ {include the point $(-1,\ 3)$}, and if $x > -1$, then $y = -2$ {exclude the point $(-1,\ -2)$}.

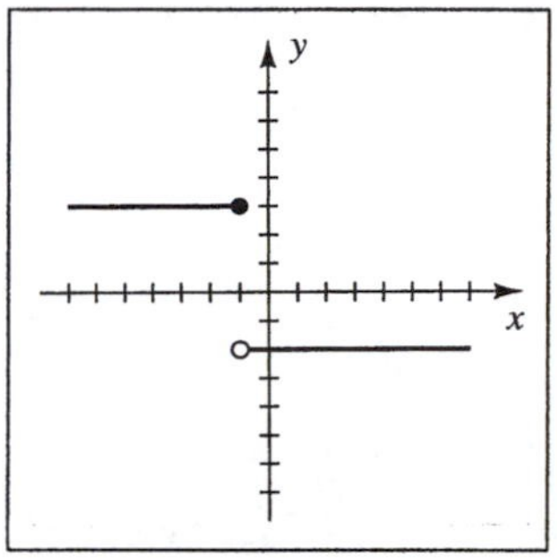


*Figure 47*

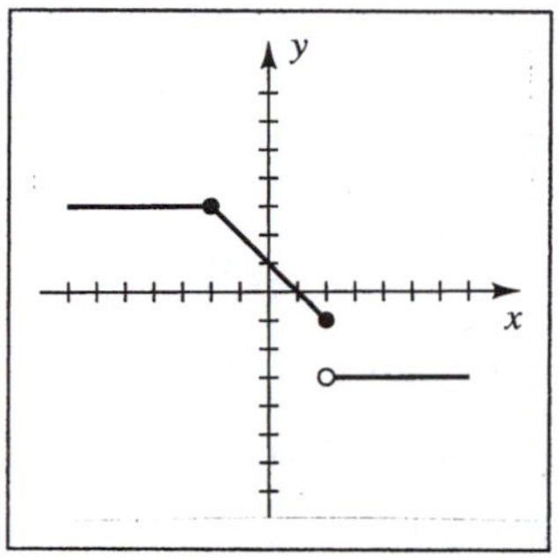


*Figure 49*

**49** $f(x) = \begin{cases} 3 & \text{if } x < -2 \\ -x+1 & \text{if } |\,x\,| \leq 2 \\ -3 & \text{if } x > 2 \end{cases}$

For the second part of the function, we have $|\,x\,| \leq 2$, or, equivalently, $-2 \leq x \leq 2$. On this part of the domain, we want to graph $f(x) = -x+1$, a line with slope $-1$ and $y$-intercept 1. Include both endpoints, $(-2,\ 3)$ and $(2,\ -1)$.

**51** $f(x) = \begin{cases} x+2 & \text{if } x \leq -1 \\ x^3 & \text{if } |\,x\,| < 1 \\ -x+3 & \text{if } x \geq 1 \end{cases}$

If $x \leq -1$, we want the graph of $y = x+2$. To determine the endpoint of this part of the graph, merely substitute $x = -1$ in $y = x+2$, obtaining $y = 1$. If $|\,x\,| < 1$, or, equivalently, $-1 < x < 1$, we want the graph of $y = x^3$. We do not include the endpoints $(-1,\ -1)$ and $(1,\ 1)$. If $x \geq 1$, we want the graph of $y = -x+3$ and include its endpoint $(1,\ 2)$.

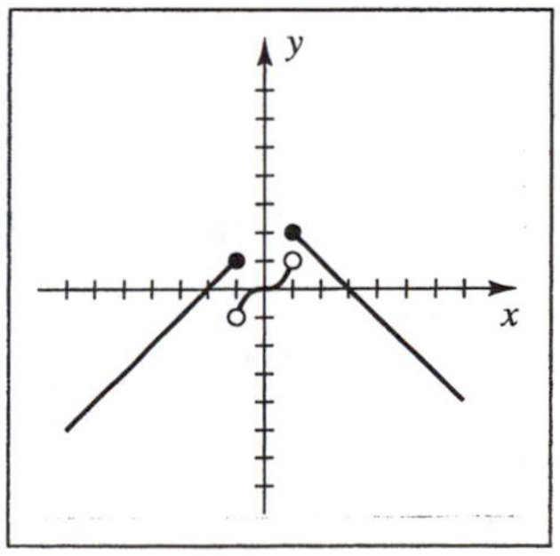


*Figure 51*

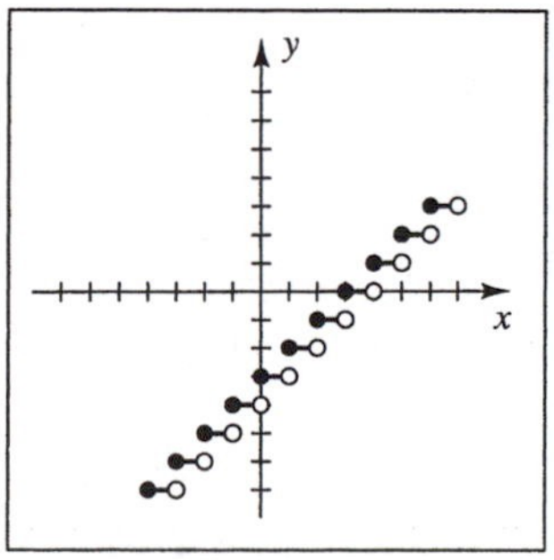


*Figure 53(a)*

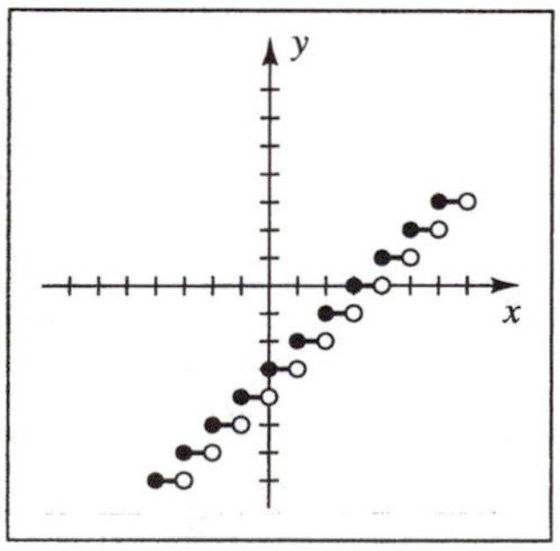


*Figure 53(b)*

[53] (a) $f(x) = [\![x-3]\!]$ • shift $g(x) = [\![x]\!]$ right 3 units {see figure on page 87}

(b) $f(x) = [\![x]\!] - 3$ • shift $g$ down 3 units, which is the same graph as in part (a).

(c) $f(x) = 2[\![x]\!]$ • vertically stretch $g$ by a factor of 2

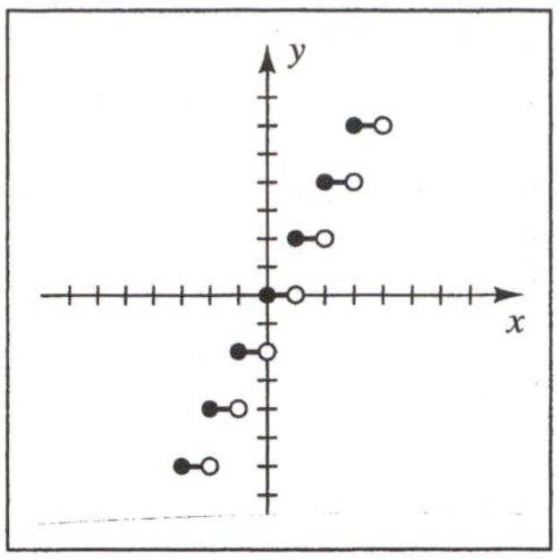

Figure 53(c)

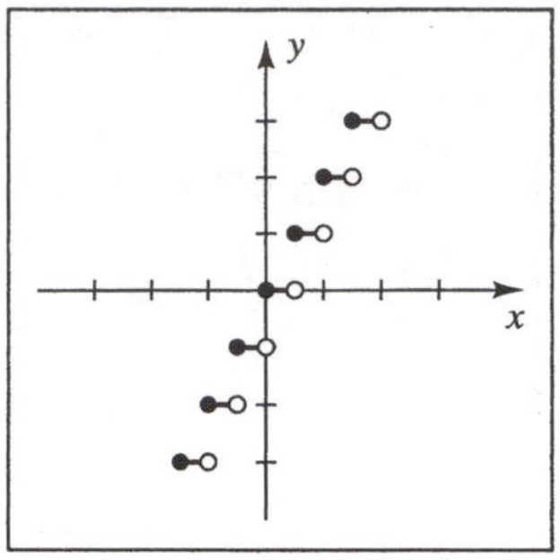

Figure 53(d)

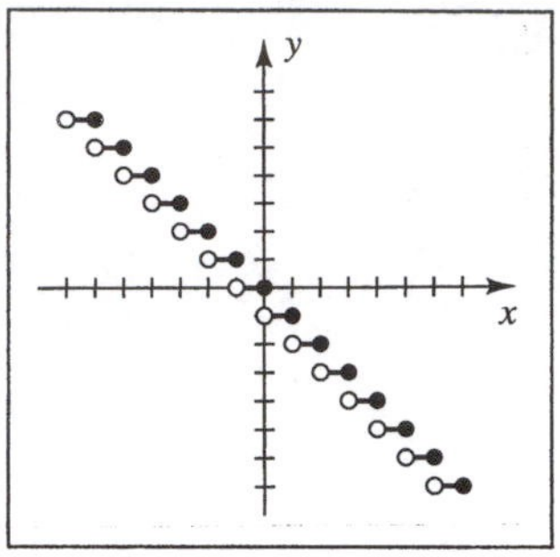

Figure 53(e)

(d) $f(x) = [\![2x]\!]$ • horizontally compress $g$ by a factor of 2

Alternatively, we could determine the pattern of "steps" for this function by finding the values of $x$ that make $f(x)$ change from 0 to 1, then from 1 to 2, etc. If $2x = 0$, then $x = 0$, and if $2x = 1$, then $x = \frac{1}{2}$.

Thus, the function will equal 0 from $x = 0$ to $x = \frac{1}{2}$ and then jump to 1 at $x = \frac{1}{2}$.

If $2x = 2$, then $x = 1$. The pattern is established: each step will be $\frac{1}{2}$ unit long.

(e) $f(x) = [\![-x]\!]$ • reflect $g$ through the $y$-axis

[55] A question you can ask to help determine if a relationship is a function is "If $x$ is a particular value, can I find a unique $y$-value?" In this case, if $x$ was 16, then $16 = y^2 \Rightarrow y = \pm 4$. Since we cannot find a unique $y$-value, this is not a function. Graphically, {see Figure 3 in §3.2 in the text} given any $x$-value greater than 0, there are two points on the graph and a vertical line intersects the graph in more than one point.

[57] Reflect each portion of the graph that is below the $x$-axis through the $x$-axis.

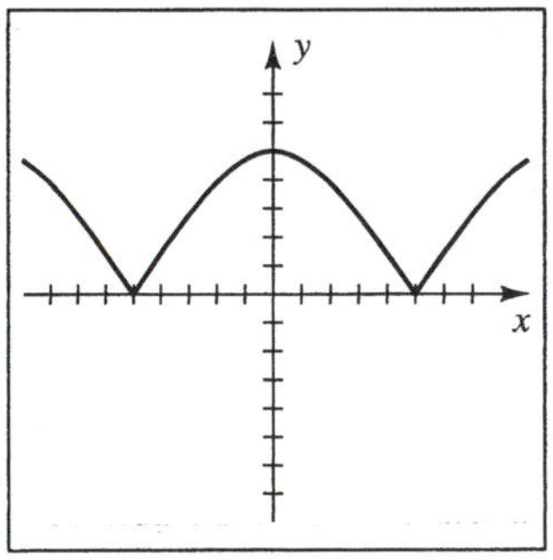

Figure 57

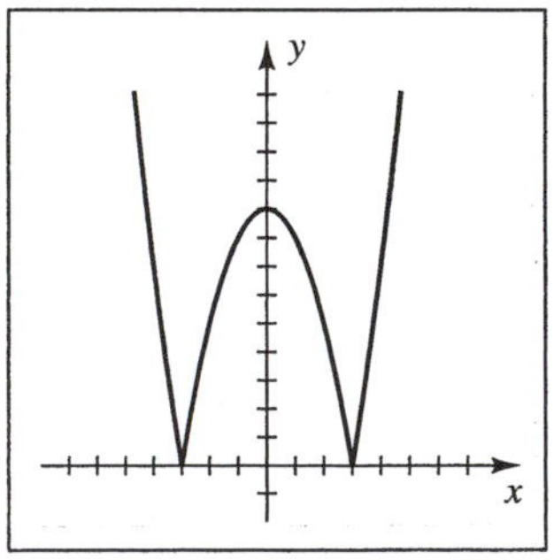

Figure 59

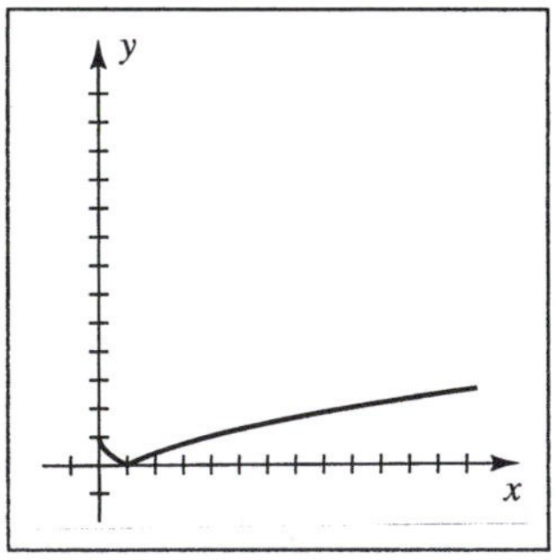

Figure 61

[59] $y = |\,9 - x^2\,|$ • First sketch $y = 9 - x^2$,

then reflect the portions of the graph below the $x$-axis through the $x$-axis.

**61** $y = |\sqrt{x} - 1|$ • First sketch $y = \sqrt{x} - 1$, which is the graph of the square root function shifted down 1 unit. Then reflect the portions of the graph below the $x$-axis through the $x$-axis. See *Figure 61* on the preceding page.

**63** (a) For $y = -2f(x)$, multiply the $y$-coordinates by $-2$. So the given range, $[-4, 8]$, becomes $[-16, 8]$ {note that we changed the order so that $-16$ appears first—listing the range as $[8, -16]$ is incorrect}. The domain {$x$-coordinates} remains the same. $D = [-2, 6]$, $R = [-16, 8]$

(b) For $y = f(\frac{1}{2}x)$, multiply the $x$-coordinates by 2.
The range {$y$-coordinates} remains the same. $D = [-4, 12]$, $R = [-4, 8]$

(c) For $y = f(x - 3) + 1$, add 3 to the $x$-coordinates and add 1 to the $y$-coordinates.
$D = [1, 9]$, $R = [-3, 9]$

(d) For $y = f(x + 2) - 3$, subtract 2 from the $x$-coordinates and subtract 3 from the $y$-coordinates. $D = [-4, 4]$, $R = [-7, 5]$

(e) For $y = f(-x)$, multiply all $x$-coordinates by $-1$. $D = [-6, 2]$, $R = [-4, 8]$

(f) For $y = -f(x)$, multiply all $y$-coordinates by $-1$. $D = [-2, 6]$, $R = [-8, 4]$

(g) $y = f(|x|)$ • Graphically, we can reflect all points with positive $x$-coordinates through the $y$-axis, so the domain $[-2, 6]$ becomes $[-6, 6]$. Algebraically, we are replacing $x$ with $|x|$, so $-2 \leq x \leq 6$ becomes $-2 \leq |x| \leq 6$, which is equivalent $|x| \leq 6$, or, equivalently, $-6 \leq x \leq 6$. The range stays the same because of the given assumptions: $f(2) = 8$ and $f(6) = -4$; that is, the full range is taken on for $x \geq 0$. Note that the range could not be determined if $f(-2)$ was equal to 8. $D = [-6, 6]$, $R = [-8, 4]$

(h) $y = |f(x)|$ • The points with $y$-coordinates having values from $-4$ to 0 will have values from 0 to 4, so the range will be $[0, 8]$. $D = [-2, 6]$, $R = [0, 8]$

**65** If $x \leq 20{,}000$, then $T(x) = 0.15x$. If $x > 20{,}000$, then the tax is 15% of the first 20,000, which is 3000, plus 20% of the amount over 20,000, which is $(x - 20{,}000)$. We may summarize and simplify as follows:

$$T(x) = \begin{cases} 0.15x & \text{if } x \leq 20{,}000 \\ 3000 + 0.20(x - 20{,}000) & \text{if } x > 20{,}000 \end{cases}$$

$$= \begin{cases} 0.15x & \text{if } x \leq 20{,}000 \\ 3000 + 0.20x - 4{,}000 & \text{if } x > 20{,}000 \end{cases}$$

$$= \begin{cases} 0.15x & \text{if } x \leq 20{,}000 \\ 0.20x - 1000 & \text{if } x > 20{,}000 \end{cases}$$

[67] The author receives \$1.20 on the first 10,000 copies, \$1.50 on the next 5000, and \$1.80 on each additional copy.

$$R(x) = \begin{cases} 1.20x & \text{if } 0 \le x \le 10{,}000 \\ 12{,}000 + 1.50(x - 10{,}000) & \text{if } 10{,}000 < x \le 15{,}000 \\ 19{,}500 + 1.80(x - 15{,}000) & \text{if } x > 15{,}000 \end{cases}$$

$$= \begin{cases} 1.20x & \text{if } 0 \le x \le 10{,}000 \\ 1.50x - 3000 & \text{if } 10{,}000 < x \le 15{,}000 \\ 1.80x - 7500 & \text{if } x > 15{,}000 \end{cases}$$

[69] Assign ABS$(1.3x + 2.8)$ to $Y_1$ and $1.2x + 5$ to $Y_2$. The viewing rectangle $[-10, 50, 10]$ by $[-10, 50, 10]$ shows intersection points at exactly $-3.12$ and $-22$. The solution of $|1.3x + 2.8| < 1.2x + 5$ is the interval $(-3.12, 22)$.

[71] Assign ABS$(1.2x^2 - 10.8)$ to $Y_1$ and $1.36x + 4.08$ to $Y_2$. The standard viewing rectangle $[-15, 15]$ by $[-10, 10]$ shows intersection points at exactly $-3$ and at approximately 1.87 and 4.13. The solution is $(-\infty, -3) \cup (-3, 1.87) \cup (4.13, \infty)$.

[73] $f(x) = (0.5x^3 - 4x - 5)$ and $g(x) = (0.5x^3 - 4x - 5) + 4 = f(x) + 4$, so the graph of $g$ can be obtained by shifting the graph of $f$ upward a distance of 4.

[−12, 12] by [−8, 8]

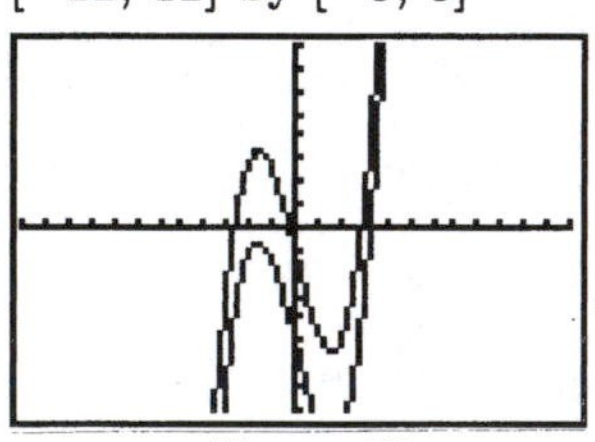

Figure 73

[−12, 12] by [−8, 8]

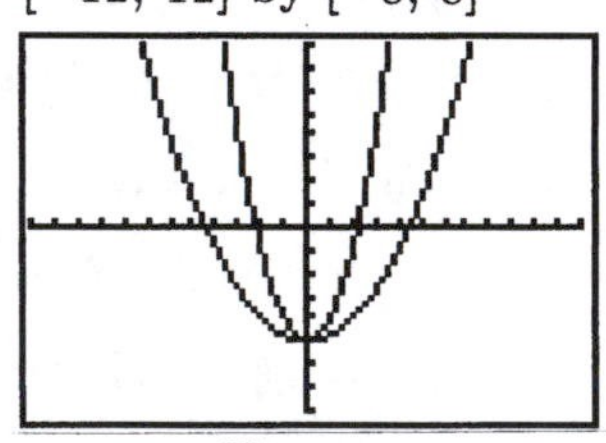

Figure 75

[75] $f(x) = x^2 - 5$ and $g(x) = (\frac{1}{2}x)^2 - 5$, so $g(x) = f(\frac{1}{2}x)$ and the graph of $g$ can be obtained by stretching the graph of $f$ horizontally by a factor of 2.

[77] $f(x) = x^3 - 5x$ and $g(x) = |f(x)|$, so the graph of $g$ is the same as the graph of $f$ if $f$ is nonnegative. If $f(x) < 0$, then the graph of $f$ will be reflected through the $x$–axis.

[−12, 12] by [−8, 8]

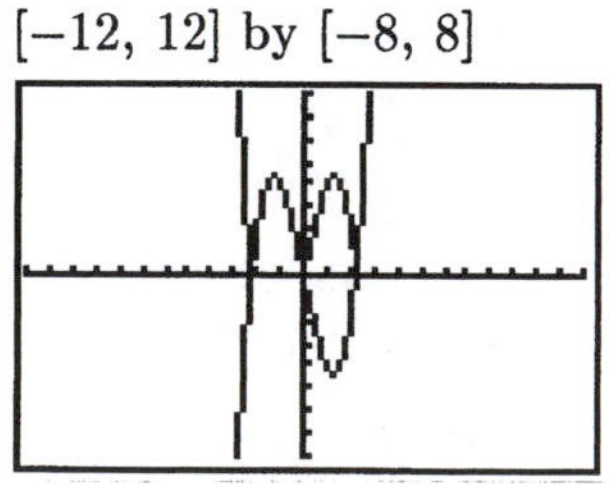

Figure 77

79 (a) Option I gives $C_1 = 4(\$29.95) + \$0.25(500 - 200) = 119.80 + 75.00 = \$194.80$.

Option II gives $C_2 = 4(\$39.95) + \$0.15(500) = 159.80 + 75.00 = \$234.80$.

(b) Let $x$ represent the mileage. The cost function for Option I is the piecewise linear function

$$C_1(x) = \begin{cases} 119.80 & \text{if } 0 \le x \le 200 \\ 119.80 + 0.25(x - 200) & \text{if } x > 200 \end{cases}$$

Option II is the linear function $C_2(x) = 159.8 + 0.15x$ for $x \ge 0$.

(c) Let $C_1 = \mathrm{Y}_1$ and $C_2 = \mathrm{Y}_2$.

Table $\mathrm{Y}_1 = 119.80 + 0.25(x - 200)*(x > 200)$ and $\mathrm{Y}_2 = 159.80 + 0.15x$

| $x$ | $\mathrm{Y}_1$ | $\mathrm{Y}_2$ | $x$ | $\mathrm{Y}_1$ | $\mathrm{Y}_2$ |
|---|---|---|---|---|---|
| 100 | 119.8 | 174.8 | 700 | 244.8 | 264.8 |
| 200 | 119.8 | 189.8 | 800 | 269.8 | 279.8 |
| 300 | 144.8 | 204.8 | 900 | 294.8 | 294.8 |
| 400 | 169.8 | 219.8 | 1000 | 319.8 | 309.8 |
| 500 | 194.8 | 234.8 | 1100 | 344.8 | 324.8 |
| 600 | 219.8 | 249.8 | 1200 | 369.8 | 339.8 |

(d) From the table, we see that the options are equal in cost for $x = 900$ miles. Option I is preferable if $x \in [0, 900)$ and Option II is preferable if $x > 900$.

## 3.6 Exercises

1 Using the standard equation of a parabola with a vertical axis having vertex $V(-3, 1)$, we have $y = a[x - (-3)]^2 + 1$. The coefficient $a$ determines whether the parabola opens upward {$a$ positive} or opens downward {$a$ negative}. If $|a| > 1$, then the parabola is narrower {steeper} than the graph of $y = x^2$; if $|a| < 1$, then the parabola is wider {flatter} than the graph of $y = x^2$. Simplifying the equation gives us $y = a(x + 3)^2 + 1$.

3 We can use the standard equation of a parabola with a vertical axis.

$$V(0, -3) \;\Rightarrow\; y = a(x - 0)^2 - 3 \;\Rightarrow\; y = ax^2 - 3.$$

5 The approach shown here {like Solution 1 in Example 2} requires us to *factor out the leading coefficient.*

$$\begin{aligned} f(x) &= -x^2 - 4x - 8 && \{\text{given}\} \\ &= -(x^2 + 4x + \underline{\quad}) - 8 + \underline{\quad} && \{\text{factor out } -1 \text{ from } -x^2 - 4x\} \\ &= -(x^2 + 4x + \underline{4}) - 8 + \underline{4} && \{\text{complete the square for } x^2 + 4x\} \\ &= -(x + 2)^2 - 4 && \{\text{equivalent equation}\} \end{aligned}$$

[7] The approach shown here { like Solution 2 in Example 2 } requires us to *divide both sides by the leading coefficient*—remember to multiply both sides by the same coefficient in the end. Either method is fine.

$f(x) = 2x^2 - 12x + 22$ { given }

$\frac{1}{2}f(x) = (x^2 - 6x + \underline{\quad}) + 11 - \underline{\quad}$ { divide the equation by 2 }

$= (x^2 - 6x + \underline{\ 9\ }) + 11 - \underline{\ 9\ }$ { complete the square for $x^2 - 6x$ }

$= (x-3)^2 + 2$ { equivalent equation }

$2 \cdot \frac{1}{2}f(x) = 2 \cdot \left[(x-3)^2 + 2\right]$ { multiply the equation by 2 }

$f(x) = 2(x-3)^2 + 4$ { the desired standard form }

[9] $f(x) = -3x^2 - 6x - 5$ { divide by $-3$ and proceed as in Exercise 7 } $\Rightarrow$

$-\frac{1}{3}f(x) = x^2 + 2x + \underline{\ 1\ } + \frac{5}{3} - \underline{\ 1\ } = (x+1)^2 + \frac{2}{3}$ $\Rightarrow$

$-3 \cdot -\frac{1}{3}f(x) = -3 \cdot \left[(x+1)^2 + \frac{2}{3}\right]$ $\Rightarrow$ $\underline{f(x) = -3(x+1)^2 - 2}$

[11] $f(x) = -\frac{3}{4}x^2 + 9x - 34$ { divide by $-\frac{3}{4}$ and proceed as in Exercise 7 } $\Rightarrow$

$-\frac{4}{3}f(x) = x^2 - 12x + \frac{136}{3} = x^2 - 12x + \underline{\ 36\ } + \frac{136}{3} - \underline{\ 36\ } = (x-6)^2 + \frac{28}{3}$ $\Rightarrow$

$f(x) = (-\frac{3}{4})(x-6)^2 + (-\frac{3}{4}) \cdot (\frac{28}{3})$ $\Rightarrow$ $\underline{f(x) = -\frac{3}{4}(x-6)^2 - 7}$

[13] (a) $x^2 - 4x = 0$ { $a = 1$, $b = -4$, $c = 0$ } $\Rightarrow$ $x = \dfrac{4 \pm \sqrt{16 - 0}}{2} = \dfrac{4 \pm 4}{2} = 0, 4$

(b) Using the theorem for locating the vertex of a parabola with $y = x^2 - 4x$ gives us $x$-coordinate $-\dfrac{b}{2a} = -\dfrac{-4}{2(1)} = 2$. The parabola opens upward since $a = 1 > 0$, so there is a minimum value of $y$. To find the minimum, substitute 2 for $x$ in $f(x) = x^2 - 4x$ to get $f(2) = -4$.

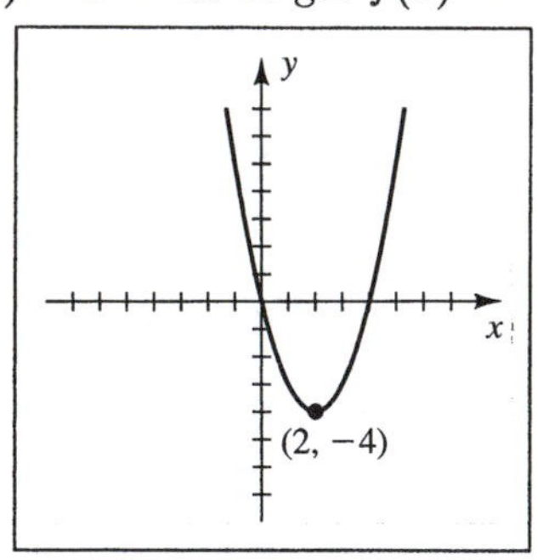


*Figure 13*

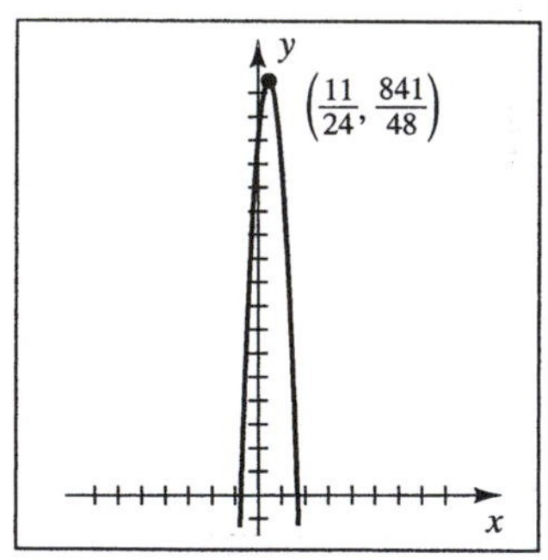


*Figure 15*

[15] (a) $f(x) = -12x^2 + 11x + 15 = 0$ $\Rightarrow$ $x = \dfrac{-11 \pm \sqrt{121 + 720}}{-24} = \dfrac{-11 \pm 29}{-24} = -\dfrac{3}{4}, \dfrac{5}{3}$

(b) The $x$-coordinate of the vertex is given by $x = -\dfrac{b}{2a} = -\dfrac{11}{2(-12)} = \dfrac{11}{24}$.

Note that this value is easily obtained from part (a). The $y$-coordinate of the vertex is then $f(\frac{11}{24}) = \frac{841}{48} \approx 17.52$. This is a maximum since $a = -12 < 0$.

17 (a) $9x^2 + 24x + 16 = 0 \Rightarrow x = \dfrac{-24 \pm \sqrt{576 - 576}}{18} = \dfrac{-24}{18} = -\dfrac{4}{3}$

(b) $f(x) = 9x^2 + 24x + 16 \Rightarrow -\dfrac{b}{2a} = -\dfrac{24}{2(9)} = -\frac{4}{3}$.

$f(-\frac{4}{3}) = 0$ is a minimum since $a > 0$.

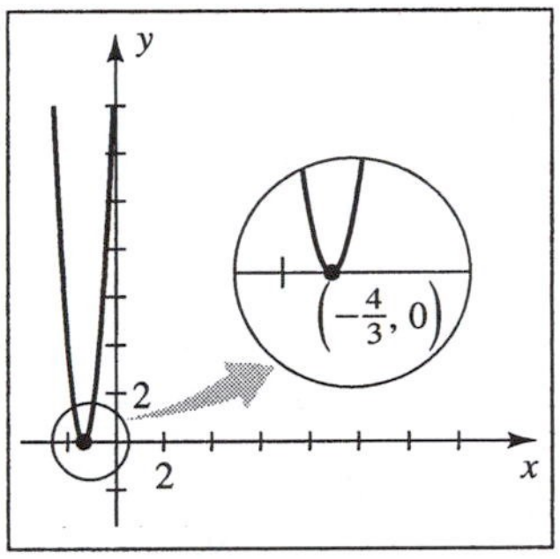

Figure 17

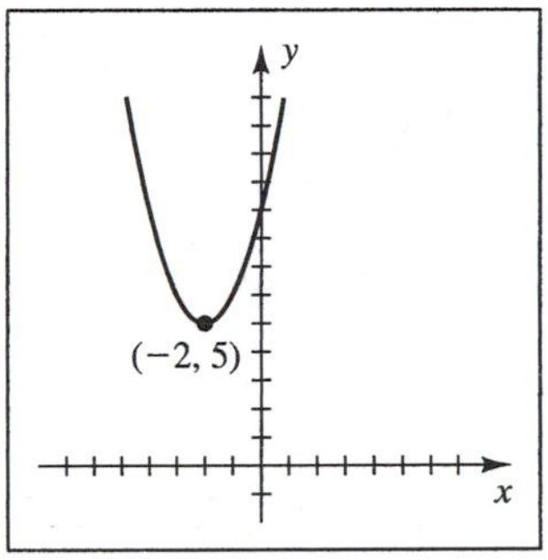

Figure 19

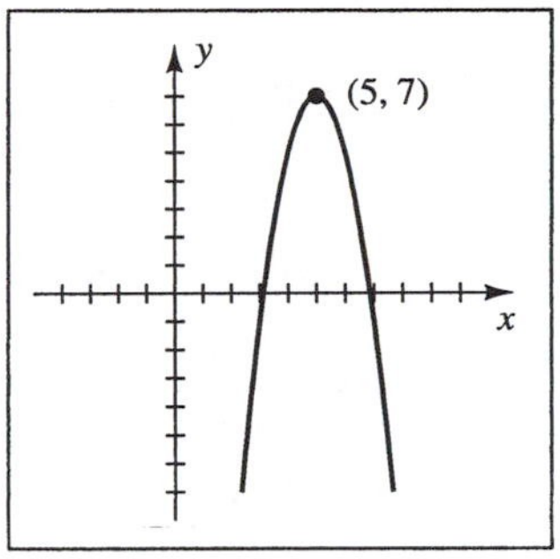

Figure 21

19 (a) $f(x) = x^2 + 4x + 9 = 0 \Rightarrow x = \dfrac{-4 \pm \sqrt{16 - 36}}{2} = \dfrac{-4 \pm \sqrt{-20}}{2} = -2 \pm \sqrt{5}\, i.$

The imaginary part indicates that there are *no* $x$-intercepts.

(b) $-\dfrac{b}{2a} = -\dfrac{4}{2(1)} = -2.$ $f(-2) = 5$ is a minimum since $a = 1 > 0$.

21 (a) $-2x^2 + 20x - 43 = 0 \Rightarrow x = \dfrac{-20 \pm \sqrt{400 - 344}}{-4} = 5 \pm \frac{1}{2}\sqrt{14} \approx 6.87,\ 3.13$

(b) $f(x) = -2x^2 + 20x - 43 \Rightarrow -\dfrac{b}{2a} = -\dfrac{20}{2(-2)} = 5.$

$f(5) = 7$ is a maximum since $a < 0$.

23 $V(4, -1) \Rightarrow y = a(x-4)^2 - 1$. Substituting 0 for $x$ and 1 for $y$ gives us

$1 = a(0-4)^2 - 1 \Rightarrow 2 = 16a \Rightarrow a = \frac{1}{8}$. Hence, $y = \frac{1}{8}(x-4)^2 - 1$.

25 $V(-2, 4) \Rightarrow y = a(x+2)^2 + 4$. $x = 1, y = 0 \Rightarrow 0 = a(1+2)^2 + 4 \Rightarrow$

$-4 = 9a \Rightarrow a = -\frac{4}{9}$. Hence, $y = -\frac{4}{9}(x+2)^2 + 4$.

27 From the figure, the $x$-intercepts are $-2$ and 4, so the equation must have the form $y = a[x - (-2)](x-4) = a(x+2)(x-4)$. To find the value of $a$, use the point (2, 4). $4 = a(2+2)(2-4) \Rightarrow 4 = a(4)(-2) \Rightarrow -8a = 4 \Rightarrow a = -\frac{1}{2}$, so the equation is $y = -\frac{1}{2}(x+2)(x-4)$.

29 $V(0, -2) \Rightarrow (h, k) = (0, -2)$. $x = 3, y = 25 \Rightarrow 25 = a(3-0)^2 - 2 \Rightarrow$

$27 = 9a \Rightarrow a = 3$. Hence, $y = 3(x-0)^2 - 2$, or $y = 3x^2 - 2$.

31 $V(3, 5) \Rightarrow y = a(x-3)^2 + 5$ $(*)$. If the $x$-intercept is 0, then the point (0, 0) is on the parabola. Substituting $x = 0$ and $y = 0$ into $(*)$ gives us $0 = a(0-3)^2 + 5 \Rightarrow$

$-5 = 9a \Rightarrow a = -\frac{5}{9}$. Hence, $y = -\frac{5}{9}(x-3)^2 + 5$.

**33** Since the $x$-intercepts are $-3$ and 5, the $x$-coordinate of the vertex of the parabola is 1 {the average of the $x$-intercept values}. Since the highest point has $y$-coordinate 4, the vertex is (1, 4). $V(1, 4) \Rightarrow y = a(x-1)^2 + 4$. We can use the point $(-3, 0)$ since there is an $x$-intercept of $-3$. $x = -3, y = 0 \Rightarrow 0 = a(-3-1)^2 + 4 \Rightarrow$

$-4 = 16a \Rightarrow a = -\frac{1}{4}$. Hence, $y = -\frac{1}{4}(x-1)^2 + 4$.

**35** Let $d$ denote the distance between the parabola and the line.

$$d = (\text{parabola}) - (\text{line}) = (-2x^2 + 4x + 3) - (x - 2) = -2x^2 + 3x + 5.$$

This relation is quadratic and the $x$-value of its maximum value is

$$-\frac{b}{2a} = -\frac{3}{2(-2)} = \frac{3}{4}. \text{ Thus, maximum } d = -2\left(\frac{3}{4}\right)^2 + 3\left(\frac{3}{4}\right) + 5 = \frac{49}{8} = 6.125.$$

Note that the maximum value *of the parabola* is $f(1) = 5$,

which is not the same as the maximum value of the distance $d$.

***Note:*** We may find the vertex of a parabola in the following problems using either:

(1) the complete the square method,

(2) the formula method, or

(3) the fact that the vertex lies halfway between the $x$-intercepts.

**37** For $D(h) = -0.058h^2 + 2.867h - 24.239$, the vertex is located at

$h = \frac{-b}{2a} = \frac{-2.867}{2(-0.058)} \approx 24.72$ km. Since $a < 0$, this will produce a maximum value.

**39** Since the $x$-intercepts of $y = cx(21 - x)$ are 0 and 21, the maximum will occur

halfway between them, that is, when the infant weighs 10.5 lb.

**41** (a) $s(t) = -16t^2 + 144t + 100$ will be a maximum when $t = \frac{-b}{2a} = \frac{-144}{2(-16)} = \frac{9}{2}$.

$s(\frac{9}{2}) = -16(\frac{9}{2})^2 + 144(\frac{9}{2}) + 100 = -324 + 648 + 100 = 424$ ft.

(b) When $t = 0$, $s(t) = 100$ ft, which is the height of the building.

**43** Let $x$ and $40 - x$ denote the numbers, and their product $P$ is $x(40 - x)$.

$P$ has zeros at 0 and 40 and is a maximum (since $a < 0$) when $x = \frac{0 + 40}{2} = 20$.

The product will be a maximum when both numbers are 20.

**45** (a) The 1000 ft of fence is made up of 3 sides of length $x$ and 4 sides of length $y$.

To express $y$ as a function of $x$, we need to solve $3x + 4y = 1000$ for $y$.

$3x + 4y = 1000 \Rightarrow 4y = 1000 - 3x \Rightarrow y = 250 - \frac{3}{4}x$.

(b) Using the value of $y$ from part (a), $A = xy = x(250 - \frac{3}{4}x) = -\frac{3}{4}x^2 + 250x$.

(c) $A$ will be a maximum when $x = \frac{-b}{2a} = \frac{-250}{2(-3/4)} = \frac{500}{3} = 166\frac{2}{3}$ ft. Using part (a)

to find the corresponding value of $y$, $y = 250 - \frac{3}{4}(\frac{500}{3}) = 250 - 125 = 125$ ft.

[47] The parabola has vertex $V(\frac{9}{2}, 3)$. Hence, the equation has the form $y = a(x - \frac{9}{2})^2 + 3$. Using the point (9, 0) {or (0, 0)}, we have $0 = a(9 - \frac{9}{2})^2 + 3 \Rightarrow a = -\frac{4}{27}$.

Thus, the path may be described by $y = -\frac{4}{27}(x - \frac{9}{2})^2 + 3$.

[49] (a) Since the vertex is at (0, 10), an equation for the parabola is $y = ax^2 + 10$.

The points (200, 90) and (−200, 90) are on the parabola.

Substituting (200, 90) for $(x, y)$ yields $90 = a(200)^2 + 10 \Rightarrow a = \frac{80}{40{,}000} = \frac{1}{500}$.

Hence, $y = \frac{1}{500}x^2 + 10$.

(b) The cables are spaced 40 ft apart. Using $y = \frac{1}{500}x^2 + 10$ with $x = 40$, 80, 120, and 160, we get $y = \frac{66}{5}, \frac{114}{5}, \frac{194}{5}$, and $\frac{306}{5}$, respectively.

There is one cable of length 10 ft and 2 cables of each of the other lengths.

Thus, the total length is $10 + 2\left(\frac{66}{5} + \frac{114}{5} + \frac{194}{5} + \frac{306}{5}\right) = 282$ ft.

[51] An equation describing the doorway is $y = ax^2 + 9$. Since the doorway is 6 feet wide at the base, $x = 3$ when $y = 0 \Rightarrow 0 = a(3)^2 + 9 \Rightarrow 0 = 9a + 9 \Rightarrow 9a = -9 \Rightarrow a = -1$. Thus, the equation is $y = -x^2 + 9$. To fit an 8 foot high box through the doorway, we must find $x$ when $y = 8$. If $y = 8$, then $8 = -x^2 + 9 \Rightarrow x^2 = 1 \Rightarrow x = \pm 1$. Hence, the box can only be $1 - (-1) = 2$ feet wide.

[53] Let $x$ denote the number of pairs of shoes that are ordered. If $x < 50$, then the amount $A$ of money that the company makes is $40x$. If $50 \le x \le 600$, then each pair of shoes is discounted $0.04x$, so the price per pair is $40 - 0.04x$, and the amount of money that the company makes is $(40 - 0.04x)x$. In piecewise form, we have

$$A(x) = \begin{cases} 40x & \text{if } x < 50 \\ (40 - 0.04x)x & \text{if } 50 \le x \le 600 \end{cases}$$

The maximum value of the first part of $A$ is (\$40)(49) = \$1960. For the second part of $A$, $A = -0.04x^2 + 40x$ has a maximum when $x = \frac{-b}{2a} = \frac{-40}{2(-0.04)} = 500$ pairs.

$A(500) = 10{,}000 > 1960$, so $x = 500$ produces a maximum for both parts of $A$.

**55** (a) Let $y$ denote the number of \$1 decreases in the monthly charge.

$R(y) = (\#\text{ of customers})(\text{monthly charge per customer})$

$= (5000 + 500y)(20 - y)$

$= 500(10 + y)(20 - y)$

Now let $x$ denote the monthly charge, which is $20 - y$.

$R$ becomes $500[10 + (20 - x)]\,(x) = 500x(30 - x)$.

(b) $R$ has $x$-intercepts at 0 and 30, and must have its vertex halfway between them at $x = 15$.

Note that this gives us $y = 20 - 15 = 5$, and we have $5000 + 500(5) = 7500$ customers for a revenue of $500(10 + 5)(20 - 5) = \$112{,}500$.

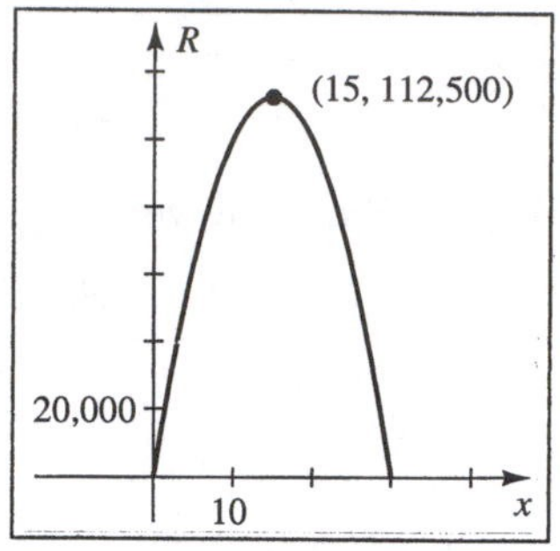

*Figure 55*

**57** From the graph, there are three points of intersection.

Their coordinates are approximately (−0.57, 0.64), (0.02, −0.27), and (0.81, −0.41).

[−3, 3] by [−2, 2]

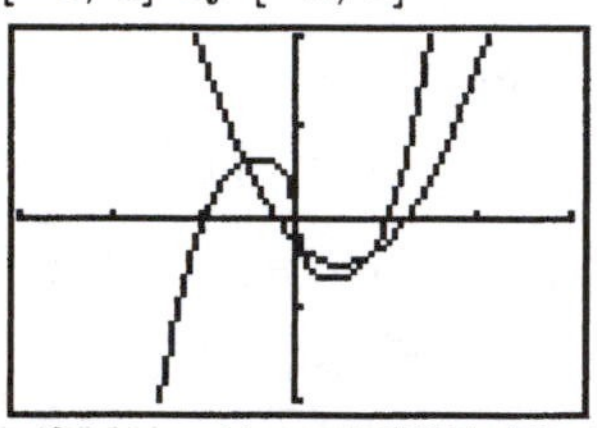
*Figure 57*

[−8, 4] by [−1, 7]

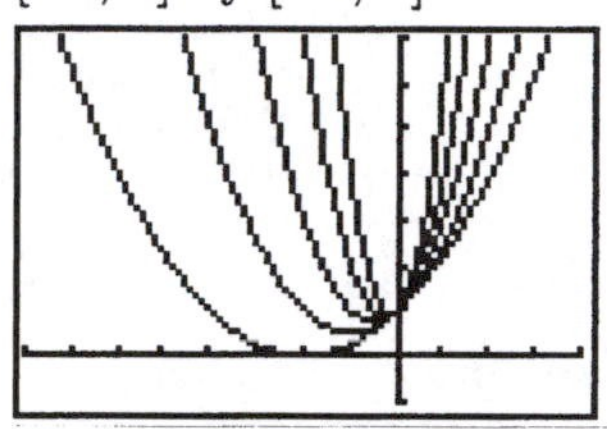
*Figure 59*

**59** Since $a > 0$, all parabolas open upward. From the graph, we can see that smaller values of $a$ result in the parabola opening wider while larger values of $a$ result in the parabola becoming narrower.

**61** (a) Let January correspond to 1, February to 2, ... , and December to 12.

(b) Let $f(x) = a(x - h)^2 + k$. The vertex appears to occur near (7, 0.8) Thus, $h = 7$ and $k = 0.8$. Using trial and error, a reasonable value for $a$ is 0.17. Thus, let $f(x) = 0.17(x - 7)^2 + 0.8$. {From the TI-83 Plus, the quadratic regression equation is $y \approx 0.1476x^2 - 1.9951x + 7.8352$.}

(c) $f(4) = 2.33$, compared to the actual value of 2.4 in.

[0.5, 12] by [0, 8]

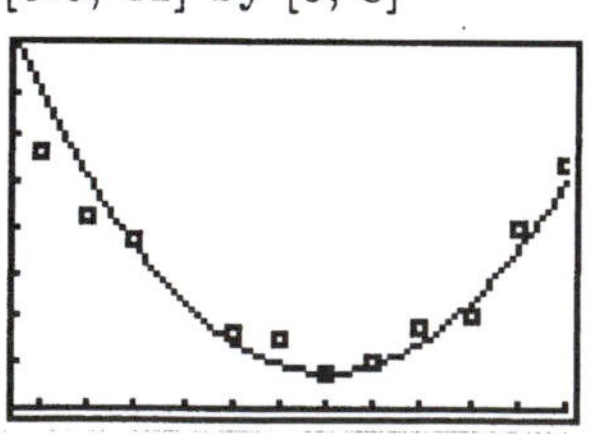
*Figure 61*

[−800, 800, 100] by [−100, 200, 100]

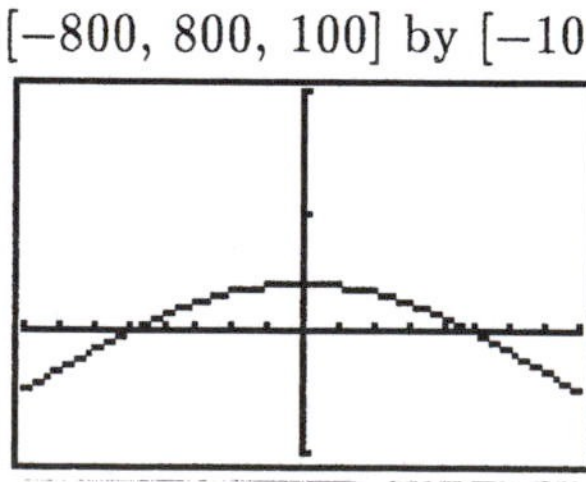
*Figure 63*

63 (a) The equation of the line passing through $A(-800,\ -48)$ and $B(-500,\ 0)$ is $y = \frac{4}{25}x + 80$. The equation of the line passing through $D(500,\ 0)$ and $E(800,\ -48)$ is $y = -\frac{4}{25}x + 80$. Let $y = a(x-h)^2 + k$ be the equation of the parabola passing through the points $B(-500,\ 0)$, $C(0,\ 40)$, and $D(500,\ 0)$. The vertex is located at $(0,\ 40)$ so $y = a(x-0)^2 + 40$. Since $D(500,\ 0)$ is on the graph, $0 = a(500-0)^2 + 40 \Rightarrow a = -\frac{1}{6250}$ and $y = -\frac{1}{6250}x^2 + 40$. Thus, let

$$f(x) = \begin{cases} \frac{4}{25}x + 80 & \text{if } -800 \le x < -500 \\ -\frac{1}{6250}x^2 + 40 & \text{if } -500 \le x \le 500 \\ -\frac{4}{25}x + 80 & \text{if } 500 < x \le 800 \end{cases}$$

(b) Graph the equations: (See *Figure 63.*)

$$Y_1 = (4/25{*}x + 80)/(x < -500),$$
$$Y_2 = (-1/6250{*}x\hat{\ }2 + 40)/(x \ge -500 \text{ and } x \le 500),$$
$$Y_3 = (-4/25{*}x + 80)/(x > 500)$$

65 (a) $f$ must have zeros of 0 and 150. Thus, $f(x) = a(x-0)(x-150)$. Also, $f$ will have a maximum of 100 occurring at $x = 75$. (The vertex will be midway between the zeros of $f$.) $a(75-0)(75-150) = 100 \Rightarrow a = \frac{100}{(75)(-75)} = -\frac{4}{225}$. $f(x) = -\frac{4}{225}(x)(x-150) = -\frac{4}{225}x^2 + \frac{8}{3}x$.

(c) The value of $k$ affects both the distance and the height traveled by the object. The distance and height decrease by a factor of $\frac{1}{k}$ when $k > 1$ and increase by a factor of $\frac{1}{k}$ when $0 < k < 1$.

[0, 180, 50] by [0, 120, 50]

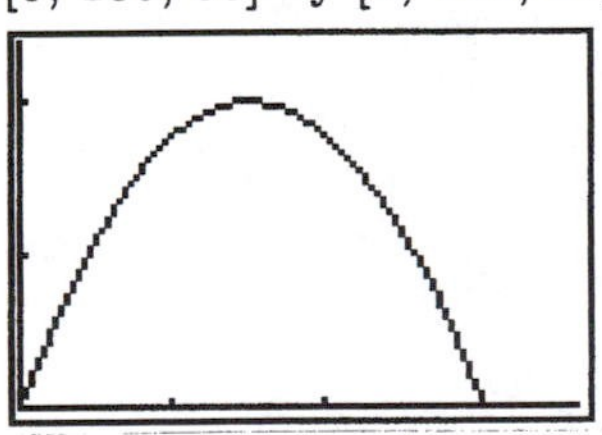

*Figure 65(b)*

[0, 600, 50] by [0, 400, 50]

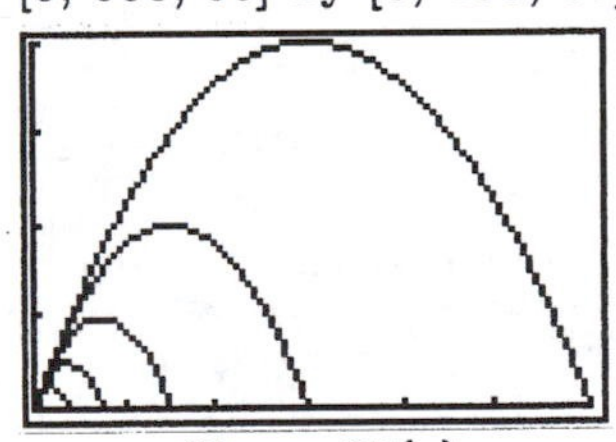

*Figure 65(c)*

## 3.7 Exercises

1 (a) $(f+g)(3) = f(3) + g(3) = 6 + 9 = 15$

(b) $(f-g)(3) = f(3) - g(3) = 6 - 9 = -3$

(c) $(fg)(3) = f(3) \cdot g(3) = 6 \cdot 9 = 54$

(d) $(f/g)(3) = \dfrac{f(3)}{g(3)} = \frac{6}{9} = \frac{2}{3}$

3 (a) $(f+g)(x) = f(x)+g(x) = (x^2+2)+(2x^2-1) = 3x^2+1$;

$(f-g)(x) = f(x)-g(x) = (x^2+2)-(2x^2-1) = 3-x^2$;

$(fg)(x) = f(x)\cdot g(x) = (x^2+2)\cdot(2x^2-1) = 2x^4+3x^2-2$;

$\left(\frac{f}{g}\right)(x) = \frac{f(x)}{g(x)} = \frac{x^2+2}{2x^2-1}$

(b) The domain of $f+g$, $f-g$, and $fg$ is the set of all real numbers, $\mathbb{R}$.

(c) The domain of $f/g$ is the same as in (b), except we must exclude the zeros of $g$.

$2x^2-1=0 \Rightarrow x^2=\frac{1}{2} \Rightarrow x = \pm\sqrt{\frac{1}{2}} = \pm\frac{\sqrt{2}}{2}$ or $\pm\frac{1}{2}\sqrt{2} \Rightarrow$

Hence, the domain of $f/g$ is all real numbers except $\pm\frac{1}{2}\sqrt{2}$.

5 (a) $(f+g)(x) = f(x)+g(x) = \sqrt{x+5}+\sqrt{x+5} = 2\sqrt{x+5}$;

$(f-g)(x) = f(x)-g(x) = \sqrt{x+5}-\sqrt{x+5} = 0$;

$(fg)(x) = f(x)\cdot g(x) = \sqrt{x+5}\cdot\sqrt{x+5} = x+5$;

$\left(\frac{f}{g}\right)(x) = \frac{f(x)}{g(x)} = \frac{\sqrt{x+5}}{\sqrt{x+5}} = 1$

(b) The radicand, $x+5$, must be nonnegative; that is $x+5 \geq 0 \Rightarrow x \geq -5$.

Thus, the domain of $f+g$, $f-g$, and $fg$ is $[-5, \infty)$.

(c) Now the radicand must be positive {can't have zero in the denominator}.

Thus, the domain of $f/g$ is $(-5, \infty)$.

7 (a) $(f+g)(x) = f(x)+g(x) = \frac{2x}{x-4}+\frac{x}{x+5} = \frac{2x(x+5)+x(x-4)}{(x-4)(x+5)} = \frac{3x^2+6x}{(x-4)(x+5)}$;

$(f-g)(x) = f(x)-g(x) = \frac{2x}{x-4}-\frac{x}{x+5} = \frac{2x(x+5)-x(x-4)}{(x-4)(x+5)} = \frac{x^2+14x}{(x-4)(x+5)}$;

$(fg)(x) = f(x)\cdot g(x) = \frac{2x}{x-4}\cdot\frac{x}{x+5} = \frac{2x^2}{(x-4)(x+5)}$;

$\left(\frac{f}{g}\right)(x) = \frac{f(x)}{g(x)} = \frac{2x/(x-4)}{x/(x+5)} = \frac{2(x+5)}{x-4}$

(b) The domain of $f$ is $\mathbb{R}-\{4\}$ and the domain of $g$ is $\mathbb{R}-\{-5\}$. The intersection of these two domains, $\mathbb{R}-\{-5, 4\}$, is the domain of the three functions.

(c) To determine the domain of the quotient $f/g$, we also exclude any values that make the denominator $g$ equal to zero. Hence, we exclude $x=0$ and the domain of the quotient is all real numbers except $-5$, $0$, and $4$, that is, $\mathbb{R}-\{-5, 0, 4\}$.

9 (a) $(f\circ g)(x) = f(g(x)) = f(-x^2) = 2(-x^2)-1 = -2x^2-1$

(b) $(g\circ f)(x) = g(f(x)) = g(2x-1) = -(2x-1)^2 = -(4x^2-4x+1) = -4x^2+4x-1$

(c) $(f\circ f)(x) = f(f(x)) = f(2x-1) = 2(2x-1)-1 = (4x-2)-1 = 4x-3$

(d) $(g\circ g)(x) = g(g(x)) = g(-x^2) = -(-x^2)^2 = -(x^4) = -x^4$

*Note:* Let $h(x) = (f \circ g)(x) = f(g(x))$ and $k(x) = (g \circ f)(x) = g(f(x))$.

$h(-2)$ and $k(3)$ could be worked two ways, as in Example 3(c) in the text.

**11** (a) $h(x) = f(3x+7) = 2(3x+7) - 5 = 6x + 9$

(b) $k(x) = g(2x-5) = 3(2x-5) + 7 = 6x - 8$

(c) Using the result from part (a), $h(-2) = 6(-2) + 9 = -12 + 9 = -3$.

(d) Using the result from part (b), $k(3) = 6(3) - 8 = 18 - 8 = 10$.

**13** (a) $h(x) = f(5x) = 3(5x)^2 + 4 = 75x^2 + 4$

(b) $k(x) = g(3x^2+4) = 5(3x^2+4) = 15x^2 + 20$

(c) $h(-2) = 75(-2)^2 + 4 = 300 + 4 = 304$

(d) $k(3) = 15(3)^2 + 20 = 135 + 20 = 155$

**15** (a) $h(x) = f(2x-1) = 2(2x-1)^2 + 3(2x-1) - 4 = 8x^2 - 2x - 5$

(b) $k(x) = g(2x^2+3x-4) = 2(2x^2+3x-4) - 1 = 4x^2 + 6x - 9$

(c) $h(-2) = 8(-2)^2 - 2(-2) - 5 = 32 + 4 - 5 = 31$

(d) $k(3) = 4(3)^2 + 6(3) - 9 = 36 + 18 - 9 = 45$

**17** (a) $h(x) = f(2x^3 - 5x) = 4(2x^3 - 5x) = 8x^3 - 20x$

(b) $k(x) = g(4x) = 2(4x)^3 - 5(4x) = 128x^3 - 20x$

(c) $h(-2) = 8(-2)^3 - 20(-2) = -64 + 40 = -24$

(d) $k(3) = 128(3)^3 - 20(3) = 3456 - 60 = 3396$

**19** (a) $h(x) = f(-7) = |-7| = 7$ (b) $k(x) = g(|x|) = -7$

(c) $h(-2) = 7$ since $h(\text{any value}) = 7$ (d) $k(3) = -7$ since $k(\text{any value}) = -7$

**21** (a) $h(x) = f(\sqrt{x+2}) = (\sqrt{x+2})^2 - 3(\sqrt{x+2}) = x + 2 - 3\sqrt{x+2}$. The domain of $f \circ g$ is the set of all $x$ in the domain of $g$, $x \geq -2$, such that $g(x)$ is in the domain of $f$. Since the domain of $f$ is $\mathbb{R}$, any value of $g(x)$ is in its domain. Thus, the domain is all $x$ such that $x \geq -2$. Note that *both* $g(x)$ *and* $f(g(x))$ are defined for $x$ in $[-2, \infty)$.

(b) $k(x) = g(x^2 - 3x) = \sqrt{(x^2 - 3x) + 2} = \sqrt{x^2 - 3x + 2}$. The domain of $g \circ f$ is the set of all $x$ in the domain of $f$, $\mathbb{R}$, such that $f(x)$ is in the domain of $g$. Since the domain of $g$ is $x \geq -2$, we must solve $f(x) \geq -2$.

$$x^2 - 3x \geq -2 \Rightarrow x^2 - 3x + 2 \geq 0 \Rightarrow (x-1)(x-2) \geq 0$$

| Resulting sign: | ⊕ | ⊖ | ⊕ |
|---|---|---|---|
| Sign of $x-2$: | − | − | + |
| Sign of $x-1$: | − | + | + |
| $x$ values: | 1 | 2 | |

*Diagram 21*

From the sign diagram, $(x-1)(x-2) \geq 0 \Rightarrow x \in (-\infty, 1] \cup [2, \infty)$. (continued)

Thus, the domain is all $x$ such that $x \in (-\infty, 1] \cup [2, \infty)$. Note that *both* $f(x)$ *and* $g(f(x))$ are defined for $x$ in $(-\infty, 1] \cup [2, \infty)$.

**23** (a) $h(x) = f(\sqrt{3x}) = (\sqrt{3x})^2 - 4 = 3x - 4$.

Domain of $g = [0, \infty)$. Domain of $f = \mathbb{R}$. Since $g(x)$ is always in the domain of $f$, the domain of $f \circ g$ is the same as the domain of $g$, $[0, \infty)$. Note that *both* $g(x)$ *and* $f(g(x))$ are defined for $x$ in $[0, \infty)$.

(b) $k(x) = g(x^2 - 4) = \sqrt{3(x^2 - 4)} = \sqrt{3x^2 - 12}$.

Domain of $f = \mathbb{R}$. Domain of $g = [0, \infty)$.

$f(x) \geq 0 \Rightarrow x^2 - 4 \geq 0 \Rightarrow x^2 \geq 4 \Rightarrow |x| \geq 2 \Rightarrow x \in (-\infty, -2] \cup [2, \infty)$.

Note that *both* $f(x)$ *and* $g(f(x))$ are defined for $x$ in $(-\infty, -2] \cup [2, \infty)$.

**25** (a) $h(x) = f(\sqrt{x+5}) = \sqrt{\sqrt{x+5} - 2}$.

Domain of $g = [-5, \infty)$. Domain of $f = [2, \infty)$.

$g(x) \geq 2 \Rightarrow \sqrt{x+5} \geq 2 \Rightarrow x + 5 \geq 4 \Rightarrow x \geq -1$ or $x \in [-1, \infty)$.

(b) $k(x) = g(\sqrt{x-2}) = \sqrt{\sqrt{x-2} + 5}$. Domain of $f = [2, \infty)$.

Domain of $g = [-5, \infty)$. $f(x) \geq -5 \Rightarrow \sqrt{x-2} \geq -5$. This is always true since the result of a square root is nonnegative. The domain is $[2, \infty)$.

**27** (a) $h(x) = f(\sqrt{x^2 - 16}) = \sqrt{3 - \sqrt{x^2 - 16}}$.

Domain of $g = (-\infty, -4] \cup [4, \infty)$. Domain of $f = (-\infty, 3]$.

$g(x) \leq 3 \Rightarrow \sqrt{x^2 - 16} \leq 3 \Rightarrow x^2 - 16 \leq 9 \Rightarrow x^2 \leq 25 \Rightarrow x \in [-5, 5]$.

But $|x| \geq 4$ from the domain of $g$, so the domain of $f \circ g$ is $[-5, -4] \cup [4, 5]$.

(b) $k(x) = g(\sqrt{3-x}) = \sqrt{(\sqrt{3-x})^2 - 16} = \sqrt{3 - x - 16} = \sqrt{-x - 13}$.

Domain of $f = (-\infty, 3]$. Domain of $g = (-\infty, -4] \cup [4, \infty)$.

$f(x) \geq 4$ $\{f(x)$ cannot be less than $0\}$ $\Rightarrow$

$\sqrt{3-x} \geq 4 \Rightarrow 3 - x \geq 16 \Rightarrow x \leq -13$.

**29** (a) $h(x) = f\left(\dfrac{2x-5}{3}\right) = \dfrac{3\left(\dfrac{2x-5}{3}\right) + 5}{2} = \dfrac{2x - 5 + 5}{2} = \dfrac{2x}{2} = x$.

Domain of $g = \mathbb{R}$. Domain of $f = \mathbb{R}$. All values of $g(x)$ are in the domain of $f$.

Hence, the domain of $f \circ g$ is $\mathbb{R}$.

(b) $k(x) = g\left(\dfrac{3x+5}{2}\right) = \dfrac{2\left(\dfrac{3x+5}{2}\right) - 5}{3} = \dfrac{3x + 5 - 5}{3} = \dfrac{3x}{3} = x$.

Domain of $f = \mathbb{R}$. Domain of $g = \mathbb{R}$. All values of $f(x)$ are in the domain of $g$.

Hence, the domain of $g \circ f$ is $\mathbb{R}$.

**31** (a) $h(x) = f\left(\frac{1}{x^3}\right) = \left(\frac{1}{x^3}\right)^2 = \frac{1}{x^6}$. Domain of $g = \mathbb{R} - \{0\}$. Domain of $f = \mathbb{R}$.

All values of $g(x)$ are in the domain of $f$. Hence, the domain of $f \circ g$ is $\mathbb{R} - \{0\}$.

(b) $k(x) = g(x^2) = \frac{1}{(x^2)^3} = \frac{1}{x^6}$. Domain of $f = \mathbb{R}$. Domain of $g = \mathbb{R} - \{0\}$.

All values of $f(x)$ are in the domain of $g$ except for 0.

Since $f$ is 0 when $x$ is 0, the domain of $f \circ g$ is $\mathbb{R} - \{0\}$.

**33** (a) $h(x) = f\left(\frac{x-3}{x-4}\right) = \dfrac{\frac{x-3}{x-4} - 1}{\frac{x-3}{x-4} - 2} \cdot \frac{x-4}{x-4} = \frac{x-3-1(x-4)}{x-3-2(x-4)} = \frac{1}{5-x}$.

Domain of $g = \mathbb{R} - \{4\}$. Domain of $f = \mathbb{R} - \{2\}$. $g(x) \neq 2 \;\Rightarrow$

$\frac{x-3}{x-4} \neq 2 \;\Rightarrow\; x - 3 \neq 2x - 8 \;\Rightarrow\; x \neq 5$. The domain is $\mathbb{R} - \{4, 5\}$.

(b) $k(x) = g\left(\frac{x-1}{x-2}\right) = \dfrac{\frac{x-1}{x-2} - 3}{\frac{x-1}{x-2} - 4} \cdot \frac{x-2}{x-2} = \frac{x-1-3(x-2)}{x-1-4(x-2)} = \frac{-2x+5}{-3x+7}$.

Domain of $f = \mathbb{R} - \{2\}$. Domain of $g = \mathbb{R} - \{4\}$. $f(x) \neq 4 \;\Rightarrow$

$\frac{x-1}{x-2} \neq 4 \;\Rightarrow\; x - 1 \neq 4x - 8 \;\Rightarrow\; x \neq \frac{7}{3}$. The domain is $\mathbb{R} - \{2, \frac{7}{3}\}$.

**35** $(f \circ g)(x) = f(g(x)) = f(x+3) = (x+3)^2 - 2$.

$(f \circ g)(x) = 0 \;\Rightarrow\; (x+3)^2 - 2 = 0 \;\Rightarrow\; (x+3)^2 = 2 \;\Rightarrow\; x + 3 = \pm\sqrt{2} \;\Rightarrow$

$x = -3 \pm \sqrt{2}$

**37** (a) $(f \circ g)(6) = f(g(6)) = f(8) = 5$ (b) $(g \circ f)(6) = g(f(6)) = g(7) = 6$

(c) $(f \circ f)(6) = f(f(6)) = f(7) = 6$ (d) $(g \circ g)(6) = g(g(6)) = g(8) = 5$

(e) $(f \circ g)(9) = f(g(9)) = f(4)$, but $f(4)$ cannot be determined from the table.

**39** $(D \circ R)(x) = D(R(x)) = D(20x) =$

$$\sqrt{400 + (20x)^2} = \sqrt{400 + 400x^2} = \sqrt{400(1+x^2)} = \sqrt{400}\,\sqrt{x^2+1} = 20\sqrt{x^2+1}$$

**41** We need to examine $(fg)(-x)$—if we obtain $(fg)(x)$, then $fg$ is an even function, whereas if we obtain $-(fg)(x)$, then $fg$ is an odd function.

$(fg)(-x) = f(-x)g(-x)$ {definition of the product of two functions}

$= -f(x)g(x)$ { $f$ is odd, so $f(-x) = -f(x)$; $g$ is even, so $g(-x) = g(x)$ }

$= -(fg)(x)$ {definition of the product of two functions}

Since $(fg)(-x) = -(fg)(x)$, $fg$ is an odd function.

**43** (ROUND2 ∘ SSTAX)(437.21) = ROUND2(SSTAX(437.21))

= ROUND2(0.0715 · 437.21)

= ROUND2(31.260515) = 31.26

[45] $r = 6t$ and $A = \pi r^2 \Rightarrow A = \pi(6t)^2 = 36\pi t^2$ ft$^2$.

[47] The formula for the volume of a cone is $V = \frac{1}{3}\pi r^2 h = \frac{1}{3}\pi r^3$ {since $h = r$} $\Rightarrow$

$$r^3 = \frac{3V}{\pi} \Rightarrow r = \sqrt[3]{\frac{3V}{\pi}}.\quad V = 243\pi t \Rightarrow r = \sqrt[3]{\frac{3 \cdot 243\pi t}{\pi}} = \sqrt[3]{729t} = 9\sqrt[3]{t} \text{ ft.}$$

[49] Let $l$ denote the length of the rope. At $t = 0$, $l = 20$. At $t = 1$, $l = 25$.

At time $t$, $l = 20 + 5t$, *not* just $5t$. We have a right triangle with sides 20, $h$, and $l$.

$h^2 + 20^2 = l^2 \Rightarrow$

$$h = \sqrt{(20 + 5t)^2 - 20^2} = \sqrt{25t^2 + 200t} = \sqrt{25(t^2 + 8t)} = 5\sqrt{t^2 + 8t}.$$

[51] From Exercise 77 of Section 3.4, $d(x) = \sqrt{90{,}400 + x^2}$. The distance $x$ of the plane from the control tower is 500 feet plus 150 feet per second, that is, $x(t) = 500 + 150t$.

Thus, $d(t) = \sqrt{90{,}400 + (500 + 150t)^2} = \sqrt{90{,}400 + 250{,}000 + 150{,}000t + 22{,}500t^2}$

$$= \sqrt{22{,}500t^2 + 150{,}000t + 340{,}400} = 10\sqrt{225t^2 + 1500t + 3404}.$$

[53] $y = (x^2 + 3x)^{1/3}$ • Suppose you were to find the value of $y$ if $x$ was equal to 3. Using a calculator, you might compute the value of $x^2 + 3x$ first, and then raise that result to the $\frac{1}{3}$ power. Thus, we would choose $y = u^{1/3}$ and $u = x^2 + 3x$.

[55] For $y = \dfrac{1}{(x-3)^4}$, choose $u = x - 3$ and $y = 1/u^4 = u^{-4}$.

[57] For $y = (x^4 - 2x^2 + 5)^5$, choose $u = x^4 - 2x^2 + 5$ and $y = u^5$.

[59] For $y = \dfrac{\sqrt{x+4} - 2}{\sqrt{x+4} + 2}$, there is not a "simple" choice for $y$ as in previous exercises.

One choice for $u$ is $u = x + 4$. Then $y$ would be $\dfrac{\sqrt{u} - 2}{\sqrt{u} + 2}$.

Another choice for $u$ is $u = \sqrt{x+4}$. Then $y$ would be $\dfrac{u-2}{u+2}$.

[61] $(f \circ g)(x) = f(g(x)) = f(x^3 + 1) = \sqrt{x^3 + 1} - 1$. We will multiply this expression by

$\dfrac{\sqrt{x^3+1}+1}{\sqrt{x^3+1}+1}$, treating it as though it was one factor of a difference of two squares.

$$(\sqrt{x^3+1} - 1) \times \frac{\sqrt{x^3+1}+1}{\sqrt{x^3+1}+1} = \frac{(\sqrt{x^3+1})^2 - 1^2}{\sqrt{x^3+1}+1} = \frac{x^3}{\sqrt{x^3+1}+1}$$

Thus, $(f \circ g)(0.0001) = f(g(10^{-4})) \approx \dfrac{(10^{-4})^3}{2} = 5 \times 10^{-13}$.

[63] *Note:* If we relate this problem to the composite function $g(f(u))$, then $Y_2$ can be considered to be $f$, $Y_1$ {which can be considered to be $u$} is the function of $x$ we are substituting into $f$, and $Y_3$ {which can be considered to be $g$} is accepting $Y_2$'s output values as its input values. Hence, $Y_3$ is a function of a function of a function.

(a) $y = -2f(x)$; $Y_1 = x$, $Y_2 = 3\sqrt{(Y_1+2)(6-Y_1)} - 4$, graph $Y_3 = -2Y_2$
{turn off $Y_1$ and $Y_2$, leaving only $Y_3$ on}; $D = [-2, 6]$, $R = [-16, 8]$

*Note:* The graphs often do not show the correct endpoints—you need to change the viewing rectangle or zoom in to actually view them on the screen.

$[-12, 12, 2]$ by $[-16, 8, 2]$

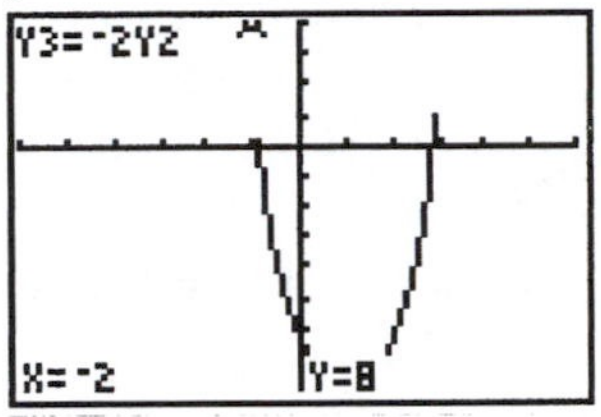


*Figure 63(a)*

$[-12, 12, 2]$ by $[-16, 8, 2]$

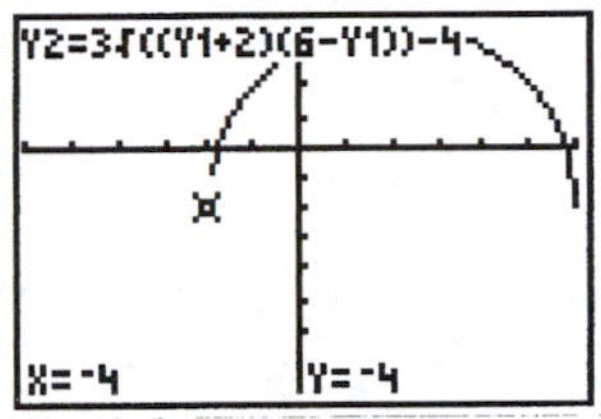


*Figure 63(b)*

(b) $y = f(\frac{1}{2}x)$; $Y_1 = 0.5x$, graph $Y_2$; $D = [-4, 12]$, $R = [-4, 8]$

(c) $y = f(x-3)+1$; $Y_1 = x-3$, graph $Y_3 = Y_2 + 1$; $D = [1, 9]$, $R = [-3, 9]$

$[-12, 12, 2]$ by $[-6, 10, 2]$

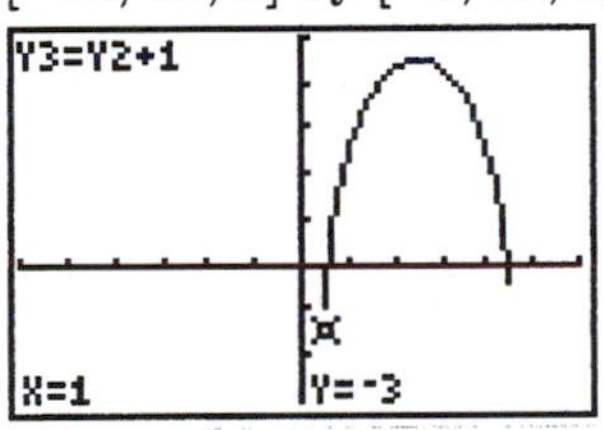


*Figure 63(c)*

$[-12, 12, 2]$ by $[-8, 8, 2]$

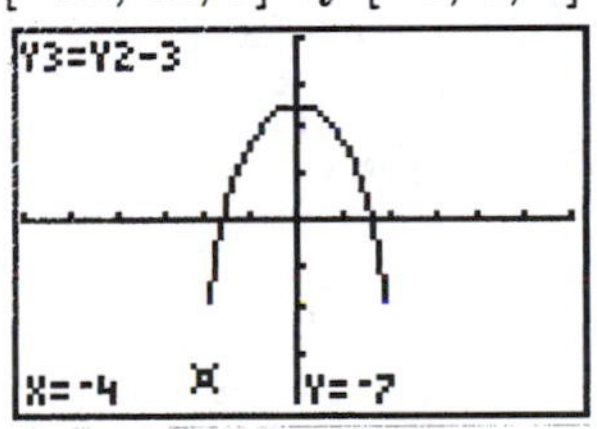


*Figure 63(d)*

(d) $y = f(x+2)-3$; $Y_1 = x+2$, graph $Y_3 = Y_2 - 3$; $D = [-4, 4]$, $R = [-7, 5]$

(e) $y = f(-x)$; $Y_1 = -x$, graph $Y_2$; $D = [-6, 2]$, $R = [-4, 8]$

$[-12, 12, 2]$ by $[-8, 8, 2]$

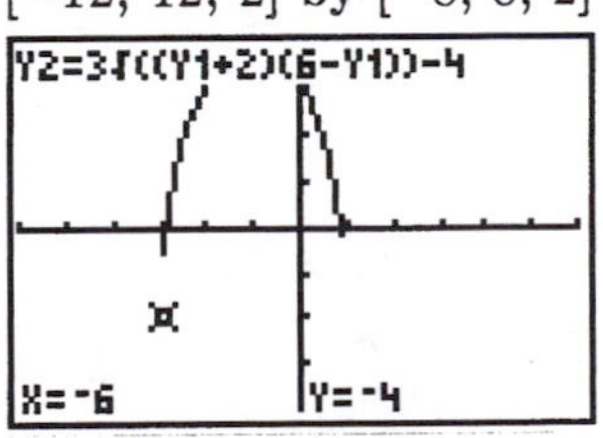


*Figure 63(e)*

$[-12, 12, 2]$ by $[-8, 8, 2]$

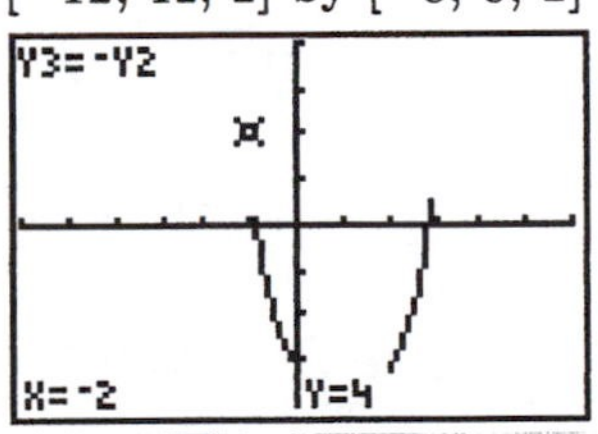


*Figure 63(f)*

(f) $y = -f(x)$; $Y_1 = x$, graph $Y_3 = -Y_2$; $D = [-2, 6]$, $R = [-8, 4]$

(g) $y = f(\,|\,x\,|\,)$; $Y_1 = \text{abs}\,x$, graph $Y_2$; $D = [-6, 6]$, $R = [-4, 8]$

[−12, 12, 2] by [−6, 10, 2]

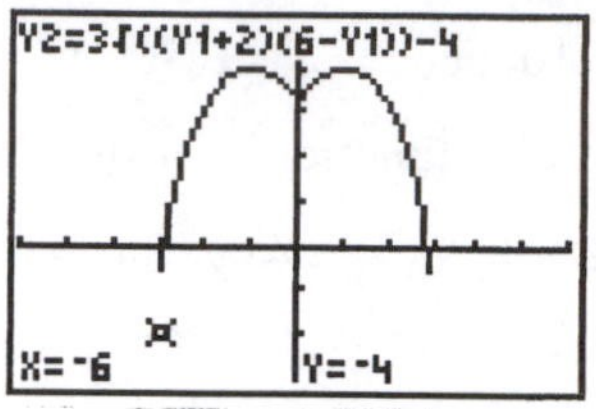


Figure 63(g)

[−4, 8] by [−2, 10]

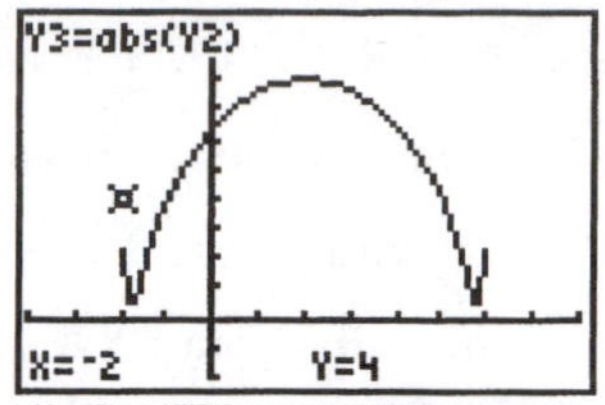


Figure 63(h)

(h) $y = \,|\,f(x)\,|\,$; $Y_1 = x$, graph $Y_3 = \text{abs}\,Y_2$; $D = [-2, 6]$, $R = [0, 8]$

## Chapter 3 Review Exercises

1 If $y/x < 0$, then $y$ and $x$ must have opposite signs, and hence,

the set consists of all points in quadrants II and IV.

3 (a) $P(-5, 9)$, $Q(-8, -7)$ $\Rightarrow$

$$d(P, Q) = \sqrt{[-8-(-5)]^2 + (-7-9)^2} = \sqrt{9+256} = \sqrt{265}.$$

(b) $P(-5, 9)$, $Q(-8, -7)$ $\Rightarrow$ $M_{PQ} = \left(\dfrac{-5+(-8)}{2}, \dfrac{9+(-7)}{2}\right) = (-\tfrac{13}{2}, 1).$

(c) Let $R = (x, y)$. $Q = M_{PR}$ $\Rightarrow$

$$(-8, -7) = \left(\frac{-5+x}{2}, \frac{9+y}{2}\right) \Rightarrow -8 = \frac{-5+x}{2} \text{ and } -7 = \frac{9+y}{2} \Rightarrow$$

$$-5+x = -16 \text{ and } 9+y = -14 \Rightarrow x = -11 \text{ and } y = -23 \Rightarrow R = (-11, -23).$$

5 With $P(a, 1)$ and $Q(-2, a)$, $d(P, Q) < 3$ $\Rightarrow$ $\sqrt{(-2-a)^2 + (a-1)^2} < 3$ $\Rightarrow$ $4 + 4a + a^2 + a^2 - 2a + 1 < 9$ $\Rightarrow$ $2a^2 + 2a - 4 < 0$ $\Rightarrow$ $a^2 + a - 2 < 0$ $\Rightarrow$ $(a+2)(a-1) < 0$. Using *Diagram 5*,

we see that $-2 < a < 1$ will assure us that $d(P, Q) < 3$.

| Resulting sign: | ⊕ | ⊖ | ⊕ |
|---|---|---|---|
| Sign of $a+2$: | − | + | + |
| Sign of $a-1$: | − | − | + |
| $a$ values: | −2 | 1 | |

Diagram 5

7 The center of the circle is the midpoint of $A(8, 10)$ and $B(-2, -14)$.

$M_{AB} = \left(\dfrac{8+(-2)}{2}, \dfrac{10+(-14)}{2}\right) = (3, -2)$. The radius of the circle is

$$\tfrac{1}{2} \cdot d(A, B) = \tfrac{1}{2}\sqrt{(-2-8)^2 + (-14-10)^2} = \tfrac{1}{2}\sqrt{100+576} = \tfrac{1}{2} \cdot 26 = 13.$$

An equation is $(x-3)^2 + (y+2)^2 = 13^2 = 169$.

[9] $C(11, -5)$, $D(-8, 6)$ $\Rightarrow$ $m_{CD} = \dfrac{6-(-5)}{-8-11} = \dfrac{11}{-19} = -\dfrac{11}{19}$.

[11] (a) $6x + 2y + 5 = 0$ $\Leftrightarrow$ $y = -3x - \frac{5}{2}$. Using the same slope, $-3$, with $A(\frac{1}{2}, -\frac{1}{3})$,

we have $y + \frac{1}{3} = -3(x - \frac{1}{2})$ $\Rightarrow$ $6y + 2 = -18x + 9$ $\Rightarrow$ $18x + 6y = 7$.

(b) Using the negative reciprocal of $-3$ for the slope,

$y + \frac{1}{3} = \frac{1}{3}(x - \frac{1}{2})$ $\Rightarrow$ $6y + 2 = 2x - 1$ $\Rightarrow$ $2x - 6y = 3$.

[13] The radius of the circle is the distance from the line $x = 4$ to the $x$-value of the

center $C(-5, -1)$; $r = 4 - (-5) = 9$. An equation is $(x + 5)^2 + (y + 1)^2 = 81$.

[15] $P(4, -3)$ with $m = 5$ $\Rightarrow$ $y + 3 = 5(x - 4)$ $\Rightarrow$ $y + 3 = 5x - 20$ $\Rightarrow$ $5x - y = 23$.

[16] $A(-1, 2)$ and $B(3, -4)$ $\Rightarrow$ $M_{AB} = (1, -1)$ and $m_{AB} = -\frac{3}{2}$. We want the

equation of the line through $(1, -1)$ with slope $\frac{2}{3}$ {the negative reciprocal of $-\frac{3}{2}$}.

$y + 1 = \frac{2}{3}(x - 1)$ $\Rightarrow$ $3y + 3 = 2x - 2$ $\Rightarrow$ $2x - 3y = 5$.

[17] $x^2 + y^2 - 12y + 31 = 0$ $\Rightarrow$ $x^2 + y^2 - 12y + \underline{\ 36\ } = -31 + \underline{\ 36\ }$ $\Rightarrow$ $x^2 + (y - 6)^2 = 5$.

$C(0, 6)$; $r = \sqrt{5}$

[19] (a) $f(x) = \dfrac{x}{\sqrt{x+3}}$ $\Rightarrow$ $f(1) = \dfrac{1}{\sqrt{4}} = \dfrac{1}{2}$ (b) $f(-1) = -\dfrac{1}{\sqrt{2}}$ (c) $f(0) = \dfrac{0}{\sqrt{3}} = 0$

(d) $f(-x) = \dfrac{-x}{\sqrt{-x+3}} = -\dfrac{x}{\sqrt{3-x}}$ (e) $-f(x) = -1 \cdot f(x) = -\dfrac{x}{\sqrt{x+3}}$

(f) $f(x^2) = \dfrac{x^2}{\sqrt{x^2+3}}$ (g) $[f(x)]^2 = \left(\dfrac{x}{\sqrt{x+3}}\right)^2 = \dfrac{x^2}{x+3}$

[21] $f(x) = \dfrac{-2(x^2 - 20)(5 - x)}{(6 - x^2)^{4/3}}$ $\Rightarrow$ $f(4) = \dfrac{(-)(-)(+)}{(+)} = +$. $f(4)$ is positive.

Note that the factor $-2$ always contributes one negative sign to the quotient and that

the denominator is always positive if $x \neq \pm\sqrt{6}$.

[22] (a) $3x - 4 \geq 0$ $\Rightarrow$ $x \geq \frac{4}{3}$; $D = [\frac{4}{3}, \infty)$.

Since $y$ is the result of a square root, $y \geq 0$; $R = [0, \infty)$.

(b) $D =$ All real numbers except $-3$.

Since $y$ is the square of the nonzero term $\dfrac{1}{x+3}$, $y > 0$; $R = (0, \infty)$.

[23] $\dfrac{f(a+h) - f(a)}{h} = \dfrac{[-(a+h)^2 + (a+h) + 5] - [-a^2 + a + 5]}{h} =$

$\dfrac{-a^2 - 2ah - h^2 + a + h + 5 + a^2 - a - 5}{h} = \dfrac{-2ah - h^2 + h}{h} = \dfrac{h(-2a - h + 1)}{h} =$

$-2a - h + 1$

[25] $f(x) = ax + b$ is the desired form. $a = \text{slope} = \dfrac{7-2}{3-1} = \dfrac{5}{2}$. $f(x) = \frac{5}{2}x + b$ $\Rightarrow$

$f(1) = \frac{5}{2} + b$, but $f(1) = 2$, so $\frac{5}{2} + b = 2$, and $b = -\frac{1}{2}$. Thus, $f(x) = \frac{5}{2}x - \frac{1}{2}$.

[27] $x + 5 = 0 \;\Leftrightarrow\; x = -5$, a vertical line; $x$-intercept $-5$; $y$-intercept: None

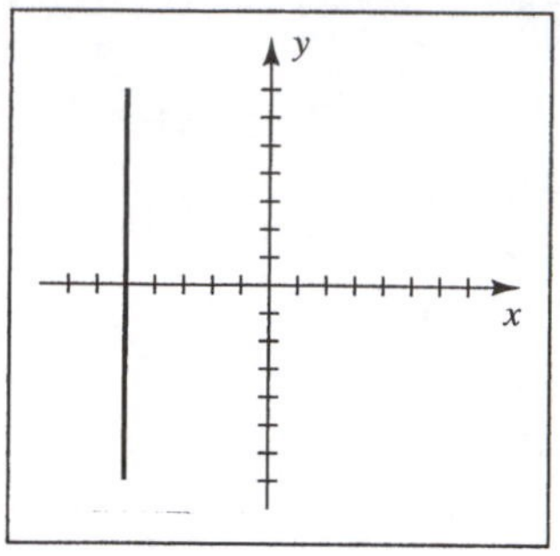

Figure 27

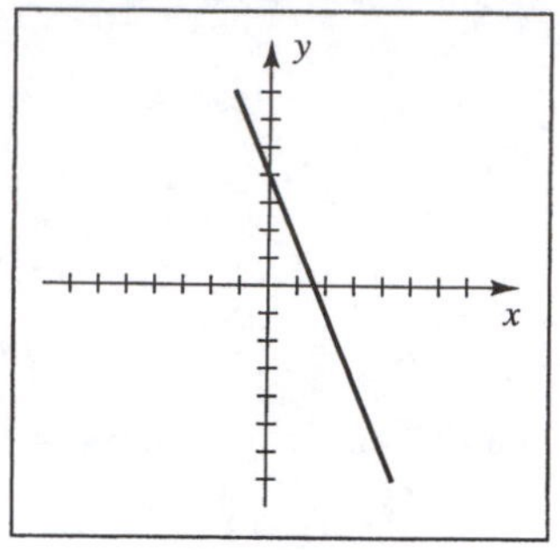

Figure 29

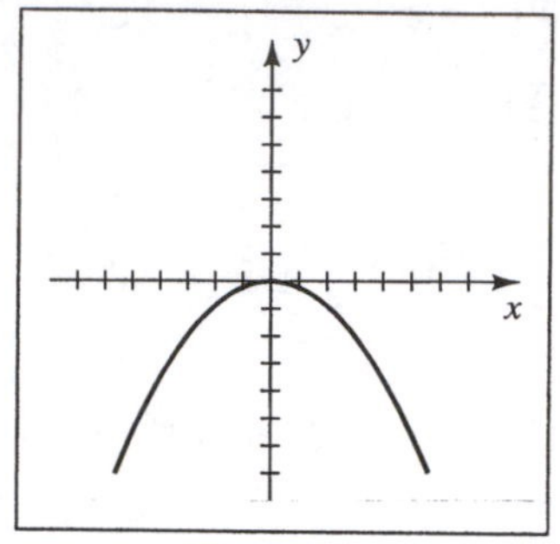

Figure 31

[29] $2y + 5x - 8 = 0 \;\Leftrightarrow\; y = -\frac{5}{2}x + 4$, a line with slope $-\frac{5}{2}$ and $y$-intercept 4.

$x$-intercept: $y = 0 \;\Rightarrow\; 2(0) + 5x - 8 = 0 \;\Rightarrow\; 5x = 8 \;\Rightarrow\; x = 1.6$

[31] $9y + 2x^2 = 0 \;\Leftrightarrow\; y = -\frac{2}{9}x^2$, a parabola opening down; $x$- and $y$-intercept 0

[33] $y = \sqrt{1 - x}$ • The radicand must be nonnegative for the radical to be defined.

$1 - x \geq 0 \;\Rightarrow\; 1 \geq x$, or equivalently, $x \leq 1$.

The domain is $(-\infty, 1]$ and the range is $[0, \infty)$.

$x$-intercept: $y = 0 \;\Rightarrow\; 0 = \sqrt{1 - x} \;\Rightarrow$

$0 = 1 - x \;\Rightarrow\; x = 1$

$y$-intercept: $x = 0 \;\Rightarrow\; y = \sqrt{1 - 0} = 1$

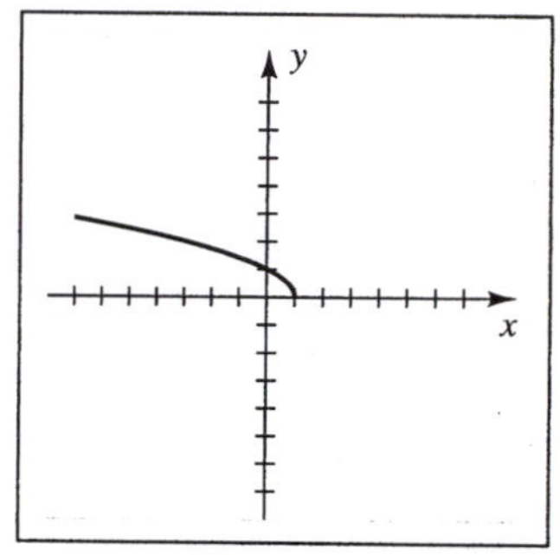

Figure 33

[35] $y^2 = 16 - x^2 \;\Leftrightarrow\; x^2 + y^2 = 16$; $x$-intercepts $\pm 4$; $y$-intercepts $\pm 4$

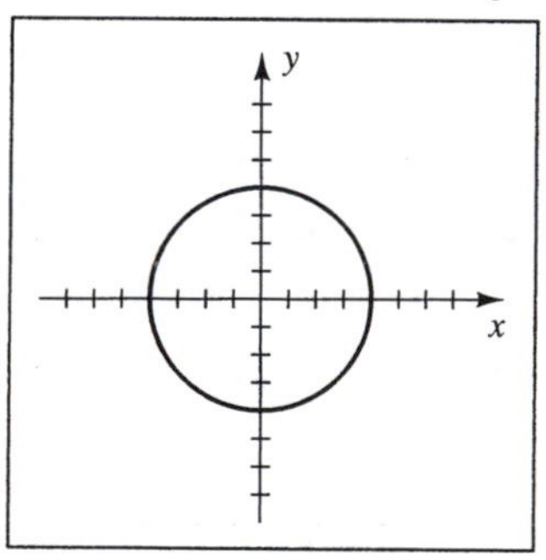

Figure 35

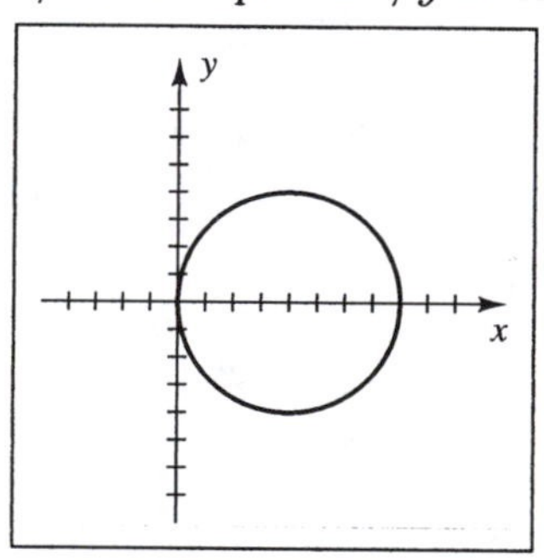

Figure 37

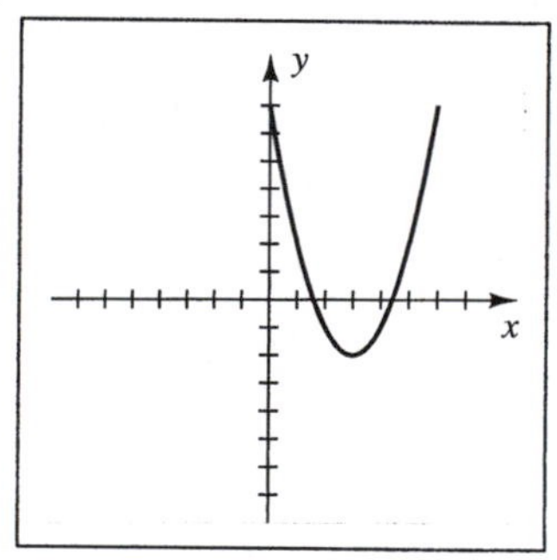

Figure 39

[37] $x^2 + y^2 - 8x = 0 \;\Leftrightarrow\; x^2 - 8x + \underline{16} + y^2 = \underline{16} \;\Leftrightarrow\; (x - 4)^2 + y^2 = 16$;

a circle with center $C(4, 0)$ and radius $r = \sqrt{16} = 4$. $x$-intercepts:

$y = 0 \;\Rightarrow\; x^2 - 8x = 0 \;\Rightarrow\; x(x - 8) = 0 \;\Rightarrow\; x = 0$ and 8, $y$-intercept: 0

[39] $y = (x - 3)^2 - 2$ is a parabola that opens upward and has vertex $(3, -2)$.

$x$-intercepts: $y = 0 \;\Rightarrow\; 0 = (x - 3)^2 - 2 \;\Rightarrow\; (x - 3)^2 = 2 \;\Rightarrow$

$x - 3 = \pm\sqrt{2} \;\Rightarrow\; x = 3 \pm \sqrt{2}$

$y$-intercept: $x = 0 \;\Rightarrow\; y = (0 - 3)^2 - 2 = 9 - 2 = 7$

**41** The radius of the large circle is 3 and the radius of the small circle is 1. Since the center of the small circle is 4 units from the origin and lies on the line $y = x$, we must have $x^2 + y^2 = 4^2 \Rightarrow x^2 + x^2 = 16 \Rightarrow 2x^2 = 16 \Rightarrow x^2 = 8 \Rightarrow x = \sqrt{8}$. The center of the small circle is $(\sqrt{8}, \sqrt{8})$.

**43** (a) The graph of $f(x) = \frac{1 - 3x}{2} = -\frac{3}{2}x + \frac{1}{2}$ is a line with slope $-\frac{3}{2}$ and $y$-intercept $\frac{1}{2}$.

(b) The function is defined for all $x$, so the domain $D$ is the set of all real numbers.

The range is the set of all real numbers, so $D = \mathbb{R}$ and $R = \mathbb{R}$.

(c) The function $f$ is decreasing on $(-\infty, \infty)$.

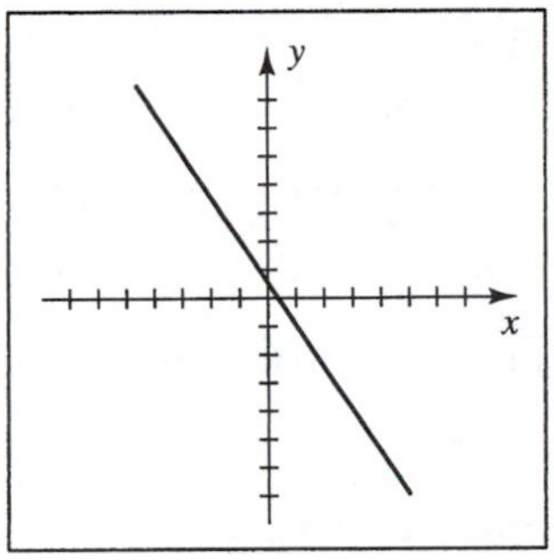


*Figure 43*

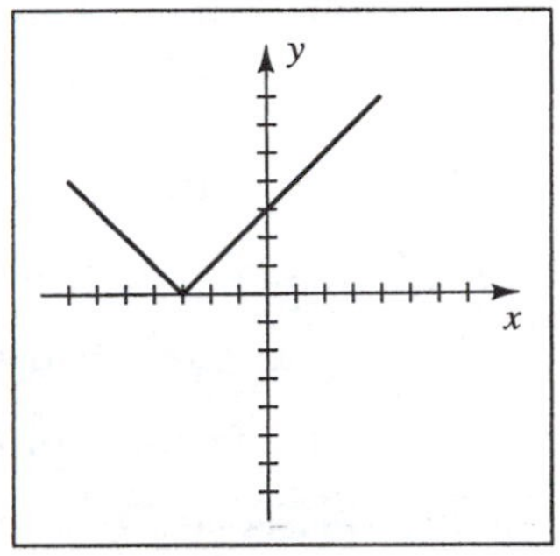


*Figure 45*

**45** (a) The graph of $f(x) = |x + 3|$ can be thought of as

the graph of $g(x) = |x|$ shifted left 3 units.

(b) The function is defined for all $x$, so the domain $D$ is the set of all real numbers.

The range is the set of all nonnegative numbers, that is, $R = [0, \infty)$.

(c) The function $f$ is decreasing on $(-\infty, -3]$ and is increasing on $[-3, \infty)$.

**47** (a) $f(x) = 1 - \sqrt{x + 1} = -\sqrt{x + 1} + 1$ • We can think of this graph as the graph of $y = \sqrt{x}$ shifted left 1 unit, reflected through the $x$-axis, and shifted up 1 unit.

(b) For $f$ to be defined, we must have $x + 1 \geq 0 \Rightarrow x \geq -1$, so $D = [-1, \infty)$.

The $y$-values of $f$ are all the values less than or equal to 1, so $R = (-\infty, 1]$.

(c) The function $f$ is decreasing on $[-1, \infty)$.

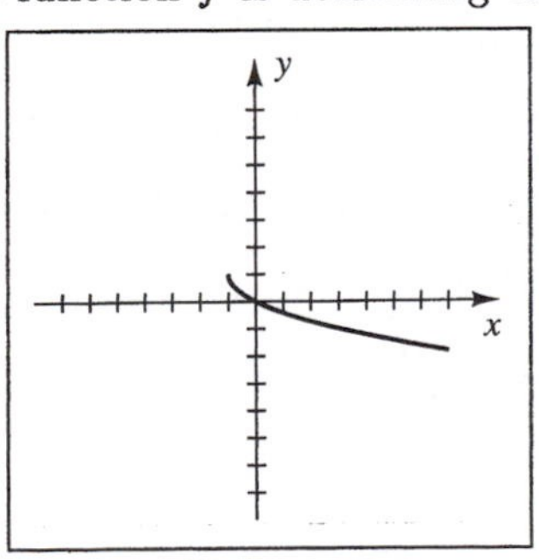


*Figure 47*

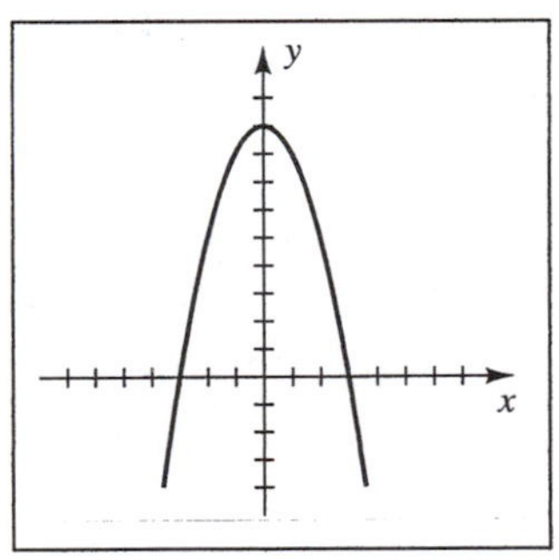


*Figure 49*

[49] (a) $f(x) = 9 - x^2 = -x^2 + 9$ • We can think of this graph as the graph of $y = x^2$ reflected through the $x$-axis and shifted up 9 units.

See *Figure 49* on the preceding page.

(b) The function is defined for all $x$, so the domain $D$ is the set of all real numbers. The $y$-values of $f$ are all the values less than or equal to 9, so $R = (-\infty, 9]$.

(c) The function $f$ is increasing on $(-\infty, 0]$ and is decreasing on $[0, \infty)$.

[51] (a) $f(x) = \begin{cases} x^2 & \text{if } x < 0 \\ 3x & \text{if } 0 \leq x < 2 \\ 6 & \text{if } x \geq 2 \end{cases}$

If $x < 0$, we want the graph of the parabola $y = x^2$. The endpoint of this part of the graph is (0, 0), but it is not included. If $0 \leq x < 2$, we want the graph of $y = 3x$, a line with slope 3 and $y$-intercept 0. Now we do include the endpoint (0, 0), but we don't include the endpoint (2, 6). If $x \geq 2$, we want the graph of the horizontal line $y = 6$ and include its endpoint (2, 6), so there are no open endpoints on the graph of $f$.

(b) The function is defined for all $x$, so the domain $D$ is the set of all real numbers. The $y$-values of $f$ are all the values greater than or equal to 0, so $R = [0, \infty)$.

(c) $f$ is decreasing on $(-\infty, 0]$, increasing on $[0, 2]$, and constant on $[2, \infty)$.

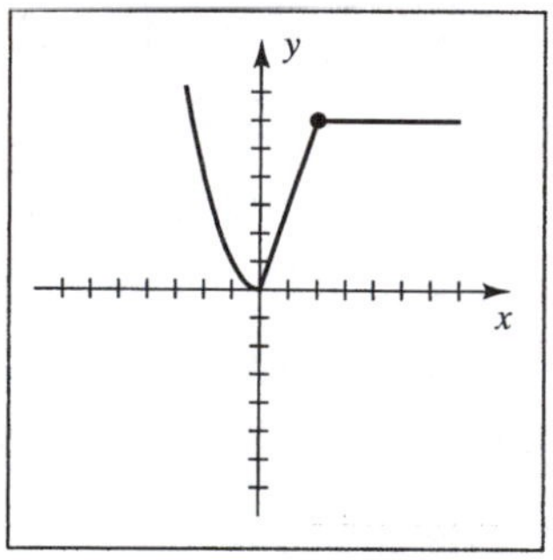


*Figure 51*

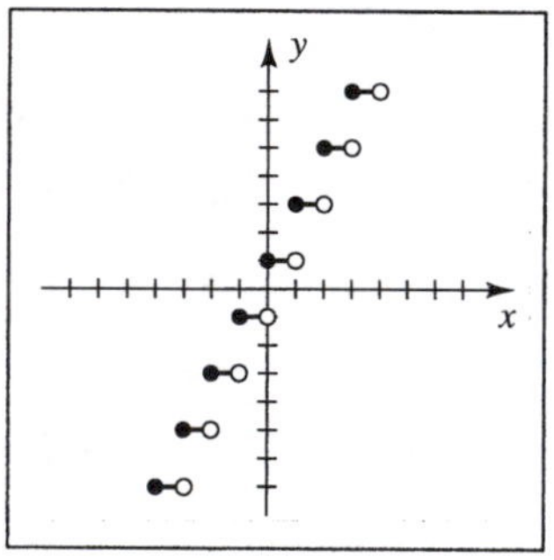


*Figure 52*

[52] (a) The "2" in front of $[\![x]\!]$ has the effect of doubling all the $y$-values of $g(x) = [\![x]\!]$. The "+1" has the effect of vertically shifting the graph of $h(x) = 2[\![x]\!]$ up 1 unit.

(b) The function is defined for all $x$, so the domain $D$ is the set of all real numbers. The range of $g(x) = [\![x]\!]$ is the set of integers, that is, $\{\ldots, -2, -1, 0, 1, 2, \ldots\}$. The range of $h(x) = 2[\![x]\!]$ is the set of even integers since we are doubling the values of $g$—that is, $\{\ldots, -4, -2, 0, 2, 4, \ldots\}$. Since $f(x) = 1 + h(x)$, the range $R$ of $f$ is $\{\ldots, -3, -1, 1, 3, \ldots\}$.

(c) The function $f$ is constant on intervals such as [0, 1), [1, 2), and [2, 3). In general, $f$ is constant on $[n, n + 1)$, where $n$ is any integer.

**53** (a) $y = \sqrt{x}$ • This is the square root function, whose graph is a half-parabola.

(b) $y = \sqrt{x+4}$ • shift $y = \sqrt{x}$ left 4 units

(c) $y = \sqrt{x}+4$ • shift $y = \sqrt{x}$ up 4 units

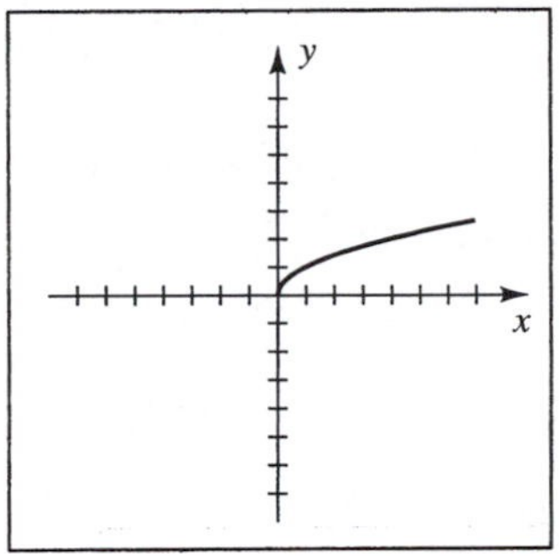

*Figure 53(a)*

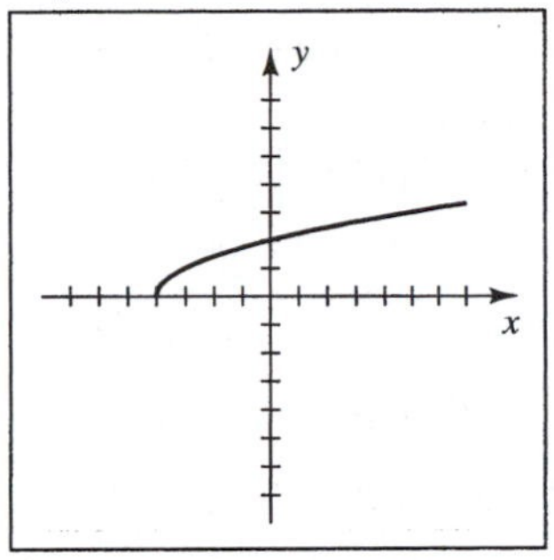

*Figure 53(b)*

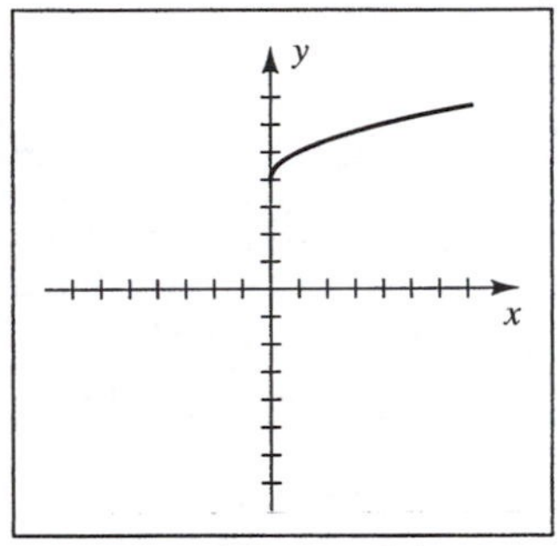

*Figure 53(c)*

(d) $y = 4\sqrt{x}$ • vertically stretch the graph of $y = \sqrt{x}$ by a factor of 4

(e) $y = \frac{1}{4}\sqrt{x}$ • vertically compress the graph of $y = \sqrt{x}$ by a factor of $1/\frac{1}{4} = 4$

(f) $y = -\sqrt{x}$ • reflect the graph of $y = \sqrt{x}$ through the $x$-axis

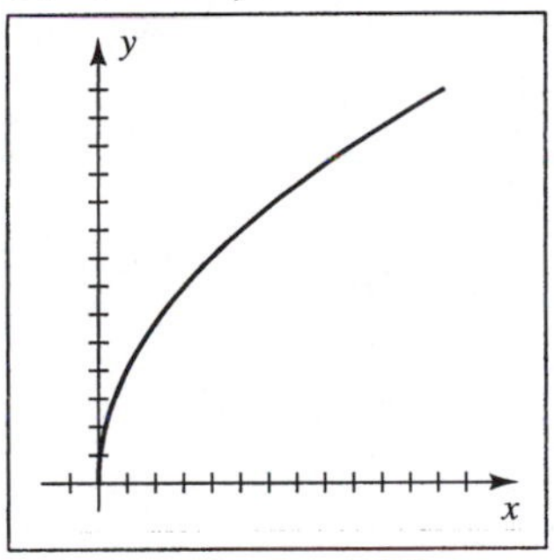

*Figure 53(d)*

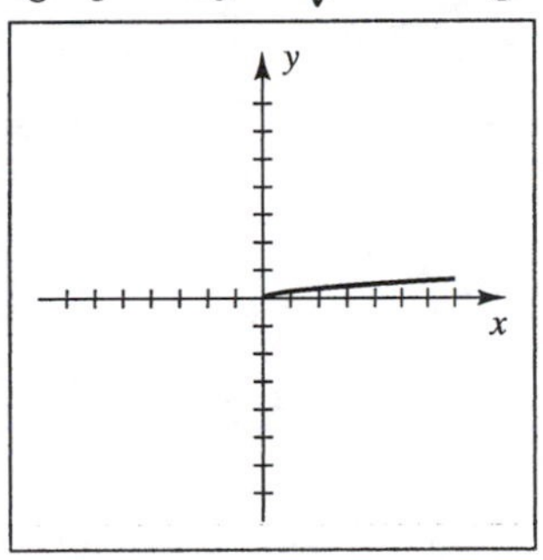

*Figure 53(e)*

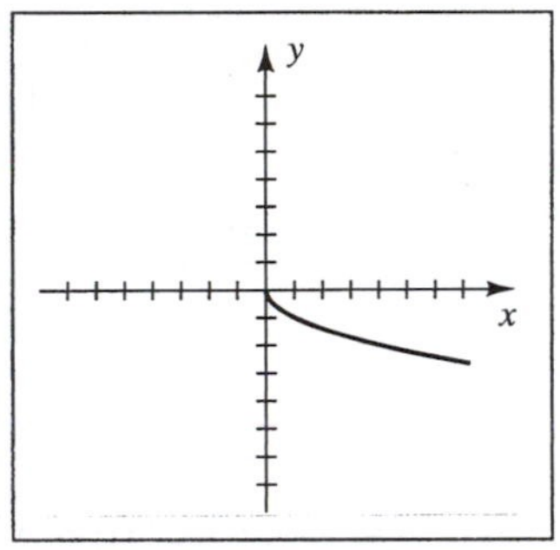

*Figure 53(f)*

**55** Using the intercept form of a line with $x$-intercept 5 and $y$-intercept $-2$, an equation is $\frac{x}{5} + \frac{y}{-2} = 1$. Multiplying by 10 gives us $\left[\frac{x}{5} + \frac{y}{-2} = 1\right] \cdot 10$, or, equivalently,

$$2x - 5y = 10.$$

**57** $V(2, -4)$ and $P(-2, 4)$ with $y = a(x-h)^2 + k \Rightarrow 4 = a(-2-2)^2 - 4 \Rightarrow$

$8 = 16a \Rightarrow a = \frac{1}{2}$. An equation is $y = \frac{1}{2}(x-2)^2 - 4$.

**58** The graph could be made by taking the graph of $y = |x|$, reflecting it through the $x$-axis $\{ y = -|x| \}$, then shifting that graph to the right 2 units $\{ y = -|x-2| \}$, and then shifting that graph down 1 unit, resulting in the graph of the equation $y = -|x-2| - 1$.

**59** $f(x) = 5x^2 + 30x + 49 \Rightarrow -\frac{b}{2a} = -\frac{30}{2(5)} = -3.$

$f(-3) = 4$ is a minimum since $a = 5 > 0$.

**61** $f(x) = -12(x+1)^2 - 37$ is in the standard form. $f(-1) = -37$ is a maximum.

[62] $f(x) = 3(x+2)(x-10)$ has $x$-intercepts at $-2$ and 10. The vertex is halfway between them at $x = 4$. $f(4) = 3 \cdot 6 \cdot (-6) = -108$ is a minimum.

[63] $f(x) = -2x^2 + 12x - 14 = -2(x^2 - 6x + \underline{\quad}) - 14 + \underline{\quad}$ {complete the square}
$= -2(x^2 - 6x + \underline{9}) - 14 + 18 = -2(x-3)^2 + 4.$

[65] The domain of $f(x) = \sqrt{4 - x^2}$ is $[-2, 2]$. The domain of $g(x) = \sqrt{x}$ is $[0, \infty)$.

(a) The domain of $fg$ is the intersection of those two domains, $[0, 2]$.

(b) The domain of $f/g$ is the same as that of $fg$,
excluding any values that make $g$ equal to 0. Thus, the domain of $f/g$ is $(0, 2]$.

[67] (a) $(f \circ g)(x) = f(g(x)) = 2(3x+2)^2 - 5(3x+2) + 1 = 18x^2 + 9x - 1$

(b) $(g \circ f)(x) = g(f(x)) = 3(2x^2 - 5x + 1) + 2 = 6x^2 - 15x + 5$

[69] (a) $(f \circ g)(x) = f(g(x)) = f(\sqrt{x-3}) = \sqrt{25 - (\sqrt{x-3})^2}$
$= \sqrt{25 - (x-3)} = \sqrt{28 - x}.$

Domain of $g(x) = \sqrt{x-3} = [3, \infty)$. Domain of $f(x) = \sqrt{25 - x^2} = [-5, 5]$.

$g(x) \leq 5$ { $g(x)$ cannot be less than 0 } $\Rightarrow$

$\sqrt{x-3} \leq 5 \;\Rightarrow\; x - 3 \leq 25 \;\Rightarrow\; x \leq 28.$ $[3, \infty) \cap (-\infty, 28] = [3, 28]$

(b) $(g \circ f)(x) = g(f(x)) = g(\sqrt{25 - x^2}) = \sqrt{\sqrt{25 - x^2} - 3}.$

Domain of $f = [-5, 5]$. Domain of $g = [3, \infty)$.

$f(x) \geq 3 \;\Rightarrow\; \sqrt{25 - x^2} \geq 3 \;\Rightarrow\; 25 - x^2 \geq 9 \;\Rightarrow\; x^2 \leq 16 \;\Rightarrow\; x \in [-4, 4].$

[71] For $y = \sqrt[3]{x^2 - 5x}$, choose $u = x^2 - 5x$ and $y = \sqrt[3]{u}$.

[73] (a) 2008 corresponds to $t = 2008 - 1948 = 60$. $d = 181 + 1.065(60) = 244.9 \approx 245$ ft

(b) $d = 265 \;\Rightarrow\; 265 = 181 + 1.065t \;\Rightarrow\; t = \dfrac{265 - 181}{1.065} \approx 78.9$ yr,
which corresponds to $79 + 1948 = 2027$ or Olympic year 2028.

[75] (a) $F = aC + b$ is the desired form. $F = 32$ when $C = 0 \;\Rightarrow\; F = aC + 32$. $F = 212$ when $C = 100 \;\Rightarrow\; 212 = 100a + 32 \;\Rightarrow\; a = \frac{180}{100} = \frac{9}{5}$ and hence, $F = \frac{9}{5}C + 32$.

(b) If $C$ increases 1°, $F$ increases $(\frac{9}{5})°$, or 1.8°.

[76] (a) $C_1(x) = \left(1.25\,\dfrac{\text{dollars}}{\text{gallon}}\right) \div \left(20\,\dfrac{\text{miles}}{\text{gallon}}\right) \cdot x \text{ miles} = \dfrac{1.25}{20}x = 0.0625x$, or $\dfrac{1}{16}x$.

(b) After the tune-up, the gasoline mileage will be 10% more than 20 mi/gal, that is,
22 mi/gal. $C_2(x) = \frac{1.25}{22}x + 50 = \frac{5}{88}x + 50 \approx 0.0568x + 50.$

(c) $C_2 < C_1 \;\Rightarrow\; \left[\frac{5}{88}x + 50 < \frac{1}{16}x\right] \cdot 16(11) \;\Rightarrow\; 10x + 8800 < 11x \;\Rightarrow$
$x > 8800$ miles.

[77] Surface area $S = 2(4)(x) + (4)(y) = 8x + 4y$. Cost $C = 2(8x) + 5(4y) = 16x + 20y$.

(a) $C = 400 \Rightarrow 16x + 20y = 400 \Rightarrow 20y = -16x + 400 \Rightarrow y = -\frac{4}{5}x + 20$

(b) $V = lwh = (y)(4)(x) = 4xy = 4x(-\frac{4}{5}x + 20)$

[79] (a) $V = (10 \text{ ft}^3 \text{ per minute})(t \text{ minutes}) = 10t$

(b) The height and length of the bottom triangular region are in the proportion 6–60, or 1–10, and the length is 10 times the height. When $0 \le h \le 6$, the volume is $V = (\text{cross sectional area})(\text{pool width}) = \frac{1}{2}bh(40) = \frac{1}{2}(10h)(h)(40) = 200h^2 \text{ ft}^3$. When $6 < h \le 9$, the triangular region is full and

$$V = 200(6)^2 + (h-6)(80)(40) = 7200 + 3200(h-6).$$

(c) **For $0 \le h \le 6$:** $10t = 200h^2 \Rightarrow h^2 = \frac{10t}{200} \Rightarrow h = \sqrt{\frac{t}{20}}$;

$$0 \le h \le 6 \Rightarrow 0 \le \sqrt{\frac{t}{20}} \le 6 \Rightarrow 0 \le \frac{t}{20} \le 36 \Rightarrow 0 \le t \le 720.$$

**For $6 < h \le 9$:** $10t = 7200 + 3200(h-6) \Rightarrow h - 6 = \frac{10t - 7200}{3200} \Rightarrow$

$$h = 6 + \frac{t - 720}{320};$$

$$6 < h \le 9 \Rightarrow 6 < 6 + \frac{t-720}{320} \le 9 \Rightarrow 0 < \frac{t-720}{320} \le 3 \Rightarrow$$

$$0 < t - 720 \le 960 \Rightarrow 720 < t \le 1680.$$

[81] (a) $\frac{y}{b} = \frac{y+h}{a} \Rightarrow ay = by + bh \Rightarrow y(a-b) = bh \Rightarrow y = \frac{bh}{a-b}$

(b) $V = \frac{1}{3}\pi a^2(y+h) - \frac{1}{3}\pi b^2 y = \frac{\pi}{3}\left[(a^2 - b^2)y + a^2 h\right] =$

$$\frac{\pi}{3}\left[(a^2 - b^2)\frac{bh}{a-b} + a^2 h\right] = \frac{\pi}{3}h\left[(a+b)b + a^2\right] = \frac{\pi}{3}h(a^2 + ab + b^2)$$

(c) $a = 6,\ b = 3,\ V = 600 \Rightarrow \frac{\pi}{3}h(6^2 + 6 \cdot 3 + 3^2) = 600 \Rightarrow h = \frac{1800}{63\pi} = \frac{200}{7\pi} \approx 9.1$ ft

[82] Let $t$ denote the time (in hr) after 1:00 P.M. If the starting point for ship B is the origin, then the locations of A and B are $-30 + 15t$ and $-10t$, respectively. Using the Pythagorean theorem, $d^2 = (-30 + 15t)^2 + (-10t)^2 = 325t^2 - 900t + 900$. The time at which the distance between the ships is minimal is the same as the time at which the square of the distance between the ships is minimal.

$$\text{Thus, } t = -\frac{b}{2a} = -\frac{-900}{2(325)} = \frac{18}{13}, \text{ or about 2:23 P.M.}$$

[83] Let $r$ denote the radius of the semicircles and $x$ the length of the rectangle.

Perimeter = half-mile $\Rightarrow 2x + 2\pi r = \frac{1}{2} \Rightarrow x = -\pi r + \frac{1}{4}$.

$A = 2rx = 2r(-\pi r + \frac{1}{4}) = -2\pi r^2 + \frac{1}{2}r$. The maximum value of $A$ occurs when

$$r = -\frac{b}{2a} = -\frac{1/2}{2(-2\pi)} = \frac{1}{8\pi} \text{ mi.} \quad x = -\pi\left(\frac{1}{8\pi}\right) + \frac{1}{4} = \frac{1}{8} \text{ mi.}$$

[85] (a) Solving $-0.016x^2 + 1.6x = \frac{1}{5}x$ for $x$ represents the intersection between the parabola and the line. $-0.08x^2 + 8x = x \Rightarrow 7x - \frac{8}{100}x^2 = 0 \Rightarrow$ $x(7 - \frac{8}{100}x) = 0 \Rightarrow x = 0, \frac{175}{2}$. The rocket lands at $(\frac{175}{2}, \frac{35}{2}) = (87.5, 17.5)$.

(b) The *difference* $d$ between the parabola and the line is to be maximized here. $d = (-0.016x^2 + 1.6x) - (\frac{1}{5}x) = -0.016x^2 + 1.4x$. $d$ obtains a maximum when $x = -\frac{b}{2a} = -\frac{1.4}{2(-0.016)} = 43.75$. The maximum height of the rocket *above the ground* is $d = -0.016(43.75)^2 + 1.4(43.75) = 30.625$ units.

## Chapter 3 Discussion Exercises

[1] Graphs of equations of the form $y = x^{p/q}$, where $x \geq 0$, and $p$ and $q$ are positive integers all pass through (0, 0) and (1, 1). If $p/q < 1$, the graph is above $y = x$ for $0 \leq x \leq 1$ and below $y = x$ for $x \geq 1$. The closer $p/q$ is to 1, the closer $y = x^{p/q}$ is to $y = x$. If $p/q > 1$, the graph is below $y = x$ for $0 \leq x \leq 1$ and above $y = x$ for $x \geq 1$.

[3] For the graph of $g(x) = \sqrt{f(x)}$, where $f(x) = ax^2 + bx + c$, consider 2 cases:

**(1) ($a > 0$)** If $f$ has 0 or 1 $x$-intercept(s), the domain of $g$ is $\mathbb{R}$ and its range is $[\sqrt{k}, \infty)$, where $k$ is the $y$-value of the vertex of $f$. If $f$ has 2 $x$-intercepts (say $x_1$ and $x_2$ with $x_1 < x_2$), then the domain of $g$ is $(-\infty, x_1] \cup [x_2, \infty)$ and its range is $[0, \infty)$. The general shape is similar to the v-shape of the graph of $y = \sqrt{a}\,|\,x\,|$.

**(2) ($a < 0$)** If $f$ has no $x$-intercepts, there is no graph of $g$. If $f$ has 1 $x$-intercept, the graph of $g$ consists of that point. If $f$ has 2 $x$-intercepts, the domain of $g$ is $[x_1, x_2]$ and the range is $[0, \sqrt{k}]$. The shape of $g$ is that of the top half of an oval.

The main advantage of graphing $g$ as a composition (say $Y_1 = f$ and $Y_2 = \sqrt{Y_1}$ on a graphing calculator) is to observe the relationship between the range of $f$ and the domain of $g$.

[5] The expression $2x + h + 6$ represents the slope of the line between the points $P(x, f(x))$ and $Q(x + h, f(x + h))$. If $h = 0$, then $2x + 6$ represents the slope of the tangent line at the point $P(x, f(x))$.

[7] The values of the $x$-intercepts (if they exist) are found by using the quadratic formula $\left(x = -\frac{b}{2a} \pm \frac{\sqrt{b^2 - 4ac}}{2a}\right)$. Hence, the distance $d$ from the axis of symmetry, $x = -\frac{b}{2a}$, to either $x$-intercept is $d = \frac{\sqrt{b^2 - 4ac}}{2\,|\,a\,|}$ and $d^2 = \frac{b^2 - 4ac}{4a^2}$. From page 217, the $y$-coordinate of the vertex is $h = c - \frac{b^2}{4a} = \frac{4ac - b^2}{4a}$, so

$$\frac{h}{d^2} = \frac{\frac{4ac - b^2}{4a}}{\frac{b^2 - 4ac}{4a^2}} = -\frac{4a^2(4ac - b^2)}{4a(b^2 - 4ac)} = -a.$$

Thus, $h = -ad^2$. Note that this relationship also reveals a connection between the discriminant $D$ and the $y$-coordinate of the vertex, namely $h = -D^2/(4a)$.

[9] $D = 0.0833x^2 - 0.4996x + 3.5491 \;\Rightarrow\; 0.0833x^2 - 0.4996x + (3.5491 - D) = 0.$

Solving for $x$ with the quadratic formula yields

$$x = \frac{0.4996 \pm \sqrt{(-0.4996)^2 - 4(0.0833)(3.5491 - D)}}{2(0.0833)}, \text{ or, equivalently,}$$

$$x = \frac{4996 \pm \sqrt{33{,}320{,}000D - 93{,}295{,}996}}{1666}.$$

From *Figure 9* (a graph of $D$), we see that $3 \le x \le 15$ corresponds to the right half of the parabola. Hence, we choose the plus sign in the equation for $x$.

[−20, 20, 2] by [0, 40, 2]

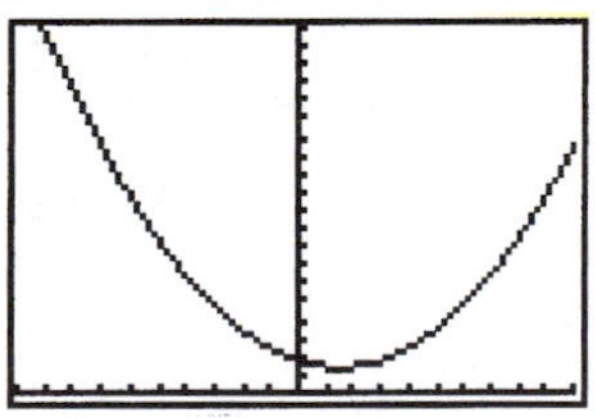

*Figure 9*

# Chapter 4: Polynomial and Rational Functions

## 4.1 Exercises

[1] $f(x) = 2x^3 + c$ • The graph of $y = g(x) = 2x^3$ is the graph of $y = x^3$ stretched vertically by a factor of 2.

(a) The effect of the "3" for $c$ is to shift the graph of $g$ *up* 3 units.

(b) The effect of the "−3" for $c$ is to shift the graph of $g$ *down* 3 units.

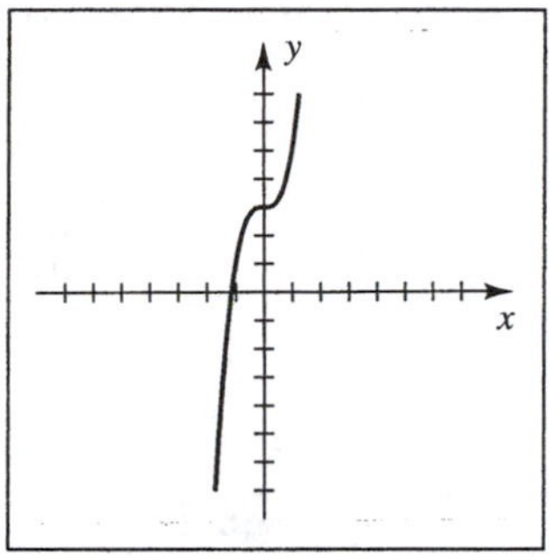


*Figure 1(a)*

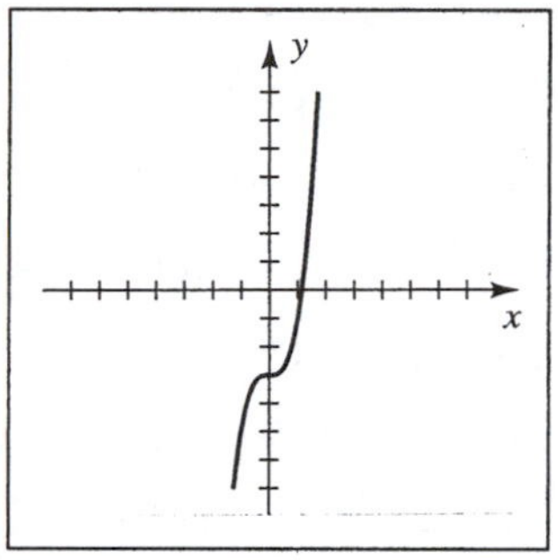


*Figure 1(b)*

[3] $f(x) = ax^3 + 2$ • The "+2" will shift each graph up 2 units.

(a) The effect of the "2" for $a$ is to vertically stretch $g(x) = x^3 + 2$ by a factor of 2 and make it appear "steeper."

(b) The effect of the "$\frac{1}{3}$" for $a$ is to vertically compress $g(x) = x^3$ by a factor of $1/(1/3) = 3$, making it appear "flatter." The "−" reflects the graph of $h(x) = \frac{1}{3}x^3$ through the $x$-axis.

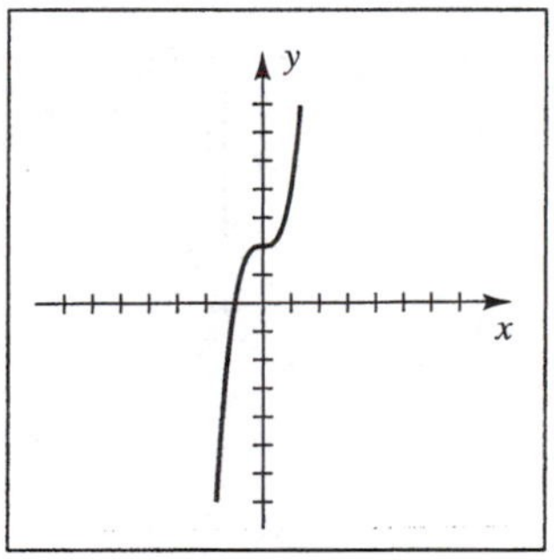


*Figure 3(a)*

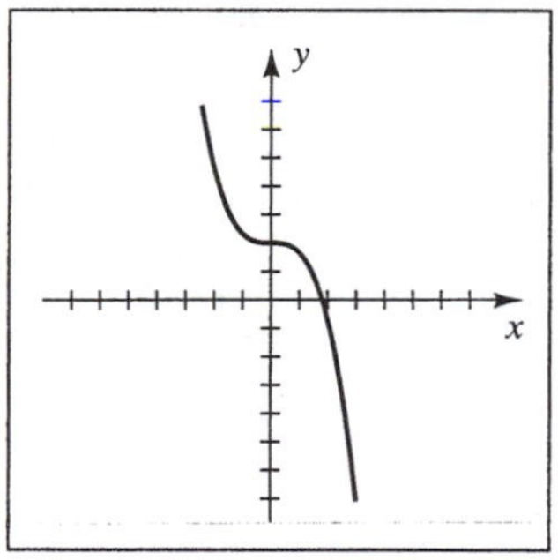


*Figure 3(b)*

[5] We need to show that there is a sign change between $f(3)$ and $f(4)$ for $f(x) = x^3 - 4x^2 + 3x - 2$.

$$f(3) = 27 - 36 + 9 - 2 = -2 \quad \text{and} \quad f(4) = 64 - 64 + 12 - 2 = 10.$$

Since $f(3) = -2 < 0$ and $f(4) = 10 > 0$, the intermediate value theorem for polynomial functions assures us that $f$ takes on every value between −2 and 10 in the interval [3, 4], namely, 0.

[7] As in Exercise 5, with $f(x) = -x^4 + 3x^3 - 2x + 1$, $a = 2$, $b = 3$,

we have $f(2) = 5 > 0$ and $f(3) = -5 < 0$.

[9] As in Exercise 5, with $f(x) = x^5 + x^3 + x^2 + x + 1$, $a = -\frac{1}{2}$, $b = -1$,

we have $f(-\frac{1}{2}) = \frac{19}{32} > 0$ and $f(-1) = -1 < 0$.

[11] (a) The graph has $x$-intercepts at $-1$, 1, and 2—so $(x+1)$, $(x-1)$, and $(x-2)$ are factors in its equation. It is the graph of a cubic. **C**

(b) The graph has $x$-intercepts at $-1$, 1, and 2. It is the graph of a quartic. **D**

(c) The graph has $x$-intercepts at 0 and 2. It is the graph of a cubic with a negative leading coefficient. **B**

(d) The graph has $x$-intercepts at 0 and 2. It is the graph of a cubic with a positive leading coefficient. **A**

[13] $f(x) = \frac{1}{4}x^3 - 2 = \frac{1}{4}(x^3 - 8) = \frac{1}{4}(x-2)(x^2 + 2x + 4)$. The general shape of the graph is that of $g(x) = x^3$. To find the $x$-intercepts, set $y$ $\{f(x)\}$ equal to 0. This means either $x - 2 = 0$ or $x^2 + 2x + 4 = 0$. $x - 2 = 0 \Rightarrow x = 2$ and $x^2 + 2x + 4 = 0 \Rightarrow x = -1 \pm \sqrt{3}\,i$. The imaginary solutions mean that we have no $x$-intercepts from the factor $x^2 + 2x + 4$ and the only $x$-intercept is 2. To find the $y$-intercept, set $x$ equal to 0. $x = 0 \Rightarrow y = -2$.

The graph of $f$ lies *above* the $x$-axis for all values of $x$ greater than 2, so $f(x) > 0$ if $x > 2$. The graph of $f$ lies *below* the $x$-axis for all values of $x$ less than 2, so $f(x) < 0$ if $x < 2$.

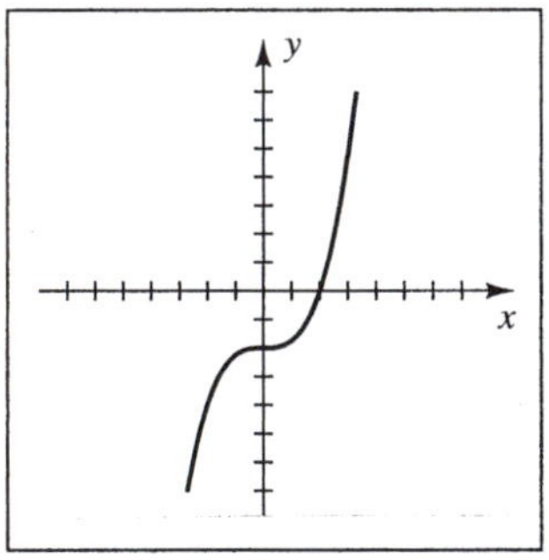


*Figure 13*

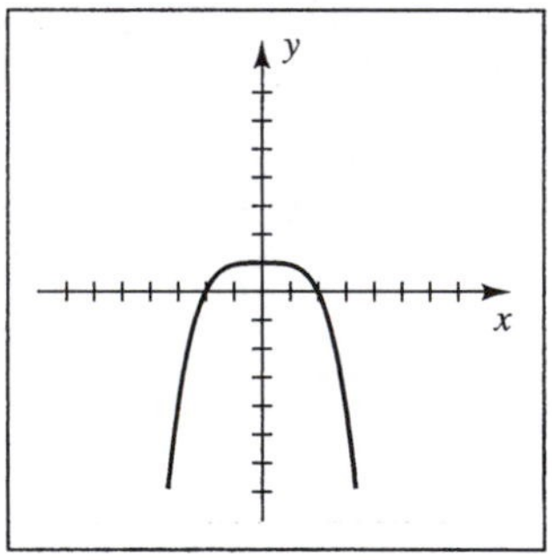


*Figure 15*

[15] $f(x) = -\frac{1}{16}x^4 + 1 = -\frac{1}{16}(x^4 - 16) = -\frac{1}{16}(x^2 + 4)(x+2)(x-2)$. The general shape of the graph is that of $g(x) = -x^4$. The $x$-intercepts are $-2$ and 2. The $y$-intercept is 1. The graph of $f$ lies *above* the $x$-axis for all values of $x$ such that $-2 < x < 2$, so we can write $f(x) > 0$ if $|\,x\,| < 2$. The graph of $f$ lies *below* the $x$-axis for all values of $x$ such that $x < -2$ or $x > 2$, so we can write $f(x) < 0$ if $|\,x\,| > 2$.

[17] $f(x) = x^4 - 4x^2 = x^2(x+2)(x-2)$. The general shape of the graph is that of $g(x) = x^4$. The $x$-intercepts are $-2$, 0, and 2. The $y$-intercept is 0. The graph of $f$ lies *above* the $x$-axis for all values of $x$ such that $x < -2$ or $x > 2$, so we can write $f(x) > 0$ if $|x| > 2$. The graph of $f$ lies *below* the $x$-axis for all values of $x$ such that $-2 < x < 0$ or $0 < x < 2$, so we can write $f(x) < 0$ if $0 < |x| < 2$. Note that $x = 0$ is excluded in the last inequality since $f(x)$ *is equal to* 0 if $x = 0$.

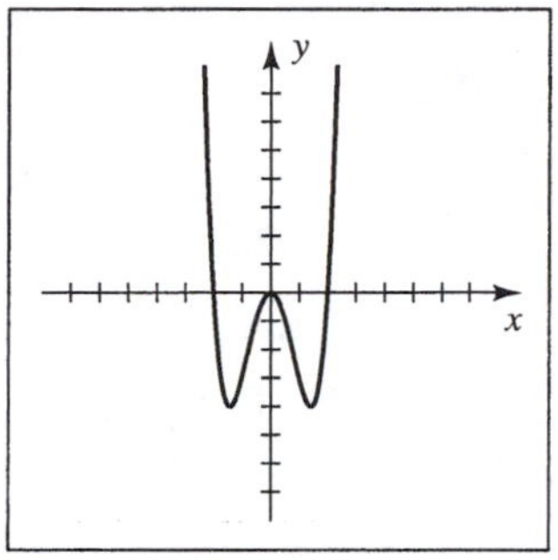


*Figure 17*

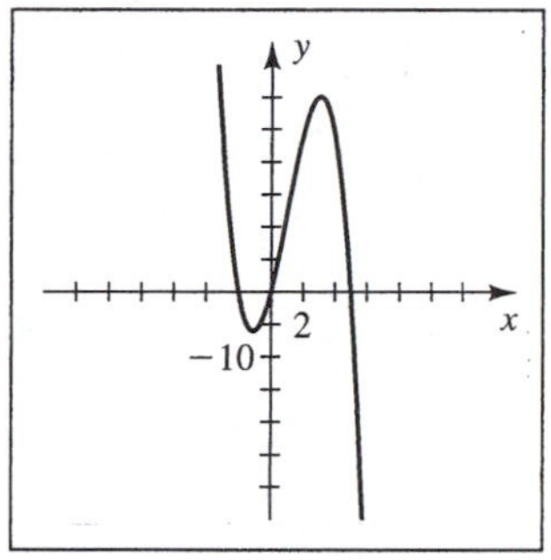


*Figure 19*

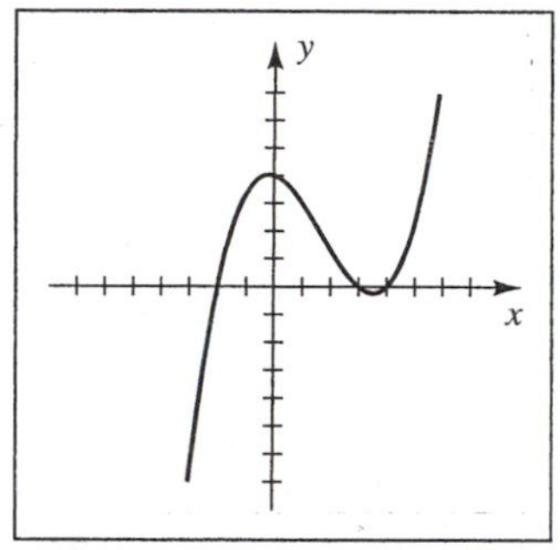


*Figure 21*

[19] $f(x) = -x^3 + 3x^2 + 10x = -x(x^2 - 3x - 10) = -x(x+2)(x-5)$. The general shape of the graph is that of $g(x) = -x^3$. The $x$-intercepts are $-2$, 0, and 5. The $y$-intercept is 0. The graph of $f$ lies *above* the $x$-axis for all values of $x$ such that $x < -2$ or $0 < x < 5$, so $f(x) > 0$ if $x < -2$ or $0 < x < 5$. The graph of $f$ lies *below* the $x$-axis for all values of $x$ such that $-2 < x < 0$ or $x > 5$, so $f(x) < 0$ if $-2 < x < 0$ or $x > 5$.

[21] $f(x) = \frac{1}{6}(x+2)(x-3)(x-4)$ • If we were to multiply the terms of $f$, we would obtain a polynomial of the form $f(x) = \frac{1}{6}x^3 + (\text{other terms})$, so the general shape of the graph is that of $g(x) = x^3$. The $x$-intercepts are $-2$, 3, and 4. The $y$-intercept is $f(0) = \frac{1}{6}(0+2)(0-3)(0-4) = \frac{1}{6}(2)(-3)(-4) = 4$.

To determine the intervals on which $f(x) > 0$ and $f(x) < 0$, you may want to refer back to the concept of a sign diagram. Below is a sign diagram for this function. Note how the sign of each region correlates to the sign of $f(x)$.

| Resulting sign: | ⊖ | ⊕ | ⊖ | ⊕ |
|---|---|---|---|---|
| Sign of $x - 4$: | $-$ | $-$ | $-$ | $+$ |
| Sign of $x - 3$: | $-$ | $-$ | $+$ | $+$ |
| Sign of $x + 2$: | $-$ | $+$ | $+$ | $+$ |
| $x$ values: | $-2$ | 3 | 4 | |

*Diagram 21*

See *Figure 21* above. The graph of $f$ lies *above* the $x$-axis for all values of $x$ such that $-2 < x < 3$ or $x > 4$, so $f(x) > 0$ if $-2 < x < 3$ or $x > 4$. The graph of $f$ lies *below* the $x$-axis for all values of $x$ such that $x < -2$ or $3 < x < 4$, so $f(x) < 0$ if $x < -2$ or $3 < x < 4$.

[23] $f(x) = x^3 + 2x^2 - 4x - 8 = x^2(x+2) - 4(x+2)$ {factor by grouping}

$= (x^2 - 4)(x+2)$ {factor out $(x+2)$}

$= (x+2)(x-2)(x+2) = (x+2)^2(x-2)$.

The general shape of the graph is that of $g(x) = x^3$. The $x$-intercepts are $-2$ and 2. The $y$-intercept is $-8$.

If an intercept value has a corresponding *linear* factor {a factor with exponent 1}, the graph will go through that intercept—that is, there will be a sign change from positive to negative or negative to positive in the function. Thus, at $x = 2$ the graph of the function "goes through" the point. If an intercept value has a corresponding quadratic factor (as in the case of $-2$ and $(x+2)^2$), then the function will not change sign. Hence, at $x = -2$ the graph of the function "touches the point and turns around."

The above discussion can be generalized as follows:

(1) If an intercept value has a corresponding factor raised to an odd power, then the function will change sign at that point.

(2) If an intercept value has a corresponding factor raised to an even power, then the function will not change sign at that point.

The graph of $f$ lies *above* the $x$-axis for all values of $x$ such that $x > 2$, so $f(x) > 0$ if $x > 2$. The graph of $f$ lies *below* the $x$-axis for all values of $x$ such that $x < -2$ or $-2 < x < 2$, so we write $f(x) < 0$ if $x < -2$ or $|x| < 2$.

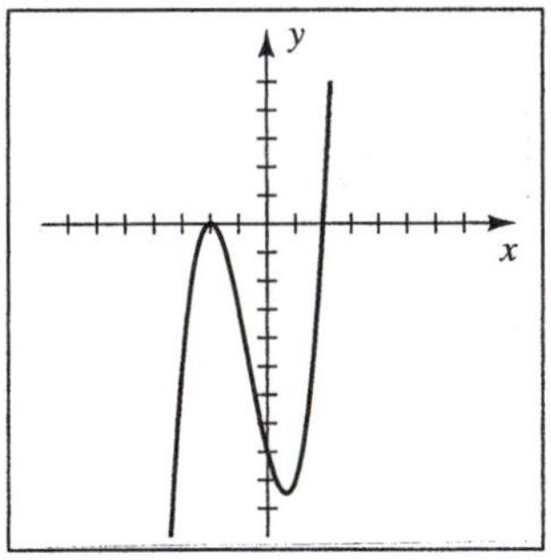


Figure 23

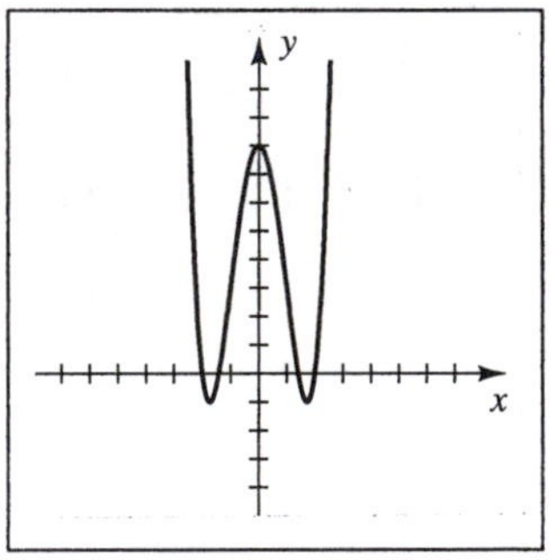


Figure 25

[25] $f(x) = x^4 - 6x^2 + 8 = (x^2 - 2)(x+2)(x-2)$. The general shape of the graph is that of $g(x) = x^4$. The $x$-intercepts are $\pm 2$ and $\pm\sqrt{2}$. The $y$-intercept is 8. The graph of $f$ lies *above* the $x$-axis for all values of $x$ such that $x < -2$ or $-\sqrt{2} < x < \sqrt{2}$ or $x > 2$, so we can write $f(x) > 0$ if $|x| > 2$ or $|x| < \sqrt{2}$. The graph of $f$ lies *below* the $x$-axis for all values of $x$ such that $-2 < x < -\sqrt{2}$ or $\sqrt{2} < x < 2$, so we can write $f(x) < 0$ if $\sqrt{2} < |x| < 2$.

[27] $f(x) = x^2(x+2)(x-1)^2(x-2)$ • If we were to multiply the terms of $f$, we would obtain a polynomial of the form $f(x) = x^6 +$ (other terms), so the general shape of the graph is that of $g(x) = x^6$. The $x$-intercepts are $-2$, 0, 1, and 2. The $y$-intercept is $f(0) = 0$.

The graph of $f$ lies *above* the $x$-axis for all values of $x$ such that $x < -2$ or $x > 2$, so we can write $f(x) > 0$ if $|\,x\,| > 2$. The graph of $f$ lies *below* the $x$-axis for all values of $x$ such that $-2 < x < 0$ or $0 < x < 1$ or $1 < x < 2$, so we can write $f(x) < 0$ if $|\,x\,| < 2$, $x \neq 0$, $x \neq 1$.

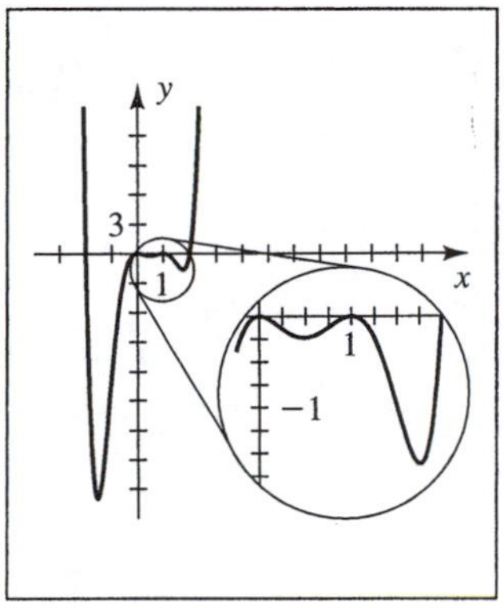


Figure 27

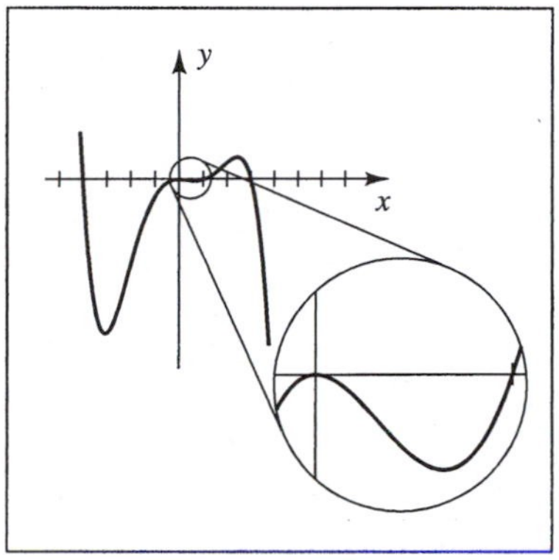


Figure 29

[29] The sign of $f(x)$ is positive on $(-\infty, -4)$, so the graph of $f$ must be above the $x$-axis on that interval. The sign of $f(x)$ is negative on $(-4, 0)$, so the graph of $f$ must cross the $x$-axis at $x = -4$ and be below the $x$-axis on the interval $(-4, 0)$. The sign of $f(x)$ is negative on $(0, 1)$, so the graph of $f$ must touch the $x$-axis at $x = 0$ and then fall below the $x$-axis on $(0, 1)$. The sign of $f(x)$ is positive on $(1, 3)$, so the graph of $f$ must cross the $x$-axis at $x = 1$ and be above the $x$-axis on that interval. The sign of $f(x)$ is negative on $(3, \infty)$, so the graph of $f$ must cross the $x$-axis at $x = 3$ and be below the $x$-axis on the interval $(3, \infty)$.

[31] (a) The graph of $f(x) = (x-a)(x-b)(x-c)$, where $a < 0 < b < c$, must have one negative zero, $a$, and two positive zeros, $b$ and $c$. The general shape is that of a cubic polynomial.

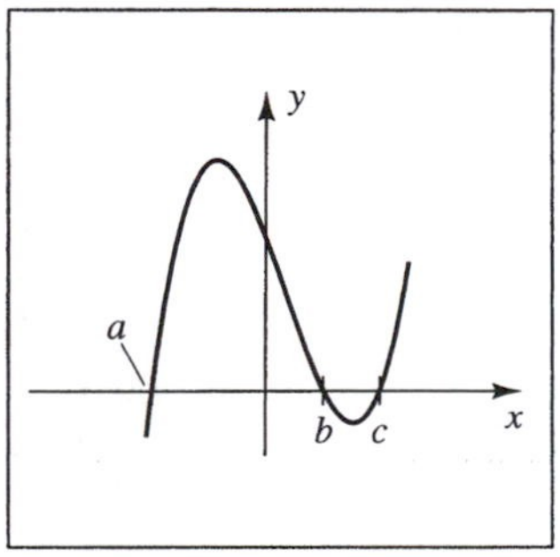


Figure 31

(b) The $y$-intercept is $f(0) = (-a)(-b)(-c) = -abc$.

(c) The solution to $f(x) < 0$ is $(-\infty, a) \cup (b, c)$; that is, where the graph of $f$ is below the $x$-axis.

(d) The solution to$f(x) \geq 0$ is $[a, b] \cup [c, \infty)$; that is, where the graph of $f$ is above or on the $x$-axis.

[33] If $f(x)$ is a polynomial such that the coefficient of every odd power of $x$ is 0, then $f(x)$ consists of only even-powered terms. If $n$ is even, then $(-x)^n = x^n$ and hence $f(-x) = f(x)$, which is the condition that must be true for $f$ to be an even function. Thus, $f$ is an even function.

[35] If a graph contains the point $(-1,\ 4)$, then $f(-1) = 4$.

For this function, $f(x) = 3x^3 - kx^2 + x - 5k$,

$$f(-1) = 3(-1)^3 - k(-1)^2 + (-1) - 5k = -3 - k - 1 - 5k = -4 - 6k.$$

Thus, $-4 - 6k$ must equal 4. Solving for $k$ yields $-4 - 6k = 4 \ \Rightarrow\ k = -\frac{4}{3}$.

[37] If one zero of $f(x) = x^3 - 2x^2 - 16x + 16k$ is 2, then $f(2) = 0$. But $f(2) = 16k - 32$, so $16k - 32 = 0 \ \Rightarrow\ 16k = 32 \ \Rightarrow\ k = 2$. Thus,

$$\begin{aligned} f(x) &= x^3 - 2x^2 - 16x + 32 && \{\text{let } k = 2\} \\ &= x^2(x-2) - 16(x-2) && \{\text{factor by grouping}\} \\ &= (x^2 - 16)(x - 2) && \{\text{factor out } (x-2)\} \\ &= (x+4)(x-4)(x-2). \end{aligned}$$

So the other two zeros of $f$ are $\pm 4$.

[39] $P(x) = \frac{1}{2}(5x^3 - 3x) = \frac{1}{2}x(5x^2 - 3)$.

$P(x) = 0 \ \Rightarrow\ x = 0$ or $x^2 = \frac{3}{5} \ \Leftrightarrow\ x = \pm\sqrt{\frac{3}{5}} = \pm\sqrt{\frac{3 \cdot 5}{5 \cdot 5}} = \pm\frac{1}{5}\sqrt{15}$.

$P(x) > 0$ on $(-\frac{1}{5}\sqrt{15},\ 0)$ and $(\frac{1}{5}\sqrt{15},\ \infty)$.

$P(x) < 0$ on $(-\infty,\ -\frac{1}{5}\sqrt{15})$ and $(0,\ \frac{1}{5}\sqrt{15})$.

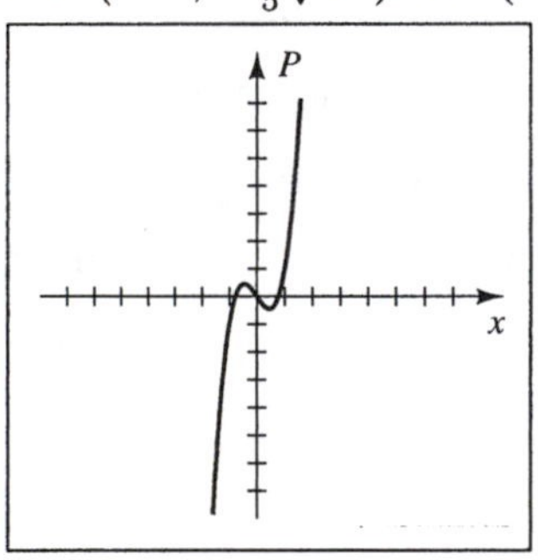


Figure 39

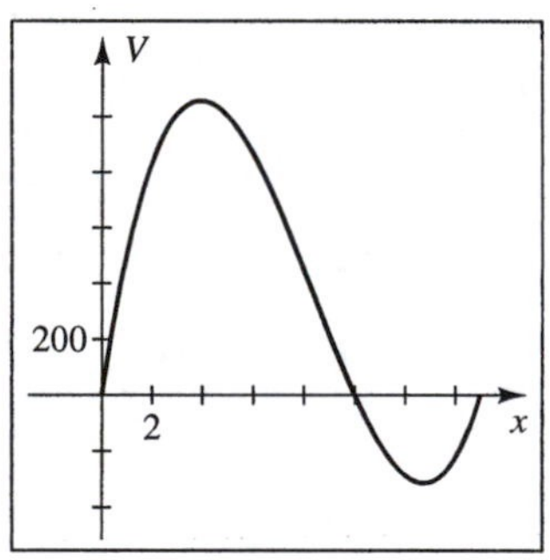


Figure 41

[41] (a) $V(x) = lwh = (30 - x - x)(20 - x - x)x = x(20 - 2x)(30 - 2x)$.

(b) $V(x) = x(20 - 2x)(30 - 2x) = x(-2)(x - 10)(-2)(x - 15) = 4x(x - 10)(x - 15)$.

The $x$-intercepts of $V$ are 0, 10, and 15. $V(x) > 0$ on $(0,\ 10)$ and $(15,\ \infty)$. Allowable values for $x$ are in $(0,\ 10)$.

[43] We will work part (b) first.

(b) $T = \frac{1}{20}t(t-12)(t-24)$ has the general shape of a cubic polynomial with zeros at 0, 12, and 24. Its sign pattern is negative, positive, negative, positive.

(a) $T = \frac{1}{20}t(t-12)(t-24) = 0 \Rightarrow t = 0, 12, 24.$ $T > 0$ for $0 < t < 12$ {6 A.M. to 6 P.M.}; $T < 0$ for $12 < t < 24$ {6 P.M. to 6 A.M.}.

(c) 12 noon corresponds to $t = 6$, $T(6) = 32.4 > 32$°F and $T(7) = 29.75 < 32$°F

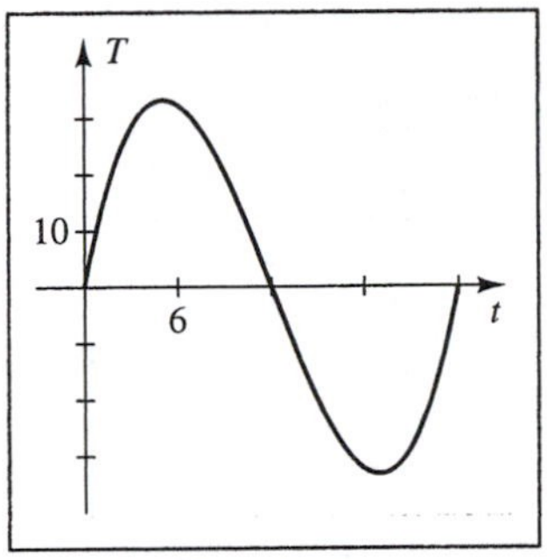


*Figure 43*

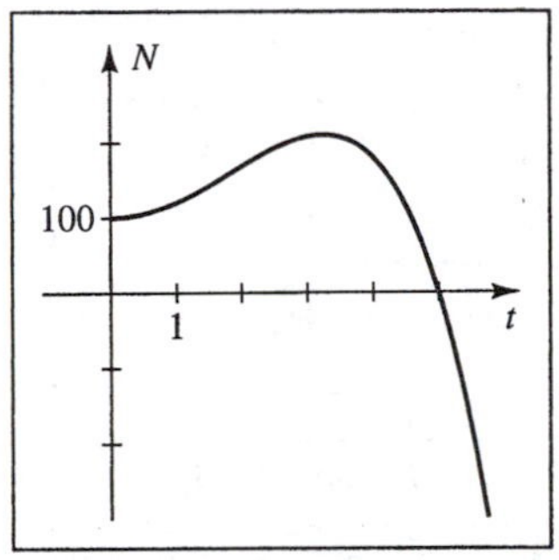


*Figure 45*

[45] (a) $N(t) = -t^4 + 21t^2 + 100$

$= -(t^4 - 21t^2 - 100)$ {factor out $-1$}

$= -(t^2 - 25)(t^2 + 4)$ {factor as a quadratic in $t^2$}

$= -(t+5)(t-5)(t^2+4)$ {difference of two squares}

$N$ has the general shape of a quartic polynomial reflected through the $t$-axis. If $t > 0$, then $N(t) > 0$ for $0 < t < 5$.

(b) The population becomes extinct when $N = 0$. This occurs after 5 years.

[47] (a) $f(x) = 2x^4$, $g(x) = 2x^4 - 5x^2 + 1$, $h(x) = 2x^4 + 5x^2 - 1$, $k(x) = 2x^4 - x^3 + 2x$

| $x$ | $f(x)$ | $g(x)$ | $h(x)$ | $k(x)$ |
|---|---|---|---|---|
| −60 | 25,920,000 | 25,902,001 | 25,937,999 | 26,135,880 |
| −40 | 5,120,000 | 5,112,001 | 5,127,999 | 5,183,920 |
| −20 | 320,000 | 318,001 | 321,999 | 327,960 |
| 20 | 320,000 | 318,001 | 321,999 | 312,040 |
| 40 | 5,120,000 | 5,112,001 | 5,127,999 | 5,056,080 |
| 60 | 25,920,000 | 25,902,001 | 25,937,999 | 25,704,120 |

(b) As $|x|$ becomes large, the function values become similar.

(c) The term with the highest power of $x$: $2x^4$.

**49** (a) $[-9, 9]$ by $[-6, 6]$

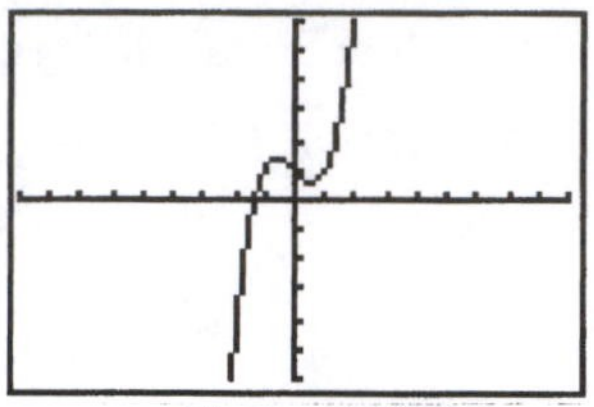

*Figure 49(1)*

$[-9, 9]$ by $[-6, 6]$

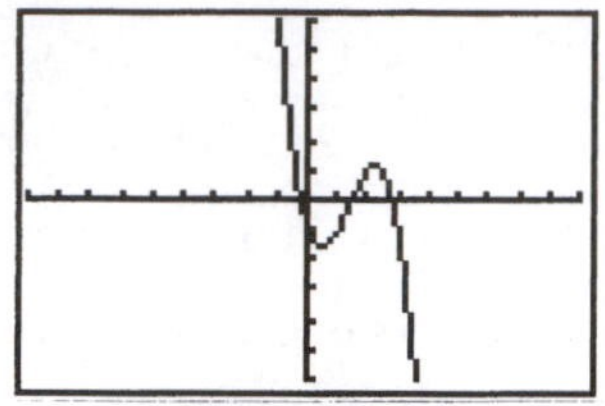

*Figure 49(2)*

$[-9, 9]$ by $[-6, 6]$

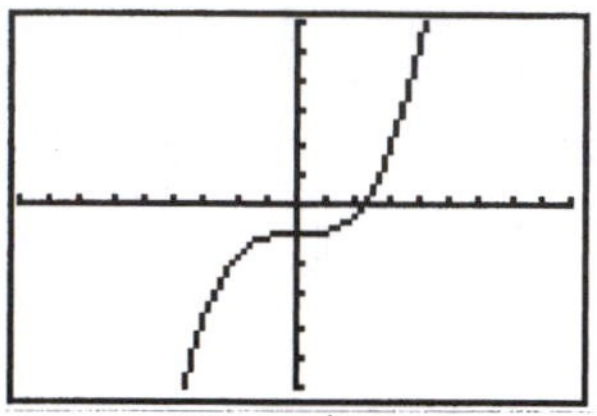

*Figure 49(3)*

$[-9, 9]$ by $[-6, 6]$

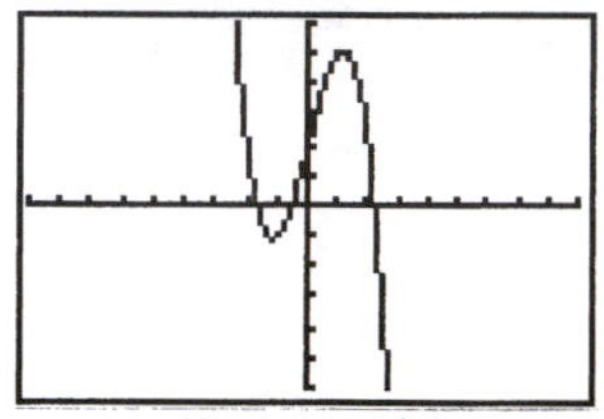

*Figure 49(4)*

(b) (1) $f(x) = x^3 - x + 1$ • As $x$ approaches $\infty$, $f(x)$ approaches $\infty$; as $x$ approaches $-\infty$, $f(x)$ approaches $-\infty$

(2) $f(x) = -x^3 + 4x^2 - 3x - 1$ • As $x$ approaches $\infty$, $f(x)$ approaches $-\infty$; as $x$ approaches $-\infty$, $f(x)$ approaches $\infty$

(3) $f(x) = 0.1x^3 - 1$ • As $x$ approaches $\infty$, $f(x)$ approaches $\infty$; as $x$ approaches $-\infty$, $f(x)$ approaches $-\infty$

(4) $f(x) = -x^3 + 4x + 2$ • As $x$ approaches $\infty$, $f(x)$ approaches $-\infty$; as $x$ approaches $-\infty$, $f(x)$ approaches $\infty$

(c) For the cubic function $f(x) = ax^3 + bx^2 + cx + d$ with $a > 0$, $f(x)$ approaches $\infty$ as $x$ approaches $\infty$ and $f(x)$ approaches $-\infty$ as $x$ approaches $-\infty$. With $a < 0$, $f(x)$ approaches $-\infty$ as $x$ approaches $\infty$ and $f(x)$ approaches $\infty$ as $x$ approaches $-\infty$.

**51** From the graph, $f(x) = x^3 + 0.2x^2 - 2.6x + 1.1$ has three zeros.

They are approximately $-1.89$, $0.49$, and $1.20$.

$[-4.5, 4.5]$ by $[-3, 3]$

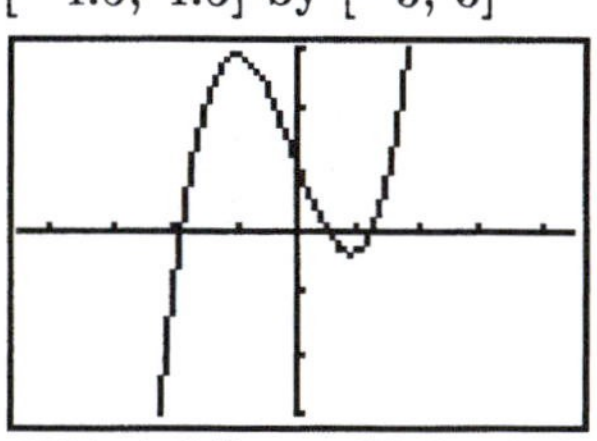

*Figure 51*

$[-4.5, 4.5]$ by $[-3, 3]$

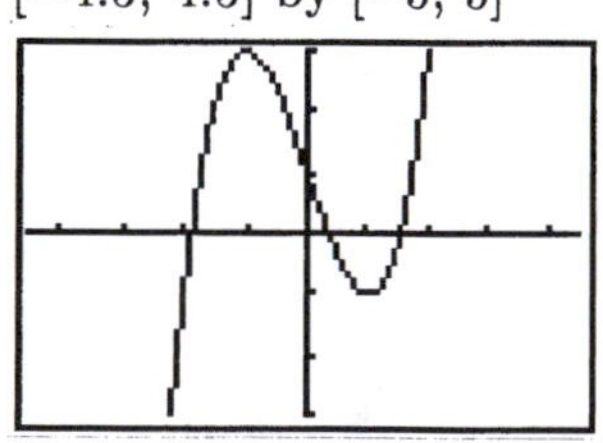

*Figure 53*

**53** From the graph, $f(x) = x^3 - 3x + 1$ has three zeros.

They are approximately $-1.88$, $0.35$, and $1.53$.

[55] If $f(x) = x^3 + 5x - 2$ and $k = 1$, then $f(x) > k$ on $(0.56, \infty)$.

Use an intersect feature to find the value 0.56.

$[-4.5, 4.5]$ by $[-3, 3]$

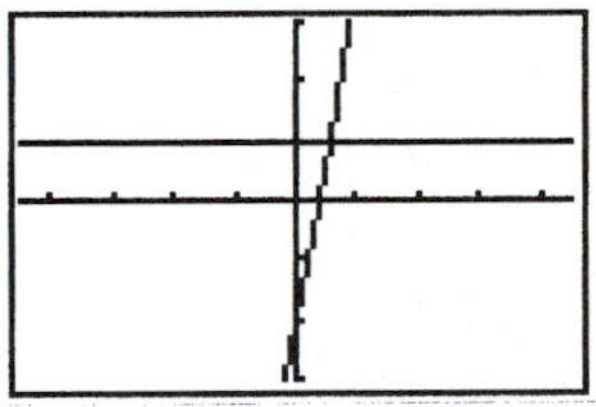

*Figure 55*

$[-4.5, 4.5]$ by $[-3, 3]$

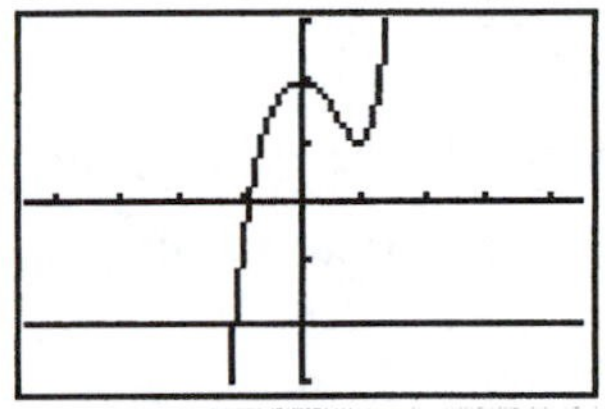

*Figure 57*

[57] If $f(x) = x^5 - 2x^2 + 2$ and $k = -2$, then $f(x) > k$ on $(-1.10, \infty)$.

[59] From the graph, there are three points of intersection.

Their coordinates are approximately $(-1.29, -0.77)$, $(0.085, 2.66)$, and $(1.36, -0.42)$.

$[-4.5, 4.5]$ by $[-2, 4]$

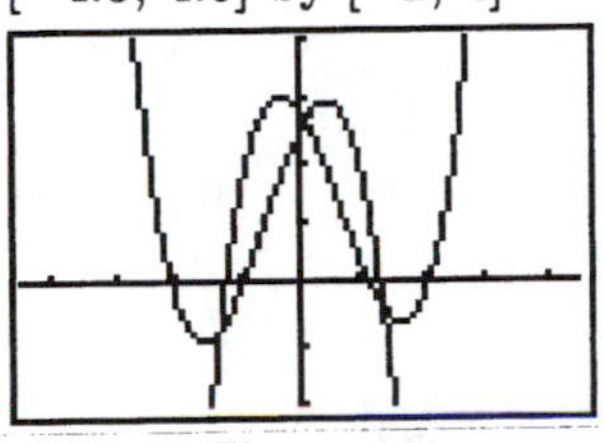

*Figure 59*

$[1975, 1995]$ by $[20, 40]$

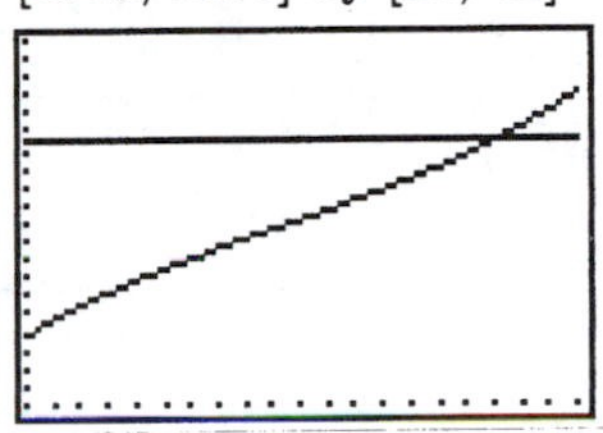

*Figure 61*

[61] (a) To enter $f$ in your graphing calculator, let $Y_1 = X - 1975$ and then assign

$$0.0014{Y_1}^3 - 0.0388{Y_1}^2 + 0.8783Y_1 + 23.82 \quad \text{to} \quad Y_2.$$

From the graph, we determine that there were approximately 23.82 million recipients in 1975 and 37.066 million recipients in 1995. The number of recipients is increasing quite rapidly.

(b) Graph $Y_3 = 34.4$ (shown in *Figure 61*). The graphs of $Y_2$ and $Y_3$ intersect when $x \approx 1991.9792 \approx 1992$. There were 34.4 million recipients in 1992.

## *4.2 Exercises*

*Note:* Refer to the illustration on page 259. We will walk through the steps to obtain the quotient and remainder for this example.

(1) Determine the quotient of the first terms of the dividend $(x^4 - 16)$ and the divisor $(x^2 + 3x + 1)$. $\dfrac{x^4}{x^2} = x^2$—this is the first term of the resulting quotient.

(2) Multiply the result in step (1), $x^2$, by the divisor, $x^2 + 3x + 1$. This product is $x^4 + 3x^3 + x^2$.

(continued)

(3) Subtract the result in step (2) from the dividend. This result is $-3x^3 - x^2$ (there is no need to worry about the lower degree terms here, namely, $-16$). A common mistake is to forget to subtract <u>all</u> of the terms and obtain (in this case) $3x^3 + x^2$.

(4) Repeat steps (1) through (3) for the remaining dividend until the degree of the remainder is less than the degree of the divisor.

[1] $f(x) = 2x^4 - x^3 - 3x^2 + 7x - 12;\ \ p(x) = x^2 - 3$ •

$$
\begin{array}{r|rrrrr}
 & & & 2x^2 - & x + & 3 \\ \cline{2-6}
x^2 - 3 & 2x^4 - & x^3 - & 3x^2 + & 7x - & 12 \\
 & 2x^4 & - & 6x^2 & & \\ \cline{2-6}
 & & -x^3 + & 3x^2 & & \\
 & & -x^3 & & + \ 3x & \\ \cline{3-6}
 & & & 3x^2 + & 4x & \\
 & & & 3x^2 & - & 9 \\ \cline{4-6}
 & & & & 4x - & 3
\end{array}
$$

First step: $\dfrac{2x^4}{x^2} = 2x^2$

★ $q(x) = 2x^2 - x + 3;\ r(x) = 4x - 3$

[3] $f(x) = 3x^3 + 2x - 4;\ p(x) = 2x^2 + 1$ •

$$
\begin{array}{r|rrrrr}
 & & & \frac{3}{2}x & & \\ \cline{2-6}
2x^2 + 1 & 3x^3 & - \ 0x^2 & + \ 2x & - & 4 \\
 & 3x^3 & & + \ \frac{3}{2}x & & \\ \cline{2-6}
 & & & \frac{1}{2}x & - & 4
\end{array}
$$

First step: $\dfrac{3x^3}{2x^2} = \frac{3}{2}x$

★ $q(x) = \frac{3}{2}x;\ r(x) = \frac{1}{2}x - 4$

[5] Note that we can't divide $f(x) = 7x + 2$ by $p(x) = 2x^2 - x - 4$ since the degree of the dividend is already less than the degree of the divisor. So the quotient is 0 and the remainder is $7x + 2$.

[7] $f(x) = 9x + 4;\ p(x) = 2x - 5$ •

$$
\begin{array}{r|rrr}
 & & & \frac{9}{2} \\ \cline{2-4}
2x - 5 & 9x & + & 4 \\
 & 9x & - & \frac{45}{2} \\ \cline{2-4}
 & & & \frac{53}{2}
\end{array}
$$

First step: $\dfrac{9x}{2x} = \frac{9}{2}$

★ $q(x) = \frac{9}{2};\ r(x) = \frac{53}{2}$

[9] We wish to find $f(2)$ by using the remainder theorem. Thus, we need to divide $f(x) = 3x^3 - x^2 + 5x - 4$ by $x - 2$ using either long division or synthetic division. Using synthetic division, we have

$$\begin{array}{r|rrrr} 2 & 3 & -1 & 5 & -4 \\ & & 6 & 10 & 30 \\ \hline & 3 & 5 & 15 & 26 \end{array}$$

The last number in the third row, 26, is our remainder. Hence, $f(2) = 26$.

[11] We divide by $x + 3$ or $x - (-3)$. Note that we include a 0 for the missing $x^3$ term in $f(x) = x^4 - 6x^2 + 4x - 8$.

$$\begin{array}{r|rrrrr} -3 & 1 & 0 & -6 & 4 & -8 \\ & & -3 & 9 & -9 & 15 \\ \hline & 1 & -3 & 3 & -5 & 7 \end{array}$$

Hence, $f(-3) = 7$.

[13] To show that $x + 3$ is a factor of $f(x) = x^3 + x^2 - 2x + 12$, we must show that $f(-3) = 0$. $f(-3) = -27 + 9 + 6 + 12 = 0$, and, hence, $x + 3$ is a factor of $f(x)$.

[15] To show that $x + 2$ is a factor of $f(x) = x^{12} - 4096$, we must show that $f(-2) = 0$. $f(-2) = (-2)^{12} - 4096 = 4096 - 4096 = 0$, and, hence, $x + 2$ is a factor of $f(x)$.

[17] $f$ has degree 3 with zeros $-2$, 0, 5 $\Rightarrow$

$f(x) = a[x - (-2)](x - 0)(x - 5)$ {let $a = 1$}

$= x(x + 2)(x - 5) = x(x^2 - 3x - 10) = x^3 - 3x^2 - 10x$

[19] $f$ has degree 4 with zeros $-2$, $\pm 1$, 4 $\Rightarrow$

$f(x) = a(x + 2)(x + 1)(x - 1)(x - 4)$ {let $a = 1$}

$= (x^2 - 1)(x^2 - 2x - 8) = x^4 - 2x^3 - 9x^2 + 2x + 8$

[21]
$$\begin{array}{r|rrrr} 2 & 2 & -3 & 4 & -5 \\ & & 4 & 2 & 12 \\ \hline & 2 & 1 & 6 & 7 \end{array}$$

The synthetic division indicates that the quotient is $2x^2 + x + 6$ and the remainder is 7.

[23]
$$\begin{array}{r|rrrr} -3 & 1 & 0 & -8 & -5 \\ & & -3 & 9 & -3 \\ \hline & 1 & -3 & 1 & -8 \end{array}$$

The synthetic division indicates that the quotient is $x^2 - 3x + 1$ and the remainder is $-8$.

[25] $3x^5 + 6x^2 + 7 = \underline{3}\,x^5 + \underline{0}\,x^4 + \underline{0}\,x^3 + \underline{6}\,x^2 + \underline{0}\,x + \underline{7}$

$$\begin{array}{r|rrrrrr} -2 & 3 & 0 & 0 & 6 & 0 & 7 \\ & & -6 & 12 & -24 & 36 & -72 \\ \hline & 3 & -6 & 12 & -18 & 36 & -65 \end{array}$$

The synthetic division indicates that the quotient is $3x^4 - 6x^3 + 12x^2 - 18x + 36$ and the remainder is $-65$.

**27** $\frac{1}{2}$

| | | | | | |
|---|---|---|---|---|---|
| $\frac{1}{2}$ | 4 | 0 | −5 | 0 | 1 |
| | | 2 | 1 | −2 | −1 |
| | 4 | 2 | −4 | −2 | 0 |

The synthetic division indicates that the quotient is $4x^3 + 2x^2 - 4x - 2$ and the remainder is 0.

**29**

| | | | | |
|---|---|---|---|---|
| 3 | 2 | 3 | −4 | 4 |
| | | 6 | 27 | 69 |
| | 2 | 9 | 23 | 73 |

The synthetic division indicates that $f(3) = 73$.

**31**

| | | | | |
|---|---|---|---|---|
| −0.2 | 0.3 | 0 | 0.04 | −0.034 |
| | | −0.06 | 0.012 | −0.0104 |
| | 0.3 | −0.06 | 0.052 | −0.0444 |

The remainder is $-0.0444$ and the remainder theorem indicates that this is $f(-0.2)$.

**33**

| | | | |
|---|---|---|---|
| $2+\sqrt{3}$ | 1 | 3 | −5 |
| | | $2+\sqrt{3}$ | $13+7\sqrt{3}$ |
| | 1 | $5+\sqrt{3}$ | $8+7\sqrt{3}$ |

The remainder is $8 + 7\sqrt{3}$ and the remainder theorem indicates that this is the value of $f(2+\sqrt{3})$.

**35** $-2$ is a zero of $f(x)$ if we can show that $f(-2) = 0$.

| | | | | | |
|---|---|---|---|---|---|
| −2 | 3 | 8 | −2 | −10 | 4 |
| | | −6 | −4 | 12 | −4 |
| | 3 | 2 | −6 | 2 | 0 |

Hence, $f(-2) = 0$ and $-2$ is a zero of $f(x)$.

**37**

| | | | | |
|---|---|---|---|---|
| $\frac{1}{2}$ | 4 | −6 | 8 | −3 |
| | | 2 | −2 | 3 |
| | 4 | −4 | 6 | 0 |

Hence, $f(\frac{1}{2}) = 0$ and $\frac{1}{2}$ is a zero of $f(x)$.

**39** $f(x) = kx^3 + x^2 + k^2x + 3k^2 + 11$, so

$$\begin{aligned} f(-2) &= k(-2)^3 + (-2)^2 + k^2(-2) + 3k^2 + 11 \\ &= -8k + 4 - 2k^2 + 3k^2 + 11 \\ &= k^2 - 8k + 15 \end{aligned}$$

The remainder, $k^2 - 8k + 15$, must be zero if $f(x)$ is to be divisible by $x + 2$.

$$k^2 - 8k + 15 = 0 \Rightarrow (k-3)(k-5) = 0 \Rightarrow k = 3, 5.$$

[41] $x-c$ will not be a factor of $f(x)=3x^4+x^2+5$ for any real number $c$ if the remainder of dividing $f(x)$ by $x-c$ is not zero for any real number $c$. The remainder is $f(c)$, or $3c^4+c^2+5$. Since $3c^4+c^2+5$ is greater than or equal to 5 for any real number $c$, it is never zero and hence, $x-c$ is never a factor of $f(x)$.

[43] $f(x)=3x^{100}+5x^{85}-4x^{38}+2x^{17}-6 \Rightarrow f(-1)=3-5-4-2-6=-14.$

{ change sign on coefficients of odd powers of $x$ }

[45] If $f(x)=x^n-y^n$ and $n$ is even, then $f(-y)=(-y)^n-(y)^n=y^n-y^n=0$.

Hence, $x+y$ is a factor of $x^n-y^n$.

[47] (a) The radius of the resulting cylinder is $x$ and the height is $y$, that is, $6-x$.

$$V=\pi r^2h=\pi x^2(6-x)$$

(b) The volume of the cylinder of radius 1 and altitude 5 is $\pi(1)^2 5=5\pi$. To determine another value of $x$ which would result in the same volume, we need to solve the equation $5\pi=\pi x^2(6-x)$. $5\pi=\pi x^2(6-x) \Rightarrow 5=6x^2-x^3 \Rightarrow$ $x^3-6x^2+5=0$. We know that $x=1$ is a solution to this equation. Synthetically dividing, we obtain

| | | | | |
|---|---|---|---|---|
| 1 | 1 | −6 | 0 | 5 |
| | | 1 | −5 | −5 |
| | 1 | −5 | −5 | 0 |

The quotient is $x^2-5x-5$. Using the quadratic formula, we see that the solutions are $x=\dfrac{5\pm\sqrt{45}}{2}$. Since $x$ must be positive in the first quadrant, $\dfrac{5+\sqrt{45}}{2}\approx 5.85$ would be an allowable value of $x$. If $x=\frac{1}{2}(5+\sqrt{45})$, then $y=6-x=6-\frac{1}{2}(5+\sqrt{45})=\frac{12}{2}-\frac{5}{2}-\frac{1}{2}\sqrt{45}=\frac{1}{2}(7-\sqrt{45})$.

The point $P$ is $P(x,\,y)=\left(\frac{1}{2}(5+\sqrt{45}),\,\frac{1}{2}(7-\sqrt{45})\right)\approx(5.85,\,0.15)$.

[49] (a) $A=lw=(2x)(y)=2x(4-x^2)=8x-2x^3$

(b) This solution is similar to the solution in Exercise 47(b).

$A=6 \Rightarrow 6=8x-2x^3 \Rightarrow x^3-4x+3=0 \Rightarrow (x-1)(x^2+x-3)=0 \Rightarrow$ $x=1$ or $x^2+x-3=0$. Solving the second equation, we get $x=\dfrac{-1\pm\sqrt{13}}{2}$.

The value $\dfrac{\sqrt{13}-1}{2}$ would be an allowable value of $x$.

The base, $2x$, would then be $\sqrt{13}-1\approx 2.61$.

51 If $f(x) = x^8 - 7.9x^5 - 0.8x^4 + x^3 + 1.2x - 9.81$,

then the remainder is $f(0.21) \approx -9.55$.

$[-1, 1]$ by $[-20, 0, 2]$

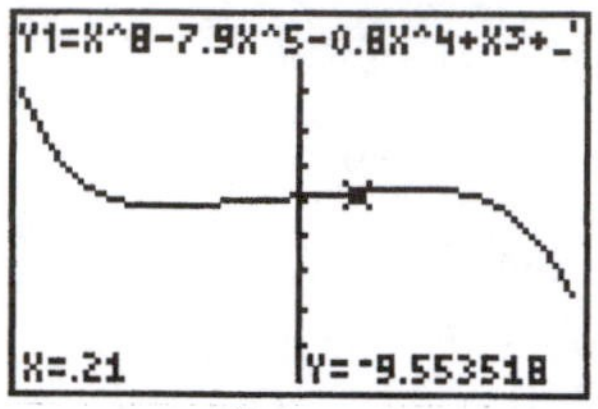


Figure 51

$[-9, 9]$ by $[-3, 9]$

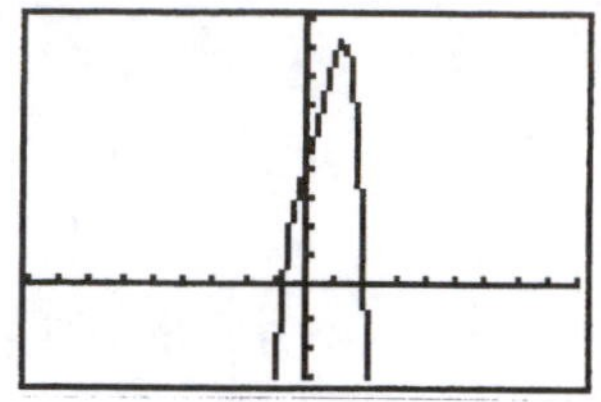

Figure 53

53 $f(1.6) = -2k^4 + 2.56k^3 + 3.2k + 4.096$. Graph $y = -2k^4 + 2.56k^3 + 3.2k + 4.096$ (that is, $y = -2x^4 + 2.56x^3 + 3.2x + 4.096$). From the graph, we see that $y = 0$ when $k \approx -0.75, 1.96$. Thus, if $k$ assumes either of these values, $f(1.6) = 0$ and $f$ will be divisible by $x - 1.6$ by the factor theorem.

## 4.3 Exercises

1 The polynomial is of the form $f(x) = a(x - (-1))(x - 2)(x - 3)$.

$f(-2) = a(-1)(-4)(-5) = 80 \Rightarrow -20a = 80 \Rightarrow a = -4$.

Hence, the polynomial is $-4(x + 1)(x - 2)(x - 3)$, or $-4x^3 + 16x^2 - 4x - 24$.

3 The polynomial is of the form $f(x) = a(x - (-4))(x - 3)(x - 0)$.

$f(2) = a(6)(-1)(2) = -36 \Rightarrow -12a = -36 \Rightarrow a = 3$.

Hence, the polynomial is $3(x + 4)(x - 3)(x)$, or $3x^3 + 3x^2 - 36x$.

5 The polynomial is of the form $f(x) = a(x - (-2i))(x - 2i)(x - 3)$.

$f(1) = a(1 + 2i)(1 - 2i)(-2) = 20 \Rightarrow -10a = 20 \Rightarrow a = -2$.

Hence, the polynomial is $-2(x + 2i)(x - 2i)(x - 3) = -2(x^2 + 4)(x - 3)$,

or $-2x^3 + 6x^2 - 8x + 24$.

7 $f(x) = a(x + 4)^2(x - 3)^2 = (x^2 + x - 12)^2 \{a = 1\} = x^4 + 2x^3 - 23x^2 - 24x + 144$

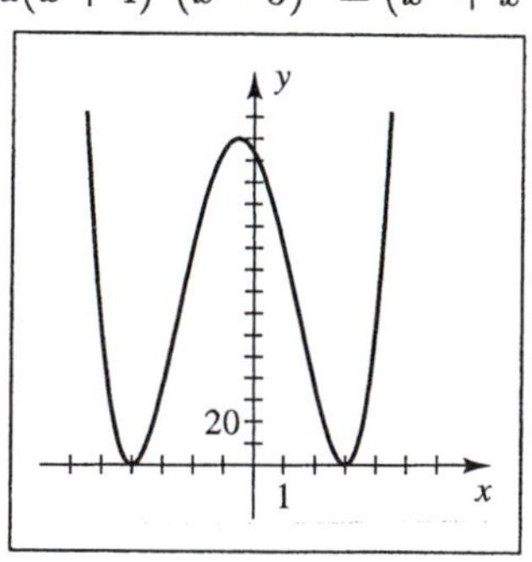


Figure 7

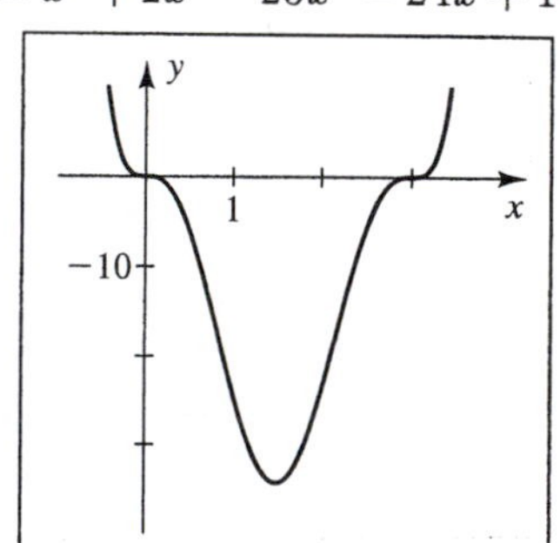


Figure 9

9 $f(x) = a(x)^3(x - 3)^3$ so $f(2) = a(8)(-1) = -8a$. But $f(2) = -24$, so $-8a = -24$, or,

$a = 3$. $f(x) = 3(x)^3(x^3 - 9x^2 + 27x - 27) = 3x^6 - 27x^5 + 81x^4 - 81x^3$.

[11] The graph has $x$-intercepts at $-1$, $\frac{3}{2}$, 3, and $f(0) = \frac{7}{2}$.

$f(x) = a(x+1)(x-\frac{3}{2})(x-3)$; $f(0) = a(1)(-\frac{3}{2})(-3) = \frac{7}{2} \Rightarrow \frac{9}{2}a = \frac{7}{2} \Rightarrow a = \frac{7}{9}$.

$$f(x) = \tfrac{7}{9}(x+1)(x-\tfrac{3}{2})(x-3)$$

[13] 3 is a zero of multiplicity one, 1 is a zero of multiplicity two, and $f(0) = 3$.

$f(x) = a(x-1)^2(x-3)$; $f(0) = a(1)(-3) = 3 \Rightarrow a = -1$. $f(x) = -1(x-1)^2(x-3)$.

[15] The values of $x$ that make $f(x) = x^2(3x+2)(2x-5)^3$ equal to zero are 0, $-\frac{2}{3}$, and $\frac{5}{2}$. The exponents associated with these zeros are 2, 1, and 3, respectively. Thus, 0 is a root of multiplicity 2, $-\frac{2}{3}$ is a root of multiplicity 1, and $\frac{5}{2}$ is a root of multiplicity 3.

[17] $f(x) = 4x^5 + 12x^4 + 9x^3 = x^3(4x^2+12x+9) = x^3(2x+3)^2$, so 0 is a root of multiplicity 3 and $-\frac{3}{2}$ is a root of multiplicity 2.

[19] $f(x) = (x^2+x-12)^3(x^2-9)^2 = [(x+4)(x-3)]^3[(x+3)(x-3)]^2$

$= (x+4)^3(x-3)^3(x+3)^2(x-3)^2 = (x+4)^3(x+3)^2(x-3)^5$

★ $-4$ (multiplicity 3); $-3$ (multiplicity 2); 3 (multiplicity 5)

[21] $f(x) = x^4 + 7x^2 - 144 = (x^2+16)(x^2-9) = (x^2+16)(x+3)(x-3)$

★ $\pm 4i$, $\pm 3$ (each of multiplicity 1)

[23] Using synthetic division,

| | | | | | |
|---|---|---|---|---|---|
| $-3$ | 1 | 7 | 13 | $-3$ | $-18$ |
| | | $-3$ | $-12$ | $-3$ | 18 |
| | 1 | 4 | 1 | $-6$ | 0 |

| | | | | |
|---|---|---|---|---|
| $-3$ | 1 | 4 | 1 | $-6$ |
| | | $-3$ | $-3$ | 6 |
| | 1 | 1 | $-2$ | 0 |

The remaining polynomial, $x^2+x-2$, can be factored as $(x+2)(x-1)$.

Hence, $f(x) = (x+3)^2(x+2)(x-1)$.

[25] *Note:* If the sum of the coefficients of the polynomial is 0, then 1 is a zero of the polynomial. Synthetically dividing 1 five times, we obtain the remaining polynomial, $x+1$. Hence, $f(x) = x^6 - 4x^5 + 5x^4 - 5x^2 + 4x - 1 = (x-1)^5(x+1)$.

*Note:* For the following exercises, let $f(x)$ denote the polynomial, P, the number of sign changes in $f(x)$, and N, the number of sign changes in $f(-x)$. The types of possible solutions are listed in the order positive, negative, nonreal complex.

[27] $f(x) = 4x^3 - 6x^2 + x - 3$. The sign pattern for coefficients is $+, -, +, -$. Since there are 3 sign changes in $f(x)$, P = 3. $f(-x) = -4x^3 - 6x^2 - x - 3$. *Note:* Simply change the sign of the coefficients of the odd-powered terms of $f(x)$ to find $f(-x)$. The sign pattern for coefficients is $-, -, -, -$. Since there are no sign changes in $f(-x)$, N = 0. If there are 3 positive solutions, then there are no negative or nonreal complex. If there is 1 positive solution, there are no negative and 2 nonreal complex solutions.

[29] $f(x) = 4x^3 + 2x^2 + 1 \Rightarrow$ P = 0. $f(-x) = -4x^3 + 2x^2 + 1 \Rightarrow$ N = 1. ★ 0, 1, 2

[31] $f(x) = 3x^4 + 2x^3 - 4x + 2 \Rightarrow$ P = 2.
$f(-x) = 3x^4 - 2x^3 + 4x + 2 \Rightarrow$ N = 2. ★ 2, 2, 0; 2, 0, 2; 0, 2, 2; 0, 0, 4

[33] $f(x) = x^5 + 4x^4 + 3x^3 - 4x + 2 \Rightarrow$ P = 2.
$f(-x) = -x^5 + 4x^4 - 3x^3 + 4x + 2 \Rightarrow$ N = 3. ★ 2, 3, 0; 2, 1, 2; 0, 3, 2; 0, 1, 4

[35] Synthetically dividing 5 into the polynomial yields a bottom row consisting of all nonnegative numbers (see below). Synthetically dividing −2 into the polynomial yields a bottom row that alternates in sign (see below).

| | | | | |
|---|---|---|---|---|
| 5 | 1 | −4 | −5 | 7 |
| | | 5 | 5 | 0 |
| | 1 | 1 | 0 | 7 |

| | | | | |
|---|---|---|---|---|
| −2 | 1 | −4 | −5 | 7 |
| | | −2 | 12 | −14 |
| | 1 | −6 | 7 | −7 |

These results indicate that the upper bound is 5 and the lower bound is −2. From the graph of $f(x) = x^3 - 4x^2 - 5x + 7$, we see that the bounds given by the theorem (5 and −2) are indeed the smallest and largest integers that are upper and lower bounds.

[37] Synthetically dividing 2 into the polynomial yields a bottom row consisting of all nonnegative numbers (see below). Synthetically dividing −2 into the polynomial yields a bottom row that alternates in sign (see below).

| | | | | | |
|---|---|---|---|---|---|
| 2 | 1 | −1 | −2 | 3 | 6 |
| | | 2 | 2 | 0 | 6 |
| | 1 | 1 | 0 | 3 | 12 |

| | | | | | |
|---|---|---|---|---|---|
| −2 | 1 | −1 | −2 | 3 | 6 |
| | | −2 | 6 | −8 | 10 |
| | 1 | −3 | 4 | −5 | 16 |

These results indicate that the upper bound is 2 and the lower bound is −2.

(continued)

From the graph of $f(x) = x^4 - x^3 - 2x^2 + 3x + 6$, we see that there are *no* real zeros. Remember, the theorem only gives us upper positive and lower negative bounds for the zeros of a polynomial—it doesn't guarantee that there are any zeros.

**39** Similar to Exercise 35 with $f(x) = 2x^5 - 13x^3 + 2x - 5$, upper bound 3, and lower bound $-3$.

| | | | | | | |
|---|---|---|---|---|---|---|
| 3 | 2 | 0 | −13 | 0 | 2 | −5 |
| | | 6 | 18 | 15 | 45 | 141 |
| | 2 | 6 | 5 | 15 | 47 | 136 |

| | | | | | | |
|---|---|---|---|---|---|---|
| −3 | 2 | 0 | −13 | 0 | 2 | −5 |
| | | −6 | 18 | −15 | 45 | −141 |
| | 2 | −6 | 5 | −15 | 47 | −146 |

**41** The zero at $x = -1$ must be of even multiplicity since the graph does not cross the $x$-axis. Since we want $f$ to have minimal degree, we will let the associated factor be $(x+1)^2$. The graph goes through the zero at $x = 1$ without "flattening out," so the associated factor is $(x-1)^1$. Because the graph flattens out at the zero at $x = 2$, we will let its associated factor be $(x-2)^3$. Thus, $f$ has the form

$$f(x) = a(x+1)^2(x-1)(x-2)^3.$$

Since the $y$-intercept is $(0, -2)$, we have $f(0) = a(1)(-1)(-8) = 8a$ and $f(0) = -2 \Rightarrow 8a = -2 \Rightarrow a = -\frac{1}{4}$. Hence, $f(x) = -\frac{1}{4}(x+1)^2(x-1)(x-2)^3$.

**43** (a) Similar to the discussion in the solution of Exercise 41, we have

$$f(x) = a(x+3)^3(x+1)(x-2)^2.$$

(b) $a = 1$ and $x = 0 \Rightarrow f(0) = 1(0+3)^3(0+1)(0-2)^2 = 1(3)^3(1)(-2)^2 = 108.$

**45** From the graph, we see that $f(x) = x^5 - 16.75x^3 + 12.75x^2 + 49.5x - 54$ has zeros of $-4$, $-2$, 1.5, and 3. There is a double root at 1.5. Since the leading coefficient of $f$ is 1, we have $f(x) = 1(x+4)(x+2)(x-1.5)^2(x-3)$.

[−5, 5] by [−150, 150, 25]

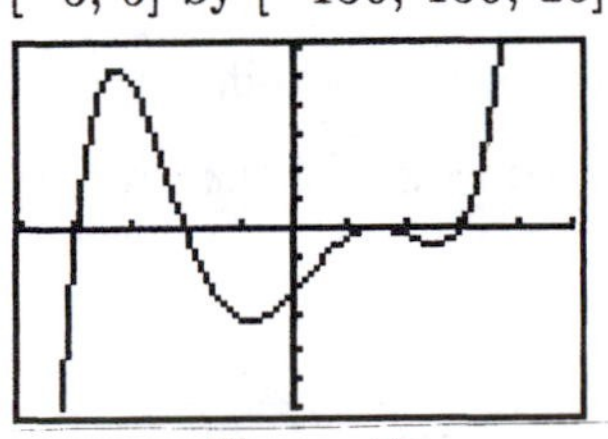

*Figure 45*

[47] Since the zeros are −2, 1, 2, and 3, the polynomial must have the form

$$f(x) = a(x+2)(x-1)(x-2)(x-3).$$

Now, $f(0) = a(2)(-1)(-2)(-3) = -12a$ and $f(0) = -24 \Rightarrow -12a = -24 \Rightarrow a = 2$.
Let $f(x) = 2(x+2)(x-1)(x-2)(x-3)$. Since $f$ has been completely determined, we must check the remaining data point(s). $f(-1) = 2(1)(-2)(-3)(-4) = -48 \neq -52$.

Thus, a fourth-degree polynomial *does not* fit the data points.

[49] Since the zeros are 2, 5.2, and 10.1, the polynomial must have the form

$$f(x) = a(x-2)(x-5.2)(x-10.1).$$

Now, $f(1.1) = a(-33.21) = -49.815 \Rightarrow a = 1.5$.
Let $f(x) = 1.5(x-2)(x-5.2)(x-10.1)$. Since $f$ has been completely determined, we must check the remaining data points. $f(3.5) = 25.245$ and $f(6.4) = -29.304$.

Thus, a third-degree polynomial *does* fit the data points.

[51] The zeros are 0, 5, 19, 24, and $f(12) = 10$. $f(t) = a(t)(t-5)(t-19)(t-24)$;
$f(12) = a(12)(7)(-7)(-12) = 10 \Rightarrow 7056a = 10 \Rightarrow a = \frac{10}{7056} = \frac{5}{3528}$.

$$f(t) = \tfrac{5}{3528}t(t-5)(t-19)(t-24)$$

[53] The graph of $f$ does not cross the $x$-axis at a zero of even multiplicity, but does cross the $x$-axis at a zero of odd multiplicity. The higher the multiplicity of a zero, the more horizontal the graph of $f$ is near that zero.

[−3, 3] by [−2, 2]

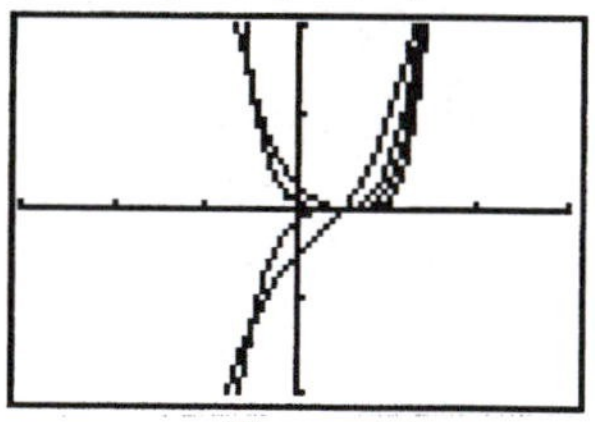

Figure 53

[−3, 3] by [−3, 1]

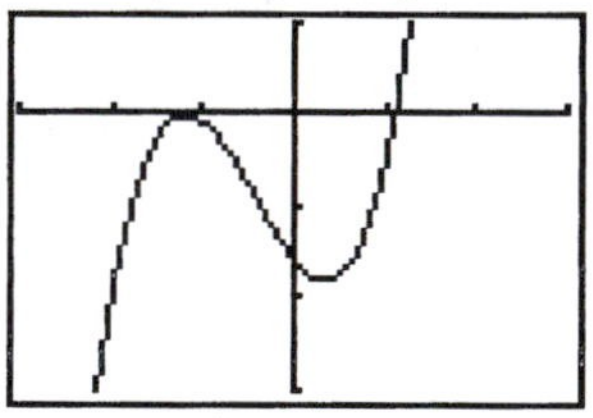

Figure 55

[55] From the graph of $f$ there are two zeros. They are −1.2 and 1.1. The zero at −1.2 has even multiplicity and the zero at 1.1 has odd multiplicity. Since $f$ has degree 3, the zero at −1.2 must have multiplicity 2 and the zero at 1.1 has multiplicity 1.

[57] From the graph of $A(t) = -\frac{1}{2400}t^3 + \frac{1}{20}t^2 + \frac{7}{6}t + 340$, we see that $A = 400$ when $t \approx 27.1$. Thus, the carbon dioxide concentration will be 400 in $1980 + 27.1 = 2007.1$, or, during the year 2007.

[0, 60, 10] by [0, 600, 100]

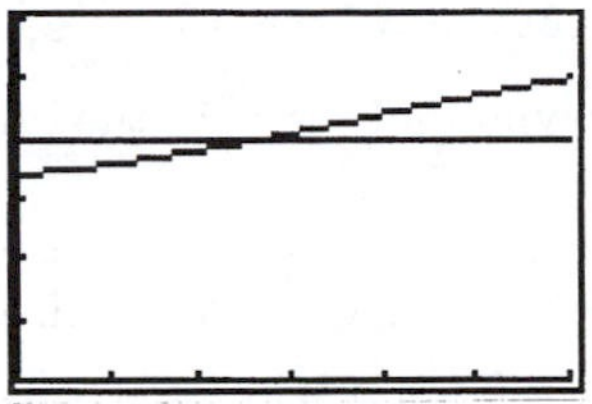

*Figure 57*

[0.5, 12.5] by [−30, 50, 10]

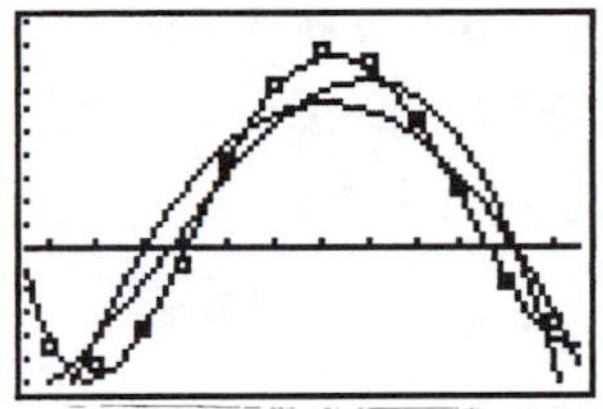

*Figure 59*

[59] (a) Graphing the data and the functions show that the best fit is $h(x)$.

(b) Since the temperature changes sign between April and May and between October and November, an average temperature of 0°F occurs when $4 \leq x \leq 5$ and $10 \leq x \leq 11$.

(c) Finding the zeros of $h$ between 1 and 12 gives us $x \approx 4.02, 10.53$.

[61] Let $r = 6$ and $k = 0.7$. Graph $Y_1 = \frac{4k}{3}\pi r^3 - \pi x^2 r + \frac{1}{3}\pi x^3 = \frac{604.8\pi}{3} - 6\pi x^2 + \frac{1}{3}\pi x^3$ and determine the positive zeros. There are two zeros located at $x \approx 7.64, 15.47$. Since the sphere floats, it will not sink deeper than twice the radius, which is 12 centimeters. Thus, the pine sphere will sink approximately 7.64 centimeters into the water.

[−20, 20, 5] by [−800, 800, 100]

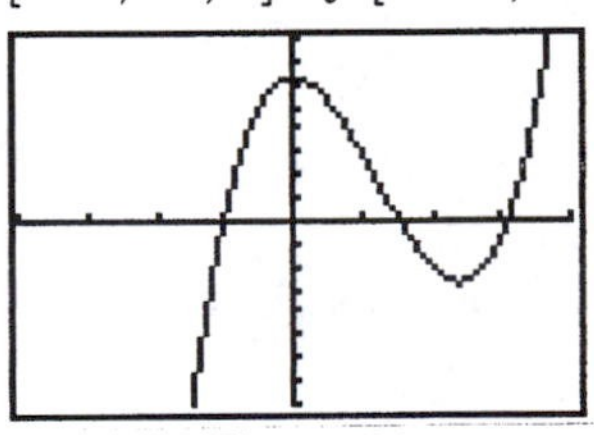

*Figure 61*

[−20, 20, 5] by [−1000, 1000, 100]

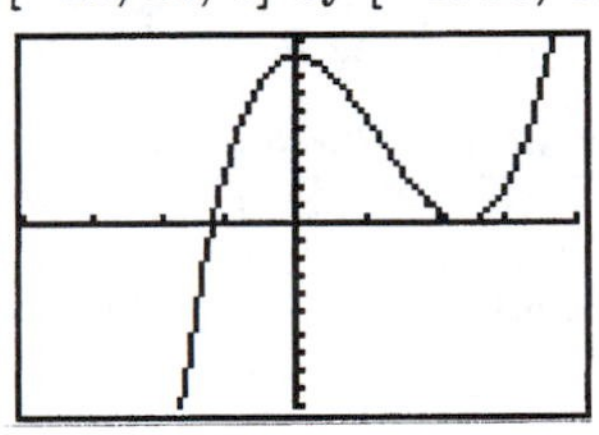

*Figure 63*

[63] Let $r = 6$ and $k = 1$. Graph $Y_1 = \frac{4k}{3}\pi r^3 - \pi x^2 r + \frac{1}{3}\pi x^3 = 288\pi - 6\pi x^2 + \frac{1}{3}\pi x^3$ and determine the positive zero. There is a zero at $x = 12$. This means that the entire sphere is just submerged. The sphere has the same density as water and neither sinks nor floats, much like a balloon filled with water.

## 4.4 Exercises

*Note:* It is helpful to remember that if $a+bi$ is a zero, then $\left[x^2-2ax+(a^2+b^2)\right]$ is the associated quadratic factor for the following exercises. (page 282 in the text)

1 Since $3+2i$ is a root, so is $3-2i$. As in the preceding note, $a=3$ and $b=-2$, and hence, $-2a=-6$ and $a^2+b^2=13$. Thus, the polynomial is of the form $\left[x-(3+2i)\right]\left[x-(3-2i)\right]=x^2-6x+13$.

3 Since 2 is a zero, $x-2$ is a factor. The associated quadratic factor for $-2\pm 5i$ is $x^2-2(-2)x+\underline{(-2)^2+(5)^2}$, or, equivalently, $x^2+4x+29$. Hence, the polynomial is of the form $(x-2)(x^2+4x+29)$.

5 Since $-1$ and 0 are zeros, $x+1$ and $x$ are factors. The associated quadratic factor for $3\pm i$ is $x^2-2(3)x+\underline{(3)^2+(1)^2}$, or, equivalently, $x^2-6x+10$. Hence, the polynomial is of the form $(x)(x+1)(x^2-6x+10)$.

7 Since $4+3i$ is a root, so is $4-3i$. The associated quadratic factor for $4\pm 3i$ is

$$x^2-2(4)x+\underline{(4)^2+(3)^2}\text{, or, equivalently, } x^2-8x+25.$$

Since $-2+i$ is a root, so is $-2-i$. The associated quadratic factor for $-2\pm i$ is

$$x^2-2(-2)x+\underline{(-2)^2+(1)^2}\text{, or, equivalently, } x^2+4x+5.$$

Hence, the polynomial is of the form $(x^2-8x+25)(x^2+4x+5)$.

9 If $-2i$ is a zero, then $2i$ is a zero. Thus, $(x-2i)(x-(-2i))=x^2-4i^2=x^2+4$ is the associated factor. Hence, the polynomial is of the form $(x)(x^2+4)(x^2-2x+2)$.

11 The constant term, 6, has positive integer divisors 1, 2, 3, 6. The leading coefficient, 1, has integer divisors $\pm 1$. By the theorem on rational zeros of a polynomial, any rational root will be of the form

$$\frac{\text{positive divisors of } 6}{\text{divisors of } 1}=\frac{1,\ 2,\ 3,\ 6}{\pm 1}\longrightarrow \pm 1,\ \pm 2,\ \pm 3,\ \pm 6.$$

Note that this quotient gives us the same possible rational roots as the one on page 284:

$$\frac{\text{factors of } a_0}{\text{factors of } a_n}=\frac{\pm 1,\ \pm 2,\ \pm 3,\ \pm 6}{\pm 1}\longrightarrow \pm 1,\ \pm 2,\ \pm 3,\ \pm 6.$$

Since $\pm 1$ are always potential solutions, they are as good as any other values to begin with. If the coefficients of the polynomial ($x^3+3x^2-4x+6$ in this exercise) sum to 0, then 1 is a zero. In this case, the sum is $1+3+(-4)+6=6\neq 0$. Thus, 1 is eliminated and we will try $-1$. If $f(x)$ is the polynomial, then $-1$ will be a zero if the coefficients of $f(-x)$ sum to 0. In this case, $f(-x)=-x^3+3x^2-10x+6$ and the sum of the coefficients is $(-1)+3+(-10)+6=-2\neq 0$. (continued)

Thus, $-1$ is not a zero. We now use synthetic division to show that $\pm 2$, $\pm 3$, and $\pm 6$ are not zeros of $f(x)$, and thus, the equation has no rational root.

[13] For $x^5 - 3x^3 + 4x^2 + x - 2 = 0$, $a_0 = -2$ and $a_n = 1$.

$$\frac{\text{factors of } a_0}{\text{factors of } a_n} = \frac{\pm 1,\ \pm 2}{\pm 1} \longrightarrow \pm 1,\ \pm 2$$

Show that $\pm 1$ and $\pm 2$ are not zeros of $f(x)$.

[15] The constant term, $-8$, has positive integer divisors 1, 2, 4, 8. The leading coefficient has integer divisors $\pm 1$. Hence, any rational root will be of the form

$$\frac{\text{positive divisors of } -8}{\text{divisors of } 1} = \frac{1,\ 2,\ 4,\ 8}{\pm 1} \longrightarrow \pm 1,\ \pm 2,\ \pm 4,\ \pm 8.$$

We now need to try to find one solution of the equation. Once we have one solution, the resulting polynomial will be a quadratic and then we can use the quadratic formula to find the other 2 solutions. As in the solution to Exercise 11, we will try $\pm 1$ first. After determining that 1 is not a solution, we try $-1$. In this case, $f(-x) = -x^3 - x^2 + 10x - 8$ and the sum of the coefficients is $(-1) + (-1) + 10 + (-8) = 0$. Thus, $-1$ is a zero and we will use synthetic division to find the remaining polynomial.

| | | | | |
|---|---|---|---|---|
| $-1$ | 1 | $-1$ | $-10$ | $-8$ |
| | | $-1$ | 2 | 8 |
| | 1 | $-2$ | $-8$ | 0 |

The remaining polynomial is $(x^2 - 2x - 8) = (x - 4)(x + 2)$.

Hence, the solutions are $-2$, $-1$, and 4.

[17] As previously noted, we first check $\pm 1$. Neither of these values are solutions so we list the possible rational roots. The values listed below are obtained by taking quotients of each number in the numerator with $\pm 1$, and then each number in the numerator with $\pm 2$, while discarding any repeat choices.

$$\frac{1,\ 2,\ 3,\ 5,\ 6,\ 10,\ 15,\ 30}{\pm 1,\ \pm 2} \longrightarrow$$

$$\pm 1,\ \pm 2,\ \pm 3,\ \pm 5,\ \pm 6,\ \pm 10,\ \pm 15,\ \pm 30,\ \pm \tfrac{1}{2},\ \pm \tfrac{3}{2},\ \pm \tfrac{5}{2},\ \pm \tfrac{15}{2}$$

Trying 2 we obtain

| | | | | |
|---|---|---|---|---|
| 2 | 2 | $-3$ | $-17$ | 30 |
| | | 4 | 2 | $-30$ |
| | 2 | 1 | $-15$ | 0 |

The remaining polynomial is $(2x^2 + x - 15) = (2x - 5)(x + 3)$.

Hence, the solutions are $-3$, 2, and $\frac{5}{2}$.

[19] $\dfrac{1,\ 2,\ 4,\ 7,\ 8,\ 14,\ 28,\ 56}{\pm 1} \longrightarrow \pm 1,\ \pm 2,\ \pm 4,\ \pm 7,\ \pm 8,\ \pm 14,\ \pm 28,\ \pm 56$. Again, $\pm 1$ are not solutions. Trying 4 and then $-7$ (using a slightly different format) we obtain

| | | | | | |
|---|---|---|---|---|---|
| 4 | 1 | 3 | −30 | −6 | 56 |
| | | 4 | 28 | −8 | −56 |
| −7 | 1 | 7 | −2 | −14 | 0 |
| | | −7 | 0 | 14 | |
| | 1 | 0 | −2 | 0 | |

The remaining polynomial is $x^2 - 2$. Its solutions are $\pm\sqrt{2}$.

Hence, the solutions are $-7$, $\pm\sqrt{2}$, and 4.

[21] We first factor out $x^2$, leaving us with the equation $x^2(6x^3 + 19x^2 + x - 6) = 0$. The $x^2$ factor indicates that the number 0 is a zero of multiplicity two. We can now concentrate on solving the equation $6x^3 + 19x^2 + x - 6 = 0$.

$$\frac{1,\ 2,\ 3,\ 6}{\pm 1,\ \pm 2,\ \pm 3,\ \pm 6} \longrightarrow \pm 1,\ \pm 2,\ \pm 3,\ \pm 6,\ \pm\tfrac{1}{2},\ \pm\tfrac{3}{2},\ \pm\tfrac{1}{3},\ \pm\tfrac{2}{3},\ \pm\tfrac{1}{6}$$

| | | | | |
|---|---|---|---|---|
| −3 | 6 | 19 | 1 | −6 |
| | | −18 | −3 | 6 |
| | 6 | 1 | −2 | 0 |

$6x^2 + x - 2 = (3x + 2)(2x - 1)$ ★ $-3,\ -\frac{2}{3},\ \frac{1}{2}$

[23] $\dfrac{1,\ 3,\ 9,\ 27}{\pm 1,\ \pm 2,\ \pm 4,\ \pm 8} \longrightarrow$

$\pm 1,\ \pm 3,\ \pm 9,\ \pm 27,\ \pm\frac{1}{2},\ \pm\frac{1}{4},\ \pm\frac{1}{8},\ \pm\frac{3}{2},\ \pm\frac{3}{4},\ \pm\frac{3}{8},\ \pm\frac{9}{2},\ \pm\frac{9}{4},\ \pm\frac{9}{8},\ \pm\frac{27}{2},\ \pm\frac{27}{4},\ \pm\frac{27}{8}$

This is a tough one because only $-\frac{3}{4}$ is a solution.

| | | | | |
|---|---|---|---|---|
| $-\frac{3}{4}$ | 8 | 18 | 45 | 27 |
| | | −6 | −9 | −27 |
| | 8 | 12 | 36 | 0 |

$8x^2 + 12x + 36 = 0$ {divide by 4} $\Rightarrow$ $2x^2 + 3x + 9 = 0$ $\Rightarrow$

$x = \dfrac{-3 \pm \sqrt{9 - 72}}{4} = -\frac{3}{4} \pm \frac{1}{4}\sqrt{63}\,i = -\frac{3}{4} \pm \frac{3}{4}\sqrt{7}\,i$ ★ $-\frac{3}{4},\ -\frac{3}{4} \pm \frac{3}{4}\sqrt{7}\,i$

[25] $f(x) = 6x^5 - 23x^4 + 24x^3 + x^2 - 12x + 4$ has zeros at $-\frac{2}{3}$, $\frac{1}{2}$, 1 (mult. 2), and 2.

Thus, $f(x) = 6(x + \frac{2}{3})(x - \frac{1}{2})(x - 1)^2(x - 2) = (3x + 2)(2x - 1)(x - 1)^2(x - 2)$.

[27] From the graph, we see that $f(x) = 2x^3 - 25.4x^2 + 3.02x + 24.75$ has zeros of approximately $-0.9$, 1.1, and 12.5. Since the leading coefficient of $f$ is 2, we have $f(x) = 2(x + 0.9)(x - 1.1)(x - 12.5)$.

[−5, 15, 5] by [−600, 100, 100]

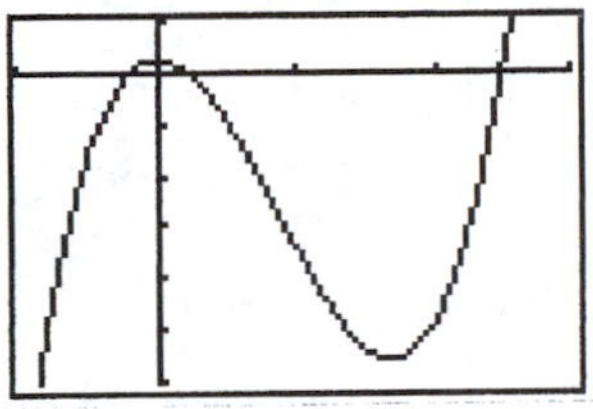

*Figure 27*

[29] No. If $i$ is a root, then $-i$ is also a root. Hence, the polynomial would have factors $x - 1$, $x + 1$, $x - i$, $x + i$ and therefore would be of degree greater than 3.

[31] Since $n$ is odd and nonreal complex zeros occur in conjugate pairs for polynomials with real coefficients, there must be at least one real zero.

[33] (a) From Exercise 41 of Section 4.1, the allowable range was $0 < x < 10$ and $V(x) = 4x(x - 10)(x - 15)$. $V = 1000 \Rightarrow 4x(x - 10)(x - 15) = 1000 \Rightarrow$ $x(x - 10)(x - 15) = 250 \Rightarrow x(x^2 - 25x + 150) - 250 = 0 \Rightarrow$ $x^3 - 25x^2 + 150x - 250 = 0$. We can show that 5 is a root, so the last equation can be written as $(x - 5)(x^2 - 20x + 50) = 0$. Using the quadratic equation gives us $(x - 5)\left[x - (10 - 5\sqrt{2})\right]\left[x - (10 + 5\sqrt{2})\right] = 0$. Discard $10 + 5\sqrt{2}$ $\{ > 10\}$.

The dimensions of the box are $x$, $20 - 2x$, and $30 - 2x$.

So the two boxes ⟦A⟧ and ⟦B⟧ having volume 1000 $\text{in}^3$ have dimensions

⟦A⟧: $5 \times \underline{20 - 2(5)} \times \underline{30 - 2(5)} = 5 \times 10 \times 20$ and

⟦B⟧: $(10 - 5\sqrt{2}) \times \underline{20 - 2(10 - 5\sqrt{2})} \times \underline{30 - 2(10 - 5\sqrt{2})} =$

$(10 - 5\sqrt{2}) \times (10\sqrt{2}) \times (10 + 10\sqrt{2})$.

(b) The surface area function is

$S(x) = (20 - 2x)(30 - 2x) + 2(x)(20 - 2x) + 2(x)(30 - 2x) = -4x^2 + 600$.

$S(5) = 500$ and $S(10 - 5\sqrt{2}) = 400\sqrt{2} \approx 565.7$, so box ⟦A⟧ has less surface area.

[35] (a) Let $x$ denote one of the sides of the triangle and $x + 1$ its hypotenuse.

Using the Pythagorean theorem, the third side $y$ is given by

$$x^2 + y^2 = (x + 1)^2 \Rightarrow y^2 = (x^2 + 2x + 1) - x^2 \Rightarrow y = \sqrt{2x + 1}.$$

Hence, the sides of the triangle are $x$, $\sqrt{2x + 1}$, and $x + 1$.

$A = \frac{1}{2}bh \Rightarrow 30 = \frac{1}{2}x\sqrt{2x + 1} \Rightarrow 60 = x\sqrt{2x + 1} \Rightarrow 60^2 = x^2(2x + 1) \Rightarrow$

$3600 = 2x^3 + x^2 \Rightarrow 2x^3 + x^2 - 3600 = 0$.

(b) There is one sign change in $f(x) = 2x^3 + x^2 - 3600$.

By Descartes' rule of signs there is one positive real root.

Synthetically dividing 13 into $f$, we obtain

| 13 | 2 | 1 | 0 | −3600 |
|---|---|---|---|---|
| | | 26 | 351 | 4563 |
| | 2 | 27 | 351 | 963 |

The numbers in the third row are nonnegative,

so 13 is an upper bound for the zeros of $f$.

(c) $2x^3 + x^2 - 3600 = 0 \Leftrightarrow (x-12)(2x^2 + 25x + 300) = 0$. The solutions of $2x^2 + 25x + 300 = 0$ are $x = -\frac{25}{4} \pm \frac{5}{4}\sqrt{71}\,i$, and hence, $x = 12$ is the only real solution. The legs of the triangle are 12 ft and 5 ft, and the hypotenuse is 13 ft.

37 (a) $\text{Volume}_{\text{total}} = \text{Volume}_{\text{cube}} + \text{Volume}_{\text{roof}}$

$$= x^3 + \tfrac{1}{2}bhx = x^3 + \tfrac{1}{2}(x)(6-x)(x) = x^3 + \tfrac{1}{2}x^2(6-x).$$

(b) $\text{Volume} = 80 \Rightarrow x^3 + \frac{1}{2}x^2(6-x) = 80 \Rightarrow x^3 + 3x^2 - \frac{1}{2}x^3 = 80 \Rightarrow$

$\frac{1}{2}x^3 + 3x^2 = 80 \Rightarrow x^3 + 6x^2 - 160 = 0 \Rightarrow (x-4)(x^2 + 10x + 40) = 0.$

The length of the side is 4 ft.

39 $x^5 + 1.1x^4 - 3.21x^3 - 2.835x^2 + 2.7x + 0.62 = -1 \Leftrightarrow$

$x^5 + 1.1x^4 - 3.21x^3 - 2.835x^2 + 2.7x + 1.62 = 0.$

The graph of $y = x^5 + 1.1x^4 - 3.21x^3 - 2.835x^2 + 2.7x + 1.62$ intersects the $x$-axis three times. The zeros at $-1.5$ and $1.2$ have even multiplicity (since the graph is tangent to the $x$-axis at these points) and the zero at $-0.5$ has odd multiplicity (since the graph crosses the $x$-axis at this point). Since the equation has degree 5, the only possibility is that the zeros at $-1.5$ and $1.2$ have multiplicity 2 and the zero at $-0.5$ has multiplicity 1. Thus, the equation has no nonreal solutions.

[−4.5, 4.5] by [−3, 3]

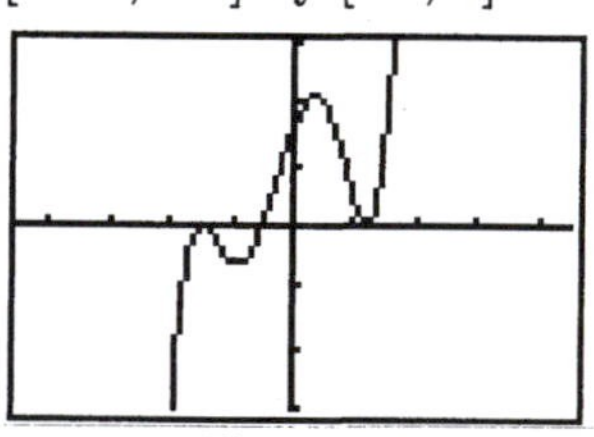

Figure 39

[−4.5, 4.5] by [−3, 3]

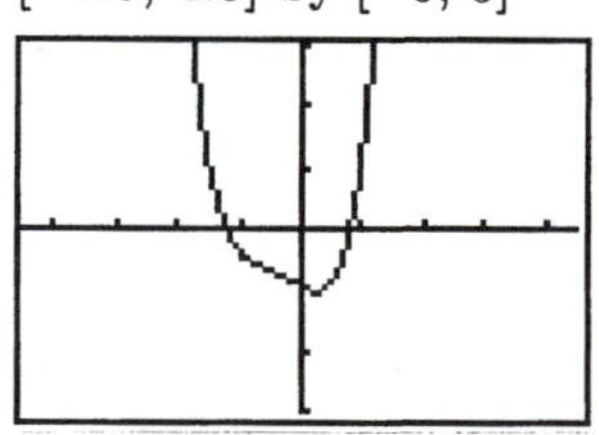

Figure 41

41 Graph $y = x^4 + 1.4x^3 + 0.44x^2 - 0.56x - 0.96$.

From the graph, zeros are located at $-1.2$ and $0.8$. Using synthetic division,

$\dfrac{x^4 + 1.4x^3 + 0.44x^2 - 0.56x - 0.96}{x + 1.2} = x^3 + 0.2x^2 + 0.2x - 0.8$ and

$\dfrac{x^3 + 0.2x^2 + 0.2x - 0.8}{x - 0.8} = x^2 + x + 1$. The zeros of $x^2 + x + 1$ are $-\frac{1}{2} \pm \frac{\sqrt{3}}{2}i$.

Thus, the solutions to the equation are $-1.2$, $0.8$, $-\frac{1}{2} \pm \frac{\sqrt{3}}{2}i$.

**43** From the graph, we see that $D(h) = 0.4$ when $h \approx 10{,}200$.

Thus, the density of the atmosphere is $0.4\ \text{kg/m}^3$ at 10,200 m.

[0, 30,000, 2000] by [0, 1.2, 0.2]

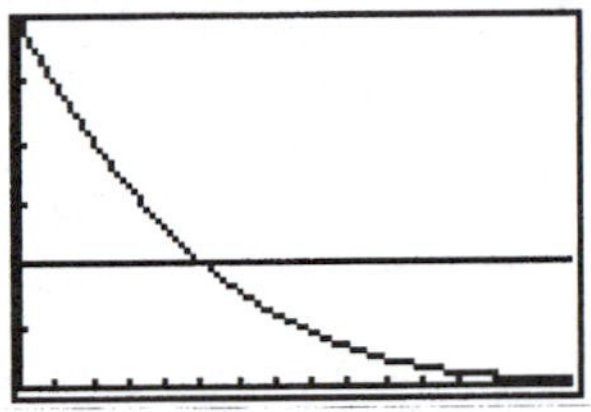

*Figure 43*

## 4.5 Exercises

*Note:* We will use the guidelines listed in the text on page 295.

Below is a summary of these guidelines.

Let $f(x) = \dfrac{a_n x^n + a_{n-1} x^{n-1} + \cdots + a_1 x + a_0}{b_k x^k + b_{k-1} x^{k-1} + \cdots + b_1 x + b_0}$, where $a_n \neq 0$ and $b_k \neq 0$.

(1) Find the $x$-intercepts {the zeros of the numerator}.

(2) Find the vertical asymptotes {the zeros of the denominator}.

(3) Find the $y$-intercept {the ratio of constant terms, $a_0/b_0$}.

(4) Find the horizontal or oblique asymptote.

(a) If $n < k$, then $y = 0$ {the $x$-axis} is the horizontal asymptote.

(b) If $n = k$,

then $y = a_n/b_k$ {the ratio of leading coefficients} is the horizontal asymptote.

(c) If $n > k$, then the asymptote is found by long division. The graph of $f$ is asymptotic to $y = q(x)$, where $q(x)$ is the quotient of the division process.

(5) Find the intersection points of the function and the asymptote found in step 3. This will help us decide *how* the function approaches the asymptote.

(6) Sketch the graph by regions, where the regions are determined by the vertical asymptotes. We may use the sign of a particular function value to help us determine if the function is positive or negative. We will not plot any points since the purpose in this section is to determine the general shape of the graph and understand the general principles involved with asymptotes.

[1] $f(x) = 4/x$ •

(a) (1) There are no $x$-intercepts since the numerator is never equal to zero.

(2) There is a vertical asymptote at $x = 0$ {the $y$-axis}.

(3) There is no $y$-intercept since the function is undefined for $x = 0$.

(4) The degree of the numerator is 0, which is less than the degree of the denominator, 1, so the horizontal asymptote is $y = 0$ {the $x$-axis}.

(5) Setting the function equal to the value of the asymptote found in step 4 gives us $4/x = 0$, which has no solutions.

(6) There is one vertical asymptote, and it separates the plane into 2 regions. For the region $x < 0$, $f(x) = 4/x < 0$—that is, the $y$-values are negative. This indicates that the graph is under the $y$-axis as in *Figure 1*. For the region $x > 0$, $f(x) > 0$, and the graph is above the $x$-axis.

(b) The domain $D$ is the set of all nonzero real numbers—that is, $\mathbb{R} - \{0\}$. The range $R$ is equal to the same set of numbers.

(c) The function is decreasing on $(-\infty, 0)$ and also on $(0, \infty)$.

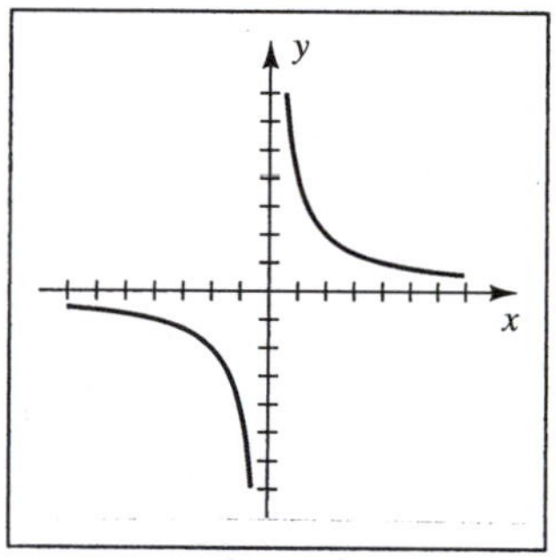


*Figure 1*

[3] $f(x) = \dfrac{-2(x+5)(x-6)}{(x-3)(x-6)} = \dfrac{-2(x+5)}{(x-3)}$ if $x \neq 6$ • Note that $x - 6$ appears in both the numerator and denominator, so there is a hole at $x = 6$. The $y$-value for the hole can be found by substituting 6 for $x$ in the rest of the function. $\dfrac{-2(6+5)}{(6-3)} = \dfrac{-2(11)}{3} = -\dfrac{22}{3}$, so the hole is at $(6, -\frac{22}{3})$. There is one other zero of the denominator, namely 3, so there is a vertical asymptote of $x = 3$. The degrees of the numerator and denominator are the same, namely 2, so the ratio of leading coefficients gives us the value of the horizontal asymptote. In this case, it is $y = \frac{-2}{1} = -2$.

[5] There is a hole at $x = -2$, so $(x+2)$ must be a factor in both the numerator and denominator. There is a vertical asymptote of $x = 1$, so $(x-1)$ must be a factor in the denominator. There is an $x$-intercept at $-3$, so $(x+3)$ must be a factor in the numerator. The horizontal asymptote is $y = 2$, so the ratio of leading coefficients must be 2. Combining this information gives us this possibility:

$$f(x) = \frac{2(x+3)(x+2)}{(x-1)(x+2)}$$

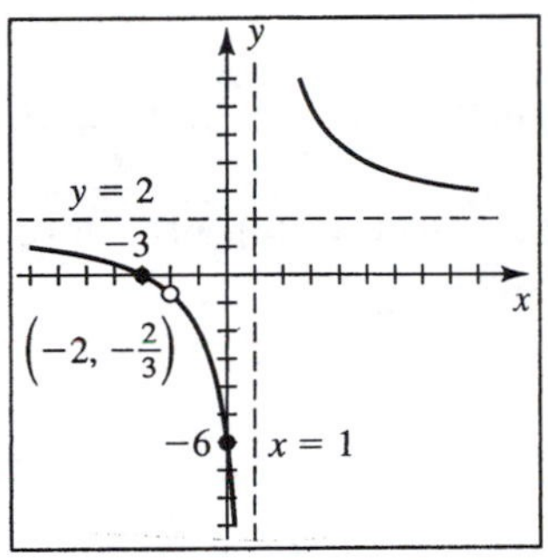

Figure 5

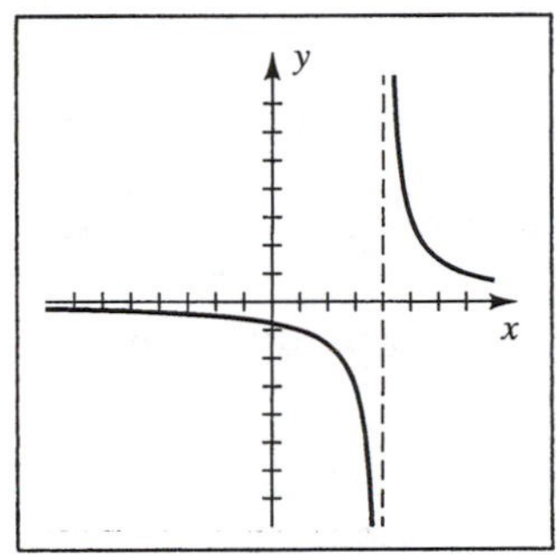

Figure 7

[7] $f(x) = \frac{3}{x-4} = 3\left(\frac{1}{x-4}\right)$ •

Rather than use the guidelines, we will think of this graph in terms of the shifting and stretching properties presented in Section 3.5. Consider the graph of $g(x) = 1/x$, which is similar to the graph in Exercise 1. Now $h(x) = 1/(x-4)$ is just $g(x-4)$, which shifts the graph of $g$ to the right by 4 units, making $x = 4$ the vertical asymptote. The effect of the 3 in the numerator is to vertically stretch the graph of $h$ by a factor of 3.

**9** $f(x) = \dfrac{-3x}{x+2}$ •

(1) Numerator $= 0 \;\Rightarrow\; -3x = 0 \;\Rightarrow\; x = 0$, the <u>only</u> $x$-intercept.

(2) Denominator $= 0 \;\Rightarrow\; x + 2 = 0 \;\Rightarrow\; x = -2$, the only vertical asymptote.

(3) $f(0) = \frac{0}{2} = 0 \;\Rightarrow\;$ 0 is the $y$-intercept. { We already knew this from step 1. }

(4) Degree of numerator $= 1 =$ degree of denominator $\Rightarrow$

$y = \frac{-3}{1}$ { ratio of leading coefficients } $= -3$ is the horizontal asymptote.

(5) Function = horizontal asymptote value $\Rightarrow\; f(x) = -3 \;\Rightarrow\; \dfrac{-3x}{x+2} = -3 \;\Rightarrow$

$-3x = -3(x+2) \;\Rightarrow\; -3x = -3x - 6 \;\Rightarrow\; 0 = -6$. This is a contradiction and indicates that there are <u>no</u> intersection points on the horizontal asymptote. Remember, the function can not intersect a vertical asymptote, but can intersect the horizontal asymptote.

(6) For the region $x < -2$, there are no $x$-intercepts and the graph must be below the horizontal asymptote $y = -3$. { If it was above $y = -3$, it would have to intersect the $x$-axis at some point, but from (1), we know that 0 is the *only* $x$-intercept. }

For the region $x > -2$, there is an $x$-intercept at 0 and the graph is above the horizontal asymptote. A common mistake is to confuse the $x$-axis with the horizontal asymptote. Remember, as $x \to \infty$ or $x \to -\infty$, $f(x)$ will get close to the horizontal or oblique asymptote, not the $x$-axis { unless the $x$-axis is the horizontal asymptote }.

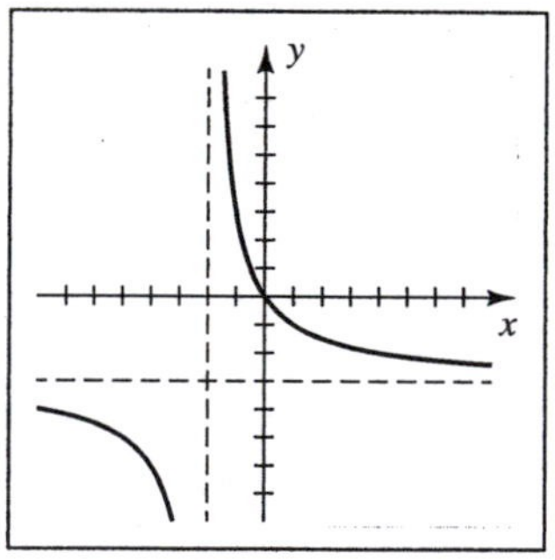


*Figure 9*

**11** $f(x) = \dfrac{4x-1}{2x+3}$ •

(1) $4x - 1 = 0 \Rightarrow 4x = 1 \Rightarrow x = \frac{1}{4}$, the only $x$-intercept.

(2) $2x + 3 = 0 \Rightarrow 2x = -3 \Rightarrow x = -\frac{3}{2}$, the only vertical asymptote.

(3) $f(0) = \frac{-1}{3} = -\frac{1}{3} \Rightarrow -\frac{1}{3}$ is the $y$-intercept.

(4) Degree of numerator $= 1 =$ degree of denominator $\Rightarrow$

$y = \frac{4}{2}$ {ratio of leading coefficients} $= 2$ is the horizontal asymptote.

(5) Function = horizontal asymptote value $\Rightarrow f(x) = 2 \Rightarrow \dfrac{4x-1}{2x+3} = 2 \Rightarrow$

$4x - 1 = 2(2x+3) \Rightarrow 4x - 1 = 4x + 6 \Rightarrow -1 = 6$. This is a contradiction and indicates that there are no intersection points on the horizontal asymptote. Remember, the function can not intersect a vertical asymptote, but can intersect the horizontal asymptote.

(6) For the region $x < -\frac{3}{2}$, there are no $x$-intercepts and the graph must be above the horizontal asymptote $y = 2$. {If it was below $y = 2$, it would have to intersect the $x$-axis at some point, but from (1), we know that $\frac{1}{4}$ is the *only* $x$-intercept.} As $x$ approaches $-\frac{3}{2}$ from the left, $y$ approaches $\infty$. As $x$ approaches $-\infty$, $y$ approaches 2 through values larger than 2.

For the region $x > -\frac{3}{2}$, there is a $y$-intercept at $-\frac{1}{3}$ and an $x$-intercept at $\frac{1}{4}$, so the graph is below the horizontal asymptote. As $x$ approaches $-\frac{3}{2}$ from the right, $y$ approaches $-\infty$. As $x$ approaches $\infty$, $y$ approaches 2 through values smaller than 2.

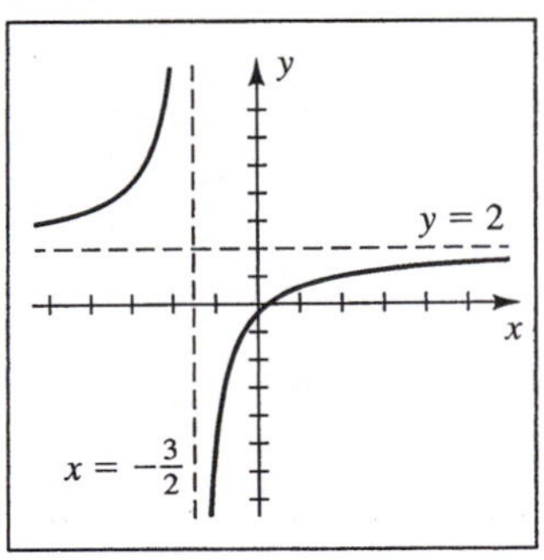

*Figure 11*

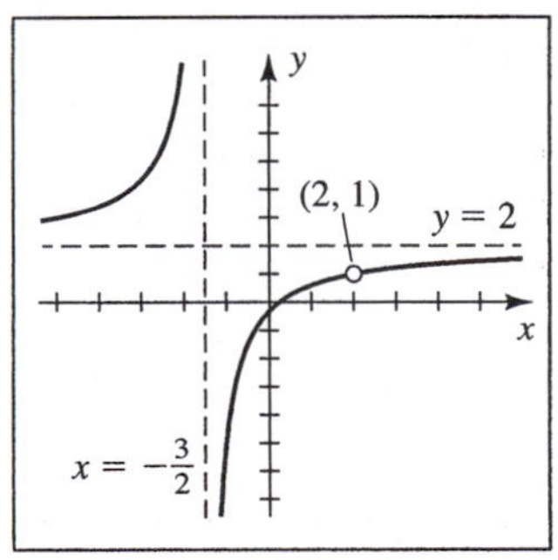

*Figure 13*

**13** $f(x) = \dfrac{(4x-1)(x-2)}{(2x+3)(x-2)} = \dfrac{4x-1}{2x+3}$ for $x \neq 2$. The reduced function is the same as the function in Exercise 11. The only difference is that this function has a hole in the graph at $x = 2$. To determine the value of $y$ when $x = 2$, substitute 2 into $\dfrac{4x-1}{2x+3}$ to get $\frac{7}{7} = 1$. There is a hole in the graph at (2, 1).

[15] $f(x) = \frac{x-2}{x^2-x-6} = \frac{x-2}{(x+2)(x-3)}$ •

(1) $x-2=0 \Rightarrow x=2$, the only $x$-intercept

(2) $(x+2)(x-3)=0 \Rightarrow x=-2$ and $x=3$, the vertical asymptotes

(3) $f(0) = \frac{-2}{-6} = \frac{1}{3} \Rightarrow \frac{1}{3}$ is the $y$-intercept

(4) Degree of numerator $= 1 < 2 =$ degree of denominator $\Rightarrow$

$y=0$ is the horizontal asymptote.

(5) We have already solved $f(x)=0$ in step 1.

Hence, we know that the function will cross the horizontal asymptote at (2, 0).

(6) For $x < -2$, we have no information and will examine $f(-5)$ { $-5$ is to the left of $-2$ }. Using the factored form of $f$ and only the signs of the values, we obtain $f(-5) = \frac{(-)}{(-)(-)} =$ { some negative number } $< 0$ since the combination of 3 negative signs will be negative. Hence, the graph will be below the horizontal asymptote.

For $-2 < x < 3$, we have an $x$-intercept at 2 and need to know what the function does on each side of 2. Choosing 0 and 2.5 for test points, we check the sign of $f(0)$ and $f(2.5)$ as above. $f(0) = \frac{(-)}{(+)(-)} > 0$ and $f(2.5) = \frac{(+)}{(+)(-)} < 0$. Hence, the function will change signs from positive to negative as it passes through 2.

For $x > 3$, $f(5) = \frac{(+)}{(+)(+)} > 0$ and the function is above the horizontal asymptote.

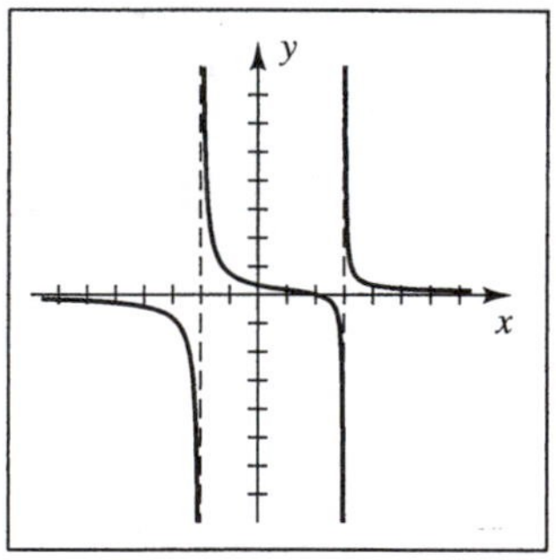


Figure 15

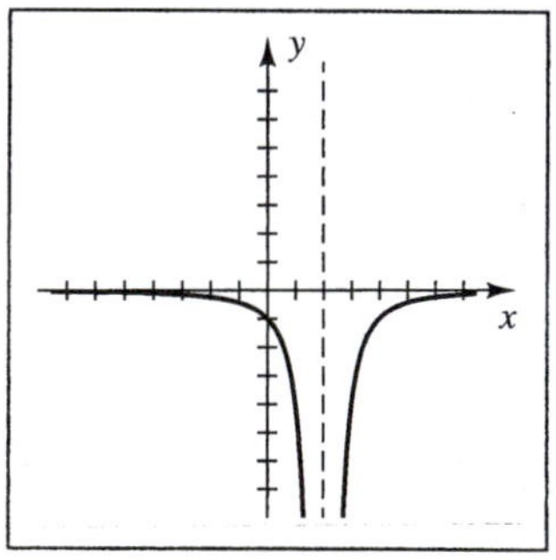


Figure 17

[17] $f(x) = \frac{-4}{(x-2)^2}$ • Note that the function is always negative since it is the quotient of a negative and a positive, provided $x \neq 2$. The $y$-intercept is $f(0) = -1$, the vertical asymptote is $x=2$, and the horizontal asymptote is $y=0$.

[19] $f(x) = \frac{x-3}{x^2-1} = \frac{x-3}{(x+1)(x-1)}$ •

(1) $x - 3 = 0 \Rightarrow x = 3$, the only $x$-intercept

(2) $(x+1)(x-1) = 0 \Rightarrow x = -1$ and $x = 1$, the vertical asymptotes

(3) $f(0) = \frac{-3}{-1} = 3 \Rightarrow 3$ is the $y$-intercept

(4) Degree of numerator $= 1 < 2 =$ degree of denominator $\Rightarrow$

$y = 0$ is the horizontal asymptote.

(5) We have already solved $f(x) = 0$ in step 1.

Hence, we know that the function will cross the horizontal asymptote at (3, 0).

(6) For $-1 < x < 1$, we have the $y$-intercept at 3. Since the function cannot intersect the $x$-axis {the only $x$-intercept is at 3}, the function remains positive and we have $f(x) \to \infty$ as $x \to 1^-$ and $f(x) \to \infty$ as $x \to -1^+$.

For $x > 1$, we have the $x$-intercept at 3 as a point to work with. The function will change sign at 3 since 3 is a zero of odd multiplicity {1}. If the function was going to merely touch the point (3, 0) and "turn around", 3 would have to be a zero of even multiplicity. Hence, it suffices to find only one sign. We choose 2 as our test point. $f(2) = \frac{(-)}{(+)(+)} < 0$. The function values are negative for $1 < x < 3$ and positive for $x > 3$. Remember, eventually $f(x)$ will be extremely close to 0 as $x \to \infty$. Make sure your sketch reflects that fact.

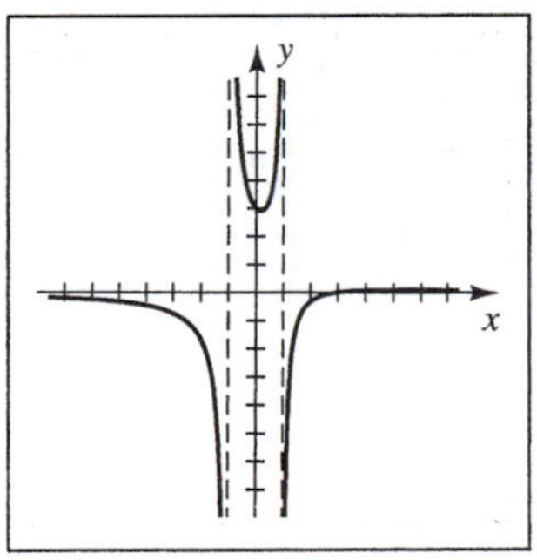


*Figure 19*

[21] $f(x) = \frac{2x^2 - 2x - 4}{x^2 + x - 12} = \frac{2(x^2 - x - 2)}{(x+4)(x-3)} = \frac{2(x+1)(x-2)}{(x+4)(x-3)}$ • As a generalization, you should always factor the function first. For steps 1–6 of the guidelines, it is convenient to use the original function for steps 3, 4, and 5, and to use the factored form for steps 1, 2, and 6. After this problem, we will concentrate on steps 5 and 6. You should be able to pick up the information for steps 1–4 by merely looking at both forms of the function { except for the oblique asymptote case }.

(1) $2(x+1)(x-2) = 0 \Rightarrow x = -1, 2$; the $x$-intercepts

(2) $(x+4)(x-3) = 0 \Rightarrow x = -4, 3$; the vertical asymptotes

(3) $f(0) = \frac{-4}{-12} = \frac{1}{3} \Rightarrow \frac{1}{3}$ is the $y$-intercept

(4) Degree of numerator $= 2 =$ degree of denominator $\Rightarrow$

$y = \frac{2}{1} = 2$ is the horizontal asymptote.

(5) $\frac{2x^2 - 2x - 4}{x^2 + x - 12} = 2 \Rightarrow 2x^2 - 2x - 4 = 2(x^2 + x - 12) \Rightarrow$

$2x^2 - 2x - 4 = 2x^2 + 2x - 24 \Rightarrow 20 = 4x \Rightarrow x = 5$. Hence, the function will intersect the horizontal asymptote only at the point (5, 2).

(6) For $x < -4$, there are no $x$-intercepts so the function is above the horizontal asymptote.

For $-4 < x < 3$, the function must go through $(-1, 0)$, $(0, \frac{1}{3})$, and $(2, 0)$. It must go "down" by both vertical asymptotes because if it went "up", it would have to cross the horizontal asymptote. But we know this only occurs at the point (5, 2) and not in this region.

For $x > 3$, we know that the graph goes up as we get close to 3 because if it went down, there would have to be an $x$-intercept, but there isn't one. The function goes through (5, 2), but then turns around { before touching the $x$-axis } and gets very close to 2 as $x$ increases.

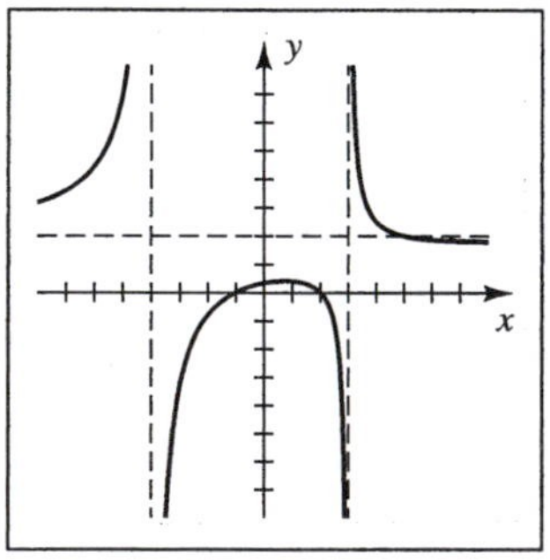


*Figure 21*

*Note:* We will let $I(x,\ y)$ denote the intersection point found in step 5.

[23] $f(x) = \dfrac{-x^2 - x + 6}{x^2 + 3x - 4} = \dfrac{-1(x^2 + x - 6)}{(x+4)(x-1)} = \dfrac{-1(x+3)(x-2)}{(x+4)(x-1)}$ •

(5) $\dfrac{-x^2 - x + 6}{x^2 + 3x - 4} = -1 \ \Rightarrow\ -x^2 - x + 6 = -x^2 - 3x + 4 \ \Rightarrow$

$2x = -2 \ \Rightarrow\ x = -1,\ I = (-1,\ -1)$

(6) For $x < -4$, the function is below $y = -1$ since there are no $x$-intercepts.

For $-4 < x < 1$, the function passes through $(-3,\ 0)$, $(-1,\ -1)$, and $(0,\ -\frac{3}{2})$. These points are known from finding information in steps 1–5.

For $x > 1$, the function passes through $(2,\ 0)$ and is always above the horizontal asymptote since it only intersects the horizontal asymptote at $(-1,\ -1)$.

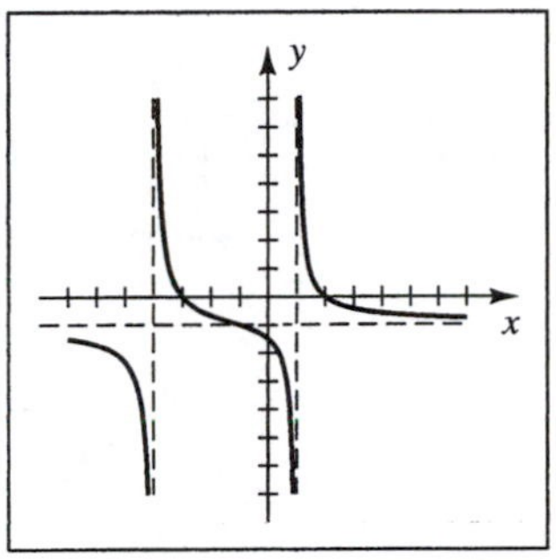

*Figure 23*

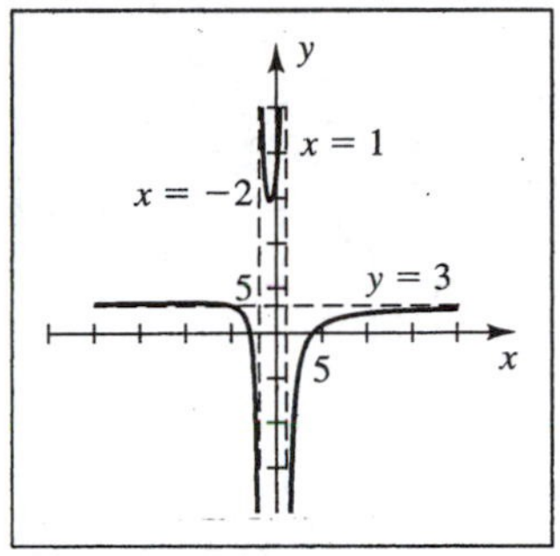

*Figure 25*

[25] $f(x) = \dfrac{3x^2 - 3x - 36}{x^2 + x - 2} = \dfrac{3(x^2 - x - 12)}{(x+2)(x-1)} = \dfrac{3(x+3)(x-4)}{(x+2)(x-1)}$ •

(5) $\dfrac{3x^2 - 3x - 36}{x^2 + x - 2} = 3 \ \Rightarrow\ 3x^2 - 3x - 36 = 3x^2 + 3x - 6 \ \Rightarrow$

$-30 = 6x \ \Rightarrow\ x = -5,\ I = (-5,\ 3)$

(6) For $x < -2$, the function passes through $(-3,\ 0)$ and $(-5,\ 3)$ and stays above the horizontal asymptote, $y = 3$, as $x \to -\infty$.

For $-2 < x < 1$, we have the $y$-intercept at 18. Since the graph doesn't intersect the horizontal asymptote in this region, the function goes up as it gets close to each vertical asymptote.

For $x > 1$, we have the $x$-intercept 4 and the function stays below the horizontal asymptote.

[27] $f(x) = \dfrac{-2x^2 + 10x - 12}{x^2 + x} = \dfrac{-2(x^2 - 5x + 6)}{(x+1)(x)} = \dfrac{-2(x-2)(x-3)}{(x+1)(x)}$ •

(2) Since $(x+1)(x) = 0 \Rightarrow x = -1$ and $0$,

we have the $y$-axis as a vertical asymptote, and hence, there is no $y$-intercept.

(5) $\dfrac{-2x^2 + 10x - 12}{x^2 + x} = -2 \Rightarrow -2x^2 + 10x - 12 = -2x^2 - 2x \Rightarrow$

$12x = 12 \Rightarrow x = 1, I = (1, -2)$

(6) For $-1 < x < 0$, $f(-\frac{1}{2}) = \dfrac{(-)(-)(-)}{(+)(-)} > 0$ and the function is above the $x$-axis.

For $x > 0$, the function goes through $(1, -2)$, $(2, 0)$, and $(3, 0)$ and gradually gets closer to $y = -3$ as $x \to \infty$.

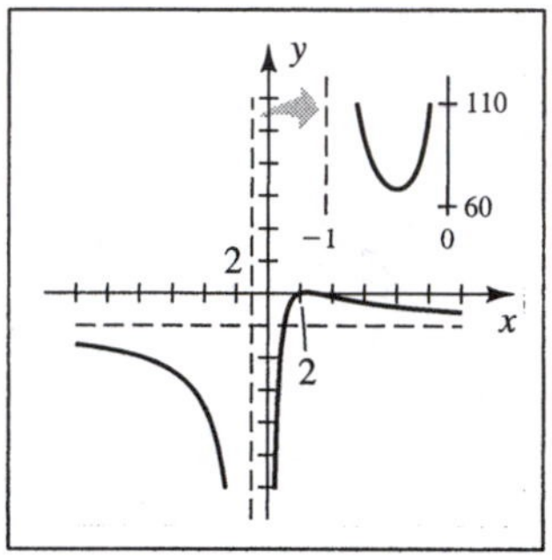


*Figure 27*

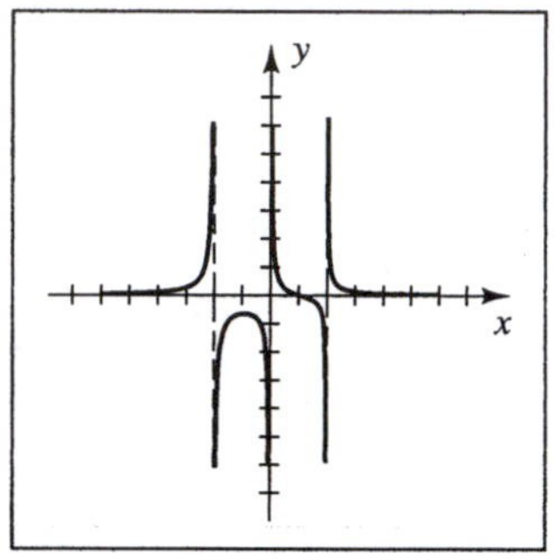


*Figure 29*

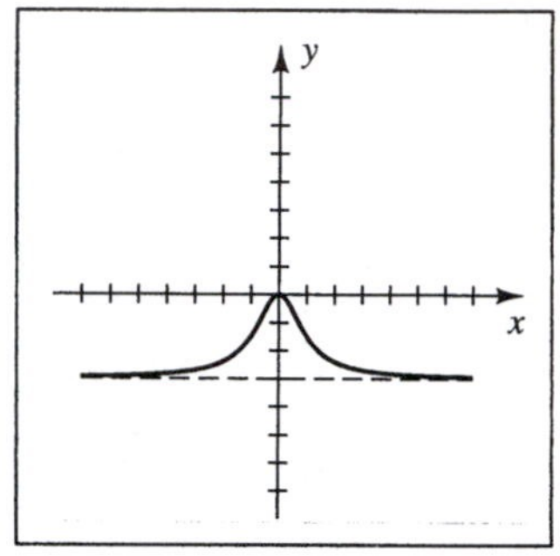


*Figure 31*

[29] $f(x) = \dfrac{x-1}{x^3 - 4x} = \dfrac{x-1}{(x)(x^2-4)} = \dfrac{x-1}{(x+2)(x)(x-2)}$ •

(5) The horizontal asymptote is $y = 0$ {the $x$-axis} and the $x$-intercept is 1, so we know that the function will cross the horizontal asymptote at $(1, 0)$.

(6) For $-2 < x < 0$, $f(-1) = \dfrac{(-)}{(+)(-)(-)} < 0$ and the function is below the $x$-axis.

For $0 < x < 1$, $f(\frac{1}{2}) = \dfrac{(-)}{(+)(+)(-)} > 0$ and the function is above the $x$-axis.

For $1 < x < 2$, $f(\frac{3}{2}) = \dfrac{(+)}{(+)(+)(-)} < 0$ and the function is below the $x$-axis.

[31] $f(x) = \dfrac{-3x^2}{x^2 + 1}$ • $f$ is an even function (See *Figure 31* above.)

(2) There are no vertical asymptotes since $x^2 + 1 \neq 0$ for any real number.

(5) $\dfrac{-3x^2}{x^2 + 1} = -3 \Rightarrow -3x^2 = -3x^2 - 3 \Rightarrow 0 = -3 \Rightarrow$

there are no intersection points of the horizontal asymptote and the function.

(6) We have the $x$-intercept at 0 and the function must get close to the horizontal asymptote, $y = -3$, as $x \to \infty$ or $x \to -\infty$.

[33] $f(x) = \dfrac{x^2 - x - 6}{x + 1} = \dfrac{(x+2)(x-3)}{x+1}$ •

(4) We could use long division to find the oblique asymptote, but since the denominator is of the from $x - c$, we will use synthetic division.

$$\begin{array}{r|rrr} -1 & 1 & -1 & -6 \\ & & -1 & 2 \\ \hline & 1 & -2 & -4 \end{array}$$

The third row indicates that $f(x) = x - 2 - \dfrac{4}{x+1}$. The expression $\dfrac{4}{x+1} \to 0$ as $x \to \pm\infty$, so $y = x - 2$ is an oblique asymptote for $f$.

(5) $\dfrac{x^2 - x - 6}{x + 1} = x - 2 \;\Rightarrow\; x^2 - x - 6 = (x-2)(x+1) \;\Rightarrow$

$x^2 - x - 6 = x^2 - x - 2 \;\Rightarrow\; -6 = -2 \;\Rightarrow$ there are no intersection points of the oblique asymptote and the function.

(6) For $x < -1$, we have the $x$-intercept at $-2$ and the function is above the oblique asymptote.

For $x > -1$, we have the points $(0, -6)$ and $(3, 0)$ and the function is below the oblique asymptote.

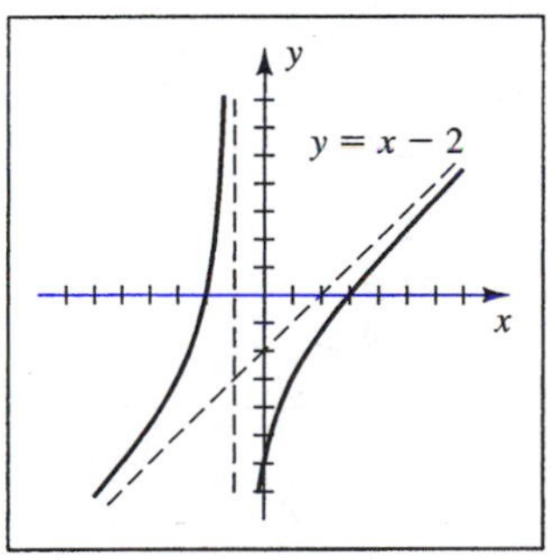


*Figure 33*

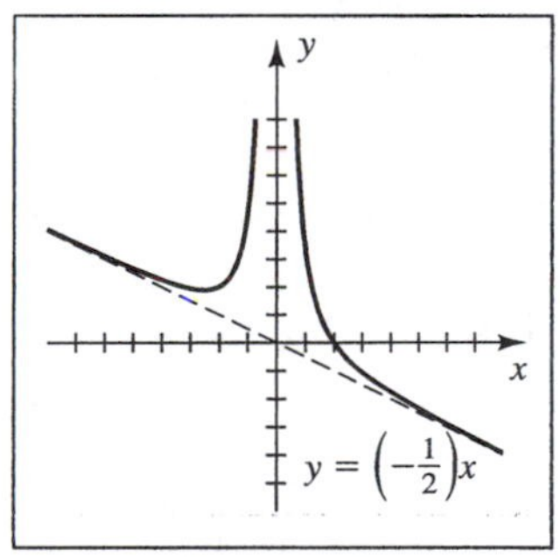


*Figure 35*

[35] $f(x) = \dfrac{8 - x^3}{2x^2} = \dfrac{(2-x)(4+2x+x^2)}{2x^2}$ •

(4) By dividing, we can write $f$ in the form $f(x) = -\frac{1}{2}x + \dfrac{8}{2x^2}$. The oblique asymptote is $y = -\frac{1}{2}x$.

(5) Function = oblique asymptote $\Rightarrow\; \dfrac{8 - x^3}{2x^2} = -\frac{1}{2}x \;\Rightarrow\; 8 - x^3 = -x^3 \;\Rightarrow$

$8 = 0 \;\Rightarrow$ there are no intersection points of the asymptote and the function.

(6) For $x < 0$, there are no $x$-intercepts, so the function is above the oblique asymptote.

For $x > 0$, we have the $x$-intercept 2, and the function is above the asymptote.

[37] $f(x) = \dfrac{2x^2 + x - 6}{x^2 + 3x + 2} = \dfrac{(x+2)(2x-3)}{(x+2)(x+1)} = \dfrac{2x-3}{x+1}$ for $x \neq -2$. We must reduce the function first and then *sketch the reduced function.* After sketching the reduced function, we need to remember to put a hole in the graph where the original function was undefined. In this case, the reduced function is $f(x) = \dfrac{2x-3}{x+1}$. We sketch this as we did the other rational functions. To determine the value of $y$ when $x = -2$, substitute $-2$ into $\dfrac{2x-3}{x+1}$ to get 7. There is a hole in the graph at $(-2, 7)$.

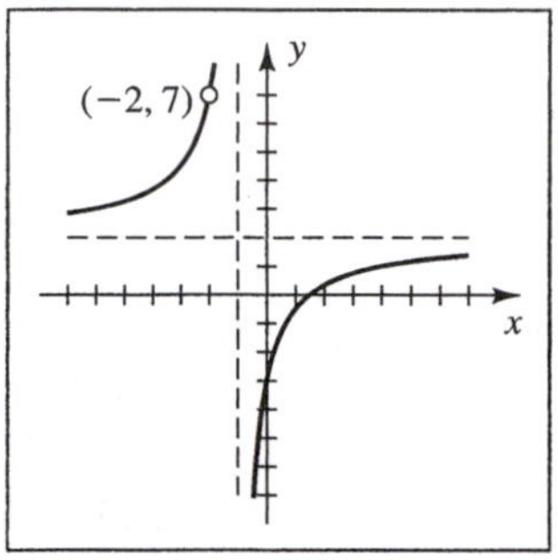

Figure 37

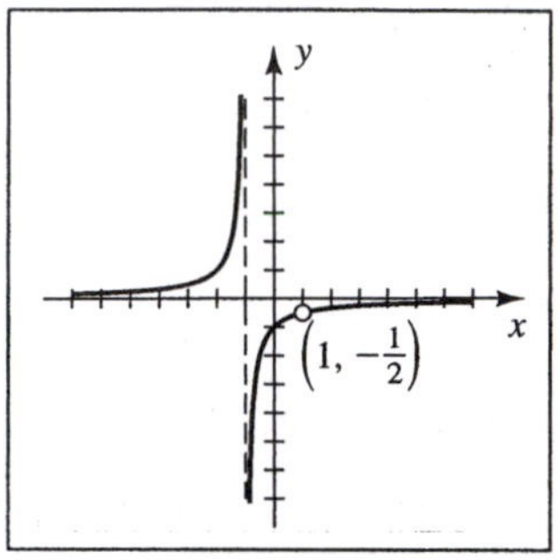

Figure 39

[39] $f(x) = \dfrac{x-1}{1-x^2} = \dfrac{x-1}{(1+x)(1-x)} = \dfrac{-1(1-x)}{(1+x)(1-x)} = \dfrac{-1}{x+1}$ for $x \neq 1$.

Substituting 1 for $x$ in $\dfrac{-1}{x+1}$ gives us a hole in the graph at $(1, -\frac{1}{2})$.

[41] $f(x) = \dfrac{x^2 + x - 2}{x+2} = \dfrac{(x+2)(x-1)}{x+2} = x - 1$ for $x \neq -2$.

Substituting $-2$ for $x$ in $x - 1$ gives us a hole in the graph {a line} at $(-2, -3)$.

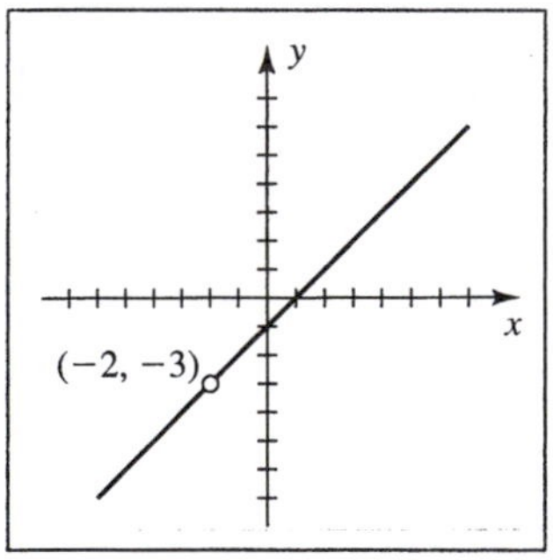

Figure 41

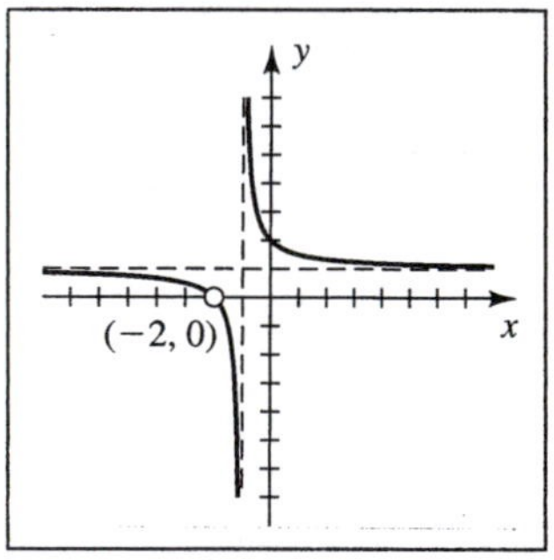

Figure 43

[43] $f(x) = \dfrac{x^2 + 4x + 4}{x^2 + 3x + 2} = \dfrac{(x+2)^2}{(x+1)(x+2)} = \dfrac{x+2}{x+1}$ for $x \neq -2$. Substituting $-2$ for $x$ in $\dfrac{x+2}{x+1}$ gives us 0, so there is a hole in the graph {on the $x$-axis} at $(-2, 0)$.

[45] The vertical asymptote of $x = 4$ implies that $x - 4$ is a factor in the denominator. Since 3 is an $x$-intercept, $x - 3$ is a factor in the numerator. The ratio of leading coefficients must be $-1$ because the horizontal asymptote is $y = -1$, and hence, we must have a factor of $-1$ in the numerator. Therefore, $f$ has the form

$$f(x) = \frac{-1(x-3)}{x-4}, \quad \text{or equivalently,} \quad f(x) = \frac{3-x}{x-4}.$$

[47] The vertical asymptotes of $x = -3$ and $x = 1$ imply that $x + 3$ and $x - 1$ are factors in the denominator. Since $-1$ is an $x$-intercept, $x + 1$ is a factor in the numerator. The hole at $x = 2$ indicates that $x - 2$ is a factor in the numerator and in the denominator. If we let $a$ denote an arbitrary coefficient, then $f$ has the form

$$f(x) = \frac{a(x+1)(x-2)}{(x-1)(x+3)(x-2)}.$$

Note that $y = 0$ is the horizontal asymptote since the degree of the numerator is less than the degree of the denominator. To determine the value of $a$, we use the fact that $f(0) = -2$.

$$f(0) = -2 \text{ and } f(0) = \frac{a(1)}{(-1)(3)} = \frac{a}{-3} \;\Rightarrow\; \frac{a}{-3} = -2 \;\Rightarrow\; a = 6.$$

Therefore, $f$ has the form $f(x) = \dfrac{6(x+1)(x-2)}{(x-1)(x+3)(x-2)}$, or equivalently, $\dfrac{6x^2 - 6x - 12}{x^3 - 7x + 6}$.

[49] (a) The radius of the outside cylinder is $(r + 0.5)$ ft and its height is $(h + 1)$ ft. Since the volume is $16\pi$ ft$^3$ and the volume is $\pi(\text{radius})^2(\text{height})$, we have

$$16\pi = \pi(r+0.5)^2(h+1) \;\Rightarrow\; h + 1 = \frac{16\pi}{\pi(r+0.5)^2} \;\Rightarrow\; h = \frac{16}{(r+0.5)^2} - 1.$$

(b) Substituting for $h$ from part (a), $V(r) = \pi r^2 h = \pi r^2\left[\dfrac{16}{(r+0.5)^2} - 1\right]$.

(c) $r$ and $h$ must both be positive. $h > 0 \;\Rightarrow$

$$\frac{16}{(r+0.5)^2} - 1 > 0 \;\Rightarrow\; \frac{16}{(r+0.5)^2} > 1 \;\Rightarrow\; 16 > (r+0.5)^2 \;\Rightarrow$$

$$(r+0.5)^2 < 16 \;\Rightarrow\; \sqrt{(r+0.5)^2} < \sqrt{16} \;\Rightarrow\; |\,r + 0.5\,| < 4 \;\Rightarrow$$

$-4 < r + 0.5 < 4 \;\Rightarrow\; -4.5 < r < 3.5$. The last inequality combined with the fact that $r > 0$ means that the excluded values are $r \le 0$ and $r \ge 3.5$.

[51] (a) Since 5 gallons of water flow into the tank each minute, $V(t) = 50 + 5t$. Since each additional gallon of water contains 0.1 lb of salt, $A(t) = 5(0.1)t = 0.5t$.

(b) $c(t) = \dfrac{A(t)}{V(t)} = \dfrac{0.5t}{50 + 5t} \cdot \dfrac{2}{2} = \dfrac{t}{10t + 100}$ lb/gal

(c) As $t \to \infty$, $c(t) \to 0.1$ lb. of salt per gal.

[53] (a) $R > S \quad \Rightarrow \quad \dfrac{4500\,S}{S+500} > S \quad \Rightarrow \quad \dfrac{4500\,S}{S+500} - S > 0 \quad \Rightarrow$

$\dfrac{4500\,S}{S+500} - \dfrac{S(S+500)}{S+500} > 0 \quad \Rightarrow \quad \dfrac{4500\,S - S^2 - 500\,S}{S+500} > 0 \quad \Rightarrow$

$\dfrac{4000\,S - S^2}{S+500} > 0 \quad \Rightarrow \quad \dfrac{S^2 - 4000\,S}{S+500} < 0$ {multiply both sides by $-1$ and reverse the direction of the inequality} $\Rightarrow \quad \dfrac{S(S-4000)}{S+500} < 0$. From *Diagram 53* and using the fact that $S > 0$, we conclude that $R > S$ when $0 < S < 4000$.

| Resulting sign: | ⊖ | ⊕ | ⊖ | ⊕ |
|---|---|---|---|---|
| $S-4000$: | $-$ | $-$ | $-$ | $+$ |
| $S$: | $-$ | $-$ | $+$ | $+$ |
| $S+500$: | $-$ | $+$ | $+$ | $+$ |
| $S$ values: | $-500$ | 0 | 4000 | |

*Diagram 53*

(b) The greatest possible number of offspring that survive to maturity is 4500, the horizontal asymptote value. 90% of 4500 is 4050.

$R = 4050 \quad \Rightarrow \quad 4050 = \dfrac{4500\,S}{S+500} \quad \Rightarrow \quad 4050S + 2{,}025{,}000 = 4500S \quad \Rightarrow$

$2{,}025{,}000 = 450S \quad \Rightarrow \quad S = 4500.$

(c) 80% of 4500 is 3600. $R = 3600 \quad \Rightarrow \quad 3600 = \dfrac{4500\,S}{S+500} \quad \Rightarrow$

$3600S + 1{,}800{,}000 = 4500S \quad \Rightarrow \quad 1{,}800{,}000 = 900S \quad \Rightarrow \quad S = 2000.$

(d) A 125% increase $\left\{\dfrac{4500-2000}{2000} \times 100\right\}$ in the number $S$ of spawners produces only a 12.5% increase $\left\{\dfrac{4050-3600}{3600} \times 100\right\}$ in the number $R$ of offspring surviving to maturity.

[55] Assign $20x^2 + 80x + 72$ to $Y_1$, $10x^2 + 40x + 41$ to $Y_2$, and $Y_1/Y_2$ to $Y_3$. Zoom-in around $(-2, -8)$ to confirm that this a low point and that there is not a vertical asymptote of $x = -2$. To determine the vertical asymptotes, graph $Y_2$ only {turn off $Y_3$}, and look for its zeros. If these values are not zeros of the numerator, then they are the values of the vertical asymptotes. No vertical asymptotes in this case.

[−9, 3] by [−9, 3]

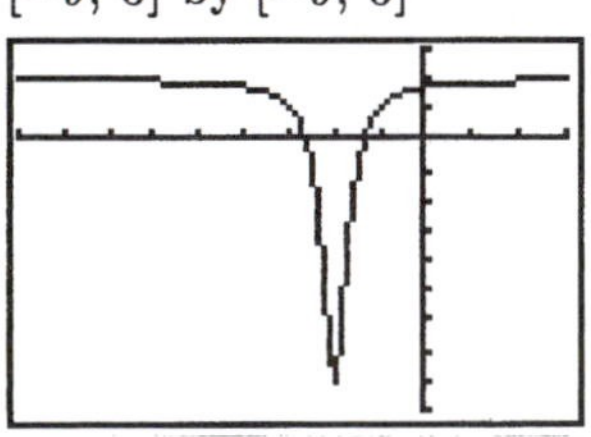

*Figure 55*

[0.7, 1.3, 0.1] by [0.8, 1.2, 0.1]

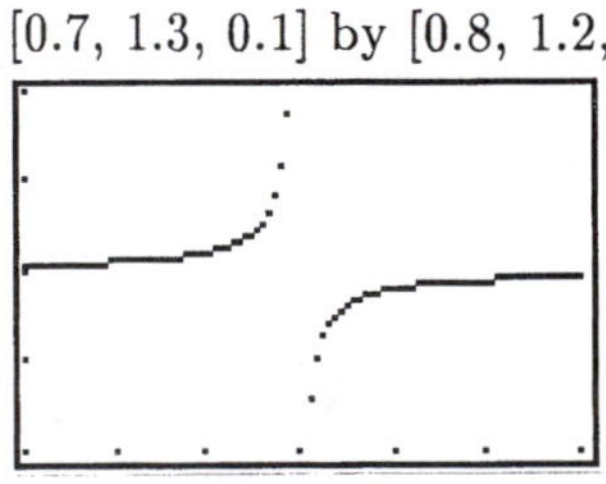

*Figure 57*

[57] $f(x) = \dfrac{(x-1)^2}{(x-0.999)^2}$ • Note that the standard viewing rectangle gives the horizontal line $y = 1$. *Figure 57* (on the preceding page) was obtained by using Dot mode. If Connected mode is used, the calculator will draw a near-vertical line at $x = 0.999$. An equation of the vertical asymptote is $x = 0.999$.

[59] (a) The graph of $g$ is the horizontal line $y = 1$ with holes at $x = 0, \pm 1, \pm 2, \pm 3$.

The TI-85/86 shows a small break in the line to indicate a hole.

(b) The graph of $h$ is the graph of $p$ with holes at $x = 0, \pm 1, \pm 2, \pm 3$.

[61] (a) GPA with additional credits = desired GPA $\Rightarrow$ $\dfrac{48(2.75) + y(4.0)}{48 + y} = x$ $\Rightarrow$

$132 + 4y = 48x + xy \Rightarrow 132 - 48x = xy - 4y \Rightarrow 132 - 48x = y(x-4) \Rightarrow$

$y = \dfrac{132 - 48x}{x - 4}$

(b)

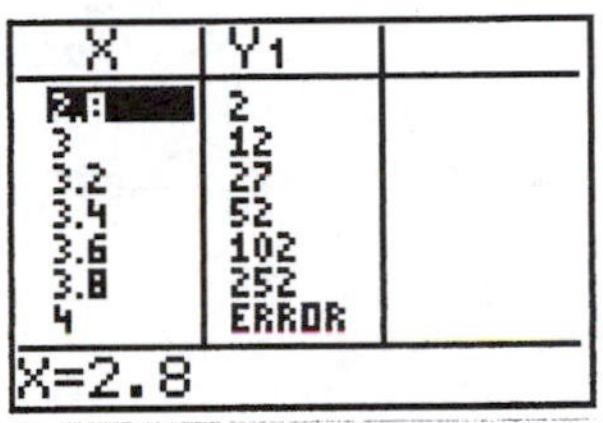

| X | Y1 |
|---|---|
| 2.8 | 2 |
| 3 | 12 |
| 3.2 | 27 |
| 3.4 | 52 |
| 3.6 | 102 |
| 3.8 | 252 |
| 4 | ERROR |

X=2.8

*Figure 61(b)*

(c) [2, 4] by [0, 1000, 100]

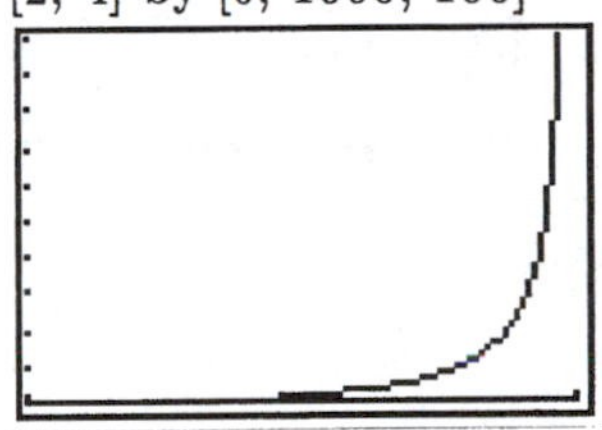

*Figure 61(c)*

(d) The vertical asymptote of the graph is $x = 4$.

(e) Regardless of the number of additional credit hours obtained at 4.0, a cumulative GPA of 4.0 is not attainable.

## *4.6 Exercises*

[1] $u$ is directly proportional to $v$ $\Rightarrow$ $u = kv$. If $v = 30$, then $u = 12$ $\Rightarrow$ $12 = k(30)$.

Solving for the constant of proportionality, $k$, we have $12 = 30k \Rightarrow k = \frac{2}{5}$.

[3] $r$ varies directly as $s$ and inversely as $t$ $\Rightarrow$ $r = k\frac{s}{t}$.

If $s = -2$ and $t = 4$, then $r = 7$ $\Rightarrow$ $7 = \dfrac{k(-2)}{4}$.

Solving for the constant of proportionality, $k$, we have $7 = \dfrac{k(-2)}{4} \Rightarrow k = -14$.

[5] $y$ is directly proportional to the square of $x$ and inversely proportional to the cube of $z$ $\Rightarrow$ $y = k\frac{x^2}{z^3}$. Substituting $x = 5$, $z = 3$, and $y = 25$ gives us $25 = \dfrac{k(25)}{27}$.

Solving for the constant of proportionality, $k$, we have $k = 27$.

[7] $z$ is directly proportional to the product of the square of $x$ and the cube of $y$ $\Rightarrow$ $z = kx^2y^3$. Substituting $x = 7$, $y = -2$, and $z = 16$ gives us $16 = k(49)(-8)$.

Solving for the constant of proportionality, $k$, we have $k = -\frac{2}{49}$.

[9] $y$ is directly proportional to $x$ and inversely proportional to the square of $z$ $\Rightarrow$ $y = k\frac{x}{z^2}$. Substituting $x = 4$, $z = 3$, and $y = 16$ gives us $16 = \frac{k(4)}{9}$.

Solving for the constant of proportionality, $k$, we have $k = 36$.

[11] $y$ is directly proportional to the square root of $x$ and inversely proportional to the cube of $z$ $\Rightarrow$ $y = k\frac{\sqrt{x}}{z^3}$. Substituting $x = 9$, $z = 2$, and $y = 5$ gives us $5 = \frac{k(3)}{8}$.

Solving for the constant of proportionality, $k$, we have $k = \frac{40}{3}$.

[13] (a) $P = kd$ (b) $118 = k(2) \Rightarrow k = 59$

(c) $P = 59(5) = 295\ \text{lb/ft}^2$

(d) $P = 59d$ is a simple linear relationship.

The graph is a line with slope 59 and $y$-intercept 0.

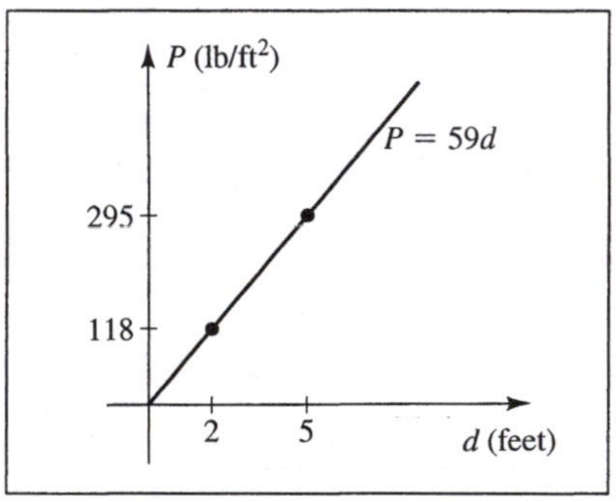


Figure 13(d)

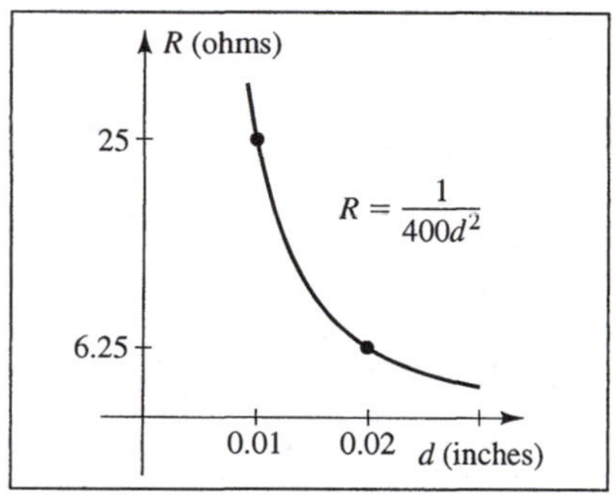


Figure 15(d)

[15] (a) $R = k\frac{l}{d^2} = \frac{kl}{d^2}$ (b) $25 = \frac{k(100)}{(0.01)^2} \Rightarrow k = \frac{1}{40{,}000}$

(c) $R = \frac{50}{(40{,}000)(0.015)^2} = \frac{50}{9}$ ohms

(d) $R = \frac{kl}{d^2} = \frac{(1/40{,}000)(100)}{d^2} = \frac{1}{400d^2}$. The graph of $R$ for $d > 0$ has a vertical asymptote of $R = 0$ and a horizontal asymptote of $d = 0$.

[17] (a) $P = k\sqrt{l}$ (b) $1.5 = k\sqrt{2} \Rightarrow k = \frac{3}{4}\sqrt{2}$

(c) $P = \frac{3}{4}\sqrt{2}(\sqrt{6}) = \frac{3}{2}\sqrt{3}$ sec

[19] (a) $T = kd^{3/2}$ (b) $365 = k(93)^{3/2} \Rightarrow k = \frac{365}{(93)^{3/2}}$

(c) $T = \frac{365}{(93)^{3/2}} \cdot (67)^{3/2} \approx 223.2$ days

[21] (a) $V = k\sqrt{L}$ (b) $35 = k\sqrt{50} \Rightarrow k = \frac{7}{2}\sqrt{2}$

(c) $V = \frac{7}{2}\sqrt{2}(\sqrt{150}) = 35\sqrt{3} \approx 60.6$ mi/hr

[23] (a) $W = kh^3$ (b) $200 = k(6)^3 \Rightarrow k = \frac{25}{27}$

(c) 5 feet 6 inches is $\frac{11}{2}$ feet. Hence, $W = \frac{25}{27}(\frac{11}{2})^3 \approx 154.1$ lb, or 154 lb.

[25] (a) $F = kPr^4 \Rightarrow P = \dfrac{F}{kr^4}$ under normal conditions.

(b) "Normal flow rates triple" means that we will use $3F$ for the flow rate $F$. "Radius increases by 10%" means that we will use $(r + 10\%r) = 1.1r$ for the radius. Hence, $3F = kP(1.1r)^4 \Rightarrow P = \dfrac{3F}{(1.1)^4 kr^4} \approx 2.05\left(\dfrac{F}{kr^4}\right)$, or about 2.05 times as hard as normal.

[27] $C = \dfrac{kDE}{Vt} \Rightarrow D = \left(\dfrac{Ct}{k}\right)\dfrac{V}{E}$, where $\dfrac{Ct}{k}$ is constant. If $V$ is twice its original value and $E$ is 0.8 of its original value (reduced by 20%), then $D$ becomes $D_1 = \left(\dfrac{Ct}{k}\right)\dfrac{2V}{0.8E}$.

Comparing $D_1$ to $D$, we have $\dfrac{D_1}{D} = \dfrac{\left(\frac{Ct}{k}\right)\frac{2V}{0.8E}}{\left(\frac{Ct}{k}\right)\frac{V}{E}} = \dfrac{2}{0.8} = 2.5 = 250\%$ of its original value. Thus, $D$ increases by 250%.

[29] The square of the distance from the origin to the point $(x, y)$ is $x^2 + y^2$. $d = \dfrac{k}{x^2 + y^2}$.
If $(x_1, y_1)$ is the new point that has density $d_1$, then

$$d_1 = \frac{k}{x_1^2 + y_1^2} = \frac{k}{(\frac{1}{3}x)^2 + (\frac{1}{3}y)^2} = \frac{k}{\frac{1}{9}x^2 + \frac{1}{9}y^2} = \frac{k}{\frac{1}{9}(x^2 + y^2)} = 9 \cdot \frac{k}{x^2 + y^2} = 9d.$$

The density $d$ is multiplied by 9.

[31] We need to examine $y/x$ for the given set of data points.

$$\frac{y}{x} = \frac{0.72}{0.6} = \frac{1.44}{1.2} = \frac{5.04}{4.2} = \frac{8.52}{7.1} = \frac{11.16}{9.3} = 1.2 \Rightarrow y = 1.2x.$$

Thus, $y$ varies directly as $x$ with constant of variation $k = 1.2$.

[33] $x^2y = -10.1$ for each data point.

Thus, $y$ varies inversely as $x^2$ with constant of variation $k = -10.1$, and $y = -\dfrac{10.1}{x^2}$.

[35] (a) $D = kS^{2.3} \Rightarrow k = \dfrac{D}{S^{2.3}}$. Using the 6 data points: $\dfrac{33}{20^{2.3}} \approx 0.0336$;

$\dfrac{86}{30^{2.3}} \approx 0.0344$; $\dfrac{167}{40^{2.3}} \approx 0.0345$; $\dfrac{278}{50^{2.3}} \approx 0.0344$; $\dfrac{414}{60^{2.3}} \approx 0.0337$; $\dfrac{593}{70^{2.3}} \approx 0.0338$.

Let $k = 0.034$. Thus, $D = 0.034S^{2.3}$.

(b) Graph the data together with $Y_1 = 0.034x^{2.3}$.

[0, 75, 10] by [0, 600, 100]

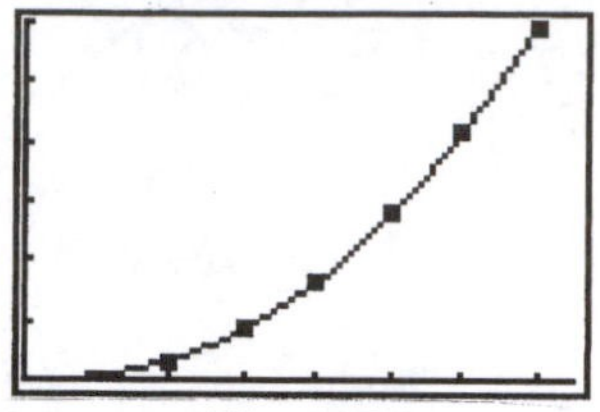

*Figure 35*

## *Chapter 4 Review Exercises*

[1] To graph $f(x) = (x+2)^3$, shift $y = x^3$ left 2 units.

$f(x) > 0$ if $x > -2$, $f(x) < 0$ if $x < -2$.

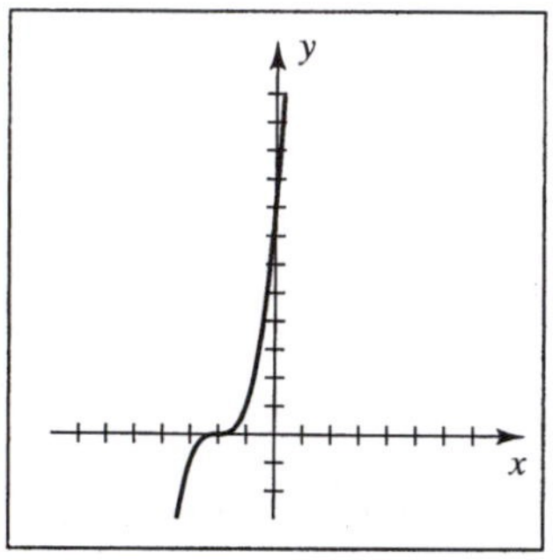


*Figure 1*

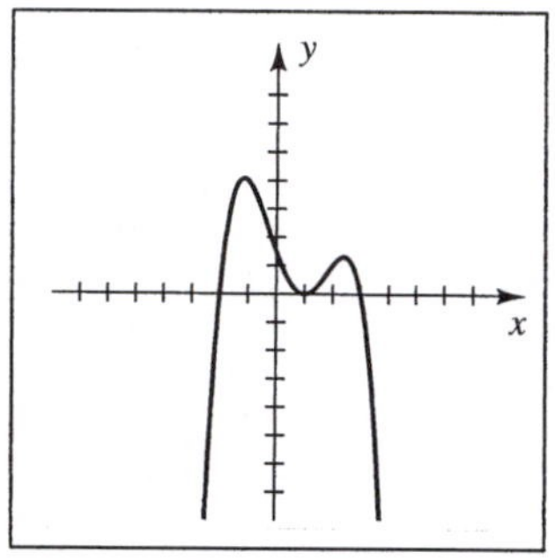


*Figure 3*

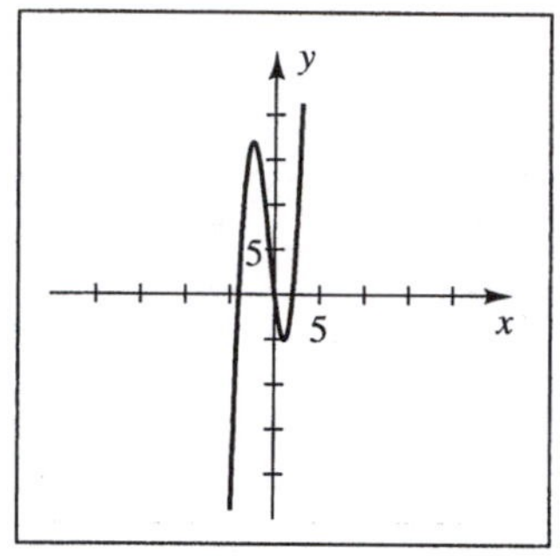


*Figure 5*

[3] $f(x) = -\frac{1}{4}(x+2)(x-1)^2(x-3)$ has zeros at $-2$, 1 (multiplicity 2), and 3.
The graph of $f$ has the general shape of $y = -x^4$.

$f(x) > 0$ if $-2 < x < 1$ or $1 < x < 3$, $f(x) < 0$ if $x < -2$ or $x > 3$.

[5] $f(x) = x^3 + 2x^2 - 8x = x(x^2 + 2x - 8) = x(x+4)(x-2)$. See *Figure 5*.
$f$ has zeros at $-4$, 0, and 2. It has the general shape of $y = x^3$.

$f(x) > 0$ if $-4 < x < 0$ or $x > 2$, $f(x) < 0$ if $x < -4$ or $0 < x < 2$.

[7] $f(0) = -9 < 100$ and $f(10) = 561 > 100$. By the intermediate value theorem for polynomial functions, $f$ takes on every value between $-9$ and 561. Hence, there is at least one real number $a$ in [0, 10] such that $f(a) = 100$.

[9] $f(x) = 3x^5 - 4x^3 + x + 5$; $p(x) = x^3 - 2x + 7$ •

$$\begin{array}{r|rrrrrr}
 & & & 3x^2 & & & +2 \\ \cline{2-7}
x^3 - 2x + 7 & 3x^5 & +0x^4 & -4x^3 & +0x^2 & +x & +5 \\
 & 3x^5 & & -6x^3 & +21x^2 & & \\ \cline{2-7}
 & & & +2x^3 & -21x^2 & & \\
 & & & +2x^3 & & -4x & +14 \\ \cline{4-7}
 & & & & -21x^2 & +5x & -9
\end{array}$$

First step: $\dfrac{3x^5}{x^3} = 3x^2$

★ $q(x) = 3x^2 + 2$; $r(x) = -21x^2 + 5x - 9$

**11** Synthetically dividing $f(x) = -4x^4 + 3x^3 - 5x^2 + 7x - 10$ by $x + 2$, we have

$$\begin{array}{r|rrrrr} -2 & -4 & 3 & -5 & 7 & -10 \\ & & 8 & -22 & 54 & -122 \\ \hline & -4 & 11 & -27 & 61 & -132 \end{array}$$

By the remainder theorem, $f(-2) = -132$.

**13** Synthetically dividing $f(x) = 6x^5 - 4x^2 + 8$ by $x + 2$, we have

$$\begin{array}{r|rrrrrr} -2 & 6 & 0 & 0 & -4 & 0 & 8 \\ & & -12 & 24 & -48 & 104 & -208 \\ \hline & 6 & -12 & 24 & -52 & 104 & -200 \end{array}$$

The synthetic division indicates that the quotient is $6x^4 - 12x^3 + 24x^2 - 52x + 104$ and the remainder is $-200$.

**15** Since $-3 + 5i$ is a zero, so is $-3 - 5i$.

$f(x) = a[x - (-3 + 5i)][x - (-3 - 5i)](x + 1) = a(x^2 + 6x + 34)(x + 1)$.

$f(1) = a(41)(2)$ and $f(1) = 4 \Rightarrow 82a = 4 \Rightarrow a = \frac{2}{41}$.

Hence, $f(x) = \frac{2}{41}(x^2 + 6x + 34)(x + 1)$.

**17** $f(x) = x^5(x + 3)^2$

{ leading coefficient is 1 }

{ 0 is a zero of multiplicity 5 }

{ $-3$ is a zero of multiplicity 2 }

$= x^5(x^2 + 6x + 9)$

$= x^7 + 6x^6 + 9x^5$

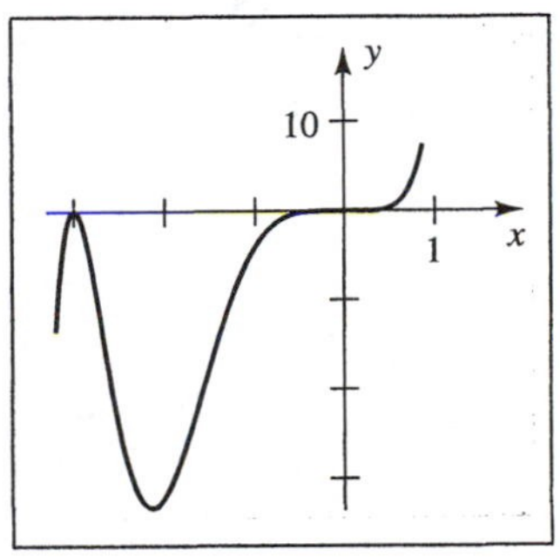


*Figure 17*

**19** $f(x) = (x^2 - 2x + 1)^2(x^2 + 2x - 3) = [(x - 1)^2]^2(x + 3)(x - 1) = (x + 3)(x - 1)^5$

★ 1 (multiplicity 5); $-3$ (multiplicity 1)

**21** (a) Let $f(x) = 2x^4 - 4x^3 + 2x^2 - 5x - 7$. Since there are 3 sign changes in $f(x)$ and 1 sign change in $f(-x) = 2x^4 + 4x^3 + 2x^2 + 5x - 7$, there are either 3 positive and 1 negative solution or 1 positive, 1 negative, and 2 nonreal complex solutions.

(b) Upper bound is 3, lower bound is $-1$

**23** Since there are only even powers, $7x^6 + 2x^4 + 3x^2 + 10 \geq 10$ for every real number $x$.

[25] $\dfrac{\text{positive divisors of } 3}{\text{divisors of } 16} = \dfrac{1,\ 3}{\pm 1,\ \pm 2,\ \pm 4,\ \pm 8,\ \pm 16} \longrightarrow$

$\pm 1,\ \pm 3,\ \pm\frac{1}{2},\ \pm\frac{3}{2},\ \pm\frac{1}{4},\ \pm\frac{3}{4},\ \pm\frac{1}{8},\ \pm\frac{3}{8},\ \pm\frac{1}{16},\ \pm\frac{3}{16}$

After a few attempts, we try $-\frac{1}{2}$.

| | | | | |
|---|---|---|---|---|
| $-\frac{1}{2}$ | 16 | $-20$ | $-8$ | 3 |
| | | $-8$ | 14 | $-3$ |
| | 16 | $-28$ | 6 | 0 |

$16x^2 - 28x + 6 = 0 \;\Rightarrow\; 8x^2 - 14x + 3 = 0 \;\Rightarrow\; (4x-1)(2x-3) = 0 \;\Rightarrow\; x = \frac{1}{4}, \frac{3}{2}.$

Thus, the solutions of $16x^3 - 20x^2 - 8x + 3 = 0$ are $x = -\frac{1}{2}, \frac{1}{4}, \frac{3}{2}$.

[27] The graph has $x$-intercepts at $-2$, 1, and 3. The equation must have the form $f(x) = a(x+2)^m(x-1)^n(x-3)^p$, where $m+n+p=6$ since $f$ is a sixth-degree polynomial. The graph goes through the $x$-axis at $m$ and $p$, so they must be odd, and it doesn't change sign at $n$, so it must be even. Since the graph flattens out at $x=-2$, $m$ must be greater than $p$. The only possibility is $m=3$, $n=2$, and $p=1$, so $f$ has the form $f(x) = a(x+2)^3(x-1)^2(x-3)$.

The $y$-intercept is 4, so

$f(0) = 4 \;\Rightarrow\; 4 = a(2)^3(-1)^2(-3) \;\Rightarrow\; -24a = 4 \;\Rightarrow\; a = -\frac{1}{6}.$

Thus, $f(x) = -\frac{1}{6}(x+2)^3(x-1)^2(x-3)$.

[29] $f(x) = \dfrac{4(x+2)(x-1)}{3(x+2)(x-5)} = \dfrac{4(x-1)}{3(x-5)}$ if $x \neq -2$ • Note that $x+2$ appears in both the numerator and denominator, so there is a hole at $x=-2$. The $y$-value for the hole can be found by substituting $-2$ for $x$ in the rest of the function. $\frac{4(-2-1)}{3(-2-5)} = \frac{4(-3)}{3(-7)} = \frac{4}{7}$, so the hole is at $(-2, \frac{4}{7})$. There is one other zero of the denominator, namely 5, so there is a vertical asymptote of $x=5$. The degrees of the numerator and denominator are the same, namely 2, so the ratio of leading coefficients gives us the value of the horizontal asymptote. In this case, it is $y = \frac{4}{3}$. There is one other zero of the numerator, namely 1, so there is an $x$-intercept of 1. If we substitute 0 for $x$, we get $\frac{4(2)(-1)}{3(2)(-5)} = \frac{4}{15}$, the $y$-intercept.

**30** $f(x) = \dfrac{-2}{(x+1)^2}$ • There is a vertical asymptote of $x = -1$. Since the numerator is negative and the denominator is positive, the functions values are negative and so $f$ is always below the $x$-axis.

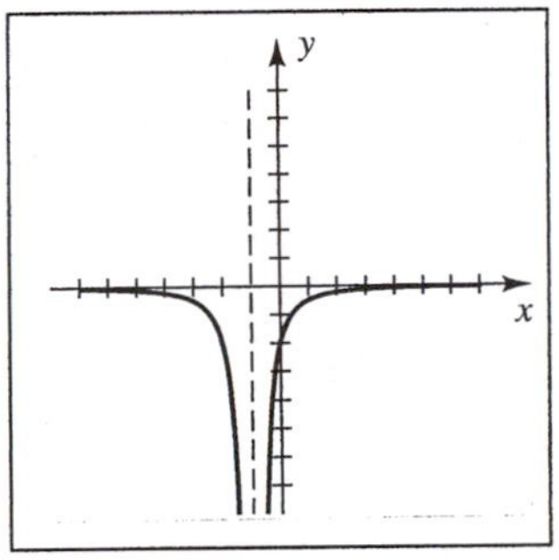


*Figure 30*

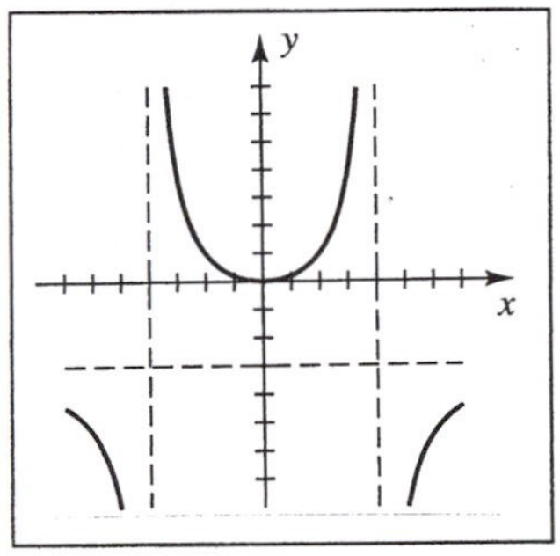


*Figure 32*

**32** $f(x) = \dfrac{3x^2}{16 - x^2} = \dfrac{3x^2}{(4+x)(4-x)}$. $f(x) = -3 \Rightarrow 3x^2 = 3x^2 - 48 \Rightarrow 0 = -48$, a contradiction $\Rightarrow$ the function does not intersect the horizontal asymptote.

**34** $f(x) = \dfrac{x^3 - 2x^2 - 8x}{-x^2 + 2x} = \dfrac{x(x^2 - 2x - 8)}{x(2-x)} = \dfrac{(x-4)(x+2)}{2-x}$ for $x \neq 0$.

Substituting $x = 0$ into $\dfrac{(x-4)(x+2)}{2-x}$ gives us $-4$. Thus, we have a hole at $(0, -4)$.

By long division, $f(x) = -x + \dfrac{-8x}{-x^2 + 2x}$, so $y = -x$ is an oblique asymptote.

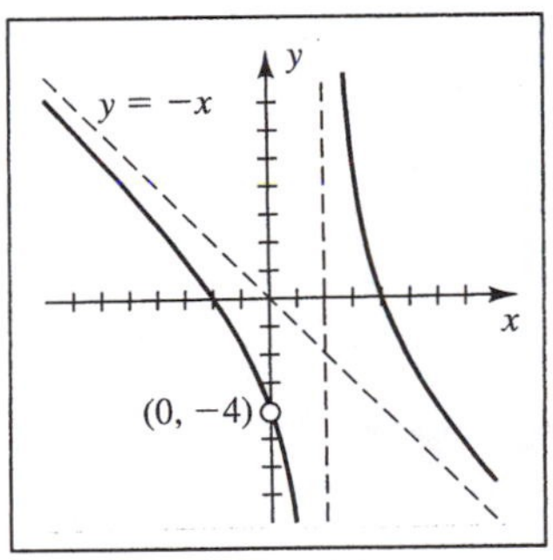


*Figure 34*

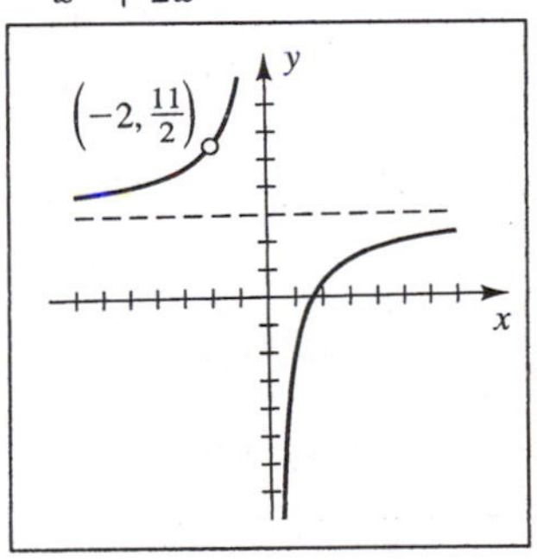


*Figure 36*

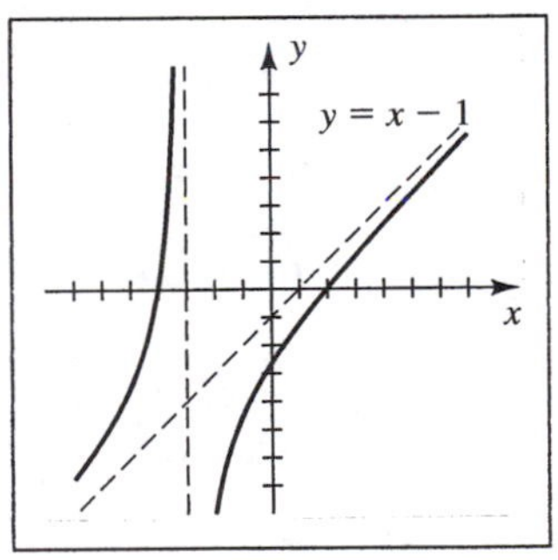


*Figure 38*

**36** $f(x) = \dfrac{3x^2 + x - 10}{x^2 + 2x} = \dfrac{(3x-5)(x+2)}{x(x+2)} = \dfrac{3x-5}{x}$ if $x \neq -2$. The horizontal asymptote is $y = 3$. The $x$-intercept is $\frac{5}{3}$ and there is no $y$-intercept since $x = 0$ is a vertical asymptote. Substituting $x = -2$ into $\dfrac{3x-5}{x}$ gives us $\dfrac{11}{2}$. Thus, we have a hole at $(-2, \frac{11}{2})$.

**38** $f(x) = \dfrac{x^2 + 2x - 8}{x+3} = \dfrac{(x+4)(x-2)}{x+3} = x - 1 - \dfrac{5}{x+3}$. The vertical asymptote is $x = -3$. The $x$-intercepts are $-4$ and 2. $y = x - 1$ is an oblique asymptote. See *Figure 38* above.

[40] There is a hole at $x = 2$, so $(x-2)$ must be a factor in both the numerator and denominator. There is a vertical asymptote of $x = -3$, so $(x+3)$ must be a factor in the denominator. There is an $x$-intercept at 5, so $(x-5)$ must be a factor in the numerator. The horizontal asymptote is $y = \frac{3}{2}$, so the ratio of leading coefficients must be $\frac{3}{2}$. Combining this information gives us this possibility:

$$f(x) = \frac{3(x-5)(x-2)}{2(x+3)(x-2)}, \quad \text{or equivalently,} \quad f(x) = \frac{3x^2 - 21x + 30}{2x^2 + 2x - 12}$$

[42] $y = \frac{k}{x^2} \Rightarrow 18 = \frac{k}{4^2} \Rightarrow k = 18 \cdot 16 = 288$, so graph $y = \frac{288}{x^2}$ for $x > 0$.

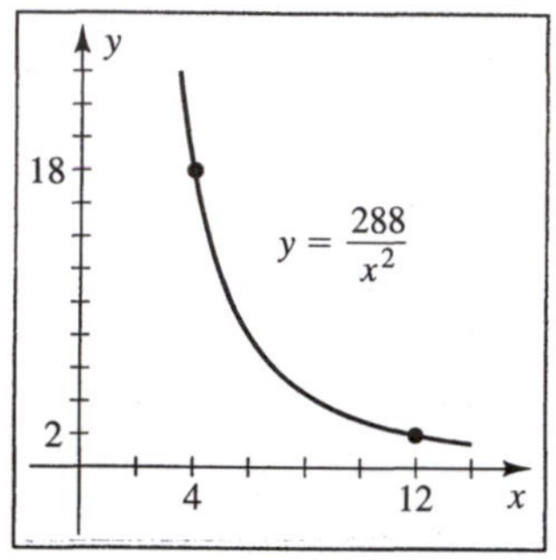


Figure 42

[43] $y = cx^2(x^2 - 4lx + 6l^2)$ •

(a) $l = 10$, $x = 10$, and $y = 2 \Rightarrow 2 = c(10)^2(10^2 - 4 \cdot 10 \cdot 10 + 6 \cdot 10^2) \Rightarrow$
$2 = c(100)(100 - 400 + 600) \Rightarrow 2 = c(100)(300) \Rightarrow$
$2 = 30{,}000c \Rightarrow c = \frac{1}{15{,}000}$

(b) $x = 6.1 \Rightarrow y \approx 0.9754 < 1$ and $x = 6.2 \Rightarrow y \approx 1.0006 > 1$.

[45] $T = \frac{1}{20}t(t-12)(t-24)$ and $T = 32 \Rightarrow \frac{1}{20}t(t-12)(t-24) = 32 \Rightarrow$
$t(t^2 - 36t + 288) = 640 \Rightarrow t^3 - 36t^2 + 288t - 640 = 0$.

Synthetically dividing $t^3 - 36t^2 + 288t - 640$ by $t - 4$ gives us

| | | | | |
|---|---|---|---|---|
| 4 | 1 | −36 | 288 | −640 |
| | | 4 | −128 | 640 |
| | 1 | −32 | 160 | 0 |

Now use the quadratic formula to solve $t^2 - 32t + 160 = 0$ for $t$.

$t = \frac{32 \pm \sqrt{32^2 - 4(1)(160)}}{2(1)} = \frac{32 \pm \sqrt{384}}{2} = \frac{32 \pm 8\sqrt{6}}{2} = 16 \pm 4\sqrt{6}$. Since $0 \le t \le 24$,

$t = 4$ (10:00 A.M.) and $t = 16 - 4\sqrt{6} \approx 6.2020$ (12:12 P.M.) are the times when the temperature was 32°F.

47 (a) $S$ is the value that is changing. As $S$ gets large, the value of $R = \frac{kS^n}{S^n + a^n}$ will approach the ratio of leading coefficients, $k/1$. Hence, an equation of the horizontal asymptote is $R = k$.

(b) $k$ is the maximum rate at which the liver can remove alcohol from the bloodstream.

48 $C(x) = \frac{20x}{101 - x}$ •

(a) $C(100) = \$2{,}000{,}000.00$ and $C(90) \approx \$163{,}636.36$

(b) Notice the great difference in cost as $x$ approaches 100. $C(50)$ is only \$19,607.84 and $C(80)$ is \$76,190.48.

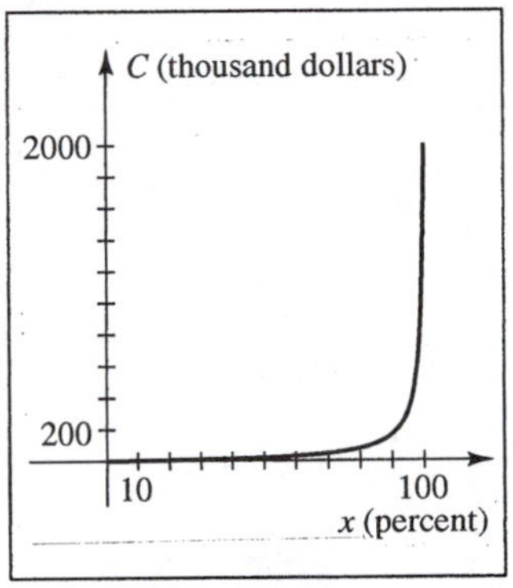

Figure 48

50 $P = kA^2v^3$; $3000 = k[\pi(5)^2]^2(20)^3 \Rightarrow k = \frac{3}{5000\pi^2}$;

$$P = \frac{3}{5000\pi^2}[\pi(5)^2]^2(30)^3 = 10{,}125 \text{ watts}$$

## *Chapter 4 Discussion Exercises*

1 **For even-degreed polynomials:** the domain is $\mathbb{R}$ and the number of $x$-intercepts ranges from *zero* to the degree of the polynomial; if the leading coefficient is positive, the range is of the form $[c, \infty)$ and the general shape has $y \to \infty$ as $|x| \to \infty$; if the leading coefficient is negative, the range is of the form $(-\infty, c]$ and the general shape has $y \to -\infty$ as $|x| \to \infty$.

**For odd-degreed polynomials:** the domain is $\mathbb{R}$, the range is $\mathbb{R}$, and the number of $x$-intercepts ranges from *one* to the degree of the polynomial; if the leading coefficient is positive, then $y \to \infty$ as $x \to \infty$ and $y \to -\infty$ as $x \to \infty$, if the leading coefficient is negative, then $y \to -\infty$ as $x \to \infty$ and $y \to \infty$ as $x \to -\infty$.

3 By long division, we obtain the quotient $2x^2 - 7x + 5$ with remainder $-6$. By synthetic division with $k = -3/2$, we obtain a bottom row of 4 $-14$ 10 $-6$. The first three numbers are twice the coefficients of the quotient and the last number is the remainder. For the factor $ax + b$, we can use synthetic division with $k = -b/a$, and obtain $a$ times the quotient and the remainder in the bottom row.

5 From your previous experience, you know that 2 points specify a first-degree polynomial (a line) and that 3 points specify a second-degree polynomial (a parabola). After working Discussion Exercise 4, you should guess that 4 points specify a third-degree polynomial, so a logical conclusion is that $n+1$ points specify an $n$-degree polynomial.

7 If the common factor is never equal to zero for any real number, then it can be canceled and has no effect on its graph. Such a factor is $x^2+1$, and an example of a function is $f(x) = \dfrac{(x^2+1)(x-1)}{(x^2+1)(x-2)}$.

9 $y = \dfrac{9x}{\sqrt{x^2+1}} \Rightarrow y\sqrt{x^2+1} = 9x \Rightarrow y^2(x^2+1) = 81x^2 \Rightarrow$

$y^2x^2 + y^2 = 81x^2 \Rightarrow y^2x^2 - 81x^2 = -y^2 \Rightarrow x^2(y^2-81) = -y^2 \Rightarrow$

$x^2 = \dfrac{y^2}{81-y^2} \Rightarrow x = \pm\sqrt{\dfrac{y^2}{81-y^2}} \Rightarrow x = \pm\dfrac{|y|}{\sqrt{81-y^2}} \Rightarrow x = \pm\dfrac{y}{\sqrt{81-y^2}}$.

From the original equation, we see that $x$ and $y$ are always both positive or both negative, so the last equation can be simplified to $x = y/\sqrt{81-y^2}$. Hence, $f^{-1}(y) = y/\sqrt{81-y^2}$ or, equivalently, $f^{-1}(x) = x/\sqrt{81-x^2}$. The denominator of $f^{-1}$ is zero for $x = \pm 9$, which are the vertical asymptotes. They are related to the horizontal asymptotes of $f$, which are $y = \pm 9$.

11 (a) $B = \dfrac{GW}{29.3 + 53.1E - 22.7C}$ •

$G = 3 \cdot 500 = 1500$, $W = 5$, $E = -0.05$, and $C = 0.95$, so

$B = \dfrac{1500(5)}{29.3 + 53.1(-0.05) - 22.7(0.95)} \approx 1476$. This tells us that a bankroll of about \$1476 would allow us to play 1500 games and be 95% confident that we will survive the 3-hour gambling session.

(b) If we use the same values for $G$, $W$, and $E$, we get $B = \dfrac{7500}{26.645 - 22.7C}$. A graph of $B$ has a vertical asymptote at $C \approx 1.17$, but that is 117% confident of surviving the 3-hour gambling session. The highest confidence value for any model like this should be 100, so the formula should have a restricted domain with it.

# Chapter 5: Inverse, Exponential, and Logarithmic Functions

## *5.1 Exercises*

[1] (a) $f$ is one-to-one and $f(4) = 5$, so $f^{-1}(5) = 4$.

(b) $g$ is *not* one-to-one since $g(1) = g(5) = 6$, so we cannot find $g^{-1}(6)$. Not possible

*Note:* To help determine if you should try to prove the function is one-to-one or look for a counterexample to show that it is not one-to-one, consider the question: "If $y$ was a particular value, could I find a unique $x$?" If the answer is yes, try to prove the function is one-to-one. Also, consider the Horizontal Line Test listed on page 321 in the text.

[3] (a) The graph is the graph of a function since it passes the Vertical Line Test. Since it passes the Horizontal Line Test, it is also one-to-one.

(b) The graph is the graph of a function, but since it doesn't pass the Horizontal Line Test, it is not one-to-one.

(c) The graph is not the graph of a function since it doesn't pass the Vertical Line Test.

[5] If $y$ was a particular value, say 5, we would have $5 = 3x - 7$. Trying to solve for $x$ would yield $12 = 3x \Rightarrow x = 4$. Since we could find a unique $x$, we will try to prove that the function is one-to-one.

**Proof** Suppose that $f(a) = f(b)$ for some numbers $a$ and $b$ in the domain. This gives us $3a - 7 = 3b - 7 \Rightarrow 3a = 3b \Rightarrow a = b$. Since $f(a) = f(b)$ implies that $a = b$, we conclude that $f$ is one-to-one.

[7] If $y$ was a particular value, say 7, we would have $7 = x^2 - 9$. Trying to solve for $x$ would yield $16 = x^2 \Rightarrow x = \pm 4$. Since we could not find a *unique* $x$, we will show that two different numbers have the same function value. Using the information already obtained, $f(4) = 7$ and $f(-4) = 7$, but $4 \neq -4$ and hence, $f$ is *not* one-to-one.

[9] Suppose $f(a) = f(b)$ with $a, b \geq 0$. $\sqrt{a} = \sqrt{b} \Rightarrow (\sqrt{a})^2 = (\sqrt{b})^2 \Rightarrow a = b$.

$f$ is one-to-one.

[11] For $f(x) = |\,x\,|$, $f(-1) = 1 = f(1)$. $f$ is *not* one-to-one.

[13] For $f(x) = \sqrt{4 - x^2}$, $f(-1) = \sqrt{3} = f(1)$. $f$ is *not* one-to-one.

[15] Suppose $f(a) = f(b)$. $\frac{1}{a} = \frac{1}{b} \Rightarrow a = b$. $f$ is one-to-one.

*Note:* For Exercises 17–20, we need to show that $f(g(x)) = x = g(f(x))$.

[17] $f(g(x)) = 3\left(\frac{x+2}{3}\right) - 2 = x + 2 - 2 = x$. $g(f(x)) = \frac{(3x-2)+2}{3} = \frac{3x}{3} = x$.

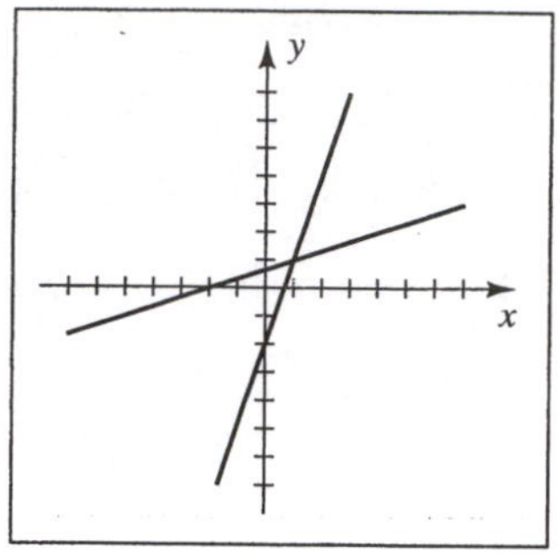


*Figure 17*

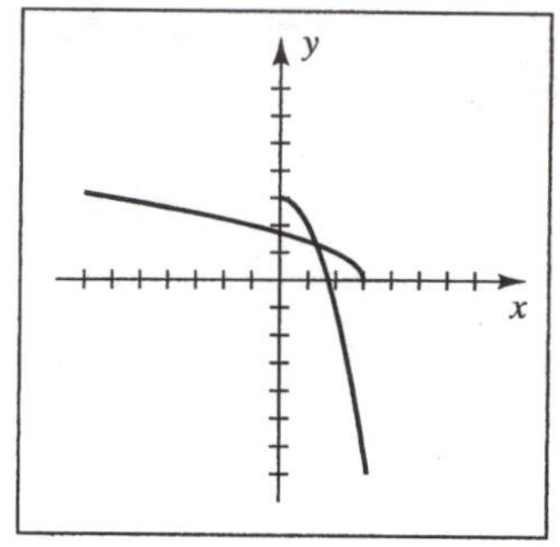


*Figure 19*

[19] $f(g(x)) = -(\sqrt{3-x})^2 + 3 = -(3-x) + 3 = x$.

$g(f(x)) = \sqrt{3 - (-x^2 + 3)} = \sqrt{x^2} = |\,x\,| = x$ {since $x \geq 0$}.

[21] $f(x) = -\frac{2}{x-1}$ • The domain of $f$ is $\mathbb{R} - \{1\}$. The horizontal asymptote of $f$ is $y = 0$ and the range of $f$ is $\mathbb{R} - \{0\}$, or equivalently, $(-\infty, 0) \cup (0, \infty)$. The domain of $f^{-1}$ is equal to the range of $f$; that is, $\mathbb{R} - \{0\}$. The range of $f^{-1}$ is equal to the domain of $f$; that is, $\mathbb{R} - \{1\}$, or equivalently, $(-\infty, 1) \cup (1, \infty)$.

[23] $f(x) = \frac{4x+5}{3x-8}$ • The domain of $f$ is $\mathbb{R} - \{\frac{8}{3}\}$. The horizontal asymptote of $f$ is $y = \frac{4}{3}$ and the range of $f$ is $\mathbb{R} - \{\frac{4}{3}\}$, or equivalently, $(-\infty, \frac{4}{3}) \cup (\frac{4}{3}, \infty)$. The domain of $f^{-1}$ is equal to the range of $f$; that is, $\mathbb{R} - \{\frac{4}{3}\}$. The range of $f^{-1}$ is equal to the domain of $f$; that is, $\mathbb{R} - \{\frac{8}{3}\}$, or equivalently, $(-\infty, \frac{8}{3}) \cup (\frac{8}{3}, \infty)$.

[25] We need to solve for $x$. $y = 3x + 5 \Rightarrow y - 5 = 3x \Rightarrow x = \frac{y-5}{3}$.

Now exchange $x$ and $y$. $y = f^{-1}(x) = \frac{x-5}{3}$

[27] $f(x) = \frac{1}{3x-2} \Rightarrow y(3x-2) = 1 \Rightarrow 3xy - 2y = 1 \Rightarrow 3xy = 2y + 1 \Rightarrow$

$x = \frac{2y+1}{3y} \Rightarrow f^{-1}(x) = \frac{2x+1}{3x}$

[29] $f(x) = \frac{3x+2}{2x-5} \Rightarrow 2xy - 5y = 3x + 2 \Rightarrow 2xy - 3x = 5y + 2 \Rightarrow$

$x(2y-3) = 5y + 2 \Rightarrow x = \frac{5y+2}{2y-3} \Rightarrow f^{-1}(x) = \frac{5x+2}{2x-3}$

[31] $f(x) = 2 - 3x^2,\ x \leq 0 \Rightarrow y + 3x^2 = 2 \Rightarrow$

$x^2 = \frac{2-y}{3} \Rightarrow x = \pm\sqrt{\frac{2-y}{3}}$ {choose minus since $x \leq 0$} $\Rightarrow f^{-1}(x) = -\sqrt{\frac{2-x}{3}}$

[33] $f(x) = 2x^3 - 5 \Rightarrow \frac{y+5}{2} = x^3 \Rightarrow x = \sqrt[3]{\frac{y+5}{2}} \Rightarrow f^{-1}(x) = \sqrt[3]{\frac{x+5}{2}}$

[35] $f(x) = \sqrt{3-x} \Rightarrow y^2 = 3-x \Rightarrow$

$x = 3 - y^2$ {Since $y \geq 0$ for $f$, $x \geq 0$ for $f^{-1}$.} $\Rightarrow f^{-1}(x) = 3 - x^2,\ x \geq 0$

[37] $f(x) = \sqrt[3]{x} + 1 \Rightarrow y - 1 = \sqrt[3]{x} \Rightarrow x = (y-1)^3 \Rightarrow f^{-1}(x) = (x-1)^3$

[39] $f(x) = x \Rightarrow y = x \Rightarrow x = y \Rightarrow f^{-1}(x) = x$

[41] $f(x) = x^2 - 6x,\ x \geq 3 \Rightarrow y = x^2 - 6x \Rightarrow x^2 - 6x - y = 0.$

This is a quadratic equation in $x$ with $a = 1$, $b = -6$, and $c = -y$.

Solving with the quadratic formula gives us

$$x = \frac{-(-6) \pm \sqrt{(-6)^2 - 4(1)(-y)}}{2(1)} = \frac{6 \pm \sqrt{36 + 4y}}{2} = \frac{6 \pm 2\sqrt{9 + y}}{2} = 3 \pm \sqrt{9 + y}.$$

Since $x \geq 3$, we choose the "+" and obtain $f^{-1}(x) = 3 + \sqrt{x + 9}$.

[43] (a) $(g^{-1} \circ f^{-1})(2) = g^{-1}(f^{-1}(2))$

$= g^{-1}(5)$ {since $f(5) = 2$}

$= 3$ {since $g(3) = 5$}

(b) $(g^{-1} \circ h)(3) = g^{-1}(h(3)) = g^{-1}(1) = -1$ {since $g(-1) = 1$}

(c) $(h^{-1} \circ f \circ g^{-1})(3) = h^{-1}(f(g^{-1}(3)))$

$= h^{-1}(f(2))$ {since $g(2) = 3$}

$= h^{-1}(-1)$

$= 5$ {since $-1 = 4 - x \Rightarrow x = 5$}

[45] (a) Remember that the domain of $f$ is the range of $f^{-1}$ and that the range of $f$ is the domain of $f^{-1}$.

(b) $D = [-1, 2]$; $R = [\frac{1}{2}, 4]$

(c) $D_1 = R = [\frac{1}{2}, 4]$; $R_1 = D = [-1, 2]$

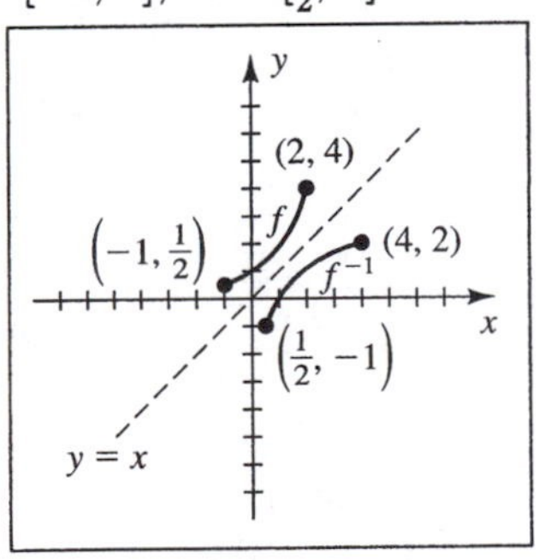


*Figure 45*

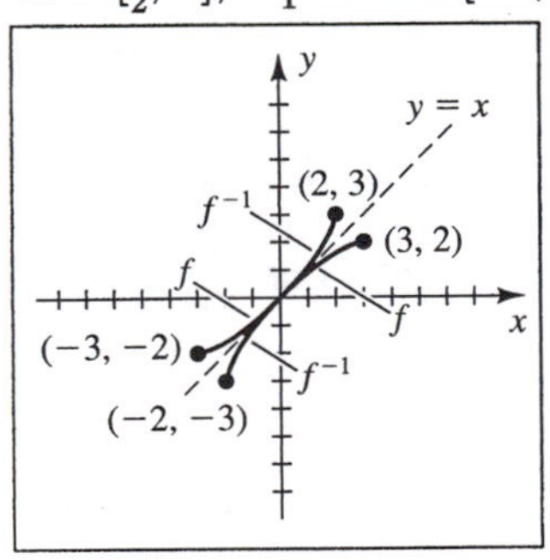


*Figure 47*

[47] (a) See *Figure 47*.

(b) $D = [-3, 3]$; $R = [-2, 2]$

(c) $D_1 = R = [-2, 2]$; $R_1 = D = [-3, 3]$

[49] (a) Since $f$ is one-to-one, an inverse exists.

If $f(x) = ax + b$, then $f^{-1}(x) = \frac{x - b}{a}$ for $a \neq 0$.

(b) No, because a constant function is not one-to-one.

51 (a) $f(x) = -x + b \Rightarrow y = -x + b \Rightarrow x = -y + b$, or $f^{-1}(x) = -x + b$.

(b) $f(x) = \frac{ax+b}{cx-a}$ for $c \neq 0 \Rightarrow y = \frac{ax+b}{cx-a} \Rightarrow cyx - ya = ax + b \Rightarrow$

$cyx - ax = ay + b \Rightarrow x(cy - a) = ay + b \Rightarrow x = \frac{ay+b}{cy-a}$, or $f^{-1}(x) = \frac{ax+b}{cx-a}$.

(c) The graph of $f$ is symmetric about the line $y = x$. Thus, $f(x) = f^{-1}(x)$.

53 From *Figure 53*, we see that $f$ is always increasing. Thus, $f$ is one-to-one.

[−6, 6] by [−4, 4]

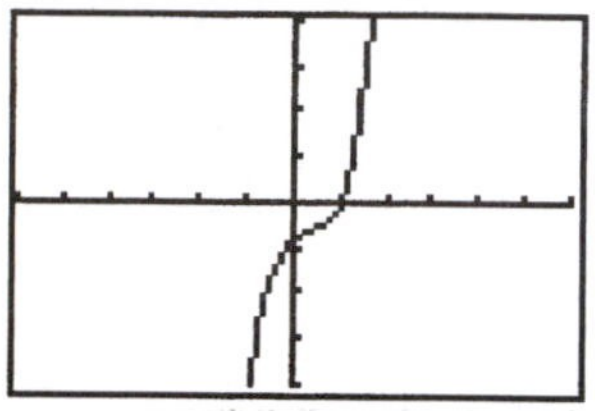

*Figure 53*

[−1, 2] by [−1, 4]

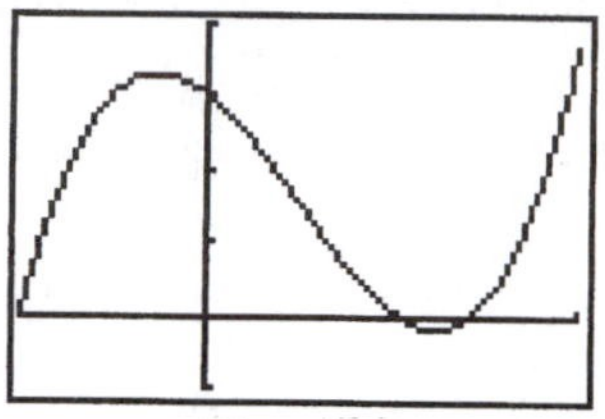

*Figure 55*

55 (a) The high point is approximately (−0.27, 3.31) and the low point is (1.22, −0.20).

Thus, $f$ decreases on [−0.27, 1.22].

(b) The function $g$ is the same as $f$ with a limited domain. The domain of $g^{-1}$ is the same as the range of $g$, that is, [−0.20, 3.31]. The range of $g^{-1}$ is the same as the domain of $g$, that is, [−0.27, 1.22].

57 The graph of $f$ will be reflected through the line $y = x$.

$y = \sqrt[3]{x-1} \Rightarrow y^3 = x - 1 \Rightarrow$

$x = y^3 + 1 \Rightarrow f^{-1}(x) = x^3 + 1$.

Graph $Y_1 = \sqrt[3]{x-1}$, $Y_2 = x^3 + 1$, and $Y_3 = x$.

[−12, 12] by [−8, 8]

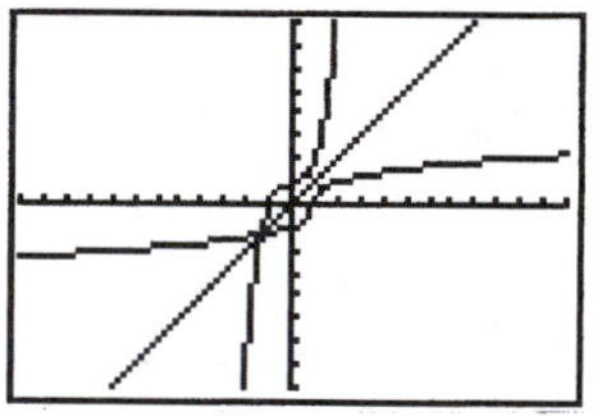

*Figure 57*

59 (a) $V(23) = 35(23) = 805 \text{ ft}^3/\text{min}$

(b) $V^{-1}(x) = \frac{1}{35}x$. Given an air circulation of $x$ cubic feet per minute, $V^{-1}(x)$ computes the maximum number of people that should be in the restaurant at one time.

(c) $V^{-1}(2350) = \frac{1}{35}(2350) \approx 67.1 \Rightarrow$ the maximum number of people is 67.

## *5.2 Exercises*

1 $7^{x+6} = 7^{3x-4} \Rightarrow x + 6 = 3x - 4 \Rightarrow 10 = 2x \Rightarrow x = 5$

3 $3^{2x+3} = 3^{(x^2)} \Rightarrow 2x + 3 = x^2 \Rightarrow x^2 - 2x - 3 = 0 \Rightarrow (x-3)(x+1) = 0 \Rightarrow$

$x = -1, 3$

[5] We need to obtain the same base on each side of the equals sign, then we can apply part (2) of the theorem about exponential functions being one-to-one.

$2^{-100x} = (0.5)^{x-4} \Rightarrow (2^{-1})^{100x} = (\frac{1}{2})^{x-4} \Rightarrow (\frac{1}{2})^{100x} = (\frac{1}{2})^{x-4} \Rightarrow$

$100x = x - 4 \Rightarrow 99x = -4 \Rightarrow x = -\frac{4}{99}$

[7] $4^{x-3} = 8^{4-x} \Rightarrow (2^2)^{x-3} = (2^3)^{4-x} \Rightarrow 2^{2x-6} = 2^{12-3x} \Rightarrow$

$2x - 6 = 12 - 3x \Rightarrow 5x = 18 \Rightarrow x = \frac{18}{5}$

[9] $4^x \cdot (\frac{1}{2})^{3-2x} = 8 \cdot (2^x)^2 \Rightarrow 2^{2x} \cdot (2^{-1})^{3-2x} = 2^3 \cdot 2^{2x} \Rightarrow$

$2^{2x-3+2x} = 2^{3+2x} \Rightarrow 2^{4x-3} = 2^{2x+3} \Rightarrow$

$4x - 3 = 2x + 3 \Rightarrow 2x = 6 \Rightarrow x = 3$

[11] (a) Let $g = f(x) = 2^x$ for reference purposes.

The graph of $g$ goes through the points $(-1, \frac{1}{2})$, $(0, 1)$, and $(1, 2)$.

(b) $f(x) = -2^x$ •

Reflect $g$ through the $x$-axis since $f$ is just $-1 \cdot 2^x$. Do not confuse this function with $(-2)^x$ — remember, the base is positive for exponential functions.

(c) $f(x) = 3 \cdot 2^x$ • vertically stretch $g$ by a factor of 3

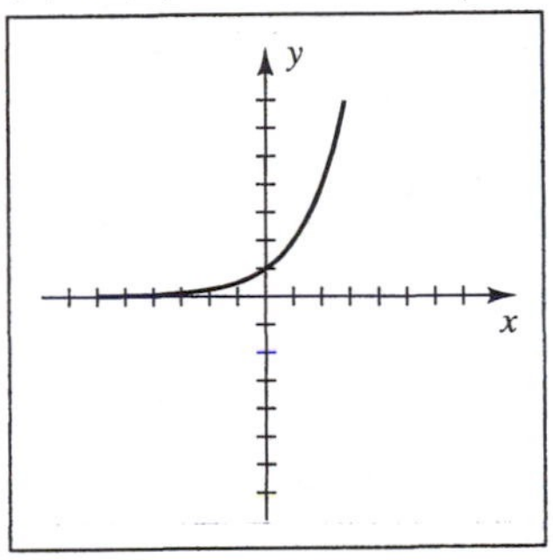

Figure 11(a)

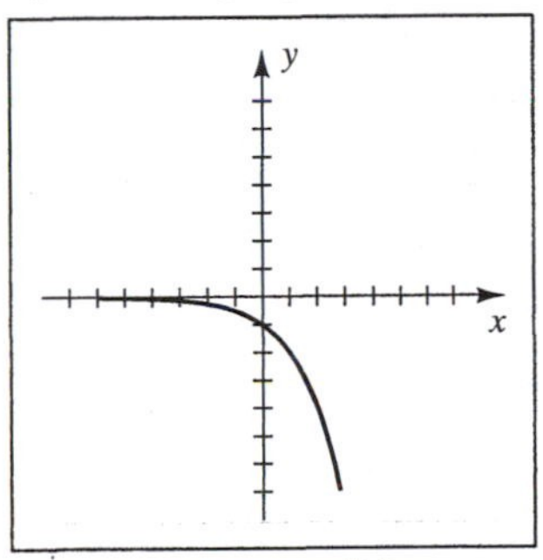

Figure 11(b)

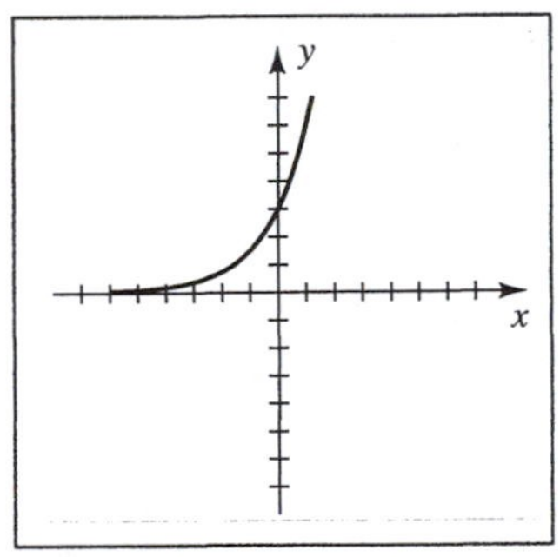

Figure 11(c)

(d) $f(x) = 2^{x+3}$ • shift $g$ left 3 units since $f$ is $g(x+3)$

(e) $f(x) = 2^x + 3$ • shift $g$ up 3 units

(f) $f(x) = 2^{x-3}$ • shift $g$ right 3 units since $f$ is $g(x-3)$

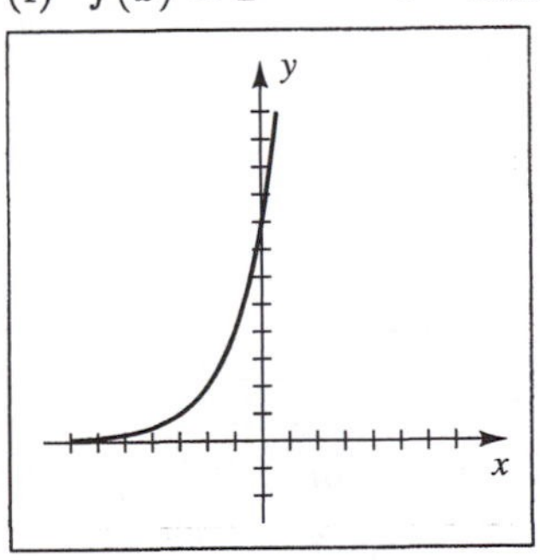

Figure 11(d)

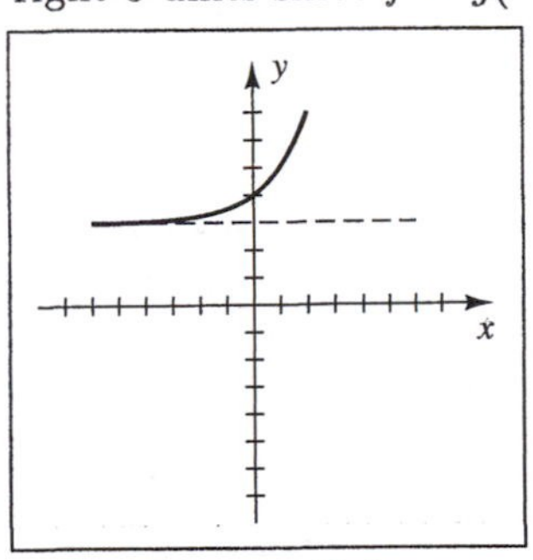

Figure 11(e)

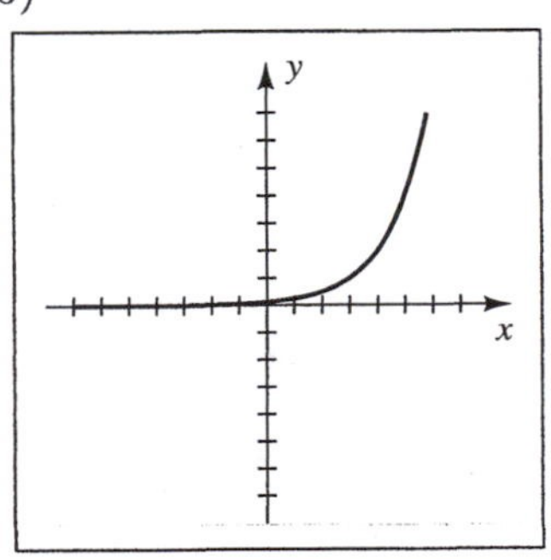

Figure 11(f)

(g) $f(x) = 2^x - 3$ • shift $g$ down 3 units

(h) $f(x) = 2^{-x}$ • reflect $g$ through the $y$-axis since $f$ is $g(-x)$

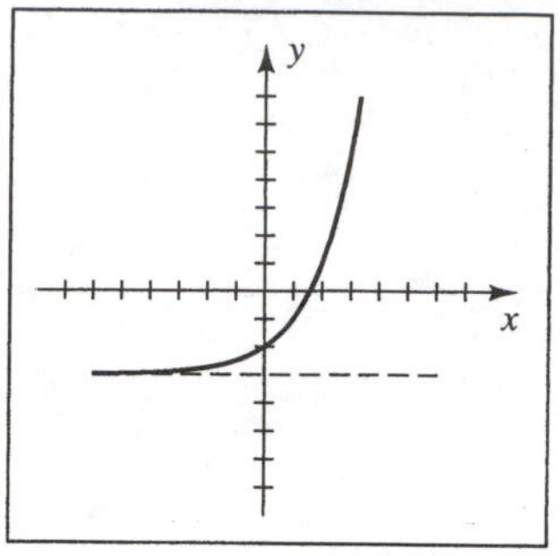

Figure 11(g)

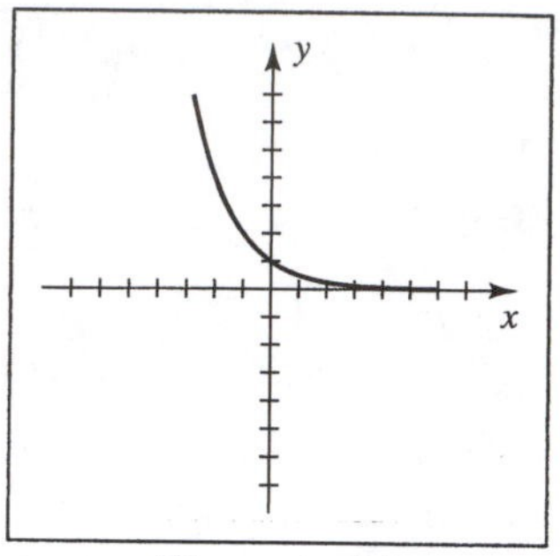

Figure 11(h)

(i) $f(x) = (\frac{1}{2})^x$ • $(\frac{1}{2})^x = (2^{-1})^x = 2^{-x}$, same graph as in part (h)

(j) $f(x) = 2^{3-x}$ • $2^{3-x} = 2^{-(x-3)}$, shift $g$ right 3 units and reflect through the line $x = 3$. Alternatively, $2^{3-x} = 2^3 2^{-x} = 8(\frac{1}{2})^x$, vertically stretch $y = (\frac{1}{2})^x$ (the graph in part (i)) by a factor of 8.

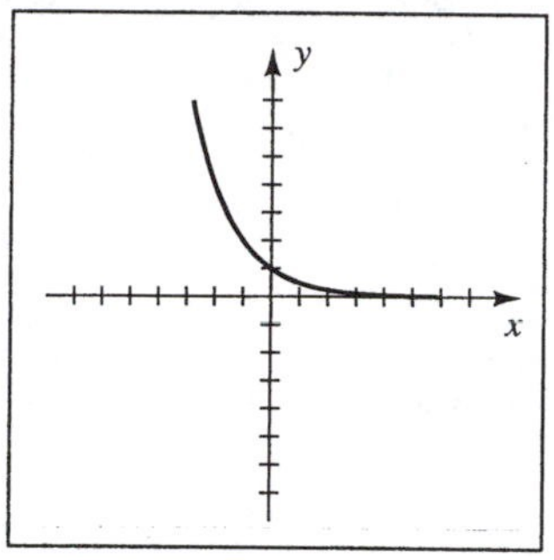

Figure 11(i)

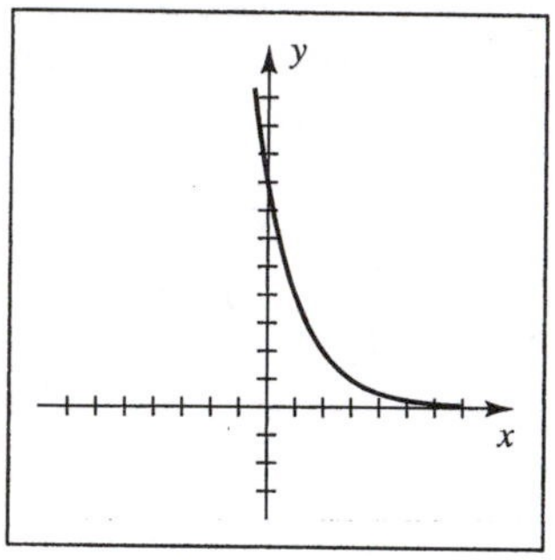

Figure 11(j)

[13] $f(x) = (\frac{2}{5})^{-x} = \left[(\frac{2}{5})^{-1}\right]^x = (\frac{5}{2})^x$ • goes through $(-1, \frac{2}{5})$, $(0, 1)$, and $(1, \frac{5}{2})$

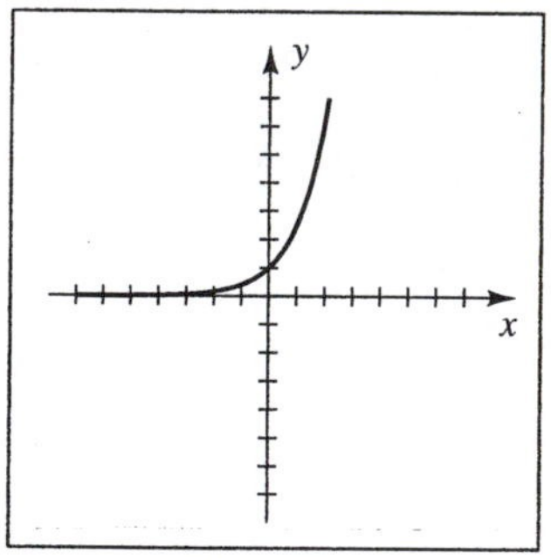

Figure 13

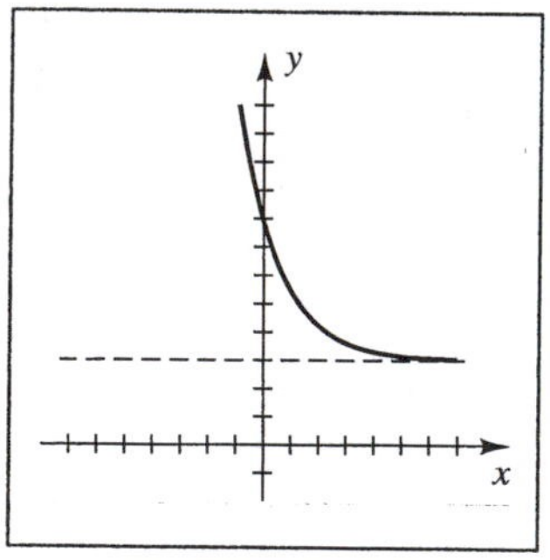

Figure 15

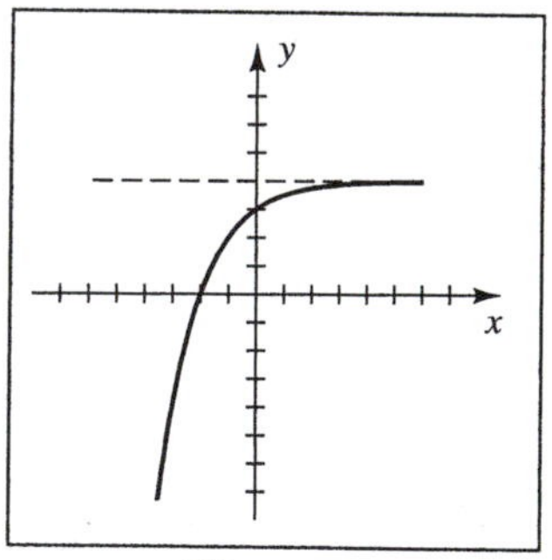

Figure 17

[15] $f(x) = 5(\frac{1}{2})^x + 3$ • vertically stretch $y = (\frac{1}{2})^x$ by a factor of 5 and shift up 3 units

[17] $f(x) = -(\frac{1}{2})^x + 4$ • reflect $y = (\frac{1}{2})^x$ through the $x$-axis and shift up 4 units

[19] $f(x) = 2^{|x|} = \begin{cases} 2^x & \text{if } x \geq 0 \\ 2^{-x} & \text{if } x < 0 \end{cases} = \begin{cases} 2^x & \text{if } x \geq 0 \\ (\frac{1}{2})^x & \text{if } x < 0 \end{cases}$

Use the portion of $y = 2^x$ with $x \geq 0$ and reflect it through the $y$-axis since $f$ is even.

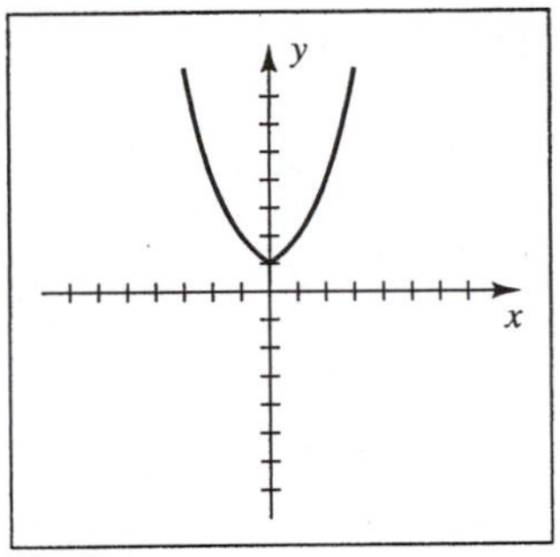

*Figure 19*

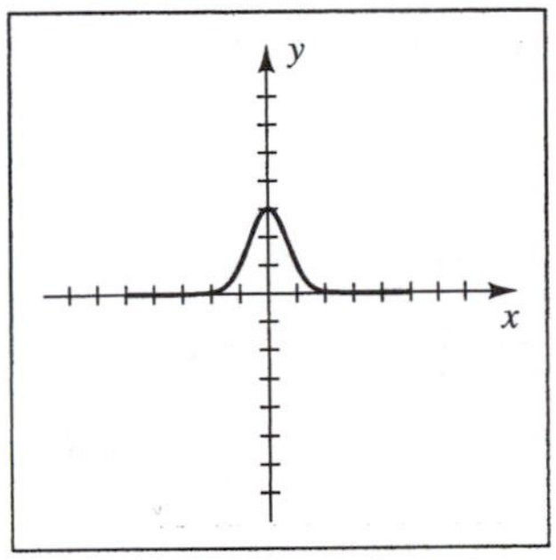

*Figure 21*

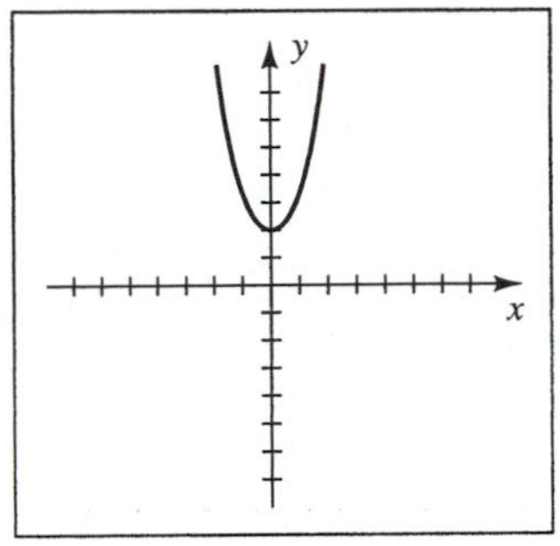

*Figure 23*

*Note:* For Exercises 21, 22, and 5 of the review exercises, refer to Example 6 in the text for the basic graph of $y = a^{-x^2} = (a^{-1})^{(x^2)} = \left(\frac{1}{a}\right)^{(x^2)}$, where $a > 1$.

[21] $f(x) = 3^{1-x^2} = 3^1 3^{-x^2} = 3(\frac{1}{3})^{(x^2)}$ • stretch $y = (\frac{1}{3})^{(x^2)}$ by a factor of 3

[23] $f(x) = 3^x + 3^{-x}$ • Adding the functions $g(x) = 3^x$ and $h(x) = 3^{-x} = (\frac{1}{3})^x$ together, we see that the $y$-intercept will be 2. If $x > 0$, $f$ looks like $y = 3^x$ since $3^x$ dominates $3^{-x}$ ($3^{-x}$ gets close to 0 and $3^x$ grows very large). If $x < 0$, $f$ looks like $y = 3^{-x}$ since $3^{-x}$ dominates $3^x$. See *Figure 23* above.

[25] The graph has a horizontal asymptote of $y = 0$, so $c = 0$ and the equation has the form $f(x) = ba^x$. The point (0, 2) is on the graph, so

$f(0) = 2 \Rightarrow 2 = ba^0 \Rightarrow 2 = b(1) \Rightarrow 2 = b$, and hence, $f(x) = 2a^x$.

Since the point $P(1, 5)$ is on the graph, $f(1) = 5$.

Thus, $5 = 2a^1 \Rightarrow a = \frac{5}{2}$, and the function is $f(x) = 2(\frac{5}{2})^x$.

[27] The graph has a horizontal asymptote of $y = -3$, so $c = -3$ and the equation has the form $f(x) = ba^x - 3$. The point $(0, -1)$ is on the graph, so

$f(0) = -1 \Rightarrow -1 = ba^0 - 3 \Rightarrow -1 = b(1) - 3 \Rightarrow 2 = b$, and hence,

$f(x) = 2a^x - 3$. Since the point $P(-1, 0)$ is on the graph, $f(-1) = 0$.

Thus, $0 = 2a^{-1} - 3 \Rightarrow 3 = \frac{2}{a} \Rightarrow 3a = 2 \Rightarrow a = \frac{2}{3}$,

and the function is $f(x) = 2(\frac{2}{3})^x - 3$.

[29] Since the $y$-intercept is 8, we must have $f(0) = 8$. $f(x) = ba^x$,

so $f(0) = 8 \Rightarrow 8 = ba^0 \Rightarrow 8 = b(1) \Rightarrow 8 = b$, and hence, $f(x) = 8a^x$.

Since the point $P(3, 1)$ is on the graph, $f(3) = 1$.

Thus, $1 = 8a^3 \Rightarrow a^3 = \frac{1}{8} \Rightarrow a = \frac{1}{2}$, and the function is $f(x) = 8(\frac{1}{2})^x$.

31 The horizontal asymptote is $y = 32$, so $f(x) = ba^{-x} + c$ has the form $f(x) = ba^{-x} + 32$. The $y$-intercept is 212, so $f(0) = 212 \Rightarrow 212 = ba^{-0} + 32 \Rightarrow 212 = b(1) + 32 \Rightarrow 180 = b \Rightarrow f(x) = 180a^{-x} + 32$. The function passes through the point $P(2, 112)$, so $f(2) = 112 \Rightarrow 112 = 180a^{-2} + 32 \Rightarrow 80 = 180a^{-2} \Rightarrow 80 = \frac{180}{a^2} \Rightarrow a^2 = \frac{180}{80} \Rightarrow a^2 = \frac{9}{4} \Rightarrow a = \frac{3}{2}$ {since $a$ must be positive, $a = -\frac{3}{2}$ is not allowed} $\Rightarrow f(x) = 180(1.5)^{-x} + 32$.

33 (a) $N(t) = 100(0.9)^t \Rightarrow N(1) = 100(0.9)^1 = 90$

(b) $N(5) = 100(0.9)^5 \approx 59$ (c) $N(10) = 100(0.9)^{10} \approx 35$

35 $f(t) = 600(3)^{t/2}$ •

(a) 8:00 A.M. corresponds to $t = 1$ and $f(1) = 600\sqrt{3} \approx 1039$.

10:00 A.M. corresponds to $t = 3$ and $f(3) = 600(3\sqrt{3}) = 1800\sqrt{3} \approx 3118$.

11:00 A.M. corresponds to $t = 4$ and $f(4) = 600(9) = 5400$.

(b) The graph of $f$ is an increasing exponential that passes through (0, 600) and the points in part (a).

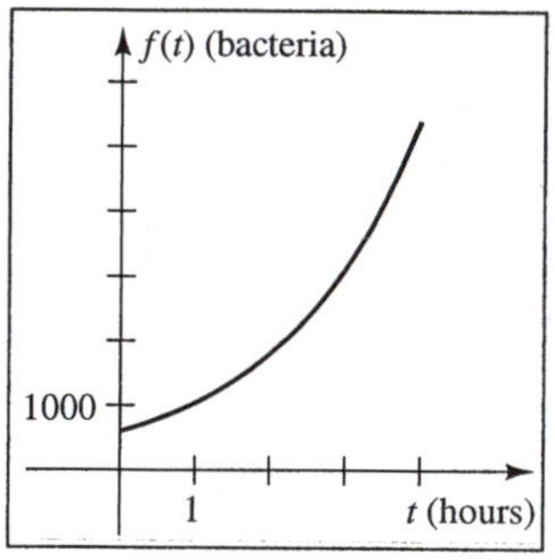

Figure 35

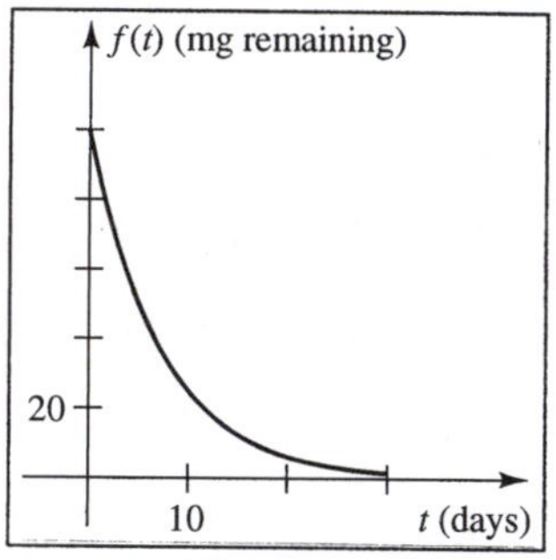

Figure 37

37 (a) $f(t) = 100(2)^{-t/5}$, so $f(5) = 100(2)^{-1} = 50$ mg, $f(10) = 100(2)^{-2} = 25$ mg, and $f(12.5) = 100(2)^{-2.5} = \frac{100}{4\sqrt{2}} = \frac{25}{2}\sqrt{2} \approx 17.7$ mg.

(b) The endpoints of this decreasing exponential are (0, 100) and (30, 1.5625).

39 A half-life of 1600 years means that when $t = 1600$, the amount remaining, $q(t)$, will be one-half the original amount—that is, $\frac{1}{2}q_0$. $q(t) = \frac{1}{2}q_0$ when $t = 1600 \Rightarrow \frac{1}{2}q_0 = q_0 2^{k(1600)} \Rightarrow 2^{-1} = 2^{1600k} \Rightarrow 1600k = -1 \Rightarrow k = -\frac{1}{1600}$.

41 Using $A = P(1 + \frac{r}{n})^{nt}$, we have $P = 1000$, $r = 0.12$, and $n = 12$.

Consider $A$ to be a function of $t$, that is, $A(t) = 1000\left(1 + \frac{0.12}{12}\right)^{12t} = 1000(1.01)^{12t}$.

(a) $A(\frac{1}{12}) = \$1010.00$ (b) $A(\frac{6}{12}) \approx \$1061.52$

(c) $A(1) \approx \$1126.83$ (d) $A(20) \approx \$10{,}892.55$

[43] $C = 10{,}000 \Rightarrow V(t) = 0.78(10{,}000)(0.85)^{t-1} = 7800(0.85)^{t-1}$

(a) $V(1) = \$7800$ (b) $V(4) \approx \$4790.18$, or $\$4790$ (c) $V(7) \approx \$2941.77$, or $\$2942$

[45] $t = 2006 - 1626 = 380$; $A = \$24(1 + 0.06/4)^{4 \cdot 380} = \$161{,}657{,}351{,}965.80$.

That's right—$161 billion!

[47] (a) Examine the pattern formed by the value $y$ in the year $n$.

| year ($n$) | value ($y$) |
|---|---|
| 0 | $y_0$ |
| 1 | $(1-a)y_0 = y_1$ |
| 2 | $(1-a)y_1 = (1-a)[(1-a)y_0] = (1-a)^2 y_0 = y_2$ |
| 3 | $(1-a)y_2 = (1-a)[(1-a)^2 y_0] = (1-a)^3 y_0 = y_3$ |

(b) $s = (1-a)^T y_0 \Rightarrow (1-a)^T = s/y_0 \Rightarrow 1 - a = \sqrt[T]{s/y_0} \Rightarrow a = 1 - \sqrt[T]{s/y_0}$

[49] (a) $r = 0.12$, $t = 30$, $L = 90{,}000 \Rightarrow$

$$k = \left(1 + \frac{0.12}{12}\right)^{12 \cdot 30} \approx 35.95 \quad \text{and} \quad M = \frac{90{,}000(0.12)k}{12(k-1)} \approx 925.75.$$

(b) Total interest paid $=$ total of payments $-$ original loan amount

$= (12 \cdot 30 \text{ payments}) \times \$925.75 - \$90{,}000 = \$243{,}270$

[51] $r = 0.15$, $t = 3$, $M = 220 \Rightarrow k = \left(1 + \frac{0.15}{12}\right)^{12 \cdot 3} \approx 1.56$.

$$M = \frac{Lrk}{12(k-1)} \Rightarrow L = \frac{12M(k-1)}{rk} = \frac{12(220)(k-1)}{(0.15)k} \approx \$6{,}346.40$$

[53] (a) $f(3) = 13^{\sqrt{3 + 1.1}} \approx 180.1206$ (b) $g(1.43) = \left(\frac{5}{42}\right)^{-1.43} \approx 20.9758$

(c) $h(1.06) = (2^{1.06} + 2^{-1.06})^{2(1.06)} \approx 7.3639$

[55] Part (b) may be interpreted as doubling an investment at 8.5%.

(a) If $y = (1.085)^x$ and $x = 40$, then $y \approx 26.13$. (b) If $y = 2$, then $x \approx 8.50$.

[0, 60, 5] by [0, 40, 5]

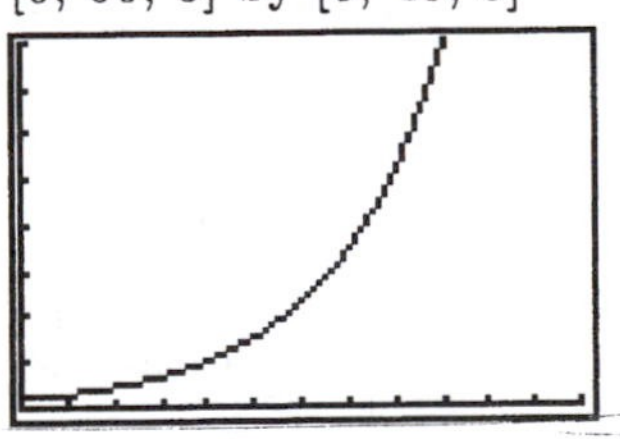

Figure 55

[−10.5, 10.5] by [−7, 7]

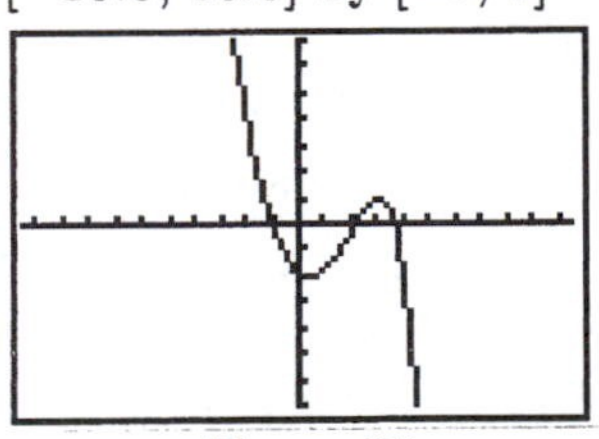

Figure 57

[57] First graph $y = 1.4x^2 - 2.2^x - 1$. The $x$-intercepts are the roots of the equation.

Using a zero or root feature, the roots are $x \approx -1.02$, $2.14$, and $3.62$.

**59** (a) $f$ is not one-to-one since

the horizontal line $y = -0.1$ intersects the graph of $f$ more than once.

(b) The only zero of $f$ is $x = 0$.

[−3, 3] by [−2, 2]

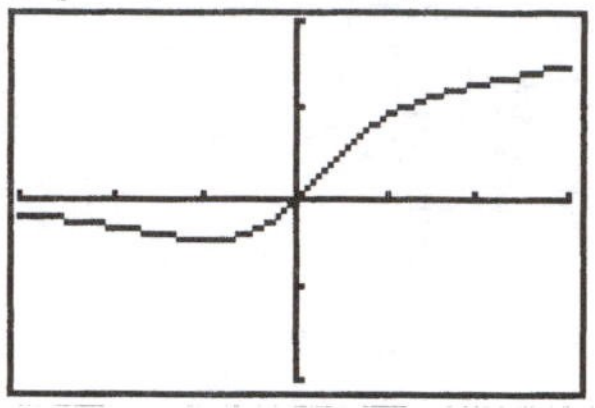

*Figure 59*

[−4, 1] by [−2, 3]

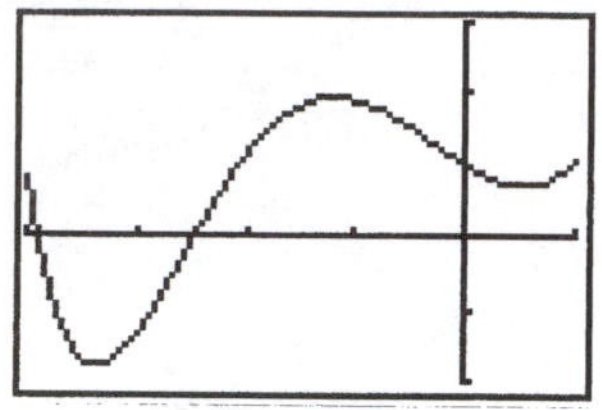

*Figure 61*

**61** (a) The low points are about $(-3.37, -1.79)$ and $(0.52, 0.71)$. The high point is approximately $(-1.19, 1.94)$. Thus, $f$ is increasing on $[-3.37, -1.19]$ and $[0.52, 1]$. $f$ is decreasing on $[-4, -3.37]$ and $[-1.19, 0.52]$.

(b) The range of $f$ on $[-4, 1]$ is approximately $[-1.79, 1.94]$.

**63** *Figure 63* is a graph of $N(t) = 1000(0.9)^t$.

By using an intersect feature, we can determine that $N = 500$ when $t \approx 6.58$ yr.

[0, 10] by [0, 1000, 100]

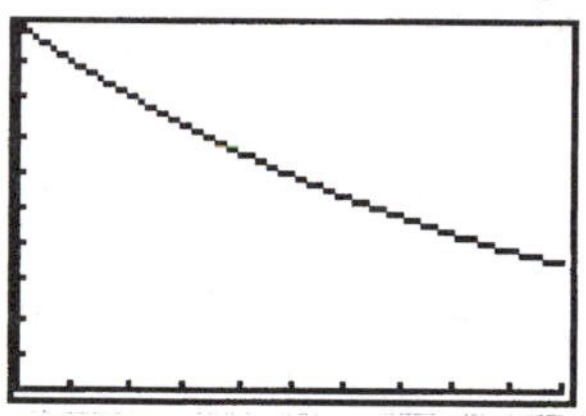

*Figure 63*

[0, 7.5] by [0, 5]

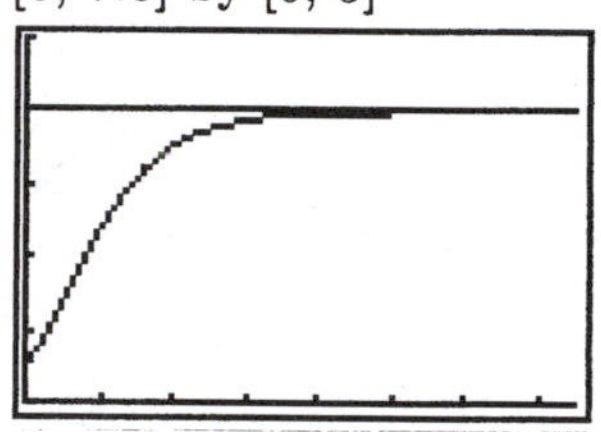

*Figure 65*

**65** Graph $y = 4(0.125)^{(0.25^x)}$. The line $y = k = 4$ is a horizontal asymptote for the Gompertz function. The maximum number of sales of the product approaches $k$.

**67** Using an intersect feature, we determine that

$$A = \frac{100\left(1+\frac{0.05}{12}\right)\left[\left(1+\frac{0.05}{12}\right)^{12n}-1\right]}{\frac{0.05}{12}}$$ and $A = 100{,}000$ intersect when $n \approx 32.8$.

[0, 40, 10] by [0, 200,000, 50,000]

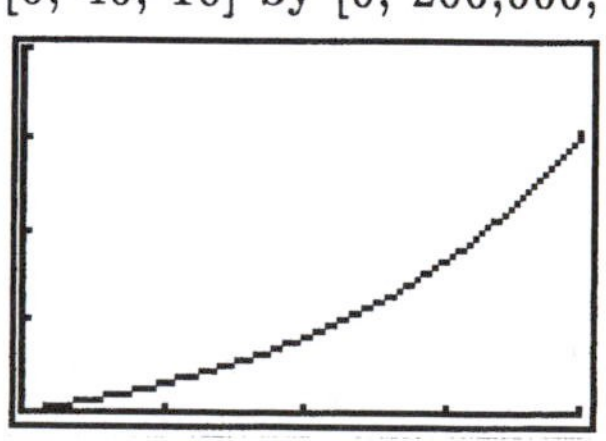

*Figure 67*

[69] (a) Let $x = 0$ correspond to 1910, $x = 20$ to 1930, ... , and $x = 85$ to 1995.

Graph the data together with the functions

(1) $f(x) = 0.809(1.094)^x$ and (2) $g(x) = 0.375x^2 - 18.4x + 88.1$.

[−10, 90, 10] by [−200, 1500, 100]

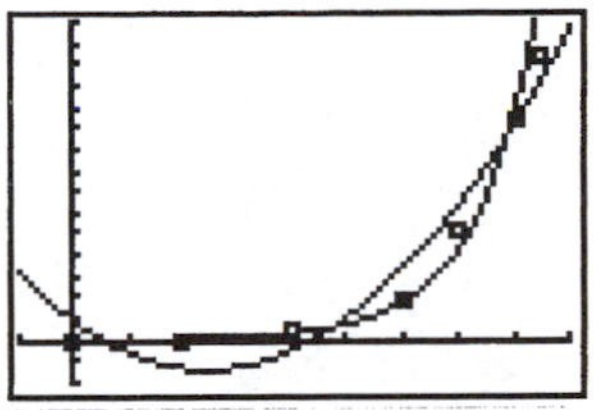

Figure 69(a)

[−10, 90, 10] by [−200, 1500, 100]

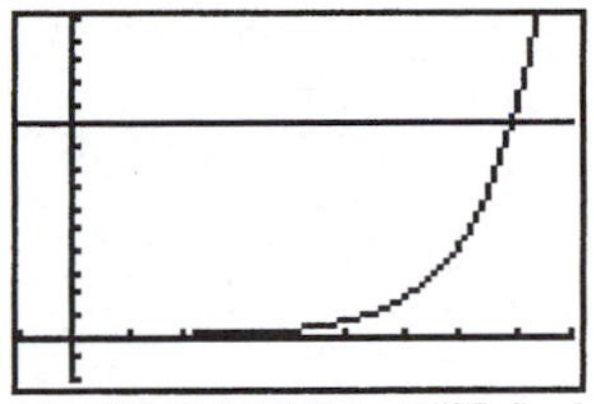

Figure 69(c)

(b) The exponential function $f$ best models the data.

(c) Graph $Y_1 = f(x)$ and $Y_2 = 1000$. The graphs intersect at $x \approx 79$, or in 1989.

[71] The model is $y = 0.03b^t$, with $t = 0$ corresponding to 1958.

The year 2002 corresponds to $t = 2002 - 1958 = 44$, so $0.37 = 0.03b^{44}$ $\Rightarrow$

$\frac{37}{3} = b^{44} \Rightarrow b = \sqrt[44]{\frac{37}{3}} \approx 1.05876$. The model is $y = 0.03(1.0588)^t$.

Using $t = 2010 - 1958 = 52$, we predict the cost of a first-class stamp in 2010 will be $0.03(1.0588)^{52} \approx 0.5842$, or 58¢.

[73] (a) $t = 1999 - 1974 = 25$ and $r/n$ is 0.0025, so $A = P\left(1 + \frac{r}{n}\right)^{nt} =$

$$\$353{,}022(1 + 0.0025)^{12 \cdot 25} = \$746{,}648.43;$$ the website gives \$1,192,971.

(b) Let $r$ denote the annual interest rate.

$$\$6{,}616{,}585 = \$353{,}022\left(1 + \frac{r}{1}\right)^{1 \cdot 25} \Rightarrow$$

$$\frac{6{,}616{,}585}{353{,}022} = (1 + r)^{25} \Rightarrow 1 + r = \sqrt[25]{\frac{6{,}616{,}585}{353{,}022}} \Rightarrow$$

$$r = \sqrt[25]{\frac{6{,}616{,}585}{353{,}022}} - 1 \approx 0.1244,$$ so $r$ is about 12.44%.

(c) In part (a), the base is constant and the variable is in the exponent, so this is an *exponential* function. In part (b), the base is variable and the exponent is constant, so this is a *polynomial* function.

## 5.3 Exercises

*Note:* Examine Figure 1 in this section to reinforce the idea that $y = e^x$ is just a special case of $y = a^x$ with $a > 1$.

[1] (a) $f(x) = e^{-x}$ • reflect $y = e^x$ through the $y$-axis

(b) $f(x) = -e^x$ • reflect $y = e^x$ through the $x$-axis

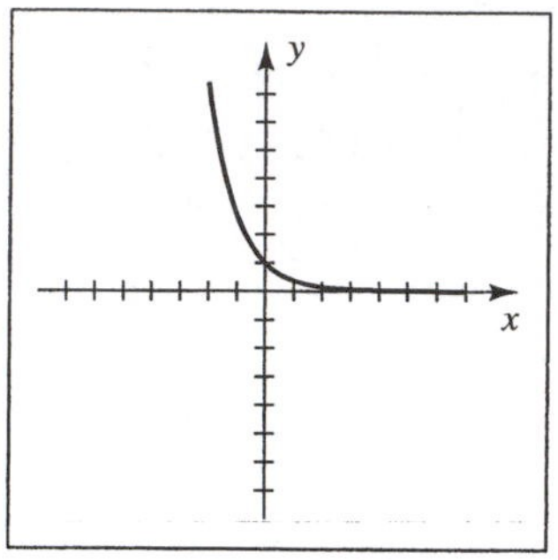


*Figure 1(a)*

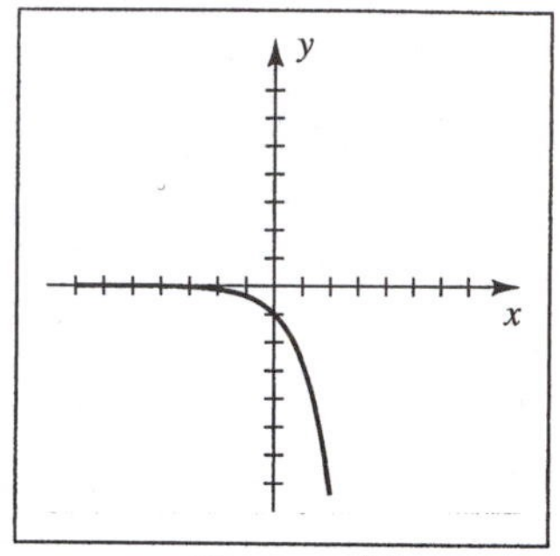


*Figure 1(b)*

[3] (a) $f(x) = e^{x+4}$ • shift $y = e^x$ left 4 units

(b) $f(x) = e^x + 4$ • shift $y = e^x$ up 4 units

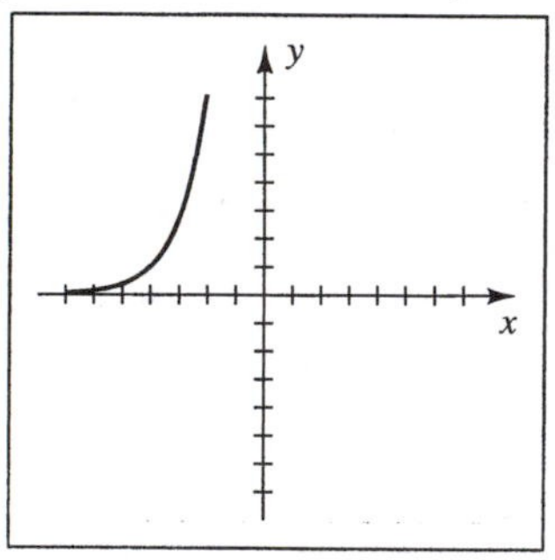


*Figure 3(a)*

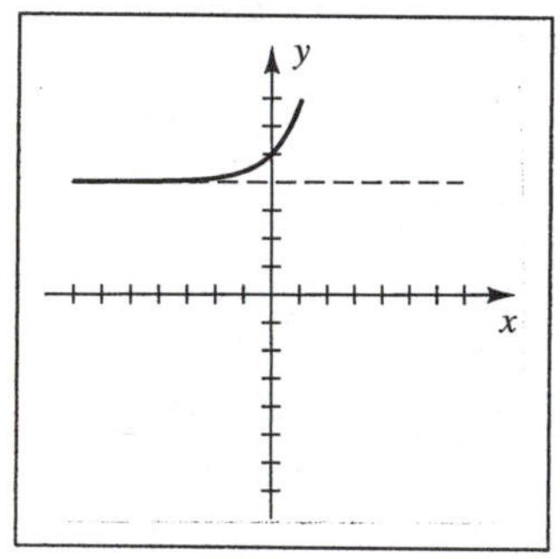


*Figure 3(b)*

[5] $A = Pe^{rt} = 1000e^{(0.0825)(5)} \approx \$1510.59$

[7] $100{,}000 = Pe^{(0.11)(18)} \Rightarrow P = \dfrac{100{,}000}{e^{1.98}} \approx \$13{,}806.92$

[9] $13{,}464 = 1000e^{(r)(20)} \Rightarrow e^{20r} = 13.464$. Using a trial and error approach on a scientific calculator or tracing and zooming or an intersect feature on a graphing calculator, we determine that $e^x \approx 13.464$ if $x \approx 2.6$. Thus, $20r = 2.6$ and $r = 0.13$ or 13%.

[11] $e^{(x^2)} = e^{7x - 12} \Rightarrow x^2 = 7x - 12 \Rightarrow x^2 - 7x + 12 = 0 \Rightarrow$

$(x-3)(x-4) = 0 \Rightarrow x = 3, 4$

[13] $xe^x + e^x = 0 \Rightarrow e^x(x+1) = 0 \Rightarrow x = -1$ {Note that $e^x \neq 0$.}

[15] $x^3(4e^{4x}) + 3x^2e^{4x} = 0 \Rightarrow x^2e^{4x}(4x+3) = 0 \Rightarrow x = -\frac{3}{4}, 0$. Note that $e^{4x} \neq 0$.

[17] When we multiply $(e^x + e^{-x})$ times $(e^x + e^{-x})$, we get

$$e^{x+x} + e^{x+(-x)} + e^{(-x)+x} + e^{(-x)+(-x)} = e^{2x} + e^0 + e^0 + e^{-2x} = e^{2x} + 2 + e^{-2x}.$$

$$\frac{(e^x + e^{-x})(e^x + e^{-x}) - (e^x - e^{-x})(e^x - e^{-x})}{(e^x + e^{-x})^2} = \frac{(e^{2x} + 2 + e^{-2x}) - (e^{2x} - 2 + e^{-2x})}{(e^x + e^{-x})^2}$$

$$= \frac{4}{(e^x + e^{-x})^2}$$

[19] $W(t) = W_0 e^{kt}$; $t = 30 \;\Rightarrow\; W(30) = 68e^{(0.2)(30)} \approx 27{,}433$ mg, or 27.43 grams

[21] The year 2010 corresponds to $t = 2010 - 1980 = 30$. Using the law of growth formula on page 348 with $q_0 = 227$ and $r = 0.007$, we have $N(t) = 227e^{0.007t}$.

Thus, $N(30) = 227e^{(0.007)(30)} \approx 280.0$ million.

[23] $N(t) = N_0 e^{-0.2t} \;\Rightarrow\; N(10) = N_0 e^{-2}$. The percentage of the original number still alive after 10 years is $\left(\dfrac{N(10)}{N_0}\right) \times 100 = \left(\dfrac{N_0 e^{-2}}{N_0}\right) \times 100 = 100e^{-2} \approx 13.5\%$.

[25] The year 2010 corresponds to $t = 2010 - 1978 = 32$.

$N(32) = 5000e^{(0.0036)(32)} = 5000e^{(0.1152)} \approx 5610.$

[27] $h = 40{,}000 \;\Rightarrow\; p = 29e^{(-0.000034)(40{,}000)} = 29e^{(-1.36)} \approx 7.44$ in.

[29] $x = 1 \;\Rightarrow\; y = 79.041 + 6.39(1) - e^{3.261 - 0.993(1)} \approx 75.77$ cm.

$x = 1 \;\Rightarrow\; R = 6.39 + 0.993e^{3.261 - 0.993(1)} \approx 15.98$ cm/yr.

[31] $2010 - 1971 = 39 \;\Rightarrow\; t = 39$ years. Using the continuously compounded interest formula with $P = 1.60$ and $r = 0.05$, we have $A = 1.60e^{(0.05)(39)} \approx \$11.25$ per hour.

[33] (a) Note here that the amount of money invested is not of interest and that we are only concerned with the percent of growth.

$$\left(1 + \tfrac{0.07}{4}\right)^{4 \cdot 1} \approx 1.0719. \quad (1.0719 - 1) \times 100\% = 7.19\%$$

(b) $e^{(0.07)(1)} \approx 1.0725$. $(1.0725 - 1) \times 100\% = 7.25\%$. The results indicate that we would receive an extra 0.06% in interest by investing our money in an account that is compounded continuously rather than quarterly. This is only an extra 6 cents on a \$100 investment, but \$600 extra on a \$1,000,000 investment (actually \$649.15 if the computations are carried beyond 0.01%).

35 The graph of $y = e^{1000x}$ nearly coincides with the negative $x$-axis and then increases so fast that it nearly coincides with the positive $y$-axis. To see the graph on a calculator, try "turning off" your axes display.

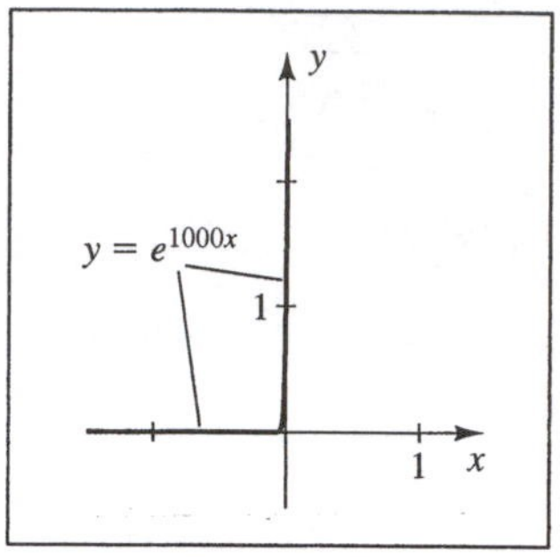


Figure 35

[0, 60, 5] by [0, 40, 5]

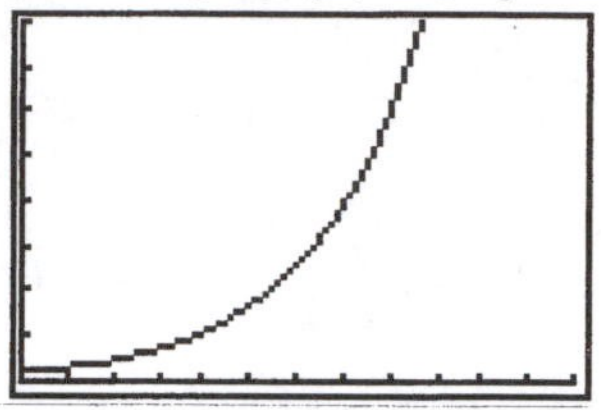

Figure 37

37 It may be of interest to compare this graph with the graph of $y = (1.085)^x$ in Exercise 55 of §5.2. Both are compounding functions with $r = 8.5\%$.
Note that $e^{0.085x} = (e^{0.085})^x \approx (1.0887)^x > (1.085)^x$ for $x > 0$.

(a) If $y = e^{0.085x}$ and $x = 40$, then $y \approx 29.96$. (b) If $y = 2$, then $x \approx 8.15$.

39 (a) As $x \to \infty$, $e^{-x} \to 0$ and $f$ will resemble $\frac{1}{2}e^x$.
As $x \to -\infty$, $e^x \to 0$ and $f$ will resemble $-\frac{1}{2}e^x$.

(b) At $x = 0$, $f(x) = 0$, and $g$ will have a vertical asymptote since $g$ is undefined (division by 0). As $x \to \infty$, $f(x) \to \infty$, and since the reciprocal of a large positive number is a small positive number, we have $g(x) \to 0$. As $x \to -\infty$, $f(x) \to -\infty$, and since the reciprocal of a large negative number is a small negative number, we have $g(x) \to 0$.

[−7.5, 7.5] by [−5, 5]

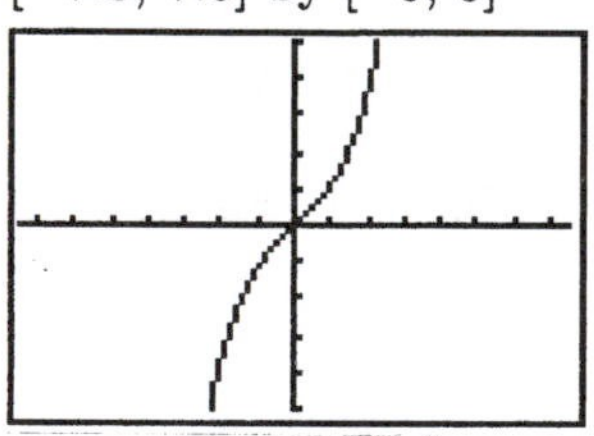

Figure 39(a)

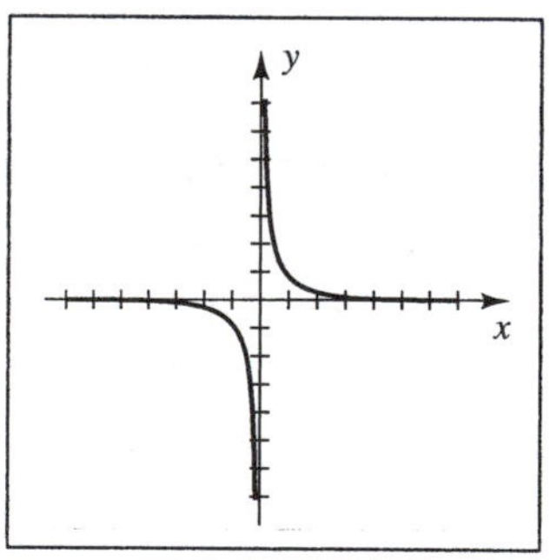


Figure 39(b)

41 (a) $f(x) = \dfrac{e^x - e^{-x}}{e^x + e^{-x}} = \dfrac{e^x - 1/e^x}{e^x + 1/e^x} \cdot \dfrac{e^x}{e^x} = \dfrac{e^{2x} - 1}{e^{2x} + 1}$. At $x = 0$, $f(x) = 0$. As $x \to \infty$, $f(x) \to 1$ since the numerator and denominator are nearly the same number. As $x \to -\infty$, $f(x) \to \dfrac{0 - 1}{0 + 1} = -1$. See *Figure 41(a)* on the next page.

(b) At $x = 0$, we will have a vertical asymptote. As $x \to \infty$, $f(x) \to 1$, and since $g$ is the reciprocal of $f$, $g(x) \to 1$. Similarly, as $x \to -\infty$, $g(x) \to -1$.

[−4.5, 4.5] by [−3, 3]

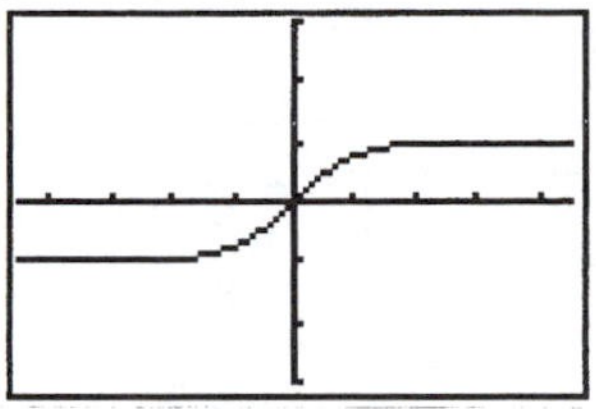

*Figure 41(a)*

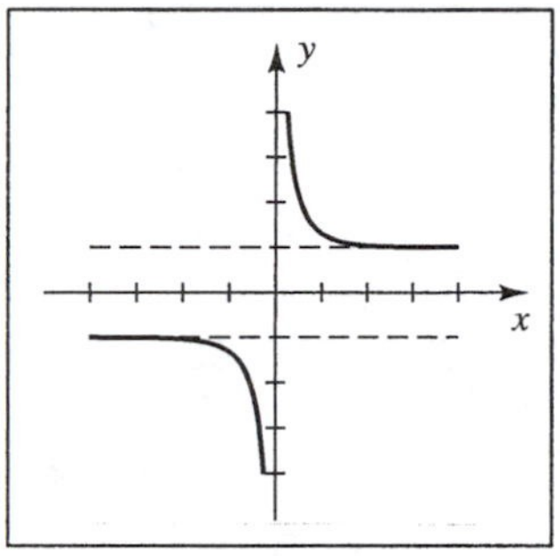


*Figure 41(b)*

**43** The approximate coordinates of the points where the graphs of $f$ and $g$ intersect are (−1.04, −0.92), (2.11, 2.44), and (8.51, 70.42). The region near the origin in *Figure 43(a)* is enhanced in *Figure 43(b)*. Thus, the solutions are $x \approx -1.04$, 2.11, and 8.51.

[−3, 11] by [−10, 80, 10]

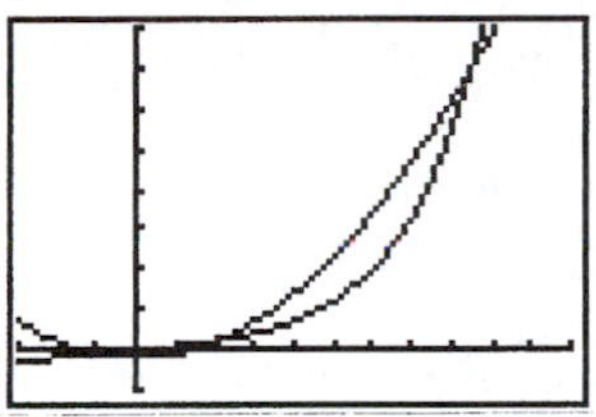

*Figure 43(a)*

[−2.26, 3.34] by [−7.14, 8.57, 10]

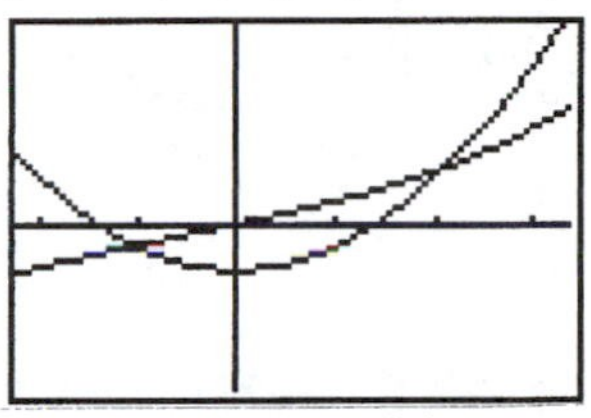

*Figure 43(b)*

**45** $f(x) = x + 1$ is a more accurate approximation to $e^x$ near $x = 0$, whereas $g(x) = 1.72x + 1$ is a more accurate approximation to $e^x$ near $x = 1$.

[0, 4.5] by [0, 3]

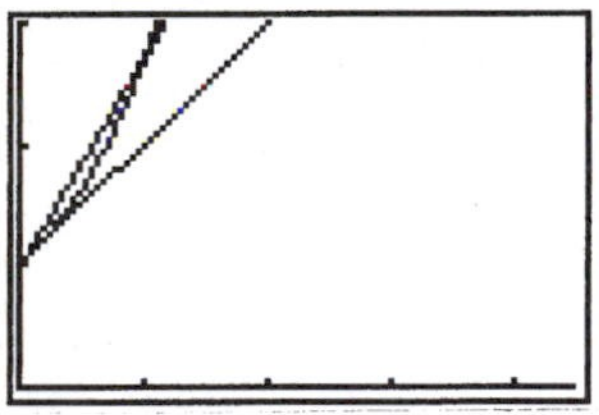

*Figure 45*

**47** From the graph, we see that $f$ has zeros at $x \approx 0.11$, 0.79, and 1.13.

[−2, 2.5] by [−1, 2]

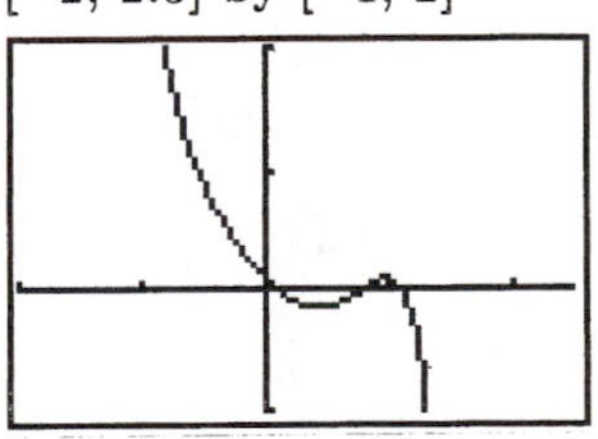

*Figure 47*

[0, 200, 50] by [0, 8]

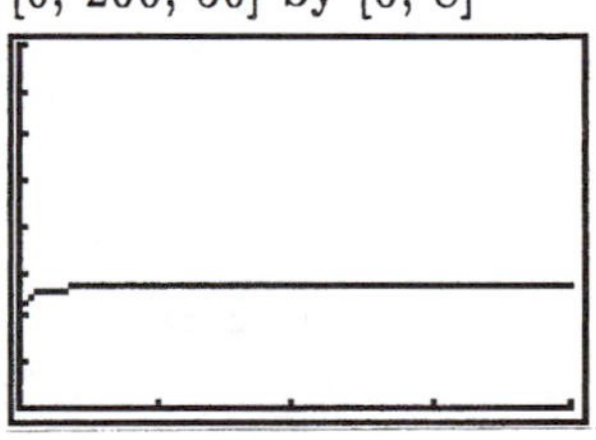

*Figure 49*

**49** From the graph, there is a horizontal asymptote of $y \approx 2.71$.

$f$ is approaching the value of $e$ asymptotically.

[51] $Y_1 = e^{-x}$ and $Y_2 = x$ intersect when $x \approx 0.567$.

[−4.5, 4.5] by [−3, 3]

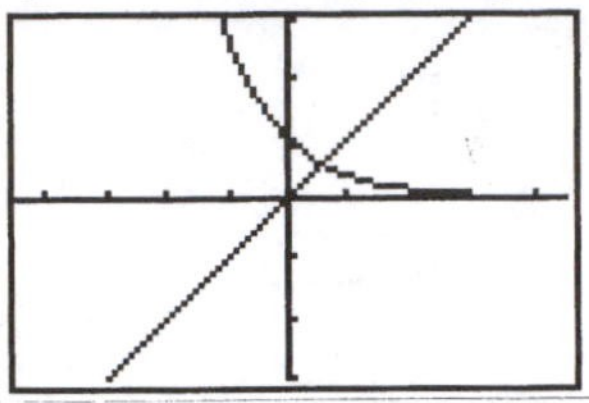

Figure 51

[−5.5, 5] by [−2, 5]

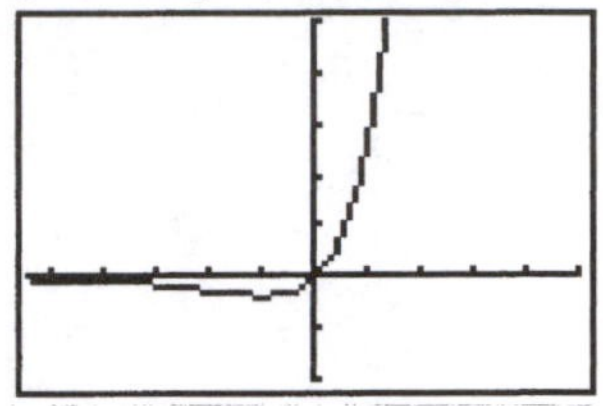

Figure 53

[53] $f$ is increasing on $[-1, \infty)$ and $f$ is decreasing on $(-\infty, -1]$.

[55] (a) When $y = 0$ and $z = 0$, the equation becomes $C = \dfrac{2Q}{\pi vab} e^{-h^2/(2b^2)}$. As $h$ increases, the exponent becomes a larger *negative* value, and hence the concentration $C$ decreases.

(b) When $z = 0$, the equation becomes $C = \dfrac{2Q}{\pi vab} e^{-y2/(2a2)} e^{-h^2/(2b^2)}$.

As $y$ increases, the concentration $C$ decreases.

[57] (a) Choose two arbitrary points that appear to lie on the curve such as (0, 1.225) and (10,000, 0.414).

$$f(0) = Ce^0 = C = 1.225 \quad \text{and} \quad f(10{,}000) = 1.225e^{10{,}000k} = 0.414.$$

To solve the last equation, graph $Y_1 = 1.225e^{10{,}000x}$ and $Y_2 = 0.414$. The graph of $Y_1$ increases so rapidly that it nearly "covers" the positive $y$-axis, so a viewing rectangle such as [−0.001, 0] by [0, 1] is needed to see the point of intersection. The graphs intersect at $x \approx -0.0001085$. Thus, $f(x) = 1.225e^{-0.0001085x}$.

(b) $f(3000) \approx 0.885$ { actual = 0.909 } and $f(9000) \approx 0.461$ { actual = 0.467 }.

[−1000, 10,100, 1000] by [0, 1.5, 0.5]

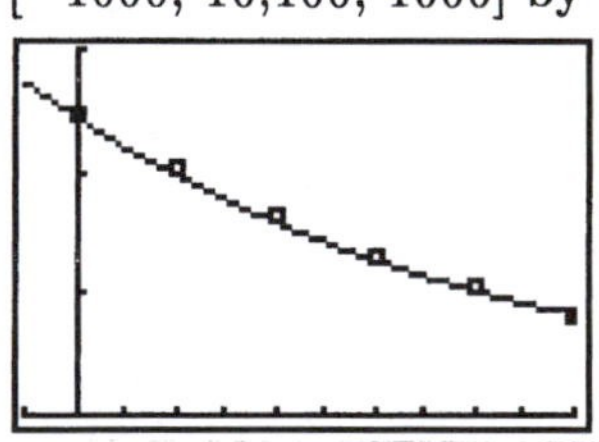

Figure 57

## 5.4 Exercises

*Note:* Exercises 1–4 are designed to familiarize the reader with the definition of $\log_a$ in this section. It is very important that you can generalize your understanding of this definition to the following case:

| |
|---|
| $\log_{\text{base}}(\text{argument}) = \text{exponent}$ *is equivalent to* $(\text{base})^{\text{exponent}} = \text{argument}$ |

Later in this section, you will also want to be able to use the following two special cases with ease:

| |
|---|
| $\log(\text{argument}) = \text{exponent}$ *is equivalent to* $(10)^{\text{exponent}} = \text{argument}$ |

| |
|---|
| $\ln(\text{argument}) = \text{exponent}$ *is equivalent to* $(e)^{\text{exponent}} = \text{argument}$ |

[1] (a) In this case, the *base* is 4, the *exponent* is 3, and the *argument* is 64. Thus,

$$4^3 = 64 \quad \text{is equivalent to} \quad \log_4 64 = 3.$$

(b) $4^{-3} = \frac{1}{64} \Leftrightarrow \log_4 \frac{1}{64} = -3$

(c) $t^r = s \Leftrightarrow \log_t s = r$

(d) $3^x = 4 - t \Leftrightarrow \log_3(4 - t) = x$

(e) In this case, the *base* is 5, the *exponent* is $7t$, and the *argument* is $\dfrac{a+b}{a}$. Thus,

$$5^{7t} = \frac{a+b}{a} \quad \text{is equivalent to} \quad \log_5 \frac{a+b}{a} = 7t.$$

(f) $(0.7)^t = 5.3 \Leftrightarrow \log_{0.7}(5.3) = t$

[3] (a) In this case, the *base* is 2, the *argument* is 32, and the *exponent* is 5. Thus,

$$\log_2 32 = 5 \quad \text{is equivalent to} \quad 2^5 = 32.$$

(b) $\log_3 \frac{1}{243} = -5 \Leftrightarrow 3^{-5} = \frac{1}{243}$

(c) $\log_t r = p \Leftrightarrow t^p = r$

(d) $\log_3(x+2) = 5 \Leftrightarrow 3^5 = (x+2)$

(e) In this case, the *base* is 2, the *argument* is $m$, and the *exponent* is $3x + 4$. Thus,

$$\log_2 m = 3x + 4 \quad \text{is equivalent to} \quad 2^{3x+4} = m.$$

(f) $\log_b 512 = \frac{3}{2} \Leftrightarrow b^{3/2} = 512$

[5] $2a^{t/3} = 5 \Leftrightarrow a^{t/3} = \frac{5}{2} \Leftrightarrow t/3 = \log_a \frac{5}{2} \Leftrightarrow t = 3\log_a \frac{5}{2}$

[7] In order to solve for $t$, we must isolate the expression containing $t$—in this case, that expression is the exponential $Ca^t$.

$$K = H - Ca^t \Rightarrow Ca^t = H - K \Rightarrow a^t = \frac{H-K}{C} \Rightarrow t = \log_a\left(\frac{H-K}{C}\right)$$

[9] In order to solve for $t$, we must isolate the expression containing $t$—in this case, that expression is the exponential $a^{Ct}$. $A = Ba^{Ct} + D \Rightarrow A - D = Ba^{Ct} \Rightarrow$

$$\frac{A-D}{B} = a^{Ct} \Rightarrow Ct = \log_a\left(\frac{A-D}{B}\right) \Rightarrow t = \frac{1}{C}\log_a\left(\frac{A-D}{B}\right).$$

The confusing step to most students in the above solution is $\frac{A-D}{B} = a^{Ct} \Rightarrow Ct = \log_a\left(\frac{A-D}{B}\right)$. This step is similar to $y = a^x \Rightarrow \log_a y = x$, except $x$ and $y$ are more complicated expressions.

[11] (a) Changing $10^5 = 100{,}000$ to logarithmic form gives us $\log_{10} 100{,}000 = 5$.

Since this is a common logarithm, we denote it as $\log 100{,}000 = 5$.

(b) $10^{-3} = 0.001 \Leftrightarrow \log 0.001 = -3$

(c) $10^x = y + 1 \Leftrightarrow \log(y+1) = x$

(d) Changing $e^7 = p$ to logarithmic form gives us $\log_e p = 7$.

Since this is a natural logarithm, we denote it as $\ln p = 7$.

(e) Changing $e^{2t} = 3 - x$ to logarithmic form gives us $\log_e(3-x) = 2t$.

Since this is a natural logarithm, we denote it as $\ln(3-x) = 2t$.

[13] (a) Remember that $\log x = 50$ is the same as $\log_{10} x = 50$.

Changing to exponential form, we have $10^{50} = x$.

(b) $\log x = 20t \; \{\log_{10} x = 20t\} \Leftrightarrow 10^{20t} = x$

(c) Remember that $\ln x = 0.1$ is the same as $\log_e x = 0.1$.

Changing to exponential form, we have $e^{0.1} = x$.

(d) $\ln w = 4 + 3x \; \{\log_e w = 4 + 3x\} \Leftrightarrow e^{4+3x} = w$

(e) $\ln(z-2) = \frac{1}{6} \; \{\log_e(z-2) = \frac{1}{6}\} \Leftrightarrow e^{1/6} = z - 2$

[15] (a) $\log_5 1 = 0$ since $5^0 = 1$ $\{\log_a 1 = 0$ for any allowable base $a\}$

(b) $\log_3 3 = 1$ $\{\log_a a = 1$ since $a^1 = a\}$

(c) Remember that you cannot take the logarithm, any base, of a negative number.

Hence, $\log_4(-2)$ is undefined.

(d) $\log_7 7^2 = 2 \; \{\log_a a^x = x\}$

(e) $3^{\log_3 8} = 8 \; \{a^{\log_a x} = x\}$

(f) $\log_5 125 = \log_5 5^3 = 3$, as in part (d)

(g) We will change the form of $\frac{1}{16}$ so that it can be written as an exponential expression with the same base as the logarithm—in this case, that base is 4.

$$\log_4 \tfrac{1}{16} = \log_4 4^{-2} = -2$$

[17] Parts (a)–(f) are direct applications of the properties in the chart on page 362.

(a) $10^{\log 3} = 3$ (b) $\log 10^5 = 5$ (c) $\log 100 = \log 10^2 = 2$

(d) $\log 0.0001 = \log 10^{-4} = -4$ (e) $e^{\ln 2} = 2$ (f) $\ln e^{-3} = -3$

For part (g), we use a property of exponents that will enable us to use the property

$$e^{\ln x} = x.$$

(g) $e^{2 + \ln 3} = e^2 e^{\ln 3} = e^2(3) = 3e^2$

[19] $\log_4 x = \log_4 (8 - x) \Rightarrow x = 8 - x$ {since the logarithm function is one-to-one} $\Rightarrow 2x = 8 \Rightarrow x = 4$. We must check to make sure that all proposed solutions do not make any of the original expressions undefined. Checking $x = 4$ in the original equation gives us $\log_4 4 = \log_4 (8 - 4)$, which is a true statement, so $x = 4$ is a valid solution.

[21] $\log_5 (x - 2) = \log_5 (3x + 7) \Rightarrow x - 2 = 3x + 7 \Rightarrow 2x = -9 \Rightarrow x = -\frac{9}{2}$. The value $x = -\frac{9}{2}$ is extraneous since it makes either of the original logarithm expressions undefined. Hence, there is no solution.

[23] $\log x^2 = \log (-3x - 2) \Rightarrow x^2 = -3x - 2 \Rightarrow x^2 + 3x + 2 = 0 \Rightarrow$ $(x + 1)(x + 2) = 0 \Rightarrow x = -1, -2$.

Checking $-1$ and $-2$, we find that both are valid solutions.

[25] $\log_3 (x - 4) = 2 \Rightarrow x - 4 = 3^2 \Rightarrow x - 4 = 9 \Rightarrow x = 13$

[27] $\log_9 x = \frac{3}{2} \Rightarrow x = 9^{3/2} = (9^{1/2})^3$ {remember, root first, power second} $= 3^3 = 27$

[29] $\ln x^2 = -2 \Rightarrow x^2 = e^{-2} \Rightarrow x^2 = \frac{1}{e^2} \Rightarrow x = \pm\sqrt{\frac{1}{e^2}} \Rightarrow x = \pm\frac{1}{e}$

[31] $e^{2 \ln x} = 9 \Rightarrow (e^{\ln x})^2 = 9 \Rightarrow x^2 = 9 \Rightarrow x = \pm 3$; $-3$ is extraneous

[33] $e^{x \ln 3} = 27 \Rightarrow (e^{\ln 3})^x = 27 \Rightarrow 3^x = 3^3 \Rightarrow x = 3$

**35** (a) $f(x) = \log_4 x$ • This graph has a vertical asymptote of $x = 0$ and goes through $(\frac{1}{4}, -1)$, $(1, 0)$, and $(4, 1)$. For reference purposes, call this $F(x)$.

(b) $f(x) = -\log_4 x$ • reflect $F$ through the $x$-axis

(c) $f(x) = 2\log_4 x$ • vertically stretch $F$ by a factor of 2

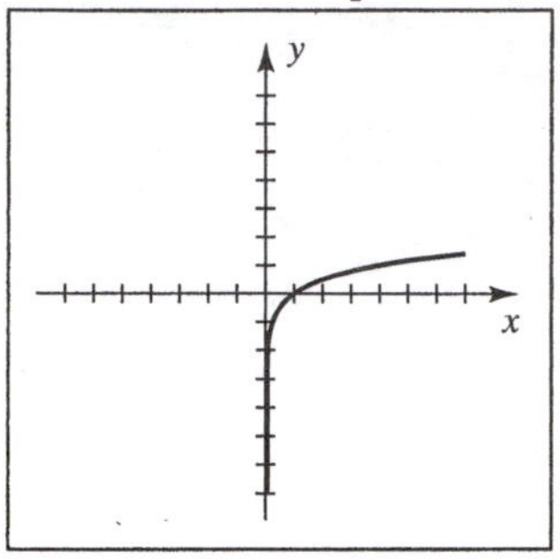


*Figure 35(a)*

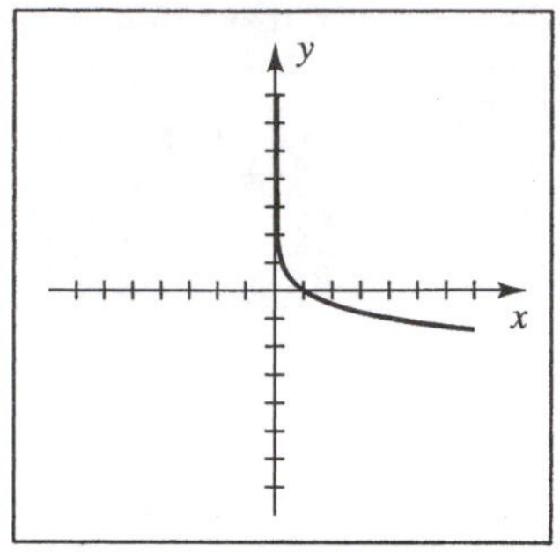


*Figure 35(b)*

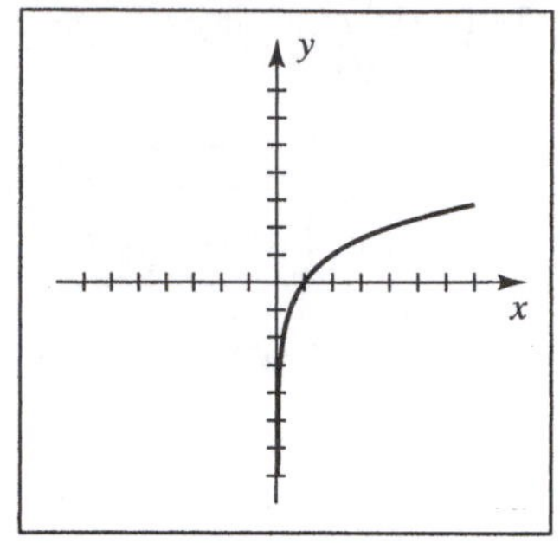


*Figure 35(c)*

(d) $f(x) = \log_4 (x + 2)$ • shift $F$ left 2 units

(e) $f(x) = (\log_4 x) + 2$ • shift $F$ up 2 units

(f) $f(x) = \log_4 (x - 2)$ • shift $F$ right 2 units

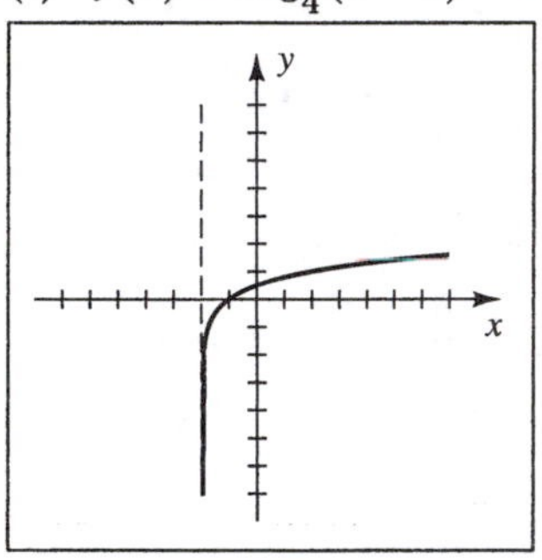


*Figure 35(d)*

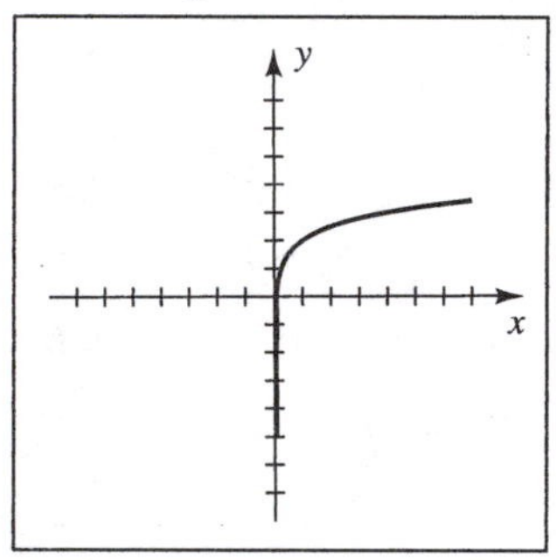


*Figure 35(e)*

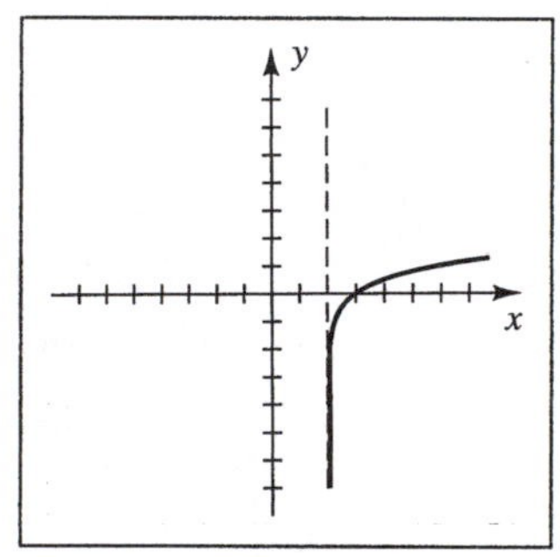


*Figure 35(f)*

(g) $f(x) = (\log_4 x) - 2$ • shift $F$ down 2 units

(h) $f(x) = \log_4 |\, x \,|$ • include the reflection of $F$ through the $y$-axis since $x$ may be positive or negative, but $\log_4 |\, x \,|$ will give the same result

(i) $f(x) = \log_4 (-x)$ • $x$ must be negative so that $-x$ is positive, reflect $F$ through the $y$-axis

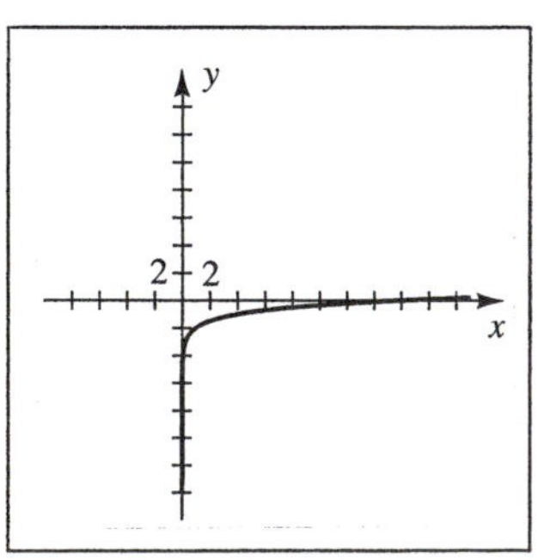


*Figure 35(g)*

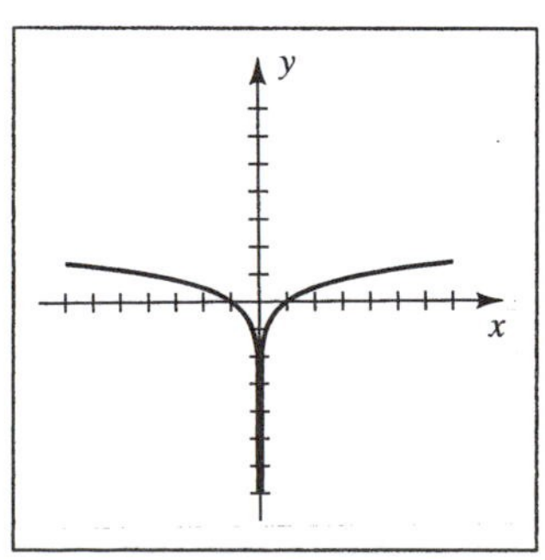


*Figure 35(h)*

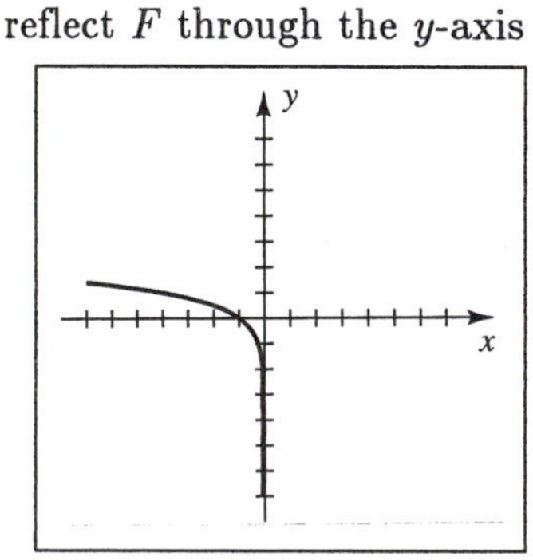


*Figure 35(i)*

(j) $f(x) = \log_4(3 - x) = \log_4[-(x - 3)]$ • Shift $F$ right 3 units and reflect through the line $x = 3$. It may be helpful to determine the domain of this function. We know that $3 - x$ must be positive for the function to be defined. Thus, $3 - x > 0 \Rightarrow 3 > x$, or, equivalently, $x < 3$.

(k) $f(x) = |\log_4 x|$ • reflect points with negative $y$-coordinates through the $x$-axis

(l) $f(x) = \log_{1/4} x$ • reflect the graph of $y = \log_4 x$ about the $x$-axis

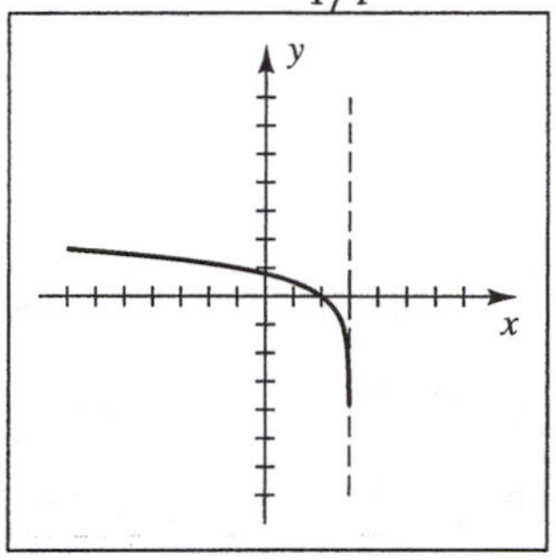

*Figure 35(j)*

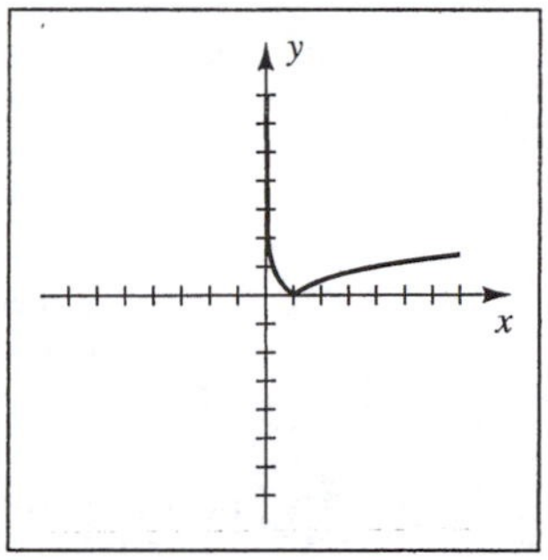

*Figure 35(k)*

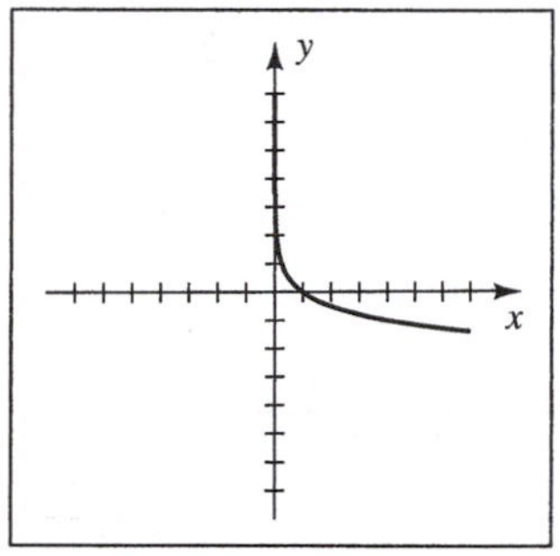

*Figure 35(l)*

[37] $f(x) = \log(x + 10)$ • This is the graph of $g(x) = \log x$ shifted 10 units to the left. There is a vertical asymptote of $x = -10$. $g$ goes through $(\frac{1}{10}, -1)$, $(1, 0)$, and $(10, 1)$, so $f$ goes through $(\frac{1}{10} - 10, -1)$, $(1 - 10, 0)$, and $(10 - 10, 1)$, that is, $(-9.9, -1)$, $(-9, 0)$, and $(0, 1)$.

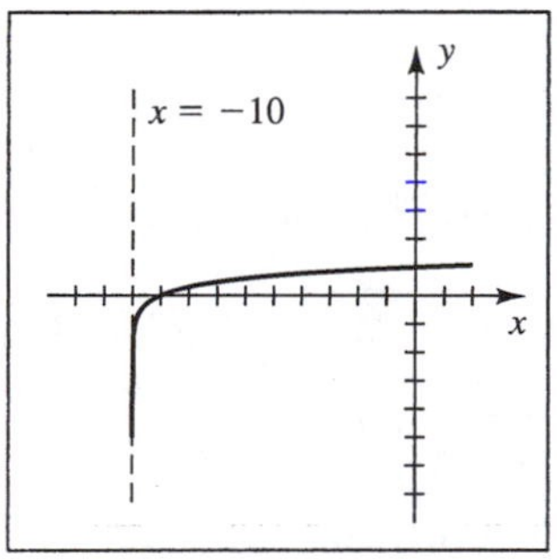

*Figure 37*

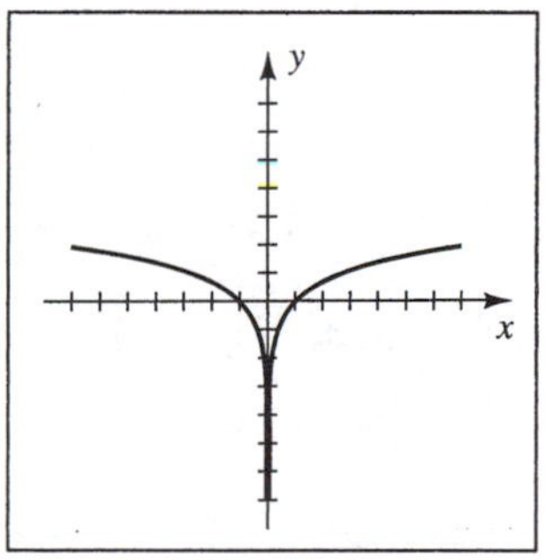

*Figure 39*

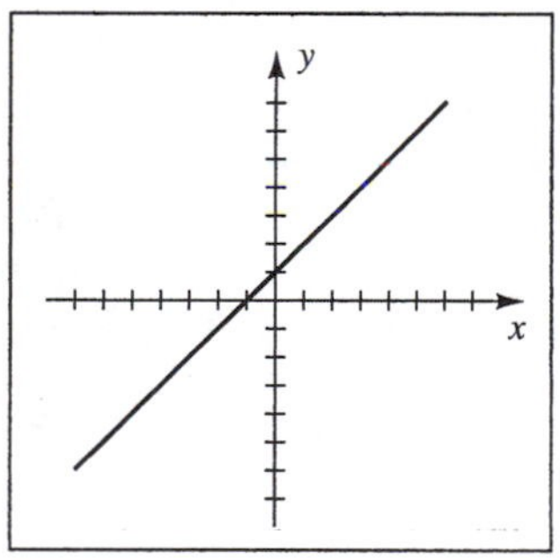

*Figure 41*

[39] $f(x) = \ln|x|$ • This is the graph of $g(x) = \ln x$ and its reflection through the $y$-axis. There is a vertical asymptote of $x = 0$. The graph of $g$ goes through $(\frac{1}{e}, -1)$, $(1, 0)$, and $(e, 1)$, so the graph of $f$ goes through $(\pm\frac{1}{e}, -1)$, $(\pm 1, 0)$, and $(\pm e, 1)$.

[41] $f(x) = \ln e + x = 1 + x$. This is the graph of a line with slope 1 and $y$-intercept 1. See *Figure 41* above.

[43] The point $(9, 2)$ is on the graph of $f(x) = \log_a x$, so $f(9) = 2 \Rightarrow 2 = \log_a 9 \Rightarrow a^2 = 9 \Rightarrow a = \pm 3$. Since the base $a$ must be positive, $a = 3$.

[45] This is the graph of $F$ reflected through the $x$-axis. ★ $f(x) = -F(x)$

[47] This is the graph of $F$ shifted right 2 units. ★ $f(x) = F(x-2)$

[49] This is the graph of $F$ shifted up 1 unit. ★ $f(x) = F(x) + 1$

[51] (a) $\log x = 3.6274 \Rightarrow x = 10^{3.6274} \approx 4240.333$, or 4240 to three significant figures

(b) $\log x = 0.9469 \Rightarrow x = 10^{0.9469} \approx 8.849$, or 8.85

(c) $\log x = -1.6253 \Rightarrow x = 10^{-1.6253} \approx 0.023697$, or 0.0237

(d) $\ln x = 2.3 \Rightarrow x = e^{2.3} \approx 9.974$, or 9.97

(e) $\ln x = 0.05 \Rightarrow x = e^{0.05} \approx 1.051$, or 1.05

(f) $\ln x = -1.6 \Rightarrow x = e^{-1.6} \approx 0.2019$, or 0.202

[53] $f(x) = 1000(1.05)^x = 1000e^{x \ln 1.05}$ (see the illustration on page 362).

This is approximately $1000e^{x\,(0.0487901642)} \approx 1000e^{0.0488x}$,

so the growth rate of $f$ is about 4.88%.

[55] $q = q_0(2)^{-t/1600} \Rightarrow \frac{q}{q_0} = 2^{-t/1600}$ {change to logarithm form} $\Rightarrow$

$-\frac{t}{1600} = \log_2(\frac{q}{q_0})$ {multiply by $-1600$} $\Rightarrow t = -1600\log_2(\frac{q}{q_0})$

[57] $I = 20e^{-Rt/L} \Rightarrow \frac{I}{20} = e^{-Rt/L} \Rightarrow \ln(\frac{I}{20}) = -\frac{Rt}{L} \Rightarrow t = -\frac{L}{R}\ln(\frac{I}{20})$

[59] $I = 10^a I_0 \Rightarrow R = \log\left(\frac{I}{I_0}\right) = \log\left(\frac{10^a I_0}{I_0}\right) = \log 10^a = a.$

Hence, for $10^2 I_0$, $10^4 I_0$, and $10^5 I_0$, the answers are: (a) 2 (b) 4 (c) 5

[61] We will find a general formula for $\alpha$ first.

$I = 10^a I_0 \Rightarrow \alpha = 10\log\left(\frac{I}{I_0}\right) = 10\log\left(\frac{10^a I_0}{I_0}\right) = 10(\log 10^a) = 10(a) = 10a.$

Hence, for $10^1 I_0$, $10^3 I_0$, and $10^4 I_0$, the answers are: (a) 10 (b) 30 (c) 40

[63] 1980 corresponds to $t = 0$ and $N(0) = 227$ million. We need to find $t$ when $N = 2 \cdot 227$. $2 \cdot 227 = 227e^{0.007t} \Rightarrow 2 = e^{0.007t} \Rightarrow \ln 2 = 0.007t \Rightarrow$

$t = \frac{\ln 2}{0.007} \Rightarrow t \approx 99$, which corresponds to the year $1980 + 99 = 2079$.

Alternatively, using the doubling time formula from page 364,

$$t = (\ln 2)/r = (\ln 2)/0.007 \approx 99.$$

[65] (a) $\ln W = \ln 2.4 + (1.84)h$ {change to exponential form} $\Rightarrow$

$W = e^{[\ln 2.4 + (1.84)h]} \Rightarrow W = e^{\ln 2.4}e^{1.84h}$ {since $e^x e^y = e^{x+y}$} $\Rightarrow$

$W = 2.4e^{1.84h}$

(b) $h = 1.5 \Rightarrow W = 2.4e^{(1.84)(1.5)} = 2.4e^{2.76} \approx 37.92$ kg

[67] (a) $p(h) = 10 \;\Rightarrow\; 10 = 14.7e^{-0.0000385h} \;\Rightarrow\; \dfrac{10}{14.7} = e^{-0.0000385h} \;\Rightarrow$

$$\ln\left(\tfrac{10}{14.7}\right) = -0.0000385h \;\Rightarrow\; h = -\tfrac{1}{0.0000385}\ln\left(\tfrac{10}{14.7}\right) \approx 10{,}007 \text{ ft.}$$

(b) At sea level, $h = 0$, and $p(0) = 14.7$. Setting $p(h)$ equal to $\frac{1}{2}(14.7)$,

and solving as in part (a), we have $h = -\frac{1}{0.0000385}\ln\left(\frac{1}{2}\right) \approx 18{,}004$ ft.

[69] (a) $t = 0 \;\Rightarrow\; W = 2600\left(1 - 0.51e^{-0.075(0)}\right)^3 = 2600(0.49)^3 \approx 305.9$ kg

(b) (1) From the graph, if $W = 1800$, $t$ appears to be about 20.

(2) Solving the equation for $t$, we have $1800 = 2600\left(1 - 0.51e^{-0.075t}\right)^3 \;\Rightarrow$

$$\tfrac{1800}{2600} = \left(1 - 0.51e^{-0.075t}\right)^3 \;\Rightarrow\; \sqrt[3]{\tfrac{9}{13}} = 1 - 0.51e^{-0.075t} \;\Rightarrow$$

$$\tfrac{51}{100}e^{-0.075t} = 1 - \sqrt[3]{\tfrac{9}{13}} \;\Rightarrow\; e^{-0.075t} = \left(\tfrac{100}{51}\right)\left(1 - \sqrt[3]{\tfrac{9}{13}}\right) \;\Rightarrow$$

$$-0.075t = \ln\left[\left(\tfrac{100}{51}\right)\left(1 - \sqrt[3]{\tfrac{9}{13}}\right)\right] \;\Rightarrow\; t = \frac{\ln\left[\left(\frac{100}{51}\right)\left(1 - \sqrt[3]{\frac{9}{13}}\right)\right]}{-0.075} \approx 19.8 \text{ yr.}$$

[71] $D = ae^{-bx}$ with $a = 5.5$ and $b = 0.10$ • The population density equals 2000 per square mile means that $D = 2$. $D = 2 \;\Rightarrow\; 5.5e^{-0.1x} = 2 \;\Rightarrow\; e^{-0.1x} = \frac{4}{11} \;\Rightarrow$

$$-0.1x = \ln\tfrac{4}{11} \;\Rightarrow\; x = -10\ln\tfrac{4}{11} \text{ mi} \approx 10.1 \text{ mi}$$

[73] Since the half-life is eight days, $A(t) = \frac{1}{2}A_0$ when $t = 8$.

Thus, $\frac{1}{2}A_0 = A_0a^{-8} \;\Rightarrow\; a^{-8} = \frac{1}{2} \;\Rightarrow\; \frac{1}{a^8} = \frac{1}{2} \;\Rightarrow$

$$a^8 = 2 \;\Rightarrow\; a = 2^{1/8} \text{ \{take the eighth root\}} \approx 1.09.$$

[75] (a) Since $\log P$ is an increasing function, increasing the population increases the walking speed. Pedestrians have faster average walking speeds in large cities.

(b) $S = 5 \;\Rightarrow\; 5 = 0.05 + 0.86\log P \;\Rightarrow\; 4.95 = 0.86\log P \;\Rightarrow$

$$\tfrac{4.95}{0.86} = \log P \;\Rightarrow\; P = 10^{4.95/0.86} \approx 570{,}000$$

[77] (a) $f(x) = \ln(x+1) + e^x \;\Rightarrow\; f(2) = \ln(2+1) + e^2 \approx 8.4877$

(b) $g(x) = \dfrac{(\log x)^2 - \log x}{4} \;\Rightarrow\; g(3.97) = \dfrac{(\log 3.97)^2 - \log 3.97}{4} \approx -0.0601$

[79] *Figure 79* shows a graph of $Y_1 = x\ln x$ and $Y_2 = 1$. By using an intersect feature, we determine that $x\ln x = 1$ when $x \approx 1.763$. Alternatively, we could graph $Y_1 = x\ln x - 1$, and find the zero of that graph.

[0, 4] by [−1, 1.67]

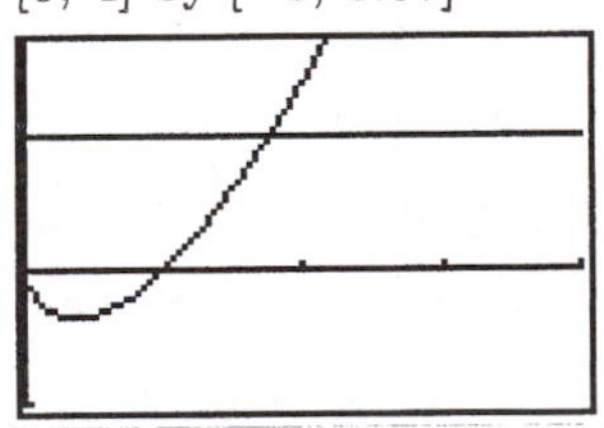

*Figure 79*

81 The domain of $f(x) = 2.2 \log(x+2)$ is $x > -2$. The domain of $g(x) = \ln x$ is $x > 0$. From *Figure 81*, we determine that $f$ intersects $g$ at about 14.90. Thus, $f(x) \geq g(x)$ on approximately (0, 14.90], not (−2, 14.90] since $g$ is not defined if $x \leq 0$. In general, the larger the base of the logarithm, the slower its graph will rise. In this case, we have base 10 and base $e \approx 2.72$, so we know that $g$ will eventually intersect $f$ even if we don't see this intersection in our first window.

[−2, 16] by [−4, 8]

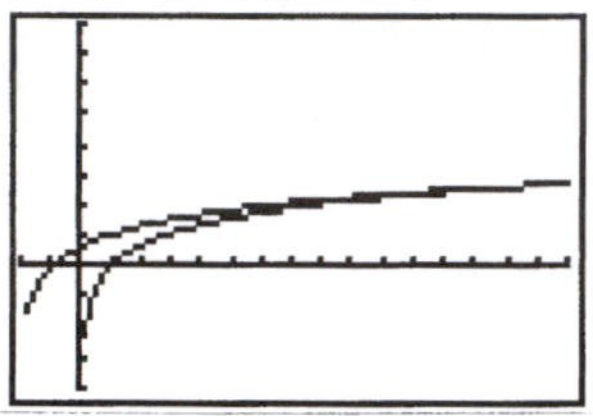

*Figure 81*

[3, 5, 0.5] by [0, 1, 0.5]

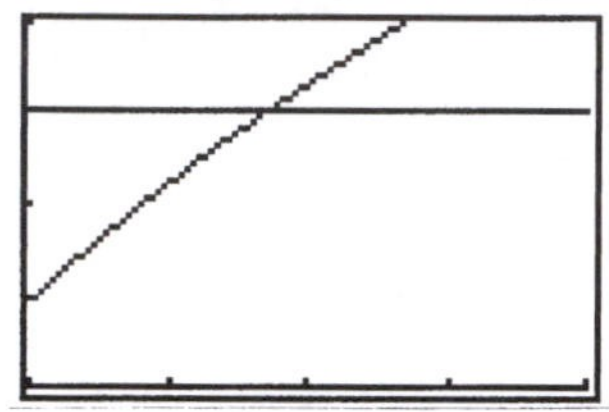

*Figure 83*

83 (a) $R = 2.07 \ln x - 2.04 = 2.07 \ln \frac{C}{H} - 2.04 = 2.07 \ln \frac{242}{78} - 2.04 \approx 0.3037 \approx 30\%$.

(b) Graph $Y_1 = 2.07 \ln x - 2.04$ and $Y_2 = 0.75$.

From the graph, $R \approx 0.75$ when $x = \frac{C}{H} \approx 3.85$.

## 5.5 Exercises

1 (a) $\log_4(xz) = \log_4 x + \log_4 z$ (b) $\log_4(y/x) = \log_4 y - \log_4 x$

(c) $\log_4 \sqrt[3]{z} = \log_4 z^{1/3} = \frac{1}{3} \log_4 z$

3 $\log_a \frac{x^3 w}{y^2 z^4} = \log_a(x^3 w) - \log_a(y^2 z^4) = (\log_a x^3 + \log_a w) - (\log_a y^2 + \log_a z^4)$

$= 3 \log_a x + \log_a w - 2 \log_a y - 4 \log_a z$

The most common mistake is to not have the minus sign in front of $4 \log_a z$. This error results from not having the parentheses in the correct place.

5 $\log \frac{\sqrt[3]{z}}{x\sqrt{y}} = \log \sqrt[3]{z} - \log x\sqrt{y} = \log z^{1/3} - \log x - \log y^{1/2} = \frac{1}{3} \log z - \log x - \frac{1}{2} \log y$

7 $\ln \sqrt[4]{\frac{x^7}{y^5 z}} = \ln x^{7/4} - \ln y^{5/4} z^{1/4} = \ln x^{7/4} - \ln y^{5/4} - \ln z^{1/4} = \frac{7}{4} \ln x - \frac{5}{4} \ln y - \frac{1}{4} \ln z$

As a generalization for exercises similar to those in 1–8, if the exponents on the variables are positive, then the sign in front of the individual logarithms will be positive if the variable was originally in the numerator and negative if the variable was originally in the denominator.

9 (a) $\log_3 x + \log_3(5y) = \log_3(x \cdot 5y) = \log_3(5xy)$

(b) $\log_3(2z) - \log_3 x = \log_3(2z/x)$ (c) $5 \log_3 y = \log_3 y^5$

[11] $2\log_a x + \frac{1}{3}\log_a(x-2) - 5\log_a(2x+3)$

$$= \log_a x^2 + \log_a(x-2)^{1/3} - \log_a(2x+3)^5 \qquad \{\text{logarithm law } 3\}$$

$$= \log_a x^2\sqrt[3]{x-2} - \log_a(2x+3)^5 \qquad \{\text{logarithm law } 1\}$$

$$= \log_a \frac{x^2\sqrt[3]{x-2}}{(2x+3)^5} \qquad \{\text{logarithm law } 2\}$$

[13] $\log(x^3y^2) - 2\log x\sqrt[3]{y} - 3\log\left(\frac{x}{y}\right) = \log(x^3y^2) - \left[\log(x\sqrt[3]{y})^2 + \log\left(\frac{x}{y}\right)^3\right]$

$$= \log(x^3y^2) - \left[\log(x^2y^{2/3}\cdot(x^3/y^3))\right]$$

$$= \log\frac{x^3y^2}{x^5y^{-7/3}} = \log\frac{y^{13/3}}{x^2}$$

[15] $\ln y^3 + \frac{1}{3}\ln(x^3y^6) - 5\ln y = \ln y^3 + \ln(xy^2) - \ln y^5 = \ln\left[(xy^5)/y^5\right] = \ln x$

[17] $\log_6(2x-3) = \log_6 12 - \log_6 3 \Rightarrow \log_6(2x-3) = \log_6\frac{12}{3} \Rightarrow 2x-3=4 \Rightarrow$

$$2x = 7 \Rightarrow x = \tfrac{7}{2}$$

[19] $2\log_3 x = 3\log_3 5 \Rightarrow \log_3 x^2 = \log_3 5^3 \Rightarrow x^2 = 125 \Rightarrow x = \pm\sqrt{125} = \pm 5\sqrt{5}$;

$-5\sqrt{5}$ is extraneous since it would make $\log_3 x$ undefined

[21] $\log x - \log(x+1) = 3\log 4 \Rightarrow \log\frac{x}{x+1} = \log 4^3 \Rightarrow \frac{x}{x+1} = 64 \Rightarrow$

$x = 64x + 64 \Rightarrow -63x = 64 \Rightarrow x = -\frac{64}{63}$. But $-\frac{64}{63}$ is extraneous since it makes an argument negative, so there is no solution for the equation.

[23] $\ln(-4-x) + \ln 3 = \ln(2-x) \Rightarrow \ln\left[(-4-x)\cdot 3\right] = \ln(2-x) \Rightarrow$

$\ln(-12-3x) = \ln(2-x) \Rightarrow -12-3x = 2-x \Rightarrow 2x = -14 \Rightarrow x = -7$.

Remember, the solution of a logarithmic equation may be negative—you must examine what happens to the original logarithm expressions. In this case, we have $\ln 3 + \ln 3 = \ln 9$, which is true.

[25] $\log_2(x+7) + \log_2 x = 3 \Rightarrow \log_2(x^2+7x) = 3 \Rightarrow x^2+7x = 2^3 \Rightarrow$

$x^2+7x-8=0 \Rightarrow (x+8)(x-1)=0 \Rightarrow x=-8,\ 1$; $-8$ is extraneous

[27] $\log_3(x+3) + \log_3(x+5) = 1 \Rightarrow \log_3\left[(x+3)(x+5)\right] = 1 \Rightarrow$

$\log_3(x^2+8x+15) = 1 \Rightarrow x^2+8x+15 = 3 \Rightarrow$

$x^2+8x+12=0 \Rightarrow (x+2)(x+6)=0 \Rightarrow x=-6,\ -2$; $-6$ is extraneous

[29] $\log(x+3) = 1 - \log(x-2) \Rightarrow \log(x+3) + \log(x-2) = 1 \Rightarrow$

$\log\left[(x+3)(x-2)\right] = 1 \Rightarrow x^2+x-6 = 10^1 \Rightarrow$

$x^2+x-16=0 \Rightarrow x = \frac{-1+\sqrt{65}}{2} \approx 3.53$; $\frac{-1-\sqrt{65}}{2} \approx -4.53$ is extraneous

**31** $\ln x = 1 - \ln(x+2) \;\Rightarrow\; \ln x + \ln(x+2) = 1 \;\Rightarrow$
$\ln[x(x+2)] = 1 \;\Rightarrow\; x^2 + 2x = e^1 \;\Rightarrow\; x^2 + 2x - e = 0 \;\Rightarrow$

$$x = \frac{-2 \pm \sqrt{4+4e}}{2} = \frac{-2 \pm 2\sqrt{1+e}}{2} = -1 \pm \sqrt{1+e}.$$

$x = -1 + \sqrt{1+e} \approx 0.93$ is a valid solution,

but $x = -1 - \sqrt{1+e} \approx -2.93$ is extraneous.

**33**

| | |
|---|---|
| $\log_3(x-2) = \log_3 27 - \log_3(x-4) - 5^{\log_5 1}$ | given |
| $\log_3(x-2) + \log_3(x-4) = \log_3 3^3 - 1$ | $5^{\log_5 1} = 1$ |
| $\log_3[(x-2)(x-4)] = 3 - 1$ | $\log_3 3^3 = 3$ |
| $\log_3(x^2 - 6x + 8) = 2$ | simplify |
| $x^2 - 6x + 8 = 3^2$ | exponential form |
| $x^2 - 6x - 1 = 0$ | subtract 9 |

Now solve $x^2 - 6x - 1 = 0$ using the quadratic formula to get
$x = (6 \pm \sqrt{40})/2 = 3 \pm \sqrt{10}$. Since $3 - \sqrt{10}$ makes the arguments of the logarithms negative, the only solution is $3 + \sqrt{10}$.

**35** $f(x) = \log_3(3x) = \log_3 3 + \log_3 x = \log_3 x + 1$ • shift $y = \log_3 x$ up 1 unit

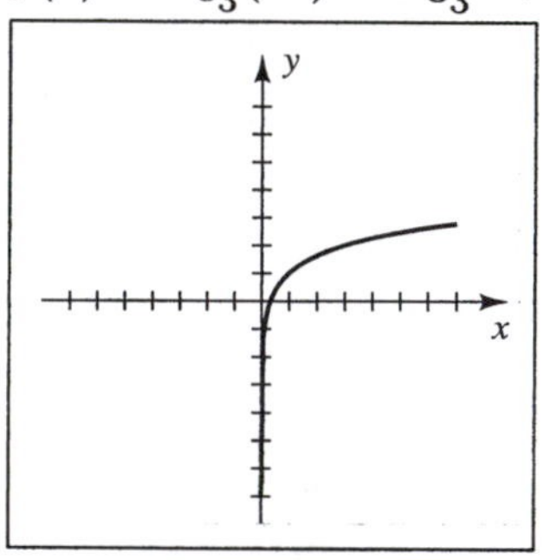

*Figure 35*

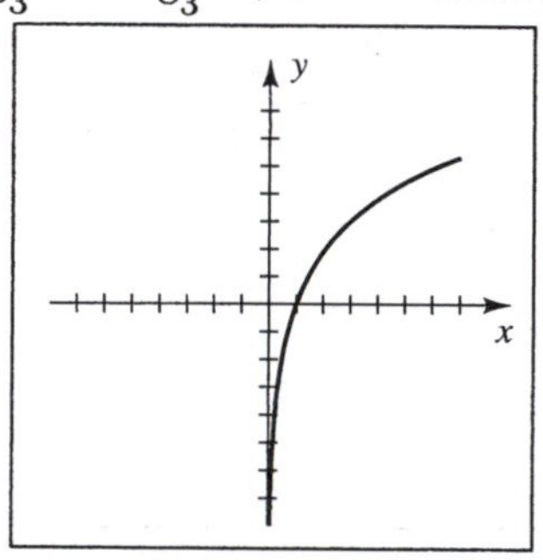

*Figure 37*

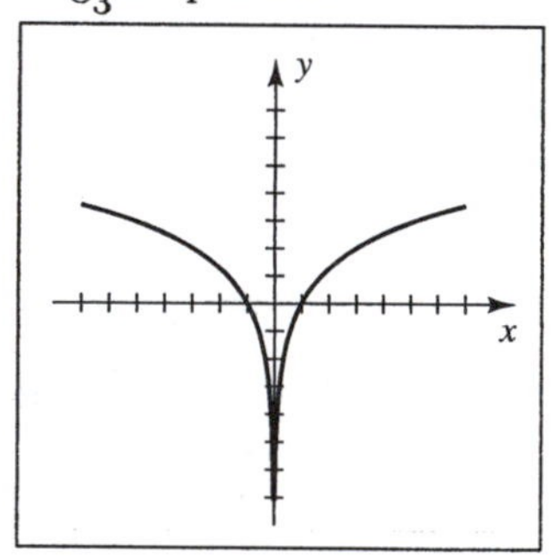

*Figure 39*

**37** $f(x) = 3\log_3 x$ • vertically stretch $y = \log_3 x$ by a factor of 3

**39** $f(x) = \log_3(x^2) = 2\log_3 x$ • Vertically stretch $y = \log_3 x$ by a factor of 2 and include its reflection through the $y$-axis since the domain of the original function, $f(x) = \log_3(x^2)$, is $\mathbb{R} - \{0\}$. Keep in mind that the laws of logarithms are established for positive real numbers, so that when we make the step $\log_3(x^2) = 2\log_3 x$, it is only for positive $x$. See *Figure 39* above.

[41] $f(x) = \log_2(x^3) = 3\log_2 x$ • vertically stretch $y = \log_2 x$ by a factor of 3

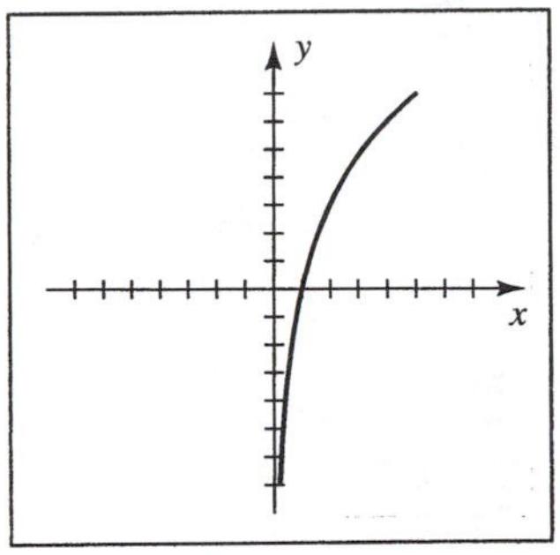

*Figure 41*

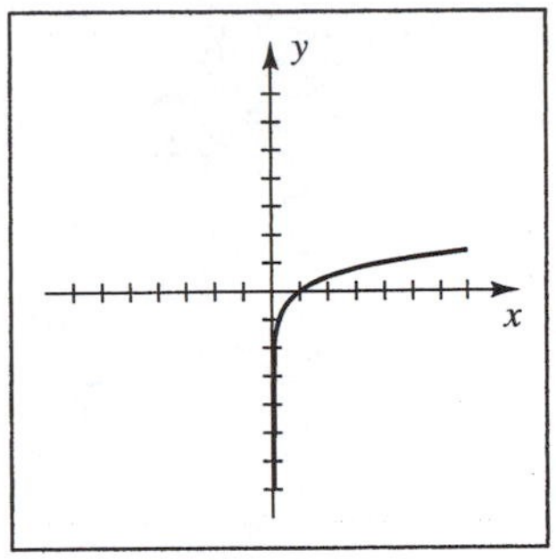

*Figure 43*

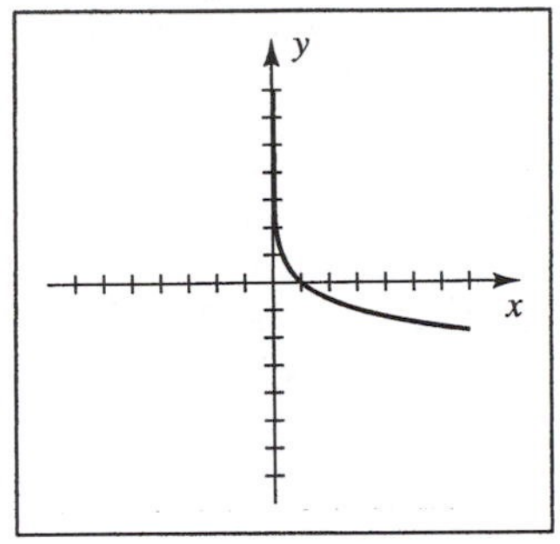

*Figure 45*

[43] $f(x) = \log_2 \sqrt{x} = \log_2 x^{1/2} = \frac{1}{2}\log_2 x$ •
vertically compress $y = \log_2 x$ by a factor of $1/(1/2) = 2$

[45] $f(x) = \log_3(\frac{1}{x}) = \log_3 x^{-1} = -\log_3 x$ • reflect $y = \log_3 x$ through the $x$-axis (See *Figure 45* above.)

[47] The values of $F(x) = \log_2 x$ are doubled and the reflection of $F$ through the $y$-axis is included. The domain of $f$ is $\mathbb{R} - \{0\}$ and $f(x) = \log_2 x^2$.

[49] $F(x) = \log_2 x$ is shifted up 3 units since (1, 0) on $F$ is (1, 3) on the graph.
Hence, $f(x) = 3 + \log_2 x = \log_2 2^3 + \log_2 x = \log_2(8x)$.

[51] $V_1 = 2$ and $V_2 = 4.5 \Rightarrow \text{db} = 20\log\frac{V_2}{V_1} = 20\log\frac{4.5}{2} \approx 7.04$, or +7.

[53] $\log y = \log b - k\log x \Rightarrow \log y = \log b - \log x^k \Rightarrow \log y = \log\frac{b}{x^k} \Rightarrow y = \frac{b}{x^k}$

[55] $c = 0.5$ and $z_0 = 0.1 \Rightarrow$

$$\begin{aligned} v &= c\ln(z/z_0) \\ &= (0.5)\ln(z/0.1) \\ &= \tfrac{1}{2}\ln(10z) \\ &= \tfrac{1}{2}(\ln 10 + \ln z) \\ &= \tfrac{1}{2}\ln 10 + \tfrac{1}{2}\ln z \approx \tfrac{1}{2}\ln z + 1.15. \end{aligned}$$

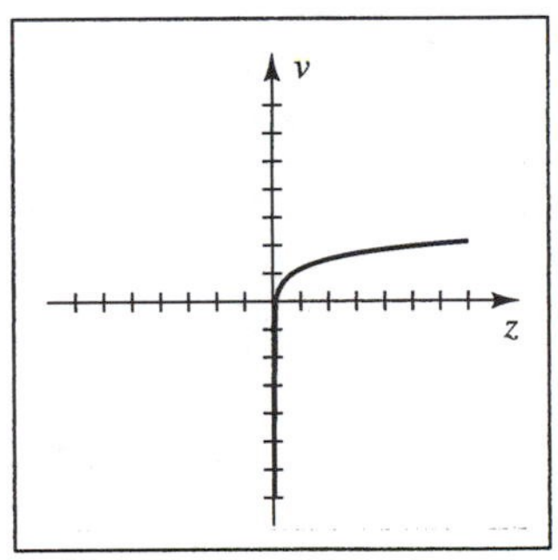

*Figure 55*

[57] (a) $R(x) = a\log(\frac{x}{x_0}) \Rightarrow R(x_0) = a\log(\frac{x_0}{x_0}) = a\log 1 = a \cdot 0 = 0$

(b) $R(2x) = a\log(\frac{2x}{x_0}) = a\log(2\cdot\frac{x}{x_0}) = a\left[\log 2 + \log(\frac{x}{x_0})\right]$

$= a\log 2 + a\log(\frac{x}{x_0}) = R(x) + a\log 2$

[59] $\ln I_0 - \ln I = kx \Rightarrow \ln\frac{I_0}{I} = kx \Rightarrow x = \frac{1}{k}\ln\frac{I_0}{I} = \frac{1}{0.39}\ln 1.12 \approx 0.29$ cm.

[61] From the graph, the coordinates of the points of intersection are approximately $(1.02, 0.48)$ and $(2.40, 0.86)$. $f(x) \geq g(x)$ on the intervals $(0, 1.02]$ and $[2.40, \infty)$.

$[0, 6]$ by $[-1, 3]$

*Figure 61*

$[0, 8]$ by $[-1.67, 3.67]$

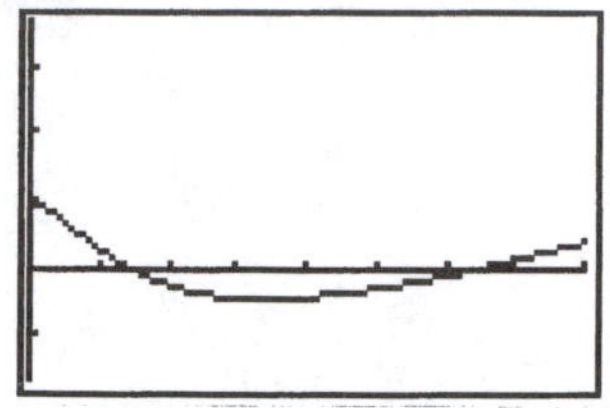

*Figure 63*

[63] Graph $y = e^{-x} - 2\log(1 + x^2) + 0.5x$ and estimate any $x$-intercepts. From the graph, we see that the roots of the equation are approximately $x \approx 1.41, 6.59$.

[65] (a) $f$ is increasing on $[0.2, 0.63]$ and $[6.87, 16]$. $f$ is decreasing on $[0.63, 6.87]$.

(b) The maximum value of $f$ is 4.61 when $x = 16$.

The minimum value of $f$ is approximately $-3.31$ when $x \approx 6.87$.

$[0.2, 16, 2]$ by $[-4.77, 5.77]$

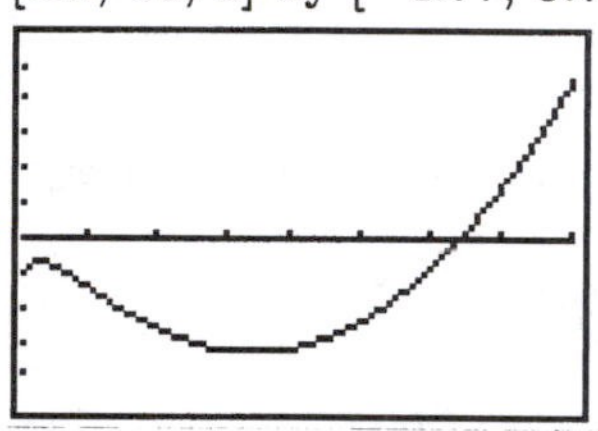

*Figure 65*

$[-5, 10]$ by $[-2, 8]$

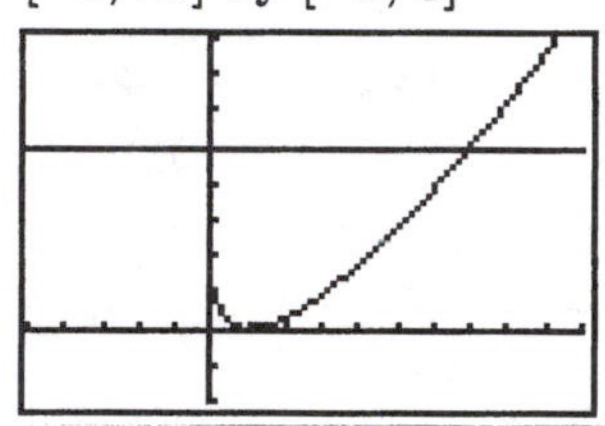

*Figure 67*

[67] Graph $Y_1 = x\log x - \log x$ and $Y_2 = 5$. The graphs intersect at $x \approx 6.94$.

[69] Let $d = x$. Graph $Y_1 = I_0 - 20\log x - kx = 70 - 20\log x - 0.076x$ and $Y_2 = 20$. At the point of intersection, $x \approx 115.3$. The distance is approximately 115 meters.

## *5.6 Exercises*

[1] (a) $5^x = 8 \Rightarrow \log 5^x = \log 8 \Rightarrow x \log 5 = \log 8 \Rightarrow x = \dfrac{\log 8}{\log 5} \approx 1.29$

(b) $5^x = 8 \Rightarrow x = \log_5 8 = \dfrac{\log 8}{\log 5}$ {by the change of base formula} $\approx 1.29$

[3] (a) $3^{4-x} = 5 \Rightarrow \log(3^{4-x}) = \log 5 \Rightarrow (4 - x)\log 3 = \log 5 \Rightarrow$

$4 - x = \dfrac{\log 5}{\log 3} \Rightarrow x = 4 - \dfrac{\log 5}{\log 3} \approx 2.54$. *Note:* The answer could also be written as $4 - \dfrac{\log 5}{\log 3} = \dfrac{4\log 3 - \log 5}{\log 3} = \dfrac{\log 81 - \log 5}{\log 3} = \dfrac{\log \frac{81}{5}}{\log 3}$.

(b) $3^{4-x} = 5 \Rightarrow 4 - x = \log_3 5 \Rightarrow x = 4 - \dfrac{\log 5}{\log 3} \approx 2.54$.

[5] $\log_5 6 = \frac{\log 6}{\log 5} \left\{ \text{or } \frac{\ln 6}{\ln 5} \right\} \approx 1.1133$

[7] $\log_9 0.2 = \frac{\log 0.2}{\log 9} \left\{ \text{or, equivalently, } \frac{\ln 0.2}{\ln 9} \right\} \approx -0.7325.$

Note that either log or ln can be used here. You should get comfortable using both.

[9] $\frac{\log_5 16}{\log_5 4} = \log_4 16 = \log_4 4^2 = 2$

[11] The steps are similar to those in Example 4. $3^{x+4} = 2^{1-3x} \Rightarrow$

$\log(3^{x+4}) = \log(2^{1-3x}) \Rightarrow (x+4)\log 3 = (1-3x)\log 2 \Rightarrow$

$x \log 3 + 4\log 3 = \log 2 - 3x \log 2 \Rightarrow x \log 3 + 3x \log 2 = \log 2 - 4\log 3 \Rightarrow$

$x(\log 3 + 3\log 2) = \log 2 - 4\log 3 \Rightarrow x = \frac{\log 2 - \log 81}{\log 3 + \log 8} \Rightarrow x = \frac{\log \frac{2}{81}}{\log 24} \approx -1.16$

[13] $2^{2x-3} = 5^{x-2} \Rightarrow \log(2^{2x-3}) = \log(5^{x-2}) \Rightarrow (2x-3)\log 2 = (x-2)\log 5 \Rightarrow$

$2x \log 2 - 3\log 2 = x \log 5 - 2\log 5 \Rightarrow 2x \log 2 - x \log 5 = 3\log 2 - 2\log 5 \Rightarrow$

$x(2\log 2 - \log 5) = \log 2^3 - \log 5^2 \Rightarrow x = \frac{\log 8 - \log 25}{\log 4 - \log 5} \Rightarrow x = \frac{\log \frac{8}{25}}{\log \frac{4}{5}} \approx 5.11$

[15] $2^{-x} = 8 \Rightarrow 2^{-x} = 2^3 \Rightarrow -x = 3 \Rightarrow x = -3$

[17] $\log x = 1 - \log(x-3) \Rightarrow \log x + \log(x-3) = 1 \Rightarrow \log(x^2 - 3x) = 1 \Rightarrow$

$x^2 - 3x = 10^1 \Rightarrow x^2 - 3x - 10 = 0 \Rightarrow (x-5)(x+2) = 0 \Rightarrow x = 5, -2.$

The solution $x = 5$ checks, but $-2$ is extraneous.

[19] $\log(x^2+4) - \log(x+2) = 2 + \log(x-2) \Rightarrow \log\left(\frac{x^2+4}{x+2}\right) - \log(x-2) = 2 \Rightarrow$

$\log\left(\frac{x^2+4}{x^2-4}\right) = 2 \Rightarrow \frac{x^2+4}{x^2-4} = 10^2 \Rightarrow x^2 + 4 = 100(x^2-4) \Rightarrow$

$x^2 + 4 = 100x^2 - 400 \Rightarrow 404 = 99x^2 \Rightarrow x^2 = \frac{404}{99} \Rightarrow$

$x = \pm\sqrt{\frac{4 \cdot 101}{9 \cdot 11}} = \pm\frac{2}{3}\sqrt{\frac{101}{11}} \approx \pm 2.02; -\frac{2}{3}\sqrt{\frac{101}{11}}$ is extraneous

[21] See Example 5 for more detail concerning this type of exercise.

$5^x + 125(5^{-x}) = 30 \{\text{multiply by } 5^x\} \Rightarrow (5^x)^2 + 125 = 30(5^x) \Rightarrow$

$(5^x)^2 - 30(5^x) + 125 = 0 \{\text{recognize as a quadratic in } 5^x \text{ and factor}\} \Rightarrow$

$(5^x - 5)(5^x - 25) = 0 \Rightarrow 5^x = 5, 25 \Rightarrow 5^x = 5^1, 5^2 \Rightarrow x = 1, 2$

[23] $4^x - 3(4^{-x}) = 8$ { multiply by $4^x$ } $\Rightarrow$ $(4^x)^2 - 3 = 8(4^x)$ $\Rightarrow$

$(4^x)^2 - 8(4^x) - 3 = 0$ { recognize as a quadratic in $4^x$ and use the quadratic

formula } $\Rightarrow$ $4^x = \dfrac{-(-8) \pm \sqrt{(-8)^2 - 4(1)(-3)}}{2(1)} = \dfrac{8 \pm \sqrt{76}}{2} = \dfrac{8 \pm 2\sqrt{19}}{2} = 4 \pm \sqrt{19}.$

Since $4^x$ is positive and $4 - \sqrt{19}$ is negative, $4 - \sqrt{19}$ is discarded.

Continuing, $4^x = 4 + \sqrt{19}$ $\Rightarrow$ $x = \log_4 (4 + \sqrt{19}) = \dfrac{\log (4 + \sqrt{19})}{\log 4}$ { use the change

of base formula to approximate } $\approx 1.53$

[25] $\log (x^2) = (\log x)^2$ $\Rightarrow$ $2 \log x = (\log x)^2$ $\Rightarrow$ $(\log x)^2 - 2 \log x = 0$ $\Rightarrow$

$(\log x)(\log x - 2) = 0$ $\Rightarrow$ $\log x = 0, 2$ $\Rightarrow$ $x = 10^0, 10^2$ $\Rightarrow$ $x = 1$ or $100$

[27] Don't confuse $\log (\log x)$ with $(\log x)(\log x)$. The first expression is

the log of the log of $x$, whereas the second expression is the log of $x$ times itself.

$\log (\log x) = 2$ $\Rightarrow$ $\log_{10} (\log x) = 2$ $\Rightarrow$ $\log x = 10^2 = 100$ $\Rightarrow$ $x = 10^{100}$

[29] $x^{\sqrt{\log x}} = 10^8$ { take the log of both sides } $\Rightarrow$ $\log\left(x^{\sqrt{\log x}}\right) = \log 10^8$ $\Rightarrow$

$\sqrt{\log x}\,(\log x) = 8$ $\Rightarrow$ $(\log x)^{1/2}(\log x)^1 = 8$ $\Rightarrow$ $(\log x)^{3/2} = 8$ $\Rightarrow$

$\left[(\log x)^{3/2}\right]^{2/3} = (8)^{2/3}$ $\Rightarrow$ $\log x = (\sqrt[3]{8})^2 = 4$ $\Rightarrow$ $x = 10{,}000$

[31] Since $e^{2x} = (e^x)^2$, we recognize $e^{2x} + 2e^x - 15$ as a quadratic in $e^x$ and factor it.

$e^{2x} + 2e^x - 15 = 0$ $\Rightarrow$ $(e^x + 5)(e^x - 3) = 0$ $\Rightarrow$ $e^x = -5, 3.$

But $e^x > 0$, so $e^x = 3$ $\Rightarrow$ $x = \ln 3.$

[33] $\log_3 x - \log_9 (x + 42) = 0$ $\Rightarrow$ $\dfrac{\ln x}{\ln 3} - \dfrac{\ln (x + 42)}{\ln 9} = 0$ { change of base theorem } $\Rightarrow$

$\dfrac{2 \ln x}{2 \ln 3} - \dfrac{\ln (x + 42)}{\ln 9} = 0$ { get a common denominator, $2 \ln 3 = \ln 3^2 = \ln 9$ } $\Rightarrow$

$\ln x^2 - \ln (x + 42) = 0$ { multiply by $\ln 9$ and change $2 \ln x$ to $\ln x^2$ } $\Rightarrow$

$\ln\left(\dfrac{x^2}{x + 42}\right) = 0$ $\Rightarrow$ $\dfrac{x^2}{x + 42} = e^0$ $\Rightarrow$ $\dfrac{x^2}{x + 42} = 1$ $\Rightarrow$ $x^2 = x + 42$ $\Rightarrow$

$x^2 - x - 42 = 0$ $\Rightarrow$ $(x - 7)(x + 6) = 0$ $\Rightarrow$ $x = 7, -6.$ $-6$ is extraneous.

*Note:* For Exercises 35–42 and 49–50 of the Chapter Review Exercises, let $D$ denote the domain of the function determined by the original equation, and $R$ its range. These are then the range and domain, respectively, of the equation listed in the answer.

**35** $y = \dfrac{10^x + 10^{-x}}{2}$ $\{D = \mathbb{R},\ R = [1, \infty)\}$ $\Rightarrow$ $2y = 10^x + 10^{-x}$

$\left\{\text{since } 10^{-x} = \frac{1}{10^x}, \text{ multiply by } 10^x \text{ to eliminate denominator}\right\}$ $\Rightarrow$

$10^{2x} - 2y\ 10^x + 1 = 0$ {treat as a quadratic in $10^x$} $\Rightarrow$

$$10^x = \frac{2y \pm \sqrt{4y^2 - 4}}{2} = y \pm \sqrt{y^2 - 1} \Rightarrow x = \log\left(y \pm \sqrt{y^2 - 1}\right)$$

**37** $y = \dfrac{10^x - 10^{-x}}{10^x + 10^{-x}}$ $\{D = \mathbb{R},\ R = (-1, 1)\}$ $\Rightarrow$ $y\,10^x + y\,10^{-x} = 10^x - 10^{-x}$ $\Rightarrow$

$y\ 10^{2x} + y = 10^{2x} - 1 \Rightarrow (y - 1)\,10^{2x} = -1 - y \Rightarrow$

$$10^{2x} = \frac{-1 - y}{y - 1} \Rightarrow 2x = \log\left(\frac{1 + y}{1 - y}\right) \Rightarrow x = \tfrac{1}{2}\log\left(\frac{1 + y}{1 - y}\right)$$

**39** $y = \dfrac{e^x - e^{-x}}{2}$ $\{D = R = \mathbb{R}\}$ $\Rightarrow$ $2y = e^x - e^{-x}$ $\Rightarrow$

$e^{2x} - 2y\,e^x - 1 = 0 \Rightarrow e^x = \dfrac{2y \pm \sqrt{4y^2 + 4}}{2} = y \pm \sqrt{y^2 + 1};$

$\sqrt{y^2 + 1} > y$, so $y - \sqrt{y^2 + 1} < 0$, but $e^x > 0$ and thus, $x = \ln\left(y + \sqrt{y^2 + 1}\right)$

**41** $y = \dfrac{e^x + e^{-x}}{e^x - e^{-x}}$ $\{D = \mathbb{R} - \{0\},\ R = (-\infty, -1) \cup (1, \infty)\}$ $\Rightarrow$

$ye^x - ye^{-x} = e^x + e^{-x} \Rightarrow ye^{2x} - y = e^{2x} + 1 \Rightarrow$

$$(y - 1)e^{2x} = y + 1 \Rightarrow e^{2x} = \frac{y + 1}{y - 1} \Rightarrow 2x = \ln\left(\frac{y + 1}{y - 1}\right) \Rightarrow x = \tfrac{1}{2}\ln\left(\frac{y + 1}{y - 1}\right)$$

**43** $f(x) = \log_2(x + 3)$ • $x = 0 \Rightarrow$ $y$-intercept $= \log_2 3 = \dfrac{\log 3}{\log 2} \approx 1.5850$

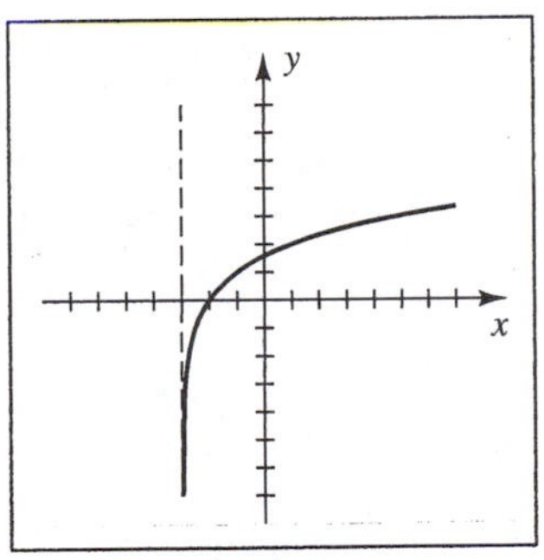


*Figure 43*

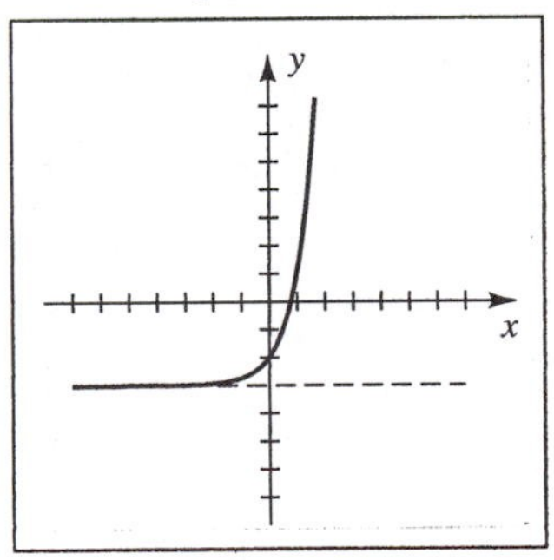


*Figure 45*

**45** $f(x) = 4^x - 3$ • $y = 0 \Rightarrow 4^x = 3 \Rightarrow$ $x$-intercept $= \log_4 3 = \dfrac{\log 3}{\log 4} \approx 0.7925$

**47** (a) vinegar: pH $\approx -\log(6.3 \times 10^{-3}) = -(\log 6.3 + \log 10^{-3}) = -(\log 6.3 - 3) =$

$3 - \log 6.3 \approx 2.2$

(b) carrots: pH $\approx -\log(1.0 \times 10^{-5}) = 5 - \log 1.0 = 5$

(c) sea water: pH $\approx -\log(5.0 \times 10^{-9}) = 9 - \log 5.0 \approx 8.3$

[49] $[\mathrm{H}^+] < 10^{-7} \Rightarrow \log[\mathrm{H}^+] < \log 10^{-7}$ { since log is an increasing function } $\Rightarrow$
$\log[\mathrm{H}^+] < -7 \Rightarrow -\log[\mathrm{H}^+] > -(-7) \Rightarrow$ pH > 7 for basic solutions;
similarly, pH < 7 for acidic solutions.

[51] Solving $A = P(1+\frac{r}{n})^{nt}$ for $t$ with $A = 2P$, $r = 0.06$, and $n = 12$ yields

$2P = P(1+\frac{0.06}{12})^{12t}$ { divide by $P$ } $\Rightarrow$

$2 = (1.005)^{12t}$ { take the ln of both sides } $\Rightarrow$

$\ln 2 = \ln(1.005)^{12t} \Rightarrow \ln 2 = 12t \ln(1.005) \Rightarrow$

$t = \dfrac{\ln 2}{12\ln(1.005)} \approx 11.58$ yr, or about 11 years and 7 months.

[53] 50% of the light reaching a depth of 13 meters corresponds to the equation

$\frac{1}{2}I_0 = I_0 c^{13}$. Solving for $c$, we have $c^{13} = \frac{1}{2} \Rightarrow c = \sqrt[13]{\frac{1}{2}} = 2^{-1/13}$.

Now letting $I = 0.01\,I_0$, $c = 2^{-1/13}$, and using the formula from Example 8,

$$x = \frac{\log(I/I_0)}{\log c} = \frac{\log[(0.01\,I_0)/I_0]}{\log 2^{-1/13}} = \frac{\log 10^{-2}}{-\frac{1}{13}\log 2} = \frac{-2}{-\frac{1}{13}\log 2} = \frac{26}{\log 2} \approx 86.4 \text{ m}.$$

[55] (a) The graph of $A = 100[1-(0.9)^t]$ is an *increasing* exponential that passes through (0, 0), (5, $\approx$ 41), and (10, $\approx$ 65).

(b) $A = 50 \Rightarrow 50 = 100[1-(0.9)^t] \Rightarrow$
$1-(0.9)^t = \frac{50}{100} \Rightarrow (0.9)^t = 0.5 \Rightarrow$
$t = \log_{0.9}(0.5) = \dfrac{\log 0.5}{\log 0.9} \approx 6.58$ min.

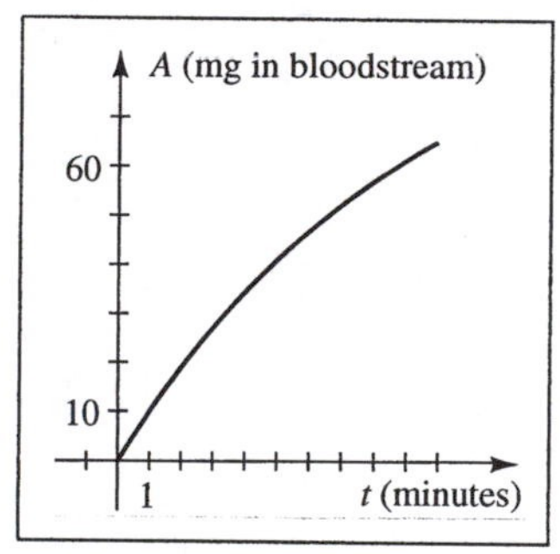


*Figure 55*

[57] (a) $F = F_0(1-m)^t \Rightarrow (1-m)^t = \dfrac{F}{F_0} \Rightarrow \log(1-m)^t = \log\left(\dfrac{F}{F_0}\right) \Rightarrow$

$t\log(1-m) = \log\left(\dfrac{F}{F_0}\right) \Rightarrow t = \dfrac{\log(F/F_0)}{\log(1-m)}$

(b) Using part (a) with $F = \frac{1}{2}F_0$ and $m = 0.00005$,

$$t = \frac{\log(F/F_0)}{\log(1-m)} = \frac{\log(\frac{1}{2}F_0/F_0)}{\log(1-0.00005)} = \frac{\log\frac{1}{2}}{\log 0.99995} \approx 13{,}863 \text{ generations}.$$

[59] (a) $t = 10 \Rightarrow h = \dfrac{120}{1+200e^{-0.2(10)}} = \dfrac{120}{1+200e^{-2}} \approx 4.28$ ft

(b) $h = 50 \Rightarrow 50 = \dfrac{120}{1+200e^{-0.2t}} \Rightarrow 1+200e^{-0.2t} = \dfrac{120}{50} \Rightarrow$

$200e^{-0.2t} = \frac{12}{5} - 1 \Rightarrow e^{-0.2t} = \frac{7}{5}\cdot\frac{1}{200} \Rightarrow e^{-0.2t} = 0.007 \Rightarrow$

$-0.2t = \ln 0.007 \Rightarrow t = \dfrac{\ln 0.007}{-0.2} \approx 24.8$ yr

[61] $\dfrac{v_0}{v_1} = \left(\dfrac{h_0}{h_1}\right)^P \Rightarrow \ln\dfrac{v_0}{v_1} = P\ln\dfrac{h_0}{h_1} \Rightarrow P = \dfrac{\ln(v_0/v_1)}{\ln(h_0/h_1)} = \dfrac{\ln(25/6)}{\ln(200/35)} \approx 0.82$

[63] If $y = c2^{kx}$ and $x = 0$, then $y = c2^0 = c = 4$. Thus, $y = 4(2)^{kx}$. Similarly, $x = 1 \Rightarrow y = 4(2)^k = 3.249 \Rightarrow 2^k = \frac{3.249}{4} \Rightarrow k = \log_2(\frac{3.249}{4}) \approx -0.300$. Thus, $y = 4(2)^{-0.3x}$. Checking the remaining two points, we see that $x = 2 \Rightarrow y \approx 2.639$ and $x = 3 \Rightarrow y \approx 2.144$. The four points lie on the graph of $y = 4(2)^{-0.3x}$ to within three-decimal-place accuracy.

[65] If $y = c\log(kx+10)$ and $x = 0$, then $y = c\log 10 = c = 1.5$. Thus, $y = 1.5\log(kx+10)$. Similarly, $x = 1 \Rightarrow y = 1.5\log(k+10) = 1.619 \Rightarrow$ $\frac{1.619}{1.5} = \log(k+10) \Rightarrow k+10 = 10^{1.619/1.5} \Rightarrow k = 10^{1.619/1.5} - 10 \approx 2.004$. Thus, $y = 1.5\log(2.004x+10)$. Checking the remaining two points, we see that $x = 2 \Rightarrow y \approx 1.720$, and $x = 3 \Rightarrow y \approx 1.807$, not 1.997. The points do not lie on the graph of $y = c\log(kx+10)$ to within three-decimal-place accuracy.

[67] $h(5.3) = \log_4 5.3 - 2\log_8(1.2\times 5.3) = \dfrac{\ln 5.3}{\ln 4} - \dfrac{2\ln 6.36}{\ln 8} \approx -0.5764$

[69] The minimum point on the graph of $y = x - \ln(0.3x) - 3\log_3 x$ is about (3.73, 0.023). So there are *no* $x$-intercepts, and hence, no roots of the equation on (0, 9).

[0, 9] by [−1, 5]

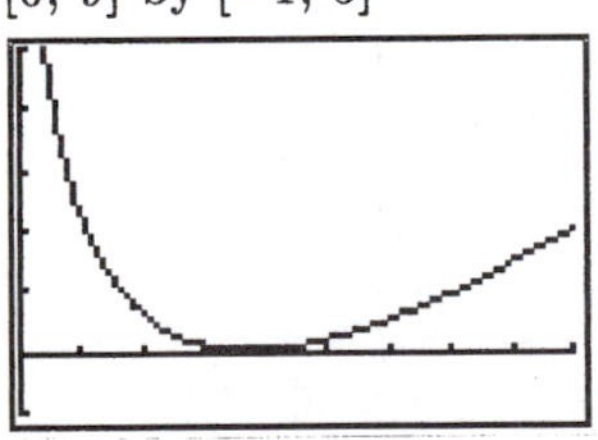

Figure 69

[−1, 17] by [−1, 11]

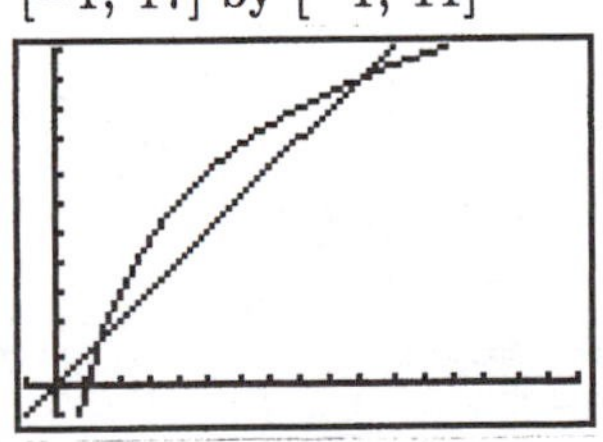

Figure 71

[71] The graphs of $f(x) = x$ and $g(x) = 3\log_2 x$ intersect at approximately (1.37, 1.37) and (9.94, 9.94). Hence, the solutions of the equation $f(x) = g(x)$ are 1.37 and 9.94.

[73] From the graph, we see that the graphs of $f$ and $g$ intersect at three points. Their coordinates are approximately $(-0.32, 0.50)$, $(1.52, -1.33)$, and $(6.84, -6.65)$. The region near the origin in *Figure 73(a)* is enhanced in *Figure 73(b)*. Thus, $f(x) > g(x)$ on $(-\infty, -0.32)$ and $(1.52, 6.84)$.

$[-5, 10]$ by $[-8, 2]$

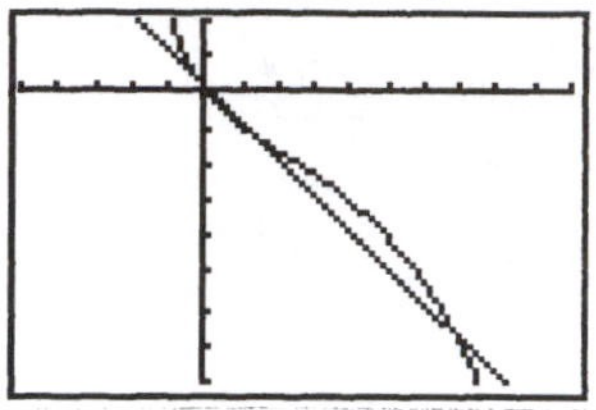

*Figure 73(a)*

$[-1.53, 2.26]$ by $[-2.92, 1.05]$

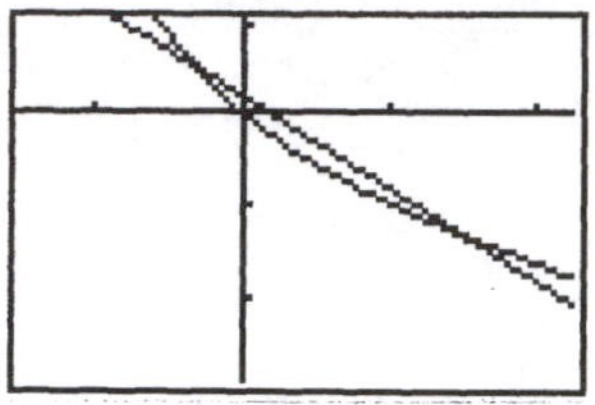

*Figure 73(b)*

[75] (1) The graph of $n(t) = 85e^{t/3}$ is increasing rapidly and soon becomes greater than 100. It is doubtful that the average score would improve dramatically without any review.

$[0, 5]$ by $[0, 200, 20]$

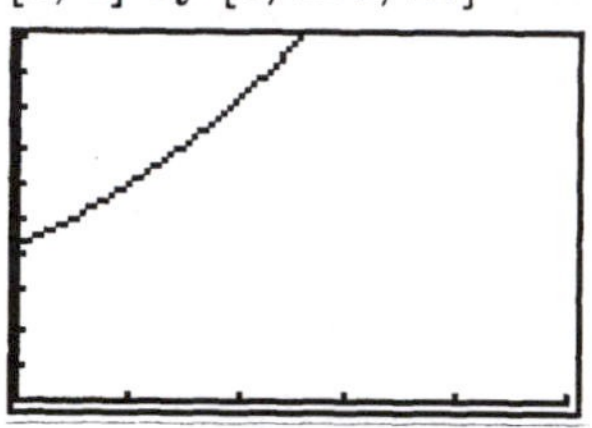

*Figure 75(1)*

$[0, 5]$ by $[0, 100, 10]$

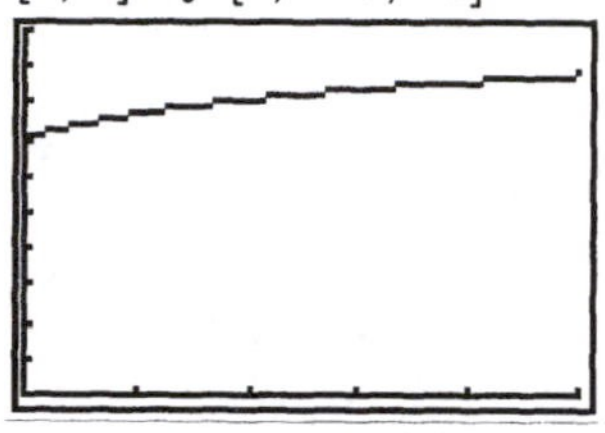

*Figure 75(2)*

(2) The graph of $n(t) = 70 + 10\ln(t+1)$ is also increasing. It is incorrect because $n(0) = 70 \neq 85$. Also, one would not expect the average score to improve without any review.

(3) The graph of $n(t) = 86 - e^t$ decreases rapidly to zero. The *average* score probably would not be zero after 5 weeks.

$[0, 5]$ by $[0, 100, 10]$

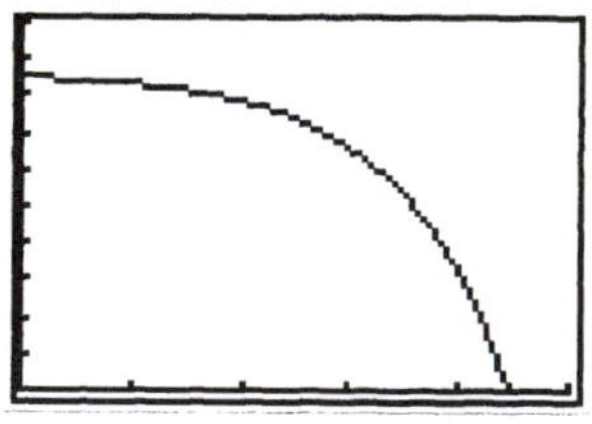

*Figure 75(3)*

$[0, 5]$ by $[0, 100, 10]$

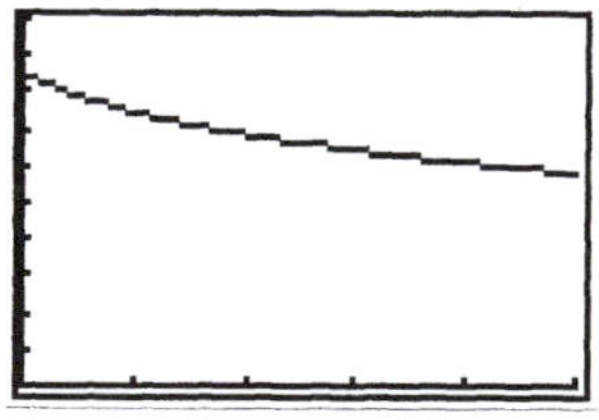

*Figure 75(4)*

(4) The graph of $n(t) = 85 - 15\ln(t+1)$ is decreasing. During the first weeks it decreases most rapidly and then starts to level off. Of the four functions, **this function seems to best model the situation.**

[77] (a) The ozone level is decreasing by 11% per year. The fraction of ozone remaining $x$ years after April 1992 is given by the function $f(x) = (0.89)^x$ {decreasing by 11% is the same as retaining 89%}. We must approximate $x$ when $f(x) = 0.5$. Using a table, $f(x) = 0.5$ when $x \approx 6$. Thus, in 1998 the ozone level would be 50% of its normal amount.

| $x$ | 1 | 2 | 3 | 4 | 5 | 6 | 7 |
|---|---|---|---|---|---|---|---|
| $f(x)$ | 0.89 | 0.79 | 0.70 | 0.63 | 0.56 | 0.50 | 0.44 |

(b) $(0.89)^t = 0.5 \Rightarrow \ln(0.89)^t = \ln 0.5 \Rightarrow t\ln 0.89 = \ln 0.5 \Rightarrow$

$t = \dfrac{\ln 0.5}{\ln 0.89} \approx 5.948$, or in 1998.

## *Chapter 5 Review Exercises*

[1] Suppose $f(a) = f(b)$. $2a^3 - 5 = 2b^3 - 5 \Rightarrow 2a^3 = 2b^3 \Rightarrow a^3 = b^3 \Rightarrow a = b$.

Thus, $f$ is a one-to-one function.

[3] $f(x) = 10 - 15x \Rightarrow 15x = 10 - y \Rightarrow$

$x = \dfrac{10 - y}{15} \Rightarrow f^{-1}(x) = \dfrac{10 - x}{15}$

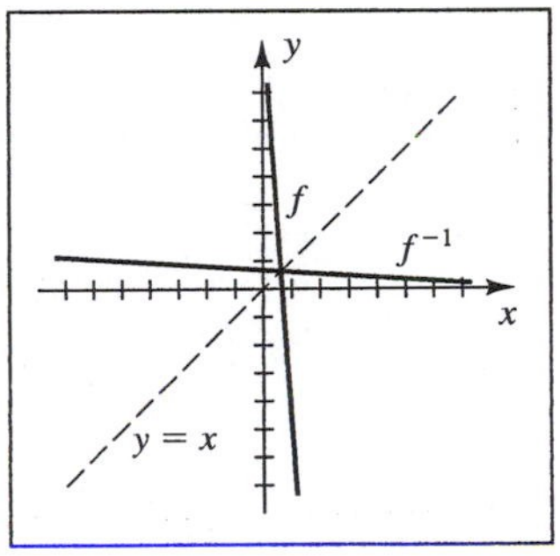


*Figure 3*

[5] (a) The point (1, 2) is on the graph, so $f(1) = 2$.

(b) $(f \circ f)(1) = f(f(1)) = f(2) = 4$.

(c) $f(2) = 4$ and $f$ is one-to-one $\Rightarrow f^{-1}(4) = 2$.

(d) $y = 4$ when $x = 2$, so $f(x) = 4 \Rightarrow x = 2$.

(e) $y > 4$ when $x > 2$, so $f(x) > 4 \Rightarrow x > 2$.

[6] Since $f$ and $g$ are one-to-one functions, we know that $f(2) = 7$, $f(4) = 2$, and $g(2) = 5$ imply that $f^{-1}(7) = 2$, $f^{-1}(2) = 4$, and $g^{-1}(5) = 2$, respectively. Part (c) is included—try part (d) and explain why you *can't* find the value.

(c) $(f^{-1} \circ g^{-1})(5) = f^{-1}(g^{-1}(5)) = f^{-1}(2) = 4$

[7] $f(x) = 3^{x+2}$ • shift $y = 3^x$ left 2 units

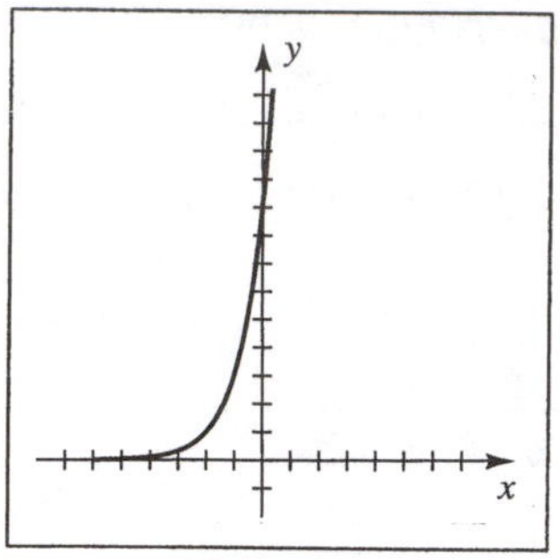

*Figure 7*

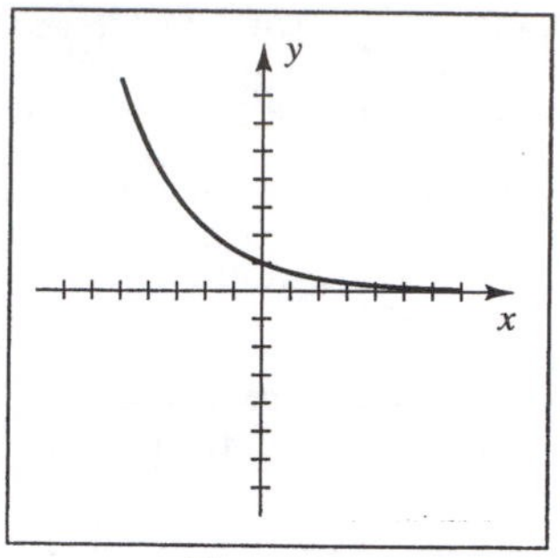

*Figure 9*

[9] $f(x) = (\frac{3}{2})^{-x} = (\frac{2}{3})^x$ • goes through $(-1, \frac{3}{2})$, $(0, 1)$, and $(1, \frac{2}{3})$

[11] $f(x) = 3^{-x^2} = (3^{-1})^{(x^2)} = (\frac{1}{3})^{(x^2)}$ • see the note in §5.2 before Exercise 21

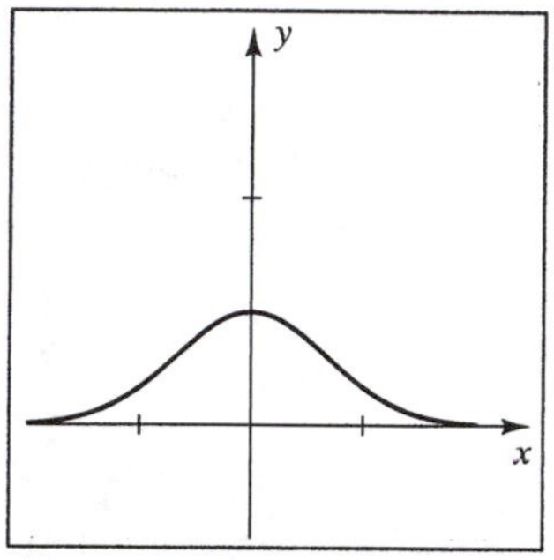

*Figure 11*

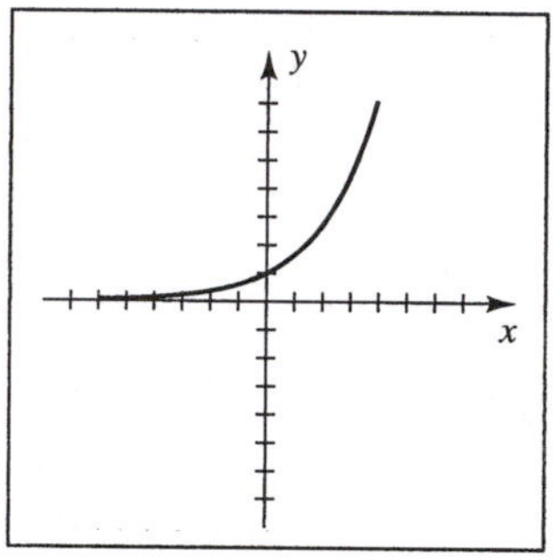

*Figure 13*

[13] $f(x) = e^{x/2} = (e^{1/2})^x \approx (1.65)^x$ • goes through $(-1, 1/\sqrt{e})$, $(0, 1)$, and $(1, \sqrt{e})$; or approximately $(-1, 0.61)$, $(0, 1)$, and $(1, 1.65)$

[15] $f(x) = e^{x-2}$ • shift $y = e^x$ right 2 units

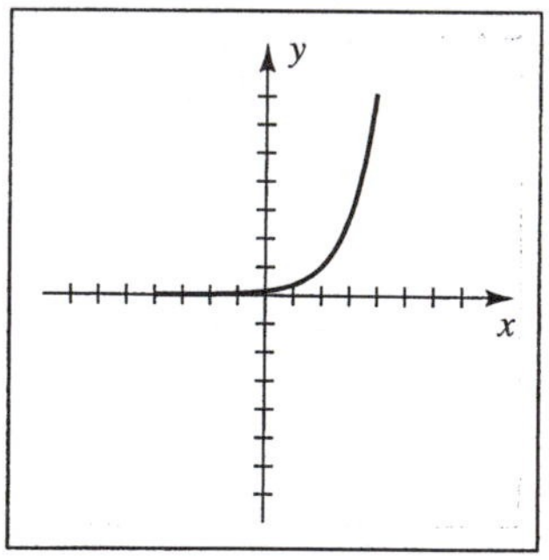

*Figure 15*

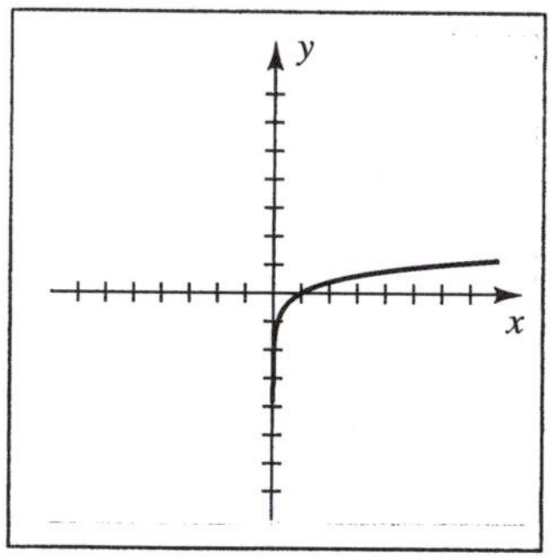

*Figure 17*

[17] $f(x) = \log_6 x$ • goes through $(\frac{1}{6}, -1)$, $(1, 0)$, and $(6, 1)$

[19] $f(x) = \log_4(x^2) = 2\log_4 x$ • stretch $y = \log_4 x$ by a factor of 2 and include its reflection through the $y$-axis since the domain of the original function is $\mathbb{R} - \{0\}$

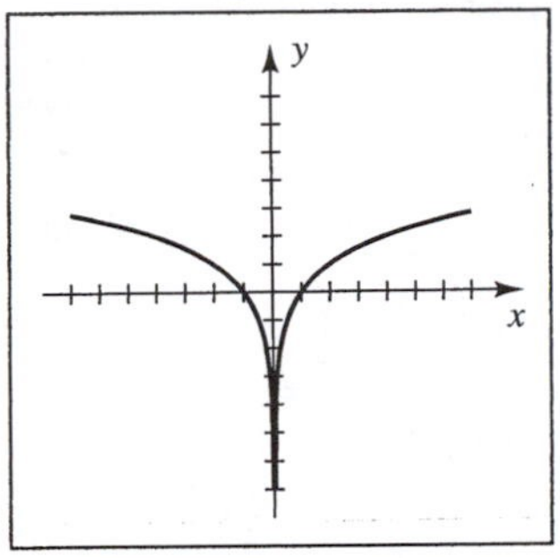


*Figure 19*

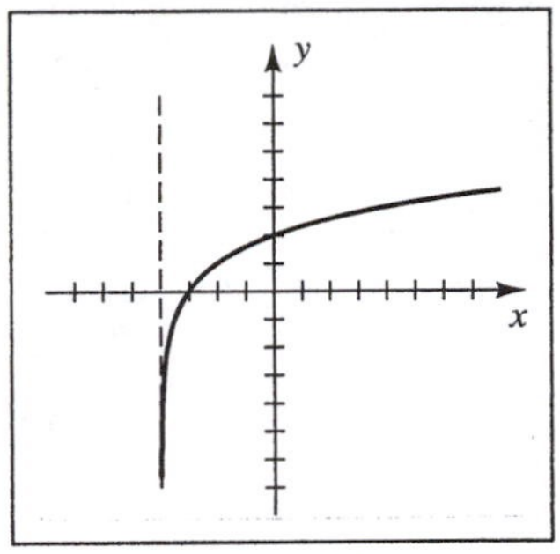


*Figure 21*

[21] $f(x) = \log_2(x+4)$ • shift $y = \log_2 x$ left 4 units

[23] (a) $\log_2 \frac{1}{16} = \log_2 2^{-4} = -4$  (b) $\log_\pi 1 = 0$  (c) $\ln e = 1$

(d) $6^{\log_6 4} = 4$  (e) $\log 1{,}000{,}000 = \log 10^6 = 6$

(f) $10^{3\log 2} = 10^{\log 2^3} = 2^3 = 8$  (g) $\log_4 2 = \log_4 4^{1/2} = \frac{1}{2}$

[25] $2^{3x-1} = \frac{1}{2} \Rightarrow 2^{3x-1} = 2^{-1} \Rightarrow 3x - 1 = -1 \Rightarrow 3x = 0 \Rightarrow x = 0$

[27] $\log\sqrt{x} = \log(x-6) \Rightarrow \sqrt{x} = x - 6 \Rightarrow (\sqrt{x})^2 = (x-6)^2 \Rightarrow$

$x = x^2 - 12x + 36 \Rightarrow x^2 - 13x + 36 = 0 \Rightarrow (x-4)(x-9) = 0 \Rightarrow$

$x = 4, 9$; 4 is extraneous

[29] $\log_4(x+1) = 2 + \log_4(3x-2) \Rightarrow \log_4(x+1) - \log_4(3x-2) = 2 \Rightarrow$

$\log_4\left(\frac{x+1}{3x-2}\right) = 2 \Rightarrow \frac{x+1}{3x-2} = 4^2 \Rightarrow x + 1 = 16(x-2) \Rightarrow$

$x + 1 = 48x - 32 \Rightarrow 33 = 47x \Rightarrow x = \frac{33}{47}$

[31] $\ln(x+2) = \ln e^{\ln 2} - \ln x \Rightarrow \ln(x+2) = \ln 2 - \ln x \Rightarrow$

$\ln(x+2) + \ln x = \ln 2 \Rightarrow \ln[x(x+2)] = \ln 2 \Rightarrow x^2 + 2x = 2 \Rightarrow$

$x^2 + 2x - 2 = 0 \Rightarrow x = \frac{-2 \pm \sqrt{4+8}}{2} = \frac{-2 \pm \sqrt{12}}{2} = \frac{-2 \pm 2\sqrt{3}}{2} = -1 + \sqrt{3}\ \{x > 0\}$

[33] $2^{5-x} = 6 \Rightarrow 5 - x = \log_2 6 \Rightarrow 5 - \log_2 6 = x \Rightarrow x = 5 - \frac{\log 6}{\log 2}$,

or equivalently, $\frac{5\log 2 - \log 6}{\log 2} = \frac{\log 2^5 - \log 6}{\log 2} = \frac{\log \frac{32}{6}}{\log 2} = \frac{\log \frac{16}{3}}{\log 2} \approx 2.415$

[35] $2^{5x+3} = 3^{2x+1} \Rightarrow \log(2^{5x+3}) = \log(3^{2x+1}) \Rightarrow$

$(5x+3)\log 2 = (2x+1)\log 3 \Rightarrow 5x\log 2 + 3\log 2 = 2x\log 3 + \log 3 \Rightarrow$

$5x\log 2 - 2x\log 3 = \log 3 - 3\log 2 \Rightarrow x(5\log 2 - 2\log 3) = \log 3 - \log 2^3 \Rightarrow$

$x = \frac{\log 3 - \log 8}{\log 32 - \log 9} = \frac{\log \frac{3}{8}}{\log \frac{32}{9}} \approx -0.773$

[37] $\log_4 x = \sqrt[3]{\log_4 x} \Rightarrow \log_4 x = (\log_4 x)^{1/3} \Rightarrow (\log_4 x)^3 = \log_4 x \Rightarrow$

$(\log_4 x)^3 - \log_4 x = 0 \Rightarrow \log_4 x\left[(\log_4 x)^2 - 1\right] = 0 \Rightarrow$

$\log_4 x = 0$ or $\log_4 x = \pm 1 \Rightarrow x = 1$ or $x = 4, \frac{1}{4}$

[39] $10^{2\log x} = 5 \Rightarrow 10^{\log x^2} = 5 \Rightarrow x^2 = 5 \Rightarrow x = \pm\sqrt{5}$; $-\sqrt{5}$ is extraneous

[41] $x^2(-2xe^{-x^2}) + 2xe^{-x^2} = 0 \Rightarrow 2xe^{-x^2}(-x^2+1) = 0 \Rightarrow x = 0, \pm 1 \left\{e^{-x^2} \neq 0\right\}$

[43] (a) $\log x^2 = \log(6-x) \Rightarrow x^2 = 6 - x \Rightarrow x^2 + x - 6 = 0 \Rightarrow$

$(x+3)(x-2) = 0 \Rightarrow x = -3, 2$

(b) $2\log x = \log(6-x) \Rightarrow \log x^2 = \log(6-x)$, which is the equation in part (a).

This equation has the same solutions provided they are in the domain.

But $-3$ is extraneous, so 2 is the only solution.

[44] (a) $\ln(e^x)^2 = 16 \Rightarrow 2\ln e^x = 16 \Rightarrow 2x = 16 \Rightarrow x = 8$

(b) $\ln e^{(x^2)} = 16 \Rightarrow x^2 = 16 \Rightarrow x = \pm 4$

[45] $\log x^4 \sqrt[3]{y^2/z} = \log(x^4 y^{2/3} z^{-1/3}) = \log x^4 + \log y^{2/3} + \log z^{-1/3}$

$= 4\log x + \frac{2}{3}\log y - \frac{1}{3}\log z$

[46] $\log(x^2/y^3) + 4\log y - 6\log\sqrt{xy} = \log\left(\frac{x^2}{y^3}\right) + \log y^4 - \log\left[(xy)^{1/2}\right]^6$

$= \log\left(\frac{x^2 \cdot y^4}{y^3}\right) - \log(x^3y^3)$

$= \log\left(\frac{x^2y}{x^3y^3}\right) = \log\left(\frac{1}{xy^2}\right)$

$= \log 1 - \log(xy^2)$

$= 0 - \log(xy^2)$

$= -\log(xy^2)$

[47] We can assume that the graph has a horizontal asymptote of $y = 0$, and use the form $f(x) = ba^x$. The point $(0, 6)$ is on the graph, so $f(0) = 6 \Rightarrow 6 = ba^0 \Rightarrow$ $6 = b(1) \Rightarrow 6 = b$, and hence, $f(x) = 6a^x$. Since the point $P(1, 8)$ is on the graph, $f(1) = 8$. Thus, $8 = 6a^1 \Rightarrow a = \frac{4}{3}$, and the function is $f(x) = 6(\frac{4}{3})^x$.

[49] $y = \frac{1}{10^x + 10^{-x}}\left\{D = \mathbb{R},\ R = (0, \tfrac{1}{2}]\right\} \Rightarrow y\ 10^x + y\ 10^{-x} = 1 \Rightarrow$

$y\ 10^{2x} + y = 10^x \Rightarrow y\ 10^{2x} - 10^x + y = 0 \Rightarrow 10^x = \frac{1 \pm \sqrt{1-4y^2}}{2y} \Rightarrow$

$$x = \log\left(\frac{1 \pm \sqrt{1-4y^2}}{2y}\right)$$

[51] (a) $x = \ln 6.6 \approx 1.89$ (b) $\log x = 1.8938 \Rightarrow x = 10^{1.8938} \approx 78.3$

(c) $\ln x = -0.75 \Rightarrow x = e^{-0.75} \approx 0.472$

[53] (a) For $y = \log_2(x+1)$, $D = (-1, \infty)$ and $R = \mathbb{R}$.

(b) $y = \log_2(x+1) \Rightarrow x = \log_2(y+1) \Rightarrow 2^x = y+1 \Rightarrow y = 2^x - 1$,

$D = \mathbb{R},\ R = (-1, \infty)$

[55] (a) $Q(t) = 2(3^t) \Rightarrow Q(0) = 2(3^0) = 2$ {in thousands}, or 2000

(b) $Q(\frac{10}{60}) = 2000(3^{1/6}) \approx 2401$; $Q(\frac{30}{60}) = 2000(3^{1/2}) \approx 3464$; $Q(1) = 2(3) = 6000$

[57] (a) $N = 64(0.5)^{t/8} = 64\left[(0.5)^{1/8}\right]^t \approx 64(0.917)^t$

(b) $N = \frac{1}{2}N_0 \Rightarrow \frac{1}{2}N_0 = N_0(0.5)^{t/8} \Rightarrow$

$(\frac{1}{2})^1 = (\frac{1}{2})^{t/8} \Rightarrow 1 = t/8 \Rightarrow t = 8$ days

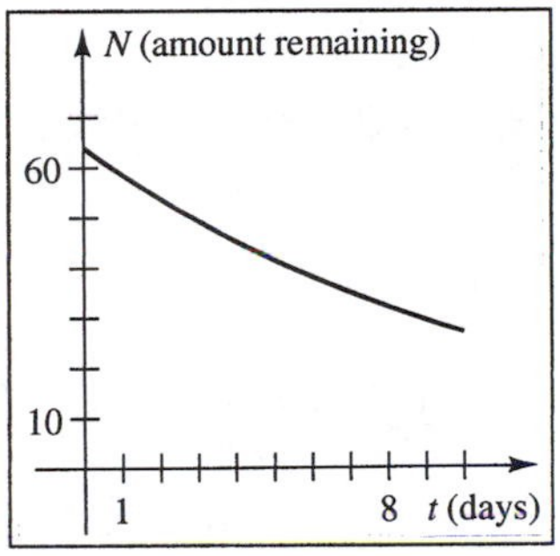


*Figure 57*

[59] (a) Using $A = Pe^{rt}$ with $A = \$35{,}000$, $P = \$10{,}000$, and $r = 11\%$, we have

$35{,}000 = 10{,}000e^{0.11t} \Rightarrow e^{0.11t} = 3.5 \Rightarrow 0.11t = \ln 3.5 \Rightarrow$

$t = \frac{1}{0.11}\ln 3.5 \approx 11.39$ yr.

(b) $2 \cdot 10{,}000 = 10{,}000e^{0.11t} \Rightarrow e^{0.11t} = 2 \Rightarrow 0.11t = \ln 2 \Rightarrow t \approx 6.30$ yr

Alternatively, using the doubling time formula, $t = (\ln 2)/r = (\ln 2)/0.11 \approx 6.30$.

[62] (a) $\alpha = 10\log\left(\frac{I}{I_0}\right) \Rightarrow \frac{\alpha}{10} = \log\left(\frac{I}{I_0}\right) \Rightarrow 10^{\alpha/10} = \frac{I}{I_0} \Rightarrow I = I_0 10^{\alpha/10}$

(b) Let $I(\alpha)$ be the intensity corresponding to $\alpha$ decibels.

$I(\alpha+1) = I_0 10^{(\alpha+1)/10} = I_0 10^{\alpha/10} 10^{1/10} = I(\alpha)\ 10^{1/10} \approx 1.26\, I(\alpha)$,

which represents a 26% increase in $I(\alpha)$.

[64] $R = 2.3\log(A + 3000) - 5.1 \Rightarrow R + 5.1 = 2.3\log(A + 3000) \Rightarrow$

$\frac{R + 5.1}{2.3} = \log(A + 3000) \Rightarrow$

$A + 3000 = 10^{(R+5.1)/2.3}$ {change to exponential form} $\Rightarrow$

$A = 10^{(R+5.1)/2.3} - 3000$

[66] $R = 4 \Rightarrow 2.3\log(A + 14{,}000) - 6.6 = 4 \Rightarrow \log(A + 14{,}000) = \frac{10.6}{2.3} \Rightarrow$

$A = 10^{106/23} - 14{,}000 \approx 26{,}615.9 \text{ mi}^2.$

[68] Substituting $v = 0$ and $m = m_1 + m_2$ in $v = -a \ln m + b$ yields

$0 = -a \ln(m_1 + m_2) + b$. Thus, $b = a \ln(m_1 + m_2)$. At burnout, $m = m_1$,

and hence, $v = -a \ln m_1 + b = -a \ln m_1 + a \ln(m_1 + m_2)$

$= a[\ln(m_1 + m_2) - \ln m_1] \Rightarrow v = a \ln\left(\frac{m_1 + m_2}{m_1}\right)$

[70] (a) $\log E = 11.4 + (1.5)R \Rightarrow E = 10^{11.4 + 1.5R}$ {merely change the form}

(b) $R = 8.4 \Rightarrow E = 10^{11.4 + 1.5(8.4)} = 10^{24}$ ergs

[72] $h = 70.228 + 5.104t + 9.222 \ln t$ and $R = 5.104 + (9.222/t)$.

$t = 2 \Rightarrow h \approx 86.8$ cm and $R = 9.715$ cm/yr.

[74] (a) $x = 4\% \Rightarrow T = -8310 \ln(0.04) \approx 26{,}749$ yr.

(b) $T = 10{,}000 \Rightarrow 10{,}000 = -8310 \ln x \Rightarrow -\frac{10{,}000}{8310} = \ln x \Rightarrow$

$x = e^{-1000/831} \approx 0.30$, or 30%.

[76] $N(t) = \frac{1}{2}N_0 \Rightarrow \frac{1}{2}N_0 = N_0(0.805)^t \Rightarrow 2^{-1} = (0.805)^t \Rightarrow$

$\ln(2^{-1}) = \ln(0.805)^t \Rightarrow -\ln 2 = t \ln(0.805) \Rightarrow$

$t = -\frac{\ln 2}{\ln(0.805)} \approx 3.196$ millennia, or 3196 yr

## Chapter 5 Discussion Exercises

[1] (a) $f(x) = -(x-1)^3 + 1 \Rightarrow$

$y = -(x-1)^3 + 1 \Rightarrow$

$(x-1)^3 = 1 - y \Rightarrow$

$x - 1 = \sqrt[3]{1-y} \Rightarrow$

$x = \sqrt[3]{1-y} + 1 \Rightarrow$

$f^{-1}(x) = \sqrt[3]{1-x} + 1$

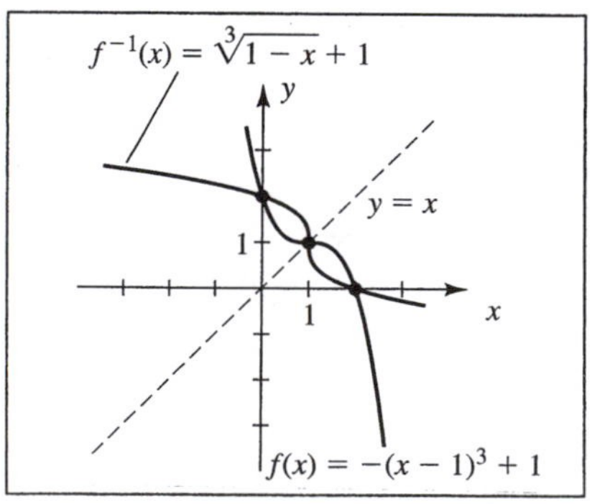


*Figure 1*

(b) If $f$ is increasing, then $f^{-1}$ moves from left to right and increases.
If $f$ is decreasing, then $f^{-1}$ moves from right to left and decreases.

(c) If $P(a,\ b)$ is a point on $f$ and $P$ is an intersection point of $f$ and $f^{-1}$, then $Q(b,\ a)$ is a point on $f^{-1}$ (since $P$ is a point on $f$) and $Q$ is a point on $f$ (since $P$ is also a point on $f^{-1}$). If $a = b$, then $P$ lies on the line $y = x$. If $a \neq b$, then $m_{PQ} = -1$, and $P$ lies on a line perpendicular to $y = x$; that is, on a line of the form $y = -x + c$.

3 (a) The $y$-intercept is $a$, so it increases as $a$ does. The graph flattens out as $a$ increases.

(b) Graph $Y_1 = \frac{a}{2}(e^{x/a} + e^{-x/a}) + (30 - a)$ on $[-20,\ 20]$ by $[30,\ 32]$ and check for $Y_1 < 32$ at $x = 20$. In this case it turns out that $a = 101$ is the smallest integer value that satisfies the conditions, so an equation is

$$y = \tfrac{101}{2}(e^{x/101} + e^{-x/101}) - 71.$$

5 (a) $x^y = y^x \;\Rightarrow\; \ln(x^y) = \ln(y^x) \;\Rightarrow\; y\ln x = x\ln y \;\Rightarrow\; \dfrac{\ln x}{x} = \dfrac{\ln y}{y}$

(b) Once you find two values of $(\ln x)/x$ that are the same (such as 0.36652), you know that the corresponding $x$-values, $x_1$ and $x_2$, satisfy the relationship $(x_1)^{x_2} \approx (x_2)^{x_1}$. In particular, when $(\ln x)/x \approx 0.36652$, we find that $x_1 \approx 2.50$ and $x_2 \approx 2.97$. Note that $2.50^{2.97} \approx 2.97^{2.50} \approx 15.20$.

(c) Note that $f(e) = \frac{1}{e}$. Any horizontal line $y = k$, with $0 < k < \frac{1}{e}$, will intersect the graph at the two points $\left(x_1,\ \dfrac{\ln x_1}{x_1}\right)$ and $\left(x_2,\ \dfrac{\ln x_2}{x_2}\right)$, where $1 < x_1 < e$ and $x_2 > e$.

7 Logarithm law 3 states that it is valid only for positive real numbers, so $y = \log_3(x^2)$ is equivalent to $y = 2\log_3 x$ only for $x > 0$. The domain of $y = \log_3(x^2)$ is $\mathbb{R} - \{0\}$, whereas the domain of $y = 2\log_3 x$ is $x > 0$.

9 There are 4 points of intersection. Listed in order of difficulty to find we have: $(-0.9999011,\ 0.00999001)$, $(-0.0001,\ 0.01)$, $(100,\ 0.01105111)$, and $(36{,}102.844,\ 4.6928 \times 10^{13})$. Exponential function values (with base $> 1$) are greater than polynomial function values (with leading term positive) for very large values of $x$.

11 $60{,}000 = 40{,}000b^5 \;\Rightarrow\; b = \sqrt[5]{1.5} \approx 1.0844717712$, or $8.44717712\%$. Mentally—it would take $70/8.5 \approx 8^+$ years to double and there would be about $40/8 = 5$ doubling periods, $2^5 = 32$ and $32 \cdot \$40{,}000 = \$1{,}280{,}000$.

Actual $= \$40{,}000(1 + 0.0844717712)^{40} = \$1{,}025{,}156.25$.

13 On the TI-83 Plus, enter the sum of the days, {0, 5168, 6728, 8136, 8407, 8735, 8857, 9010, 9274, 9631, 9666}, and the averages, {1003.16, 2002.25, 3004.46, 4003.33, 5023.55, 6010.00, 7022.44, 8038.88, 9033.23, 10,006.78, 11,014.69}, in $L_1$ and $L_2$, respectively. Use ExpReg under STAT CALC to obtain $y = ab^x$, where $a = 753.76507865478$ and $b = 1.0002468662739$. Plot the data along with the exponential function and the line $y = 20{,}000$. The functions intersect at approximately 13,282. This value corresponds to March 27, 2009. *Note:* The TI-83 has a convenient "day between dates" function for problems of this nature.

Using the first and last milestone figures and the continuously compounded interest formula gives us $A = Pe^{rt} \Leftrightarrow 11{,}014.69 = 1003.16e^{r(9666)} \Rightarrow$ $r \approx 0.0002478869045$ (daily) or about 9.05% annually. *Note:* The Dow was 100.25 on 1/12/1906. You may want to include this information and examine the differences it makes in any type of prediction.

A discussion of practical considerations should lead to mention of crashes, corrections, and the validity of any model over too long of a period of time.

15 (a) [−10, 110, 10] by [0, 1E10, 1E9]

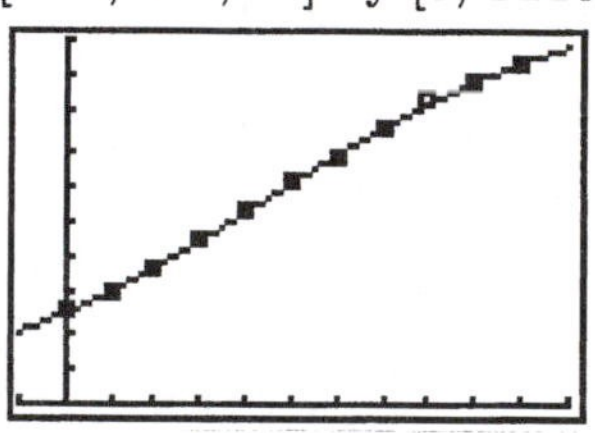

*Figure 15*

(b) Based on the shape of the curve in part (a), a logistic model is definitely more appropriate.

(c) Using the logistic model feature, we obtain $y = c/(1 + ae^{-bx})$ with

$$a = 3.567015872,\ b = 0.0281317818, \text{ and } c = 1.1355368 \times 10^{10}.$$

(d) As $x$ gets large, the term $ae^{-bx}$ gets closer to zero, so $y$ approaches $c/1 = c$.

# Chapter 6: Systems of Equations and Inequalities

## 6.1 Exercises

*Note:* The notation $E_1$ and $E_2$ refers to the first equation and the second equation.

[1] Substituting $y$ in $E_2$ into $E_1$ yields $2x-1=x^2-4 \Rightarrow x^2-2x-3=0 \Rightarrow$ $(x-3)(x+1)=0 \Rightarrow x=3, -1$. Substituting $x=3$ into $E_2$ gives us $y=5$, so $(3, 5)$ is a solution of the system. Similarly, $(-1, -3)$ is a solution.

[3] Solving $E_2$ for $x$, $x=1-2y$, and substituting into $E_1$ yields $y^2=1-(1-2y) \Rightarrow$ $y^2-2y=0 \Rightarrow y(y-2)=0 \Rightarrow y=0, 2; x=1, -3$. ★ $(1, 0), (-3, 2)$

[5] Substituting $y$ in $E_2$ into $E_1$ yields $2(4x^3)=x^2 \Rightarrow 8x^3=x^2 \Rightarrow$ $8x^3-x^2=0 \Rightarrow x^2(8x-1)=0 \Rightarrow x=0, \frac{1}{8}; y=0, \frac{1}{128}$. ★ $(0, 0), (\frac{1}{8}, \frac{1}{128})$

[7] Solving $E_1$ for $x$, $x=-2y-1$, and substituting into $E_2$ yields $2(-2y-1)-3y=12 \Rightarrow -7y=14 \Rightarrow y=-2; x=3$. ★ $(3, -2)$

[9] Solving $E_1$ for $x$, $2x=3y+1 \Rightarrow x=\frac{3}{2}y+\frac{1}{2}$, and substituting into $E_2$ yields $-6(\frac{3}{2}y+\frac{1}{2})+9y=4 \Rightarrow -9y-3+9y=4 \Rightarrow -3=4$, a contradiction.

There are *no solutions*.

[11] Solving $E_1$ for $x$, $x=5-3y$, and substituting into $E_2$ yields $(5-3y)^2+y^2=25 \Rightarrow$ $25-30y+9y^2+y^2=25 \Rightarrow 10y^2-30y=0 \Rightarrow 10y(y-3)=0 \Rightarrow$ $y=0, 3; x=5, -4$. ★ $(-4, 3), (5, 0)$

[13] Solving $E_2$ for $y$, $y=x+4$, and substituting into $E_1$ yields $x^2+(x+4)^2=8 \Rightarrow$ $x^2+x^2+8x+16=8 \Rightarrow 2x^2+8x+8=0 \Rightarrow x^2+4x+4=0 \Rightarrow$ $(x+2)^2=0 \Rightarrow x=-2; y=2$. ★ $(-2, 2)$

[15] Solving $E_2$ for $y$, $y=3x+2$, and substituting into $E_1$ yields $x^2+(3x+2)^2=9 \Rightarrow x^2+9x^2+12x+4=9 \Rightarrow 10x^2+12x-5=0 \Rightarrow$

$$x=\frac{-12\pm\sqrt{344}}{20}=\frac{-12\pm\sqrt{4\cdot 86}}{20}=\frac{-12\pm 2\sqrt{86}}{20}=\frac{-6\pm\sqrt{86}}{10}=-\frac{3}{5}\pm\frac{1}{10}\sqrt{86}.$$

Now we'll substitute the values of $x$ into $y=3x+2$ to get the corresponding values of $y$. $y=3\left(\frac{-6\pm\sqrt{86}}{10}\right)+2=\frac{-18\pm 3\sqrt{86}}{10}+\frac{20}{10}=\frac{2\pm 3\sqrt{86}}{10}=\frac{1}{5}\pm\frac{3}{10}\sqrt{86}.$

★ $(-\frac{3}{5}+\frac{1}{10}\sqrt{86}, \frac{1}{5}+\frac{3}{10}\sqrt{86}), (-\frac{3}{5}-\frac{1}{10}\sqrt{86}, \frac{1}{5}-\frac{3}{10}\sqrt{86})$

[17] Solving $E_2$ for $x$, $x=2y-4$, and substituting into $E_1$ yields $(2y-4)^2+y^2=16 \Rightarrow$ $4y^2-16y+16+y^2=16 \Rightarrow 5y^2-16y=0 \Rightarrow y(5y-16)=0 \Rightarrow$ $y=0, \frac{16}{5}; x=-4, \frac{12}{5}$. ★ $(-4, 0), (\frac{12}{5}, \frac{16}{5})$

[19] Solving $E_2$ for $x$, $x = 1 - y$, and substituting into $E_1$ yields

$(1 - y - 1)^2 + (y + 2)^2 = 10 \Rightarrow y^2 + (y^2 + 4y + 4) = 10 \Rightarrow$

$2y^2 + 4y - 6 = 0 \Rightarrow 2(y^2 + 2y - 3) = 0 \Rightarrow 2(y + 3)(y - 1) = 0 \Rightarrow$

$y = -3, 1$; $x = 4, 0$. ★ $(0, 1), (4, -3)$

[21] Substituting $y$ in $E_1$ into $E_2$ yields $20/x^2 = 9 - x^2$ {multiply by $x^2$} $\Rightarrow$

$20 = 9x^2 - x^4 \Rightarrow x^4 - 9x^2 + 20 = 0 \Rightarrow (x^2 - 4)(x^2 - 5) = 0 \Rightarrow$

$x^2 = 4, 5 \Rightarrow x = \pm 2, \pm\sqrt{5}$. If $x^2 = 4$, then $y = \frac{20}{4} = 5$.

If $x^2 = 5$, then $y = \frac{20}{5} = 4$. Thus, we get the *four* solutions $(\pm 2, 5)$ and $(\pm\sqrt{5}, 4)$.

[23] Solving $E_1$ for $y^2$, $y^2 = 4x^2 + 4$, and substituting into $E_2$ yields

$9(4x^2 + 4) + 16x^2 = 140 \Rightarrow 52x^2 = 104 \Rightarrow x^2 = 2 \Rightarrow x = \pm\sqrt{2}$; $y = \pm 2\sqrt{3}$.

There are *four* solutions. ★ $(\sqrt{2}, \pm 2\sqrt{3}), (-\sqrt{2}, \pm 2\sqrt{3})$

[25] Solving $E_1$ for $x^2$ and substituting into $E_2$ yields $(y^2 + 4) + y^2 = 12 \Rightarrow$

$2y^2 = 8 \Rightarrow y^2 = 4 \Rightarrow y = \pm 2$; $x = \pm\sqrt{8} = \pm 2\sqrt{2}$.

There are *four* solutions. ★ $(2\sqrt{2}, \pm 2), (-2\sqrt{2}, \pm 2)$

[27] Solving $E_2$ for $y$, $y = 2x + z - 9$, and substituting into $E_1$ and $E_3$ yields

$$\begin{cases} x + 2(2x + z - 9) - z = -1 \\ x + 3(2x + z - 9) + 3z = 6 \end{cases} \Rightarrow \begin{cases} 5x + z = 17 & (E_4) \\ 7x + 6z = 33 & (E_5) \end{cases}$$

Solving $E_4$ for $z$, $z = 17 - 5x$, and substituting into $E_5$ yields

$7x + 6(17 - 5x) = 33 \Rightarrow 7x + 102 - 30x = 33 \Rightarrow$

$-23x = -69 \Rightarrow x = 3$. Now $z = 17 - 5x = 17 - 5(3) = 2$ and

$y = 2x + z - 9 = 2(3) + 2 - 9 = -1$. ★ $(3, -1, 2)$

[29] Solving $E_3$ for $y$, $y = 1 - z$, and substituting into $E_2$ yields $\begin{cases} x^2 + z^2 = 5 & (E_1) \\ 2x - z = 0 & (E_4) \end{cases}$

Now solve $E_4$ for $z$, $z = 2x$, and substitute into $E_1$ yielding $x^2 + (2x)^2 = 5 \Rightarrow$

$x^2 + 4x^2 = 5 \Rightarrow 5x^2 = 5 \Rightarrow x^2 = 1 \Rightarrow x = \pm 1$;

$z = 2x = \pm 2$; $y = 1 - z = -1, 3$. ★ $(1, -1, 2), (-1, 3, -2)$

[31] Substituting $y = 4x - b$ into $y = x^2$ yields $4x - b = x^2 \Rightarrow$

$x^2 - 4x + b = 0 \Rightarrow x = \dfrac{4 \pm \sqrt{16 - 4b}}{2} = 2 \pm \sqrt{4 - b}$.

For $x$ to have 1, 2, or no values, the discriminant $4 - b$ must be equal to zero, greater than zero, or less than zero, respectively. Graphically, the line would be (a) tangent to the parabola, (b) intersect the parabola in two points, and (c) not intersect the parabola.

(continued)

(a) $4 - b = 0 \Rightarrow b = 4$.

(b) $4 - b > 0 \Rightarrow 4 > b \Rightarrow b < 4$.

(c) $4 - b < 0 \Rightarrow 4 < b \Rightarrow b > 4$.

[33] From the graph of $y = x$ and $y = 2^{-x}$, it is clear that there is a point of intersection and therefore a single solution between 0 and 1.

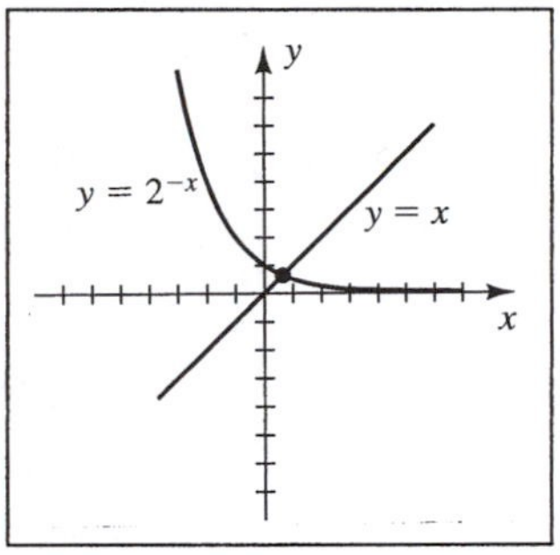


*Figure 33*

[35] The equation of the line is $y - 2 = m(x - 4)$. Substituting $y^2$ for $x$ gives us

$$y - 2 = m(y^2 - 4) \Rightarrow 0 = my^2 - y + (2 - 4m) \Rightarrow y = \frac{1 \pm \sqrt{1 - 4m(2 - 4m)}}{2m}.$$

For there to be only one value of $y$, the discriminant $1 - 4m(2 - 4m)$ must be equal to zero, i.e., $1 - 4m(2 - 4m) = 0 \Rightarrow 1 - 8m + 16m^2 = 0 \Rightarrow (4m - 1)^2 = 0 \Rightarrow m = \frac{1}{4}$. This line, $y - 2 = \frac{1}{4}(x - 4)$, or $y = \frac{1}{4}x + 1$, is tangent to the parabola.

[37] Use the fact that $(0, 3)$, $(-1, \frac{5}{3})$, and $(1, 7)$ are on the graph of $f(x) = ba^x + c$.

$$\begin{cases} f(0) = 3 \\ f(-1) = \frac{5}{3} \\ f(1) = 7 \end{cases} \Rightarrow \begin{cases} 3 = ba^0 + c & (\text{E}_1) \\ \frac{5}{3} = ba^{-1} + c & (\text{E}_2) \\ 7 = ba^1 + c & (\text{E}_3) \end{cases}$$

From $\text{E}_1$, $3 = b + c \Rightarrow c = 3 - b$. Substitute $3 - b$ for $c$ in $\text{E}_2$ and $\text{E}_3$.

$\text{E}_2$: $\frac{5}{3} = ba^{-1} + 3 - b \Rightarrow b - ba^{-1} = \frac{4}{3} \Rightarrow b\left(1 - \frac{1}{a}\right) = \frac{4}{3} \Rightarrow b\left(\frac{a-1}{a}\right) = \frac{4}{3}$

$\text{E}_3$: $7 = ba^1 + 3 - b \Rightarrow b - ba = -4 \Rightarrow b(1 - a) = -4 \Rightarrow b(a - 1) = 4$

Solving both equations for $b$ and equating the expressions gives us

$\dfrac{4a}{3(a-1)} = \dfrac{4}{a-1} \Rightarrow \frac{a}{3} = \frac{1}{1} \Rightarrow a = 3$. Substituting 3 for $a$ in $b(a - 1) = 4$ gives us $2b = 4 \Rightarrow b = 2$. Substituting 2 for $b$ in $c = 3 - b$ gives us $c = 1$, so the function is $f(x) = 2(3)^x + 1$.

[39] Using $P = 40 = 2l + 2w$ and $A = 96 = lw$, we have $\begin{cases} 2l + 2w = 40 & (\text{E}_1) \\ lw = 96 & (\text{E}_2) \end{cases}$

Solving $\text{E}_1$ for $l$, $l = 20 - w$, and substituting into $\text{E}_2$ yields $(20 - w)w = 96 \Rightarrow 20w - w^2 = 96 \Rightarrow w^2 - 20w + 96 = 0 \Rightarrow (w - 8)(w - 12) = 0 \Rightarrow w = 8, 12$; $l = 12, 8$. In either case, the rectangle is 12 inches $\times$ 8 inches.

41 (a) We have $R = aS/(S+b)$ and for 1998, let $S = 40{,}000$ and $R = 60{,}000$.

Then for 1999, let $S = 60{,}000$ and $R = 72{,}000$.

$$\begin{cases} 60{,}000 = (40{,}000a)/(40{,}000+b) \\ 72{,}000 = (60{,}000a)/(60{,}000+b) \end{cases} \Rightarrow \begin{cases} 120{,}000 + 3b = 2a & (E_1) \\ 360{,}000 + 6b = 5a & (E_2) \end{cases}$$

Solving $E_1$ for $a$ and substituting into $E_2$ yields

$$360{,}000 + 6b = 5(60{,}000 + \tfrac{3}{2}b) \Rightarrow 60{,}000 = \tfrac{3}{2}b \Rightarrow b = 40{,}000;\ a = 120{,}000.$$

(b) Now let $S = 72{,}000$ and thus $R = \dfrac{(120{,}000)(72{,}000)}{72{,}000 + 40{,}000} = \dfrac{540{,}000}{7} \approx 77{,}143$.

43 Let $R_1$ and $R_2$ equal 0. The system is then $\begin{cases} 0 = 0.01x(50 - x - y) \\ 0 = 0.02y(100 - y - 0.5x) \end{cases}$

The first equation is zero if $x = 0$ or if $50 - x - y = 0$. The second equation is zero if $y = 0$ or if $100 - y - 0.5x = 0$. Hence, there are 4 possible solutions.

**(1)** A solution is $x = 0$ and $y = 0$, or $(0,\ 0)$.

**(2)** A second solution is $x = 0$ and $100 - y - 0.5x = 0$ $\{y = 100\}$, or $(0,\ 100)$.

**(3)** A third solution is $y = 0$ and $50 - x - y = 0$ $\{x = 50\}$, or $(50,\ 0)$.

**(4)** A fourth solution occurs if $50 - x - y = 0$ and $100 - y - 0.5x = 0$.

Solve the first equation for $y$ $\{y = 50 - x\}$ and substitute into the second equation: $100 - (50 - x) - 0.5x = 0 \Rightarrow 50 = -\frac{1}{2}x \Rightarrow x = -100;\ y = 150$.

This solution is meaningless for this problem since $x$ and $y$ are nonnegative.

45 Since we have an *open* top instead of a closed top, we use $3xy$ instead of $4xy$ {as in Example 5} for the surface area formula. $\begin{cases} x^2y = 2 & \textit{Volume} \\ 2x^2 + 3xy = 8 & \textit{Surface Area} \end{cases}$

Solving $E_1$ for $y$ and substituting into $E_2$ yields $2x^2 + 3x(2/x^2) = 8 \Rightarrow$

$2x^2 + (6/x) = 8$ {multiply by $x$} $\Rightarrow 2x^3 - 8x + 6 = 0 \Rightarrow 2(x^3 - 4x + 3) = 0$.

We know that $x = 1$ is a solution of $x^3 - 4x + 3 = 0$ since the sum of the coefficients is zero. Continuing, $2(x-1)(x^2 + x - 3) = 0 \Rightarrow \{x > 0\}\ x = 1, \dfrac{-1+\sqrt{13}}{2}$.

There are two solutions: 1 ft × 1 ft × 2 ft or

$$\frac{\sqrt{13}-1}{2} \text{ ft} \times \frac{\sqrt{13}-1}{2} \text{ ft} \times \frac{8}{(\sqrt{13}-1)^2} \text{ ft} \approx 1.30 \text{ ft} \times 1.30 \text{ ft} \times 1.18 \text{ ft}.$$

47 We eliminate $n$ from the equations to determine all intersection points.

(a) $x^2 + y^2 = n^2$ and $y = n - 1$ {so $n = y + 1$} $\Rightarrow x^2 + y^2 = (y+1)^2 \Rightarrow$

$x^2 + y^2 = y^2 + 2y + 1 \Rightarrow x^2 = 2y + 1 \Rightarrow 2y = x^2 - 1 \Rightarrow y = \frac{1}{2}x^2 - \frac{1}{2}$.

The points are on the parabola $y = \frac{1}{2}x^2 - \frac{1}{2}$.

(b) As in part (a), $x^2 + y^2 = (y+2)^2 \Rightarrow x^2 = 4y + 4 \Rightarrow y = \frac{1}{4}x^2 - 1$.

49 (a) The slope of the line from $(-4,\ -3)$ to the origin is $\frac{3}{4}$ so the slope of the tangent line (which is perpendicular to the line to the origin) is $-\frac{4}{3}$. An equation of the line through $(-4,\ -3)$ with slope $-\frac{4}{3}$ is $y+3=-\frac{4}{3}(x+4)$, or, equivalently, $4x+3y=-25$. Letting $y=-50$, we find that $x=31.25$.

(b) The slope of the line from an arbitrary point $(x,\ y)$ on the circle to the origin is $\frac{y}{x}$, so the slope of the tangent line is $-\frac{x}{y}$. The line through $(0,\ -50)$ is $y+50=\left(-\frac{x}{y}\right)(x-0)$, or, equivalently, $y^2+50y=-x^2$. The equation of the circle is $x^2+y^2=25$. Substituting $x^2=25-y^2$ into $y^2+50y=-x^2$ gives us $50y=-25$, or $y=-\frac{1}{2}$. Substituting $y=-\frac{1}{2}$ into $x^2=25-y^2$ gives us $x^2=\frac{99}{4}$ and hence $x=\pm\frac{3}{2}\sqrt{11}$. The positive solution corresponds to releasing the hammer from a clockwise spin, whereas the negative solution corresponds to releasing the hammer from a counterclockwise spin as depicted in the figures. Hence, the hammer should be released at $(-\frac{3}{2}\sqrt{11},\ -\frac{1}{2})\approx(-4.975,\ -0.5)$.

51 **Graphically:** $x^2+y^2=4 \Rightarrow y=\pm\sqrt{4-x^2}$ and $x+y=1 \Rightarrow y=1-x$. Graph $Y_1=\sqrt{4-x^2}$, $Y_2=-Y_1$, and $Y_3=1-x$. There are two points of intersection at approximately $(-0.82,\ 1.82)$ and $(1.82,\ -0.82)$.

**Algebraically:** $y=1-x$ and $x^2+y^2=4 \Rightarrow x^2+(1-x)^2=4 \Rightarrow$ $x^2+1-2x+x^2=4 \Rightarrow 2x^2-2x-3=0$. Using the quadratic formula,

$x=\dfrac{2\pm\sqrt{4-4(2)(-3)}}{4}=\dfrac{1\pm\sqrt{7}}{2}$. To find the value of $y$ we use

$y=1-x \Rightarrow y=1-\dfrac{1\pm\sqrt{7}}{2}=\dfrac{1\mp\sqrt{7}}{2}$. The points of intersection are

$\left(\frac{1}{2}\pm\frac{\sqrt{7}}{2},\ \frac{1}{2}\mp\frac{\sqrt{7}}{2}\right)$. The graphical solution approximates the algebraic solution.

[−6, 6] by [−4, 4]

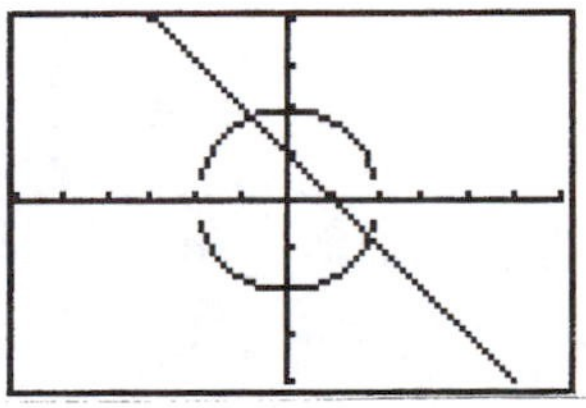

*Figure 51*

[53] After zooming-in near the region of interest in the second quadrant, we see that the cubic intersects the circle twice. Due to the symmetry, we know there are two more points of intersection in the fourth quadrant. Thus, the six points of intersection are approximately $(\mp 0.56, \pm 1.92)$, $(\mp 0.63, \pm 1.90)$, and $(\pm 1.14, \pm 1.65)$.

[−6, 6] by [−4, 4]

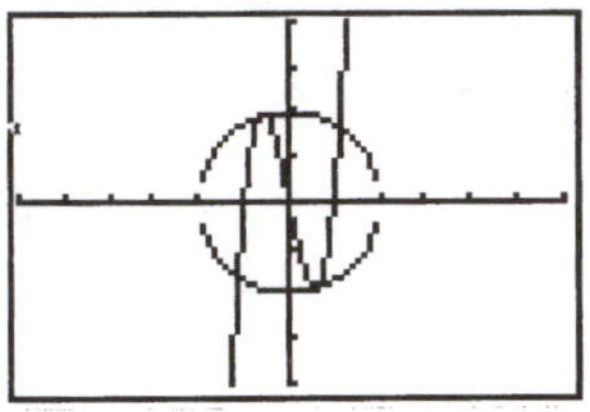

Figure 53

[−3, 3] by [−2, 2]

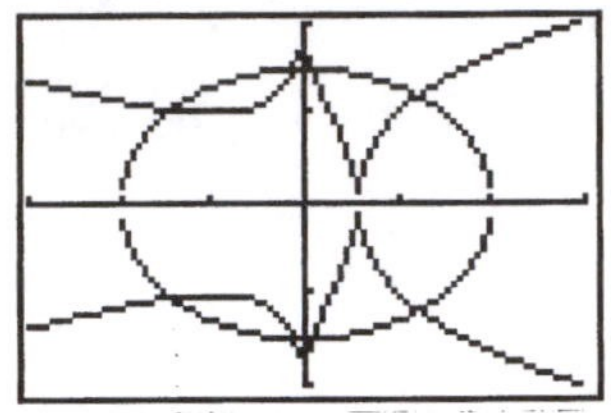

Figure 55

[55] $|x + \ln|x|| - y^2 = 0 \Rightarrow y^2 = |x + \ln|x|| \Rightarrow y = \pm\sqrt{|x + \ln|x||}$ and

$$\frac{x^2}{4} + \frac{y^2}{2.25} = 1 \Rightarrow \frac{y^2}{2.25} = 1 - \frac{x^2}{4} \Rightarrow y^2 = 2.25(1 - x^2/4) \Rightarrow$$

$y = \pm 1.5\sqrt{1 - x^2/4}$. The graph is symmetric with respect to $x$-axis. There are 8 points of intersection. Their coordinates are approximately

$$(-1.44, \pm 1.04), (-0.12, \pm 1.50), (0.10, \pm 1.50), \text{ and } (1.22, \pm 1.19).$$

[57] If $f(x) = ae^{-bx}$ and $x = 1$, then $f(1) = ae^{-b} = 0.80487 \Rightarrow a = 0.80487e^b$. When $x = 2$, $f(2) = ae^{-2b} = 0.53930 \Rightarrow a = 0.53930e^{2b}$. Let $a = y$ and $b = x$ and then graph $Y_1 = 0.80487e^x$ and $Y_2 = 0.53930e^{2x}$. The graphs intersect at approximately $(0.4004, 1.2012) = (b, a)$. Thus, $a \approx 1.2012$, $b \approx 0.4004$, and $f(x) = 1.2012e^{-0.4004x}$. The function $f$ is also accurate at $x = 3, 4$.

[0, 3] by [0, 2]

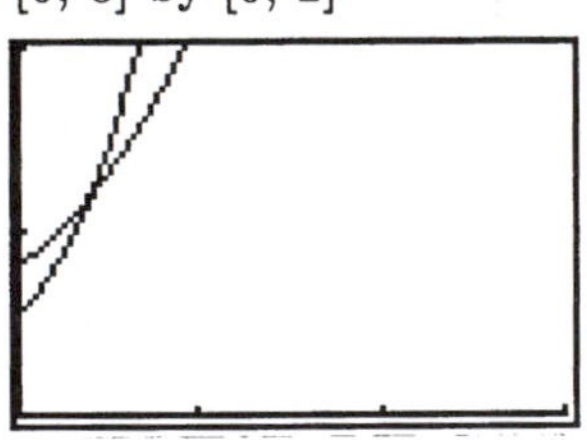

Figure 57

[0, 4] by [0, 4]

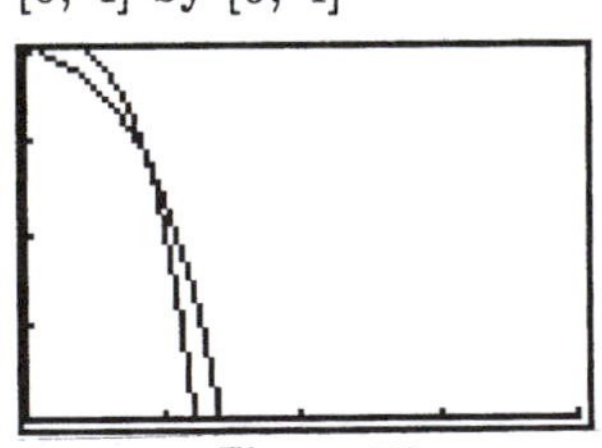

Figure 59

[59] If $f(x) = ax^2 + e^{bx}$ and $x = 2$, then $f(2) = 4a + e^{2b} = 17.2597 \Rightarrow$ $a = (17.2597 - e^{2b})/4$. When $x = 3$, $f(3) = 9a + e^{3b} = 40.1058 \Rightarrow$ $a = (40.1058 - e^{3b})/9$. Graph $Y_1 = (17.2597 - e^{2x})/4$ and $Y_2 = (40.1058 - e^{3x})/9$. The graphs intersect at approximately $(0.9002, 2.8019)$. Thus, $a \approx 2.8019$, $b \approx 0.9002$, and $f(x) = 2.8019x^2 + e^{0.9002x}$. The function $f$ is also accurate at $x = 4$.

## 6.2 Exercises

*Note:* The notation $E_1$ and $E_2$ refers to equation 1 and equation 2. $3\,E_2$ symbolizes "3 times equation 2". After the value of one variable is found, the value(s) of the other variable(s) will be stated and can be found by substituting the known value(s) back into the original equation(s).

[1] The easiest choice for multipliers is to take $-2$ times the second equation and eliminate $x$. $-2\,E_2 + E_1 \Rightarrow 4y + 3y = -16 + 2 \Rightarrow 7y = -14 \Rightarrow y = -2$; $x = 4$. Another way to write this is as follows:

$$\begin{cases} 2x + 3y = 2 & (E_1) \\ x - 2y = 8 & (E_2) \end{cases} \Rightarrow \begin{cases} 2x + 3y = 2 & (E_1) \\ -2x + 4y = -16 & (-2\,E_2) \end{cases}$$

Alternatively, if we wanted to eliminate $y$ instead of $x$, we could add 2 times the first equation to 3 times the second equation, giving us $2\,E_1 + 3\,E_2 \Rightarrow$ $4x + 3x = 4 + 24 \Rightarrow 7x = 28 \Rightarrow x = 4$; $y = -2$ ★ $(4, -2)$

[3] $3\,E_1 - 2\,E_2 \Rightarrow 15y + 14y = 48 - 48 \Rightarrow 29y = 0 \Rightarrow y = \frac{0}{29} = 0$. From the first equation, we see that $2x + 5(0) = 16 \Rightarrow 2x = 16 \Rightarrow x = 8$. ★ $(8, 0)$

[5] $2\,E_2 + E_1 \Rightarrow 2r + 3r = -8 + 3 \Rightarrow 5r = -5 \Rightarrow r = -1$. From the second equation, we see that $-1 - 2s = -4 \Rightarrow -2s = -3 \Rightarrow s = \frac{3}{2}$. ★ $(-1, \frac{3}{2})$

[7] $3\,E_1 - 5\,E_2 \Rightarrow -53y = -28 \Rightarrow y = \frac{28}{53}$. Instead of substituting into one of the equations to find the value of the other variable, it is usually easier to pick different multipliers and re-solve the system for the other variable, as in Example 4.
$7\,E_1 + 6\,E_2 \Rightarrow 53x = 76 \Rightarrow x = \frac{76}{53}$. ★ $(\frac{76}{53}, \frac{28}{53})$

[9] We will first eliminate all fractions by multiplying both sides of each equation by its lcd. $\begin{cases} 6\,E_1 \\ 3\,E_2 \end{cases} \Rightarrow \begin{cases} 2c + 3d = 30 & (E_3) \\ 3c - 2d = -3 & (E_4) \end{cases}$

$$3\,E_3 - 2\,E_4 \Rightarrow 13d = 96 \Rightarrow d = \tfrac{96}{13}.$$

$$2\,E_3 + 3\,E_4 \Rightarrow 13c = 51 \Rightarrow c = \tfrac{51}{13}.$$ ★ $(\frac{51}{13}, \frac{96}{13})$

[11] The least common multiple of 2 and 3 is 6, so we will multiply the first equation by $\sqrt{3}$ and the second equation by $\sqrt{2}$ so that the coefficients of $y$ are $\sqrt{6}$ and $-\sqrt{6}$. $\sqrt{3}\,E_1 + \sqrt{2}\,E_2 \Rightarrow 7x = 8 \Rightarrow x = \frac{8}{7}$. We will now re-solve the system to obtain a value for $y$. $2\sqrt{2}\,E_1 - \sqrt{3}\,E_2 \Rightarrow -7y = 3\sqrt{6} \Rightarrow y = -\frac{3}{7}\sqrt{6}$. ★ $(\frac{8}{7}, -\frac{3}{7}\sqrt{6})$

[13] $3\,E_1 + E_2 \Rightarrow 0 = 27$; this statement is never true so there are *no solutions.*

[15] $2\,E_1 + E_2 \Rightarrow 0 = 0$; this statement is always true so the solution is all ordered pairs $(m,\, n)$ such that $3m - 4n = 2$. We could also say that the solution is all ordered pairs $(m,\, n)$ such that $-6m + 8n = -4$. Either statement is correct.

[17] $3\,E_1 - 2\,E_2 \Rightarrow -23x = 0 \Rightarrow x = 0;\ y = 0$ ★ $(0,\, 0)$

[19] Using the hint, we let $u = 1/x$ and $v = 1/y$, and the system is $\begin{cases} 2u + 3v = -2 & (E_3) \\ 4u - 5v = 1 & (E_4) \end{cases}$

$$-2\,E_3 + E_4 \Rightarrow -11v = 5 \Rightarrow v = -\tfrac{5}{11}.$$

$$5\,E_3 + 3\,E_4 \Rightarrow 22u = -7 \Rightarrow u = -\tfrac{7}{22}.$$

Resubstituting, we have $x = 1/u = -\frac{22}{7}$ and $y = 1/v = -\frac{11}{5}$. ★ $(-\frac{22}{7},\, -\frac{11}{5})$

[21] Let $x$ denote the number of \$3.00 tickets and $y$ the number of \$4.50 tickets.

$$\begin{cases} x + y = 450 & \textit{quantity} \\ 3.00x + 4.50y = 1555.50 & \textit{value} \end{cases}$$

$E_2 - 3\,E_1 \Rightarrow 4.50y - 3y = 1555.50 - 1350 \Rightarrow 1.5y = 205.50 \Rightarrow$

$y = \dfrac{205.50}{1.5} = 137$. Since $x + y = 450$, $x = 450 - 137 = 313$.

[23] The volume for a cylinder is $\pi r^2 h$ and the volume for a cone is $\frac{1}{3}\pi r^2 h$.

The radius of the cylinder is $\frac{1}{2}$ cm. $\begin{cases} x + y = 8 & \textit{length} \\ \pi(\frac{1}{2})^2 x + \frac{1}{3}\pi(\frac{1}{2})^2 y = 5 & \textit{volume} \end{cases}$

Solving $E_1$ for $y$, $y = 8 - x$, and substituting into $E_2$ yields

$\frac{\pi}{4}x + \frac{\pi}{12}(8 - x) = 5 \Rightarrow \frac{\pi}{4}x + \frac{8\pi}{12} - \frac{\pi}{12}x = \frac{15}{3} \Rightarrow \frac{3\pi}{12}x - \frac{\pi}{12}x = \frac{15}{3} - \frac{2\pi}{3} \Rightarrow$

$\frac{\pi}{6}x = \dfrac{15 - 2\pi}{3} \Rightarrow x = \dfrac{6}{\pi}\cdot\dfrac{15 - 2\pi}{3} = \dfrac{30 - 4\pi}{\pi} = \dfrac{30}{\pi} - 4 \approx 5.55$ cm.

$y = 8 - x = 8 - \left(\dfrac{30 - 4\pi}{\pi}\right) = \dfrac{8\pi}{\pi} - \dfrac{30}{\pi} + \dfrac{4\pi}{\pi} = \dfrac{12\pi - 30}{\pi} = 12 - \dfrac{30}{\pi} \approx 2.45$ cm.

[25] The perimeter is composed of 2 sides of the rectangular portion of the table, $2l$, and 2 edges of the semicircular regions, $2\cdot\frac{1}{2}(2\pi r) = 2\pi r$. Since the radius is $\frac{1}{2}w$, the system can be represented by $\begin{cases} 2l + 2\pi(\frac{1}{2}w) = 40 & \textit{perimeter} \\ lw = 2\left[\pi(\frac{1}{2}w)^2\right] & \textit{area} \end{cases}$

Solving $E_1$ for $l$, $l = \dfrac{40 - \pi w}{2}$, and substituting into $E_2$ yields

$\left(\dfrac{40 - \pi w}{2}\right)w = \dfrac{\pi w^2}{2} \Rightarrow (40 - \pi w)w = \pi w^2 \Rightarrow 40w - \pi w^2 = \pi w^2 \Rightarrow$

$40w - 2\pi w^2 = 0 \Rightarrow 2w(20 - \pi w) = 0 \Rightarrow w = 0,\ \dfrac{20}{\pi}$.

We discard $w = 0$ and use $w = 20/\pi \approx 6.37$ ft. Thus, $l = \dfrac{40 - \pi(20/\pi)}{2} = \dfrac{20}{2} = 10$ ft.

[27] Let $x$ denote the number of adults and $y$ the number of kittens. Thus,

$(\frac{1}{2}x)$ is the number of adult females. $\begin{cases} x + y = 6000 & \textit{total} \\ y = 3(\frac{1}{2}x) & \textit{3 kittens per adult female} \end{cases}$

Substituting $y$ from $E_2$ into $E_1$ yields

$$x + \tfrac{3}{2}x = 6000 \Rightarrow \tfrac{5}{2}x = 6000 \Rightarrow x = \tfrac{2}{5} \cdot 6000 = 2400;\ y = 3600.$$

[29] Let $x$ denote the number of grams of the 35% alloy and $y$ the number of grams of the 60% alloy. $\begin{cases} x + y = 100 & \textit{quantity} \\ 0.35x + 0.60y = (0.50)(100) & \textit{quality} \end{cases}$

Multiply the second equation by 100 to obtain $35x + 60y = 5000$ $(E_3)$.

$$E_3 - 35\,E_1 \Rightarrow 25y = 1500 \Rightarrow y = 60;\ x = 40$$

[31] Let $x$ denote the speed of the plane and $y$ the speed of the wind. Use $d = rt$.

$$\begin{cases} 1200 = (x + y)(2) & \textit{with the wind} \\ 1200 = (x - y)(2\frac{1}{2}) & \textit{against the wind} \end{cases} \Rightarrow \begin{cases} 600 = x + y \\ 480 = x - y \end{cases}$$

$$E_1 + E_2 \Rightarrow 2x = 1080 \Rightarrow x = 540 \text{ mi/hr};\ y = 60 \text{ mi/hr}$$

[33] $v(t) = v_0 + at$ • $\begin{cases} v(2) = 16 \\ v(5) = 25 \end{cases} \Rightarrow \begin{cases} 16 = v_0 + 2a & (E_1) \\ 25 = v_0 + 5a & (E_2) \end{cases}$

$$E_2 - E_1 \Rightarrow 9 = 3a \Rightarrow a = 3;\ v_0 = 10$$

[35] Let $x$ denote the number of sofas produced and $y$ the number of recliners produced.

$$\begin{cases} 8x + \ 6y = 340 & \textit{labor hours} \\ 60x + 35y = 2250 & \textit{cost of materials} \end{cases}$$

$6\,E_2 - 35\,E_1 \Rightarrow 360x - 280x = 13{,}500 - 11{,}900 \Rightarrow 80x = 1600 \Rightarrow$

$$x = 20;\ y = 30$$

[37] (a) The expression $\underline{6x + 5y}$ represents the total bill for the plumber's business. This should equal the plumber's income, which is the plumber's number of hours times his/her hourly wage—that is, $(6 + 4)(x) = 10x$. The expression $\underline{4x + 6y}$ represents the total bill for the electrician's business. This should equal the electrician's income, that is, $(5 + 6)(y) = 11y$.

$$\begin{cases} 6x + 5y = 10x \\ 4x + 6y = 11y \end{cases} \Rightarrow \begin{cases} 5y = 4x \\ 4x = 5y \end{cases} \Rightarrow y = \tfrac{4}{5}x, \text{ or equivalently, } y = 0.80x$$

(b) The electrician should charge 80% of what the plumber charges—80% of \$20 per hour is \$16 per hour.

[39] Let $t = 0$ correspond to the year 1891. The average daily maximum can then be approximated by the linear equation $y_1 = 0.011t + 15.1$ and the average daily minimum by the linear equation $y_2 = 0.019t + 5.8$. We must determine $t$ when $y_1$ and $y_2$ differ by 9. $y_1 - y_2 = 9 \Rightarrow (0.011t + 15.1) - (0.019t + 5.8) = 9 \Rightarrow$ $-0.008t + 9.3 = 9 \Rightarrow -0.008t = -0.3 \Rightarrow t = 37.5$. $1891 + 37.5 = 1928.5$, or during the year 1928. Now we find the corresponding average maximum temperature: $t = 37.5 \Rightarrow y_1 = 0.011(37.5) + 15.1 = 15.5125 \approx 15.5°\text{C}$.

[41] Let $x$ and $y$ denote the amount of time in hours at the LP and SLP speeds, respectively. 5 hours and 20 minutes $= 5\frac{1}{3}$ hours, so $\dfrac{x}{5\frac{1}{3}}$ is the portion of the tape used at the LP speed. $\dfrac{y}{8}$ is the portion of the tape used at the SLP speed, and together, these fractions add up to 1 whole tape.

$$\begin{cases} x + y = 6 & \textit{total time} \\ \dfrac{x}{5\frac{1}{3}} + \dfrac{y}{8} = 1 & \textit{portions of tape} \end{cases} \Rightarrow \begin{cases} x + y = 6 & (\text{E}_1) \\ \frac{3}{16}x + \frac{1}{8}y = 1 & (\text{E}_2) \end{cases}$$

$$16\,\text{E}_2 - 2\,\text{E}_1 \Rightarrow 3x - 2x = 16 - 12 \Rightarrow x = 4;\ y = 2$$

*Note:* If you convert hours to minutes, the equations can be written as $x + y = 360$ and $x/320 + y/480 = 1$ with the solution $x = 240$ and $y = 120$.

[43] $$\begin{cases} a\,e^{3x} + b\,e^{-3x} = 0 & (\text{E}_1) \\ a\,(3e^{3x}) + b\,(-3e^{-3x}) = e^{3x} & (\text{E}_2) \end{cases}$$

Following the hint in the text, think of this system as a system of two equations in the variables $a$ and $b$. In $\text{E}_1$, we have "$e^{3x}$" times $a$, and in $\text{E}_2$, we have "$3e^{3x}$" times $a$. Hence, we will multiply $\text{E}_1$ by $-3$ and add this equation to $\text{E}_2$ to eliminate $a$ and then solve the system for $b$. $-3\,\text{E}_1 + \text{E}_2 \Rightarrow -3b\,e^{-3x} - 3b\,e^{-3x} = e^{3x} \Rightarrow$ $-6b\,e^{-3x} = e^{3x} \Rightarrow b = -\frac{1}{6}e^{6x}$. Substituting back into $\text{E}_1$ yields $a\,e^{3x} + (-\frac{1}{6}e^{6x})e^{-3x} = 0 \Rightarrow a\,e^{3x} = \frac{1}{6}e^{3x} \Rightarrow a = \frac{1}{6}$.

## 6.3 Exercises

[1] $3x - 2y < 6 \Leftrightarrow y > \frac{3}{2}x - 3$. Sketch the graph of $y = \frac{3}{2}x - 3$ with dashes. The point $(0, 0)$ is clearly on one side of the line, so we will substitute $x = 0$ and $y = 0$ into the inequality $y > \frac{3}{2}x - 3$. Checking, we have $0 > -3$, a *true* statement. Hence, we shade *all* points that are on the *same* side of the line as $(0, 0)$.

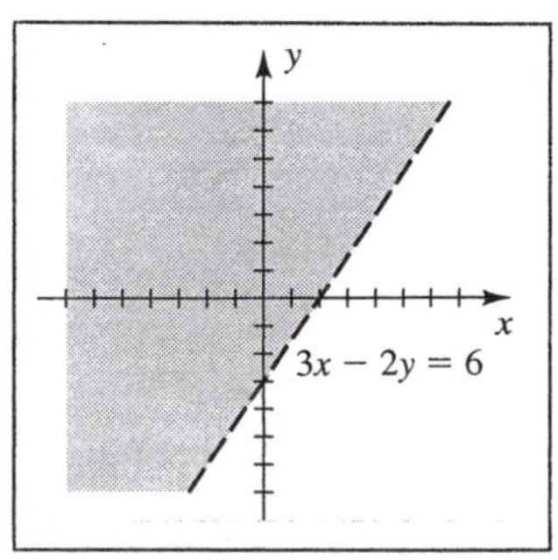


*Figure 1*

3. $2x + 3y \geq 2y + 1 \quad \Leftrightarrow \quad y \geq -2x + 1$. Sketch the graph of $y = -2x + 1$ as a solid line. Checking the test point (0, 0) in $y \geq -2x + 1$ gives us $0 \geq 1$, a *false* statement. Hence, we shade *all* points that are on the *opposite* side of the line as (0, 0).

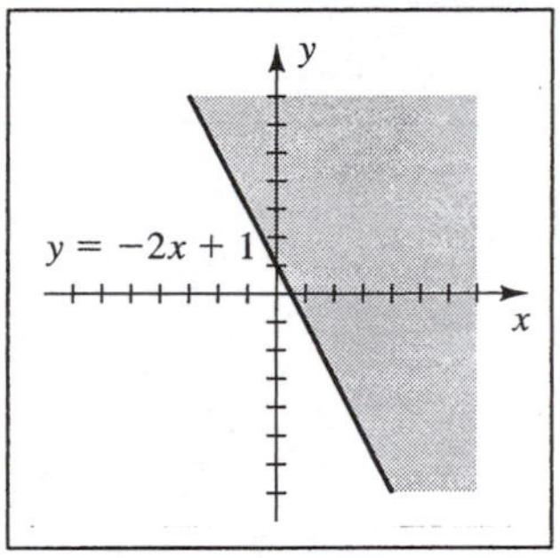


Figure 3

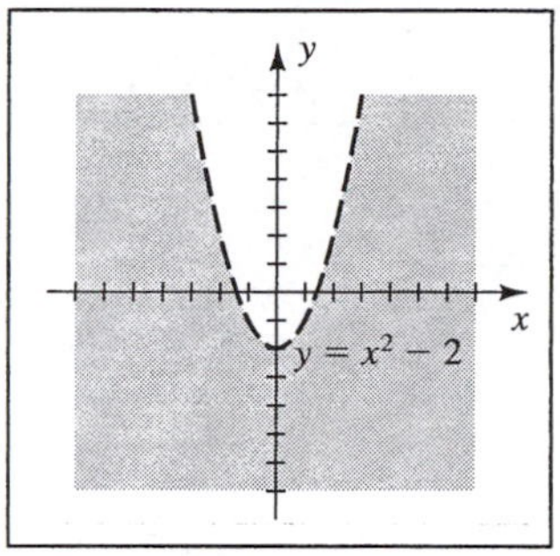


Figure 5

5. $y + 2 < x^2 \quad \Leftrightarrow \quad y < x^2 - 2$. Sketch the graph of the parabola $y = x^2 - 2$ with dashes. Checking the test point (0, 0) in $y < x^2 - 2$ gives us $0 < -2$, a *false* statement. Hence, we shade *all* points that are on the *opposite* side of the parabola as (0, 0).

7. $x^2 + 1 \leq y \quad \Leftrightarrow \quad y \geq x^2 + 1$. Sketch the graph of the parabola $y = x^2 + 1$ with a solid curve. Checking the test point (0, 0) in $y \geq x^2 + 1$ gives us $0 \geq 1$, a *false* statement. Hence, we shade *all* points that are on the *opposite* side of the parabola as (0, 0).

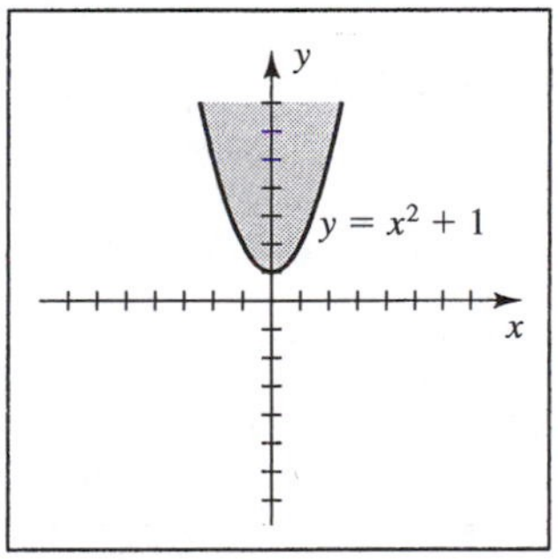


Figure 7

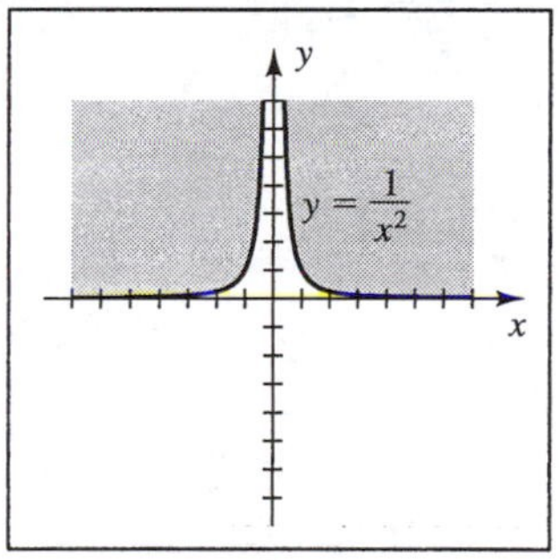


Figure 9

9. $yx^2 \geq 1 \quad \Leftrightarrow \quad y \geq 1/x^2 \; \{x \neq 0\}$. Sketch the graph of $y = 1/x^2$ with a solid curve. The point (1, 0) is clearly not on the graph. Substituting $x = 1$ and $y = 0$ into $y \geq 1/x^2$ yields $0 \geq 1$, a false statement. Hence, we do not shade the region containing (1, 0), but we do shade the regions in the first and second quadrants that are above the graph.

*Note:* The will use the notation $V$ @ $(a, b), (c, d), \ldots$ to denote the intersection point(s) of the solution region of the graph. These can be found by solving the system of *equalities* that correspond to the given system of inequalities.

[11] $\begin{cases} 3x + y < 3 \\ 4 - y < 2x \end{cases} \Leftrightarrow \begin{cases} y < -3x + 3 \\ y > -2x + 4 \end{cases}$ $V$ @ $(-1,\ 6)$

Solving the system $y = -3x + 3$ and $y = -2x + 4$ gives us the solution $(-1,\ 6)$. Testing the point $(0,\ 0)$ in both inequalities, we see that we need to shade under $y = -3x + 3$ and above $y = -2x + 4$, as shown in *Figure 11.*

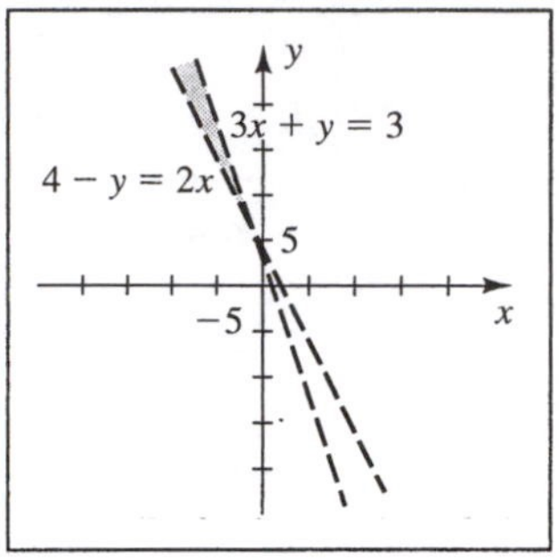

*Figure 11*

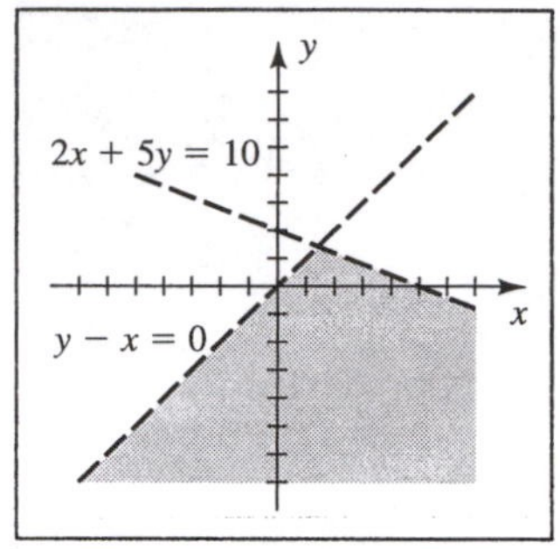

*Figure 13*

[13] $\begin{cases} y - x < 0 \\ 2x + 5y < 10 \end{cases} \Leftrightarrow \begin{cases} y < x \\ y < -\frac{2}{5}x + 2 \end{cases}$ $V$ @ $(\frac{10}{7},\ \frac{10}{7})$

Solving the system $y = x$ and $y = -\frac{2}{5}x + 2$ gives us the solution $(\frac{10}{7},\ \frac{10}{7})$.
We need to shade under $y = x$ and under $y = -\frac{2}{5}x + 2$.

[15] $\begin{cases} 3x + y \leq 6 \\ y - 2x \geq 1 \\ x \geq -2 \\ y \leq 4 \end{cases} \Leftrightarrow \begin{cases} y \leq -3x + 6 \\ y \geq 2x + 1 \\ x \geq -2 \\ y \leq 4 \end{cases}$ $V$ @ $(-2,\ -3)$, $(-2,\ 4)$, $(\frac{2}{3},\ 4)$, $(1,\ 3)$

We need to shade under $y = -3x + 6$, above $y = 2x + 1$, to the right of $x = -2$, and under $y = 4$. Include all the boundaries in the solution region.

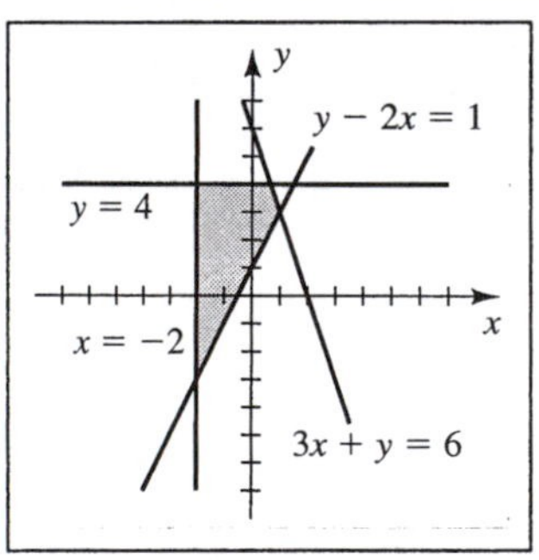

*Figure 15*

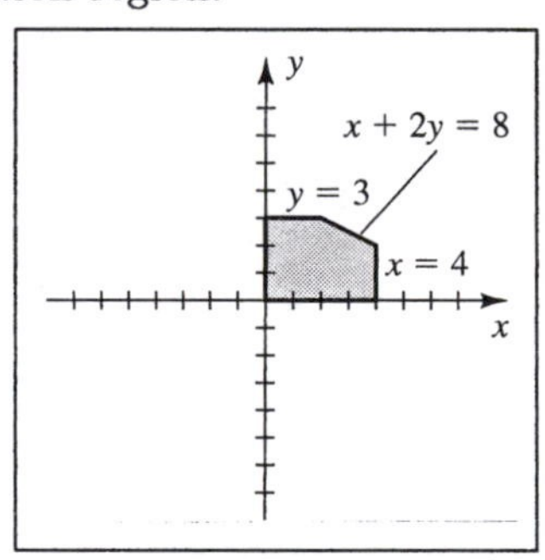

*Figure 17*

[17] $\begin{cases} x + 2y \leq 8 \\ 0 \leq x \leq 4 \\ 0 \leq y \leq 3 \end{cases}$ • $V$ @ $(0,\ 0)$, $(4,\ 0)$, $(4,\ 2)$, $(2,\ 3)$, $(0,\ 3)$

We can write $x + 2y \leq 8$ as $y \leq -\frac{1}{2}x + 4$. Shade under $y = -\frac{1}{2}x + 4$, between $x = 0$ and $x = 4$, and between $y = 0$ and $y = 3$. Include all the boundaries in the solution region.

[19] $|x| \geq 2 \Leftrightarrow x \geq 2$ or $x \leq -2$. This region is bounded on the left by the vertical line $x = 2$ and on the right by $x = -2$, including the lines.

$|y| < 3 \Leftrightarrow -3 < y < 3$. This region is bounded below by the horizontal line $y = -3$ and above by $y = 3$, excluding the lines.

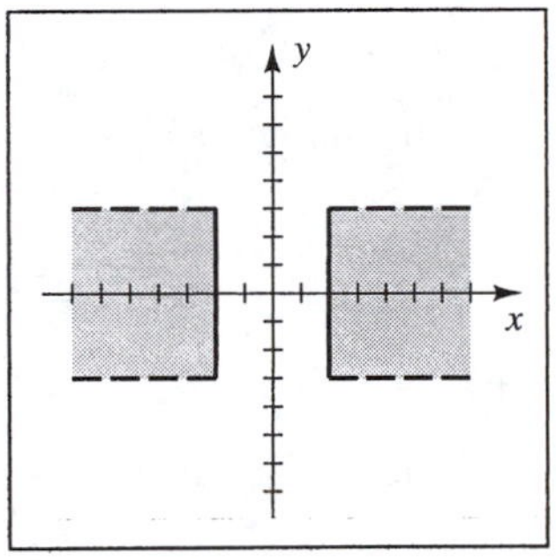


Figure 19

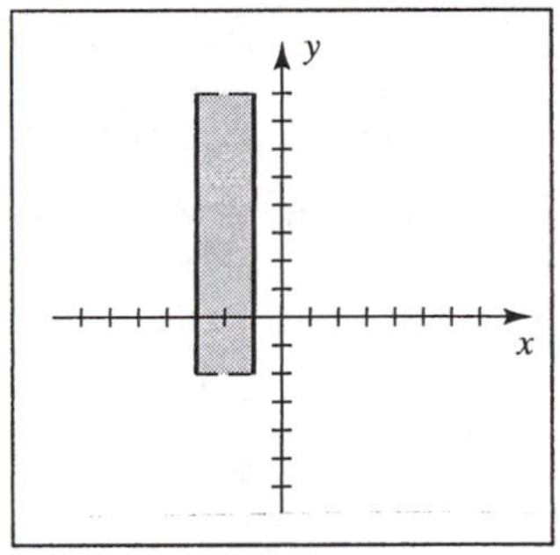


Figure 21

[21] $|x+2| \leq 1 \Leftrightarrow -1 \leq x+2 \leq 1 \Leftrightarrow -3 \leq x \leq -1$.

This region is bounded by the vertical lines $x = -3$ and $x = -1$, including the lines.

$|y-3| < 5 \Leftrightarrow -5 < y-3 < 5 \Leftrightarrow -2 < y < 8$.

This region is bounded by the horizontal lines $y = -2$ and $y = 8$, excluding the lines.

[23] $\begin{cases} x^2+y^2 \leq 4 \\ x+y \geq 1 \end{cases} \Leftrightarrow \begin{cases} x^2+y^2 \leq 2^2 \\ y \geq -x+1 \end{cases}$ $V$ @ $(\frac{1}{2} \mp \frac{1}{2}\sqrt{7}, \frac{1}{2} \pm \frac{1}{2}\sqrt{7})$

Shade inside the circle $x^2+y^2 = 4$ and above the line $y = -x+1$.

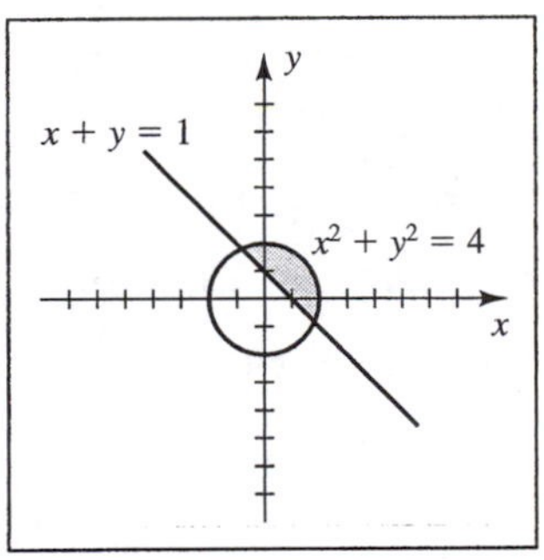


Figure 23

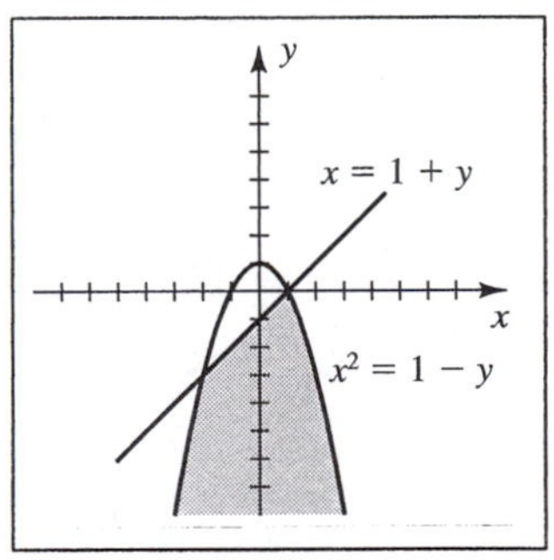


Figure 25

[25] $\begin{cases} x^2 \leq 1-y \\ x \geq 1+y \end{cases} \Leftrightarrow \begin{cases} y \leq -x^2+1 \\ y \leq x-1 \end{cases}$ $V$ @ $(-2, -3)$, $(1, 0)$

Shade under {or inside} the parabola $y = -x^2+1$ and under the line $y = x-1$.

**27** The shaded region lies between $x = 0$ and $x = 3$, including $x = 0$ and excluding $x = 3$. This region is described by $0 \leq x < 3$.

The dashed line goes through (0, 4) and (4, 0). Hence, its slope is $-1$, its $y$-intercept is 4, and an equation for it is $y = -x + 4$. Substituting $x = 0$ and $y = 0$ into $y = -x + 4$ gives us $0 = 4$. We replace $=$ by $<$ {the line is dashed} to make this statement a true statement. Hence, we have the inequality $y < -x + 4$.

The solid line goes through (0, $-4$) and (4, 0). Hence, its slope is 1, its $y$-intercept is $-4$, and an equation for it is $y = x - 4$. Substituting $x = 0$ and $y = 0$ into $y = x - 4$ gives us $0 = -4$. We replace $=$ by $\geq$ {the line is solid} to make this statement a true statement. Hence, we have the inequality $y \geq x - 4$. Thus, the graph may be described by the system $\begin{cases} 0 \leq x < 3 \\ y < -x + 4 \\ y \geq x - 4 \end{cases}$

**29** The shaded region lies *inside* the circle $x^2 + y^2 = 3^2$, so we can describe it with the inequality $x^2 + y^2 \leq 9$.

The dashed line goes through (0, 4) and (2, 0). Hence, its slope is $-2$, its $y$-intercept is 4, and an equation for it is $y = -2x + 4$. We want the region shaded *above* the line, so replace $=$ by $>$ {the line is dashed} to make this statement a true statement. Hence, we have the inequality $y > -2x + 4$.

Thus, the graph may be described by the system

$$\begin{cases} x^2 + y^2 \leq 9 \\ y > -2x + 4 \end{cases}$$

**31** The center of the circle is (2, 2) and it passes through (0, 0). The radius is the distance from (0, 0) to (2, 2), which is $\sqrt{8}$. An equation of the circle is $(x-2)^2 + (y-2)^2 = 8$. The center is clearly inside the circle, and if we substitute $x = 2$ and $y = 2$ in the equation of the circle, we get $0 = 8$. Since we want to make this a true statement and include the boundary, we use $(x-2)^2 + (y-2)^2 \leq 8$ to describe the shaded circular region and its boundary.

For the dashed line $y = x$, test the point (1, 0). This test gives us $0 = 1$, which we can make true by replacing $=$ by $<$ . Hence $y < x$ describes the shaded region under the line $y = x$. The other inequality may be found as in Exercise 27. Thus, the graph may be described by the system $\begin{cases} y < x \\ y \leq -x + 4 \\ (x-2)^2 + (y-2)^2 \leq 8 \end{cases}$

[33] For the dashed line through $(-4,\ 0)$ and $(4,\ 1)$, we can use the point-slope form to find an equation of the line. $y - 0 = \frac{1-0}{4-(-4)}(x-(-4)) \Leftrightarrow y = \frac{1}{8}x + \frac{1}{2}$. Checking the point $(0,\ 1)$ {which is in the shaded region} gives us $1 = \frac{1}{2}$, so we change $=$ to $>$ to obtain $y > \frac{1}{8}x + \frac{1}{2}$.

Similarly, for the solid line with $x$-intercept $-4$ and $y$-intercept 4, the shaded region can be described by $y \leq x + 4$.

Lastly, the solid line passing through $(4,\ 1)$ has slope $-\frac{3}{4}$, $y$-intercept 4, and its shaded region can be described by $y \leq -\frac{3}{4}x + 4$. Thus, the system is $\begin{cases} y > \frac{1}{8}x + \frac{1}{2} \\ y \leq x + 4 \\ y \leq -\frac{3}{4}x + 4 \end{cases}$

[35] Let $x$ and $y$ denote the number of sets of brand A and brand B, respectively. "Necessary to stock at least twice as many sets of brand A as of brand B" may be symbolized as $x \geq 2y$. "Necessary to have on hand at least 10 sets of brand B" may be symbolized as $y \geq 10$. Consequently, $x \geq 20$ from the first inequality. "Room for not more than 100 sets in the store" may be symbolized as $x + y \leq 100$. We now sketch the system of inequalities and shade the region that they have in common. The graph is the region bounded by the triangle with vertices $(20,\ 10)$, $(90,\ 10)$, and $(\frac{200}{3},\ \frac{100}{3})$.

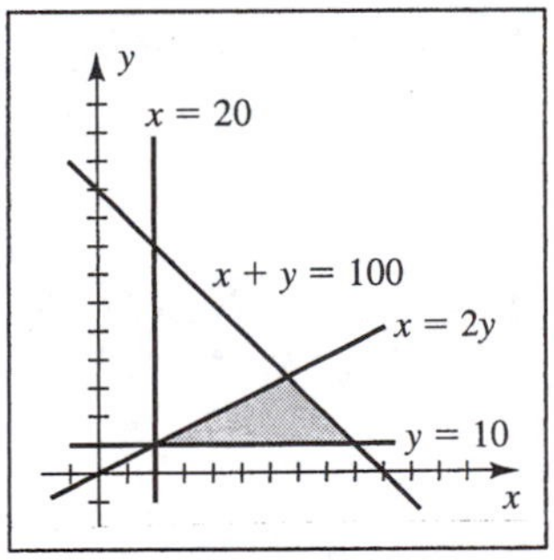


*Figure 35*

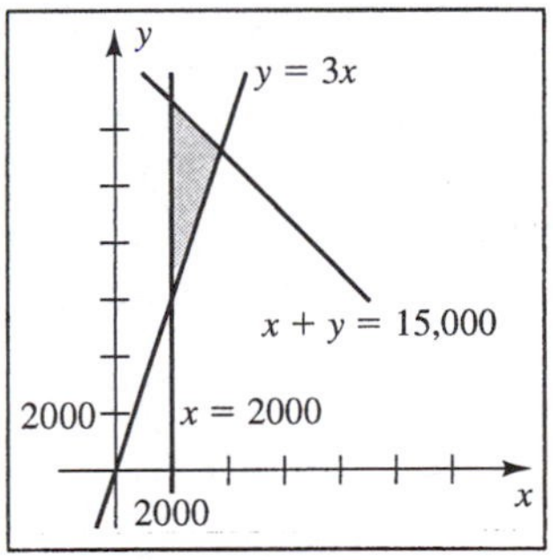


*Figure 37*

[37] If $x$ and $y$ denote the amount placed in the high-risk and low-risk investment, respectively, then a system is $x \geq 2000$, $y \geq 3x$, $x + y \leq 15{,}000$. The graph is the region bounded by the triangle with vertices $(2000,\ 6000)$, $(2000,\ 13{,}000)$, and $(3750,\ 11{,}250)$.

**39** A system is $x + y \le 9$, $y \ge x$, $x \ge 1$. To justify the condition $y \ge x$, start with

$$\frac{\text{cylinder volume}}{\text{total volume}} \ge 0.75 \Rightarrow \frac{\pi r^2 y}{\pi r^2 y + \frac{1}{3}\pi r^2 x} \ge \frac{3}{4} \Rightarrow 4\pi r^2 y \ge 3\pi r^2 y + \pi r^2 x \Rightarrow$$

$\pi r^2 y \ge \pi r^2 x \Rightarrow y \ge x$. The graph is the region bounded by the triangle with vertices (1, 1), (1, 8), and $(\frac{9}{2}, \frac{9}{2})$.

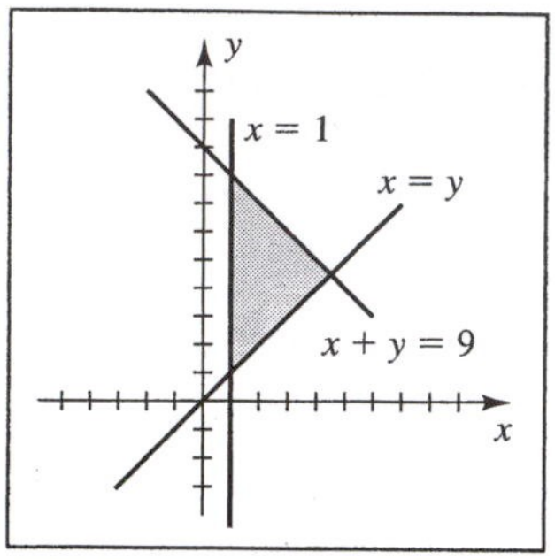

*Figure 39*

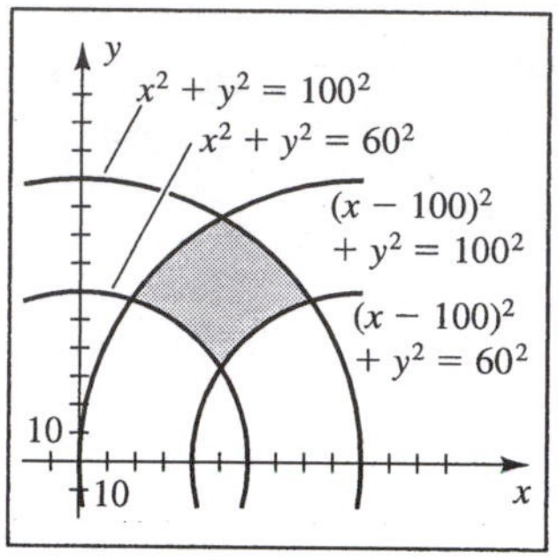

*Figure 41*

**41** If the plant is located at $(x, y)$, then a system is $(60)^2 \le x^2 + y^2 \le (100)^2$, $(60)^2 \le (x - 100)^2 + y^2 \le (100)^2$, $y \ge 0$. The graph is the region in the first quadrant that lies between the two concentric circles with center (0, 0) and radii 60 and 100, and also between the two concentric circles with center (100, 0) and radii 60 and 100.

Equating the different circle equations, we obtain the vertices of the solution region $(50, 50\sqrt{3})$, $(50, 10\sqrt{11})$, $(18, 6\sqrt{91})$, and $(82, 6\sqrt{91})$. For example, to determine where the circle $x^2 + y^2 = 100^2$ intersects the circle $(x - 100)^2 + y^2 = 60^2$, we subtract the first equation from the second to obtain $[(x - 100)^2 + y^2] - (x^2 + y^2) = 60^2 - 100^2 \Rightarrow -200x + 10{,}000 = -6400 \Rightarrow 200x = 16{,}400 \Rightarrow x = 82$. Substituting $x = 82$ into $x^2 + y^2 = 100^2$ gives us $y^2 = 100^2 - 82^2 = 3276 \Rightarrow y = \pm\sqrt{3276} = \pm 6\sqrt{91} \approx \pm 57.2$.

**43** $64y^3 - x^3 \le e^{1-2x} \Rightarrow 64y^3 \le e^{1-2x} + x^3 \Rightarrow y^3 \le \frac{1}{64}(e^{1-2x} + x^3) \Rightarrow$ $y \le \frac{1}{4}(e^{1-2x} + x^3)^{1/3}$. Graph the curve $y = \frac{1}{4}(e^{1-2x} + x^3)^{1/3}$.
The solution includes the curve and the region below it.

[−3.5, 4] by [−1, 4]

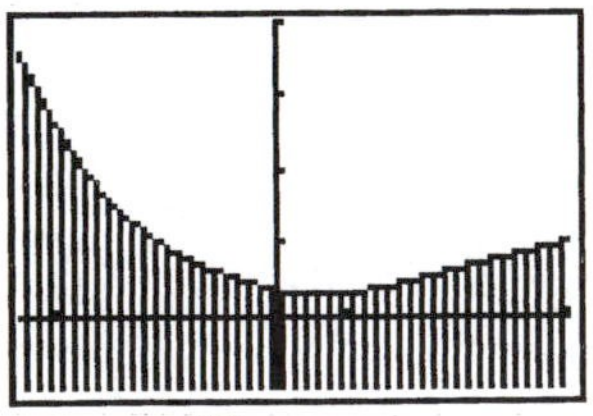
*Figure 43*

[−1.5, 1.5, 0.5] by [−1, 1, 0.5]

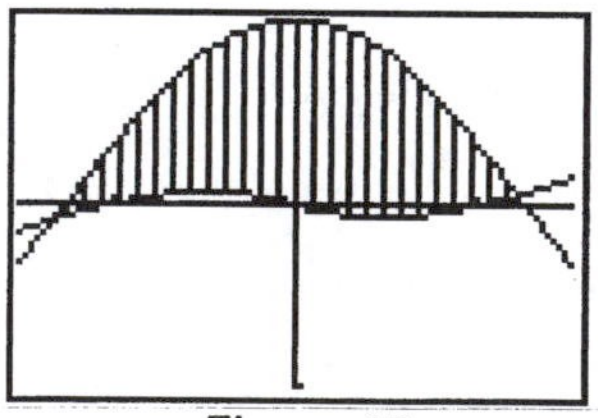
*Figure 45*

**45** $5^{1-y} \ge x^4 + x^2 + 1 \Rightarrow 1 - y \ge \log_5(x^4 + x^2 + 1) \Rightarrow y \le 1 - \log_5(x^4 + x^2 + 1)$.
$x + 3y \ge x^{5/3} \Rightarrow y \ge \frac{1}{3}(x^{5/3} - x)$. (continued)

Graph $y = 1 - \log_5(x^4 + x^2 + 1)$ $\{Y_1\}$ and $y = \frac{1}{3}(x^{5/3} - x)$ $\{Y_2\}$. The curves intersect at approximately $(-1.32, -0.09)$ and $(1.21, 0.05)$. The solution is located between the points of intersection. It includes the curves and the region below the first curve and above the second curve.

[47] $x^4 - 2x < 3y \Rightarrow y > \frac{1}{3}(x^4 - 2x)$. $x + 2y < x^3 - 5 \Rightarrow y < \frac{1}{2}(x^3 - x - 5)$. Graph $y = \frac{1}{3}(x^4 - 2x)$ and $y = \frac{1}{2}(x^3 - x - 5)$. The curves do not intersect. The solution must be above the first curve and below the second curve. Since the regions do not intersect, there is no solution.

[−4.5, 4.5] by [−3, 3]

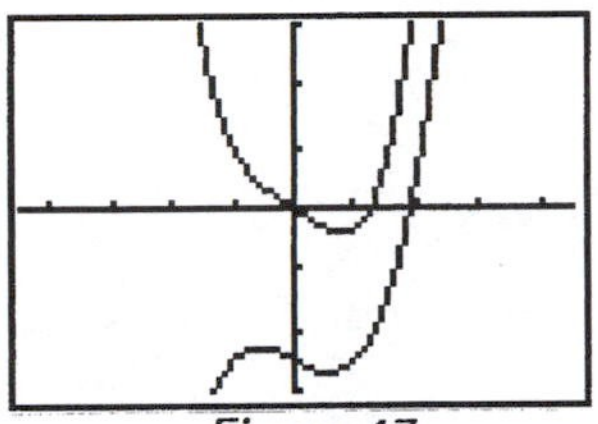

*Figure 47*

[33, 80, 5] by [0, 50, 5]

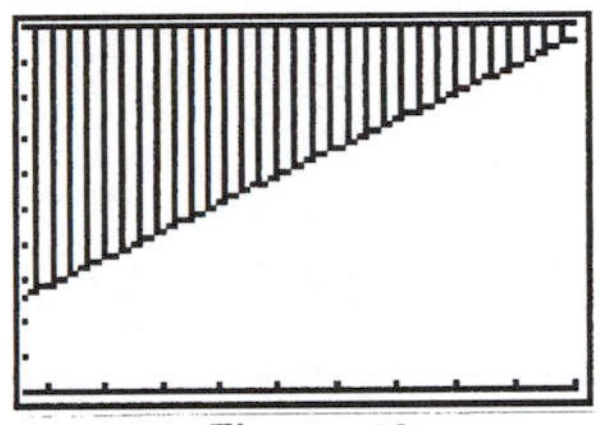

*Figure 49*

[49] (a) $29T - 39P < 450 \Rightarrow 29(37) - 39(21.2) < 450 \Rightarrow 246.2 < 450$ (true).

Yes, forests can grow.

(b) $29T - 39P < 450 \Rightarrow -39P < -29T + 450 \Rightarrow P > \frac{29}{39}T - \frac{150}{13}$.

Graph $Y_1 = \frac{29}{39}x - \frac{150}{13}$.

(c) Forests can grow whenever the point $(T, P)$ lies above the line $P = \frac{29}{39}T - \frac{150}{13}$.

## 6.4 Exercises

[1] We substitute the $x$ and $y$ values of each vertex into $C = 3x + 2y + 5$ and summarize the results in the following table. We see that there is a maximum of 27 at (6, 2), and a minimum of 9 at (0, 2).

| $(x, y)$ | (0, 2) | (0, 4) | (3, 5) | (6, 2) | (5, 0) | (2, 0) |
|---|---|---|---|---|---|---|
| $C$ | 9 ■ | 13 | 24 | 27 ■ | 20 | 11 |

[3] The region $R$ is sketched in *Figure 3* on the next page. The vertices are found by obtaining the intersection points of the system of equations corresponding to the given inequalities.

| $(x, y)$ | (0, 0) | (0, 3) | (4, 6) | (6, 3) | (5, 0) |
|---|---|---|---|---|---|
| $C$ | 0 | 3 | 18 | 21 ■ | 15 |

$C = 3x + y$;

maximum of 21 at (6, 3)

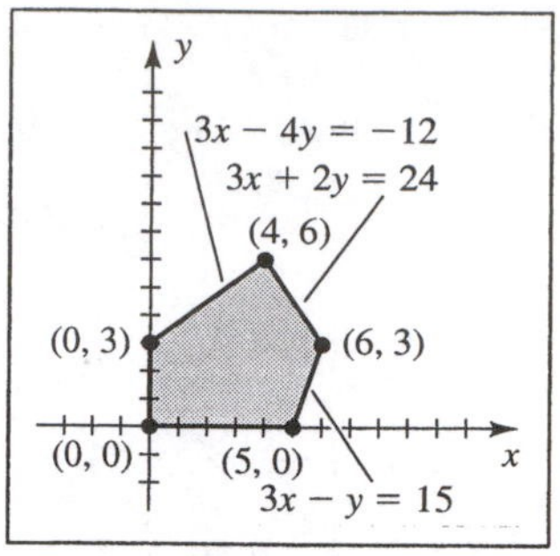


Figure 3

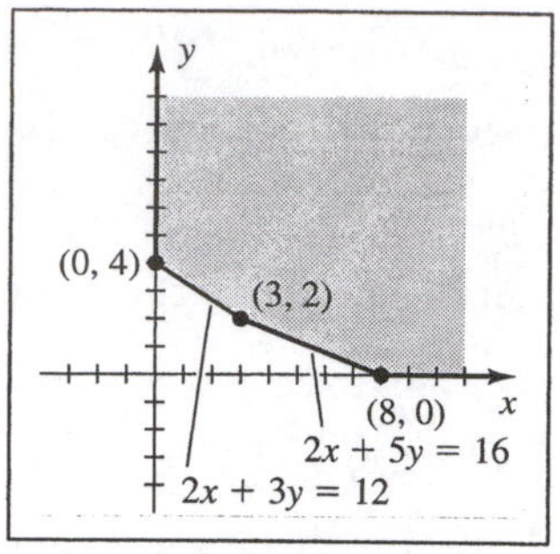


Figure 5

5

| $(x, y)$ | (8, 0) | (3, 2) | (0, 4) |
|---|---|---|---|
| $C$ | 24 | 21 ■ | 24 |

$C = 3x + 6y$;
minimum of 21 at (3, 2)

7

| $(x, y)$ | (0, 0) | (0, 4) | (2, 5) | (6, 3) | (8, 0) |
|---|---|---|---|---|---|
| $C$ | 0 | 16 | 24 ■ | 24 ■ | 16 |

As in Example 2, $C = 2x + 4y$ has the maximum value 24 for *any* point on the line segment from (2, 5) to (6, 3). To show that the last statement is true, we will substitute $6 - \frac{1}{2}x$ for $y$ in $C$. $C = 2x + 4y = 2x + 4(6 - \frac{1}{2}x) = 2x + 24 - 2x = 24$.

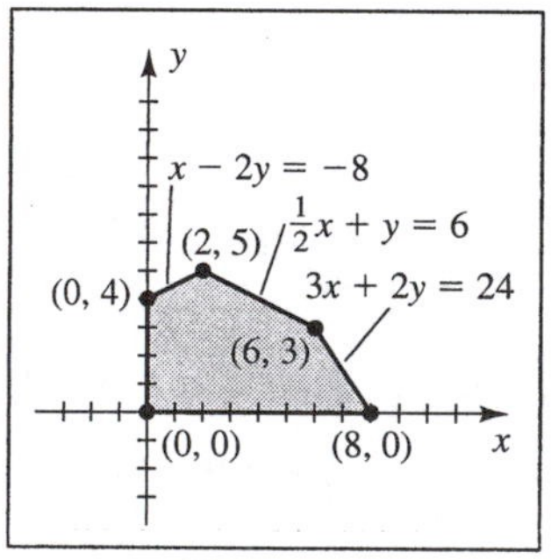


Figure 7

*Note:* **For the linear programming application exercises, the solution outline is as follows:**

(1) Define the variables used in the problem.

(2) Define the function to be maximized or minimized.

(3) List the system of inequalities that determine the solution region $R$.

(4) A table of intersection points and values of the function in (2) at those points is given. The maximum or minimum value is denoted by a ■.

(5) A summarizing statement is given.

(6) A figure is shown for the system of inequalities.

**9** Let $x$ and $y$ denote the number of oversized and standard rackets, respectively.

Profit function: $P = 15x + 8y$

$$\begin{cases} 30 \le y \le 80 \\ 10 \le x \le 30 \\ x + y \le 80 \end{cases}$$

| $(x, y)$ | (10, 30) | (30, 30) | (30, 50) | (10, 70) |
|---|---|---|---|---|
| $P$ | 390 | 690 | 850 ■ | 710 |

The maximum profit of \$850 per day occurs when 30 oversized rackets and 50 standard rackets are manufactured.

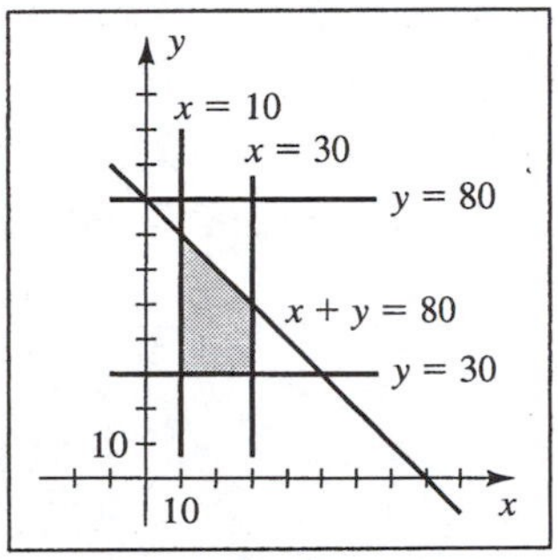

*Figure 9*

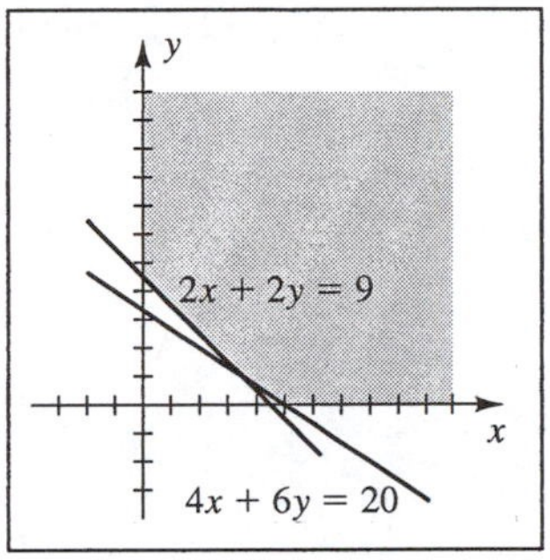

*Figure 11*

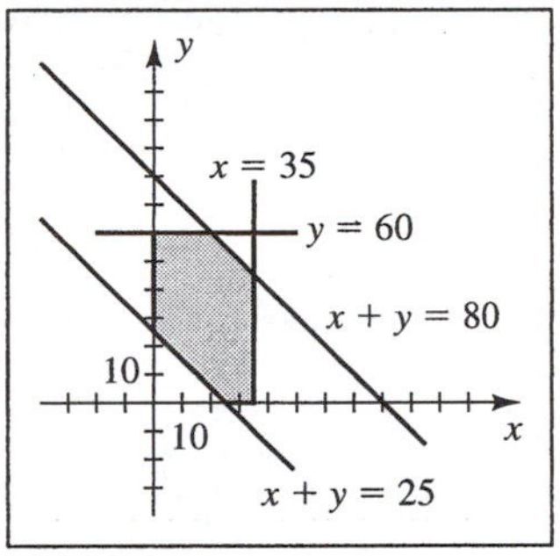

*Figure 13*

**11** Let $x$ and $y$ denote the number of pounds of S and T, respectively.

Cost function: $C = 3x + 4y$

$$\begin{cases} 2x + 2y \ge 9 & \textit{amount of } \mathrm{I} \\ 4x + 6y \ge 20 & \textit{amount of } \mathrm{G} \\ x \ge 0 \\ y \ge 0 \end{cases}$$

| $(x, y)$ | (0, 4.5) | (3.5, 1) | (5, 0) |
|---|---|---|---|
| $C$ | 18 | 14.5 ■ | 15 |

The minimum cost of \$14.50 occurs when 3.5 pounds of S and 1 pound of T are used.

**13** Let $x$ and $y$ denote the number of units sent to A and B, respectively, from $W_1$.

Cost function: $C = 12x + 10(35 - x) + 16y + 12(60 - y) = \underline{2x + 4y + 1070}$

The points are the same as those in Example 4.

$$\begin{cases} 0 \le x \le 35 \\ 0 \le y \le 60 \\ x + y \le 80 \\ x + y \ge 25 \end{cases}$$

| $(x, y)$ | (0, 25) | (0, 60) | (20, 60) | (35, 45) | (35, 0) | (25, 0) |
|---|---|---|---|---|---|---|
| $C$ | 1170 | 1310 | 1350 | 1320 | 1140 | 1120 ■ |

To minimize the shipping costs, send 25 units from $W_1$ to A and 0 from $W_1$ to B. Send 10 $\{35 - 25\}$ units from $W_2$ to A and 60 $\{60 - 0\}$ units from $W_2$ to B. See *Figure 13* above.

[15] Let $x$ and $y$ denote the number of acres planted with alfalfa and corn, respectively.

Profit function: $P = 110x - 4x - 20x + 150y - 6y - 10y = \underline{86x + 134y}$

$$\begin{cases} 4x + 6y \leq 480 & \textit{seed cost} \\ 20x + 10y \leq 1400 & \textit{labor cost} \\ x + y \leq 90 & \textit{area} \\ x,\ y \geq 0 & \end{cases}$$

| **(x, y)** | (70, 0) | (50, 40) | (30, 60) | (0, 80) | (0, 0) |
|---|---|---|---|---|---|
| **P** | 6020 | 9660 | 10,620 | 10,720 ■ | 0 |

The maximum profit of $10,720 occurs when

0 acres of alfalfa are planted and 80 acres of corn are planted.

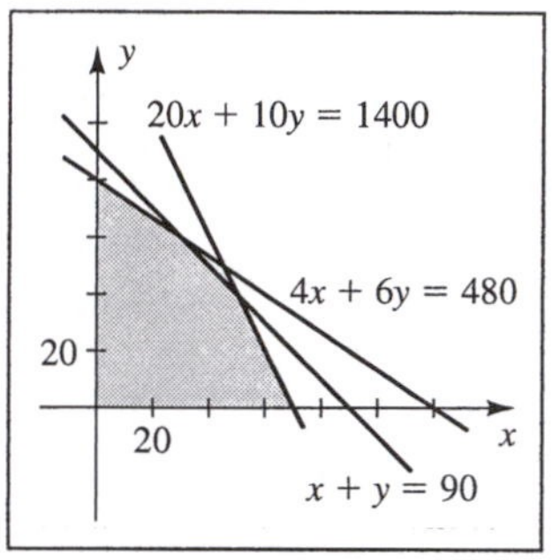

*Figure 15*

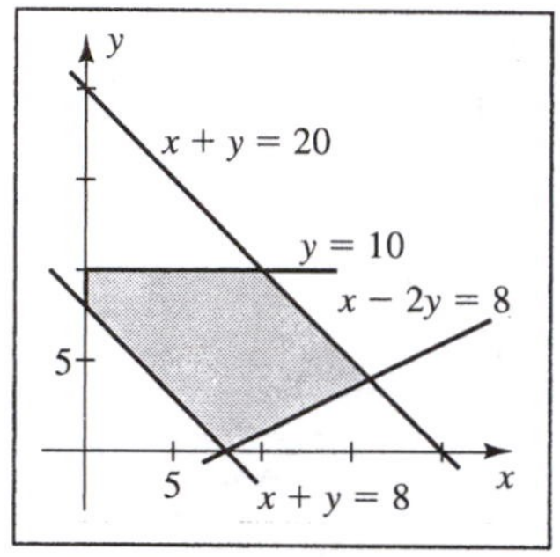

*Figure 17*

[17] Let $x$, $y$, and $z$ denote the number of ounces of X, Y, and Z, respectively.

Cost function: $C = 0.25x + 0.35y + 0.50z$

$= 0.25x + 0.35y + 0.50(20 - x - y) = \underline{10 - 0.25x - 0.15y}$

$$\begin{cases} 0.20x + 0.20y + 0.10z \geq 0.14(20) & \textit{amount of A} \\ 0.10x + 0.40y + 0.20z \geq 0.16(20) & \textit{amount of B} \\ 0.25x + 0.15y + 0.25z \geq 0.20(20) & \textit{amount of C} \end{cases} \Rightarrow \begin{cases} x + y \geq 8 \\ x - 2y \leq 8 \\ y \leq 10 \\ x + y \leq 20 \\ 0 \leq x,\ y \leq 20 \end{cases}$$

The new restrictions are found by substituting $z = 20 - x - y$ into the 3 inequalities, simplifying, and adding the last 2 inequalities.

| **(x, y)** | (8, 0) | (16, 4) | (10, 10) | (0, 10) | (0, 8) |
|---|---|---|---|---|---|
| **C** | 8.00 | 5.40 ■ | 6.00 | 8.50 | 8.80 ■■ |

The minimum cost of $5.40 requires 16 oz of X, 4 oz of Y, and 0 oz of Z.

The maximum cost of $8.80 requires 0 oz of X, 8 oz of Y, and 12 oz of Z.

[19] Let $x$ and $y$ denote the number of vans and buses purchased, respectively.

$$\begin{cases} 10{,}000x + 20{,}000y \leq 100{,}000 & \textit{purchase} \\ 100x + 75y \leq 500 & \textit{maintenance} \\ x \geq 0 \\ y \geq 0 \end{cases} \Rightarrow \begin{cases} x + 2y \leq 10 \\ 4x + 3y \leq 20 \\ x \geq 0 \\ y \geq 0 \end{cases}$$

| $(x, y)$ | (5, 0) | (2, 4) | (0, 5) | (0, 0) |
|---|---|---|---|---|
| $P$ | 75 | 130 ■ | 125 | 0 |

Passenger capacity function: $P = 15x + 25y$

The maximum passenger capacity of 130 would occur if the community purchases 2 vans and 4 buses.

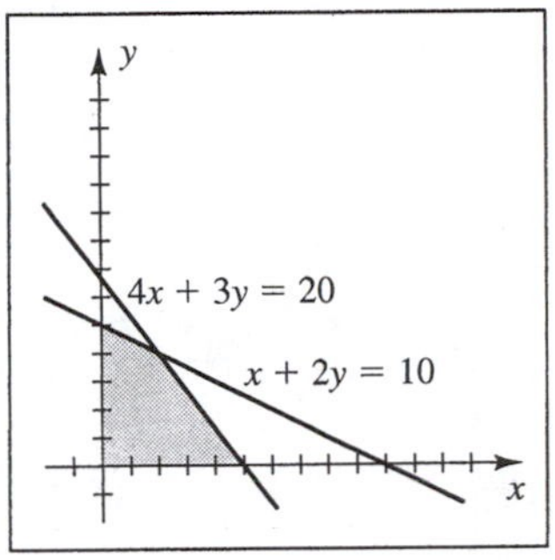

*Figure 19*

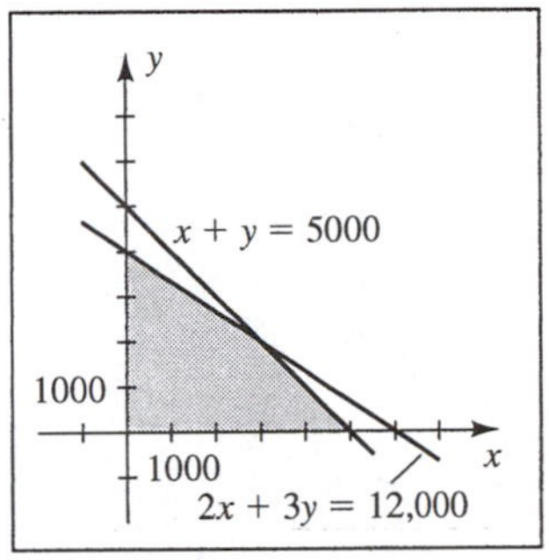

*Figure 21*

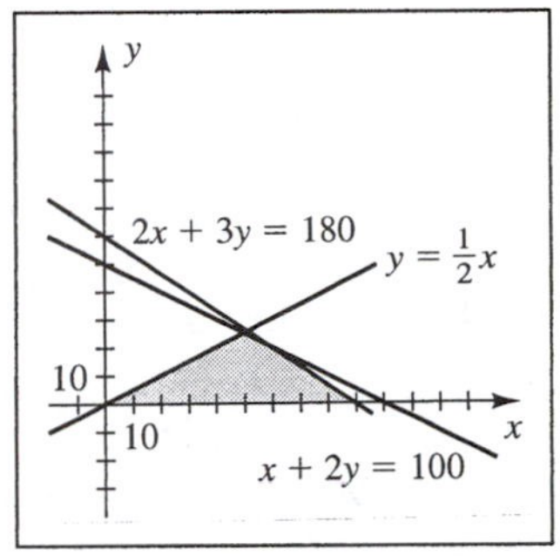

*Figure 23*

[21] Let $x$ and $y$ denote the number of trout and bass, respectively.

Pound function: $P = 3x + 4y$

$$\begin{cases} x + y \leq 5000 & \textit{number of fish} \\ 0.50x + 0.75y \leq 3000 & \textit{cost} \\ x \geq 0 \\ y \geq 0 \end{cases} \Rightarrow \begin{cases} x + y \leq 5000 \\ 2x + 3y \leq 12{,}000 \\ x \geq 0 \\ y \geq 0 \end{cases}$$

| $(x, y)$ | (5000, 0) | (3000, 2000) | (0, 4000) | (0, 0) |
|---|---|---|---|---|
| $P$ | 15,000 | 17,000 ■ | 16,000 | 0 |

The total number of pounds of fish will be a maximum of 17,000 if 3000 trout and 2000 bass are purchased.

[23] Let $x$ and $y$ denote the number of basic and deluxe units constructed, respectively.

$$\begin{cases} 300x + 600y \leq 30{,}000 & \textit{cost} \\ x \geq 2y & \textit{ratio} \\ 80x + 120y \leq 7200 & \textit{area} \\ x,\, y \geq 0 \end{cases} \Rightarrow \begin{cases} x + 2y \leq 100 \\ y \leq \frac{1}{2}x \\ 2x + 3y \leq 180 \\ x,\, y \geq 0 \end{cases}$$

| $(x, y)$ | (90, 0) | (60, 20) | (50, 25) | (0, 0) |
|---|---|---|---|---|
| $R$ | 3600 | 3900 ■ | 3875 | 0 |

Revenue function: $R = 40x + 75y$

The maximum monthly revenue of $3900 occurs if 60 basic units and 20 deluxe units are constructed. See *Figure 23* above.

## 6.5 Exercises

*Note:* Most systems are solved using the back substitution method. The solution for Exercise 7 uses the reduced echelon method. To avoid fractions, some solutions include linear combinations of rows—that is, $mR_i + nR_j$.

[1] $$\left[\begin{array}{cccc} 1 & -2 & -3 & -1 \\ 2 & 1 & 1 & 6 \\ 1 & 3 & -2 & 13 \end{array}\right] \begin{array}{l} \\ R_2 - 2R_1 \to R_2 \\ R_3 - R_1 \to R_3 \end{array}$$

$$\left[\begin{array}{cccc} 1 & -2 & -3 & -1 \\ 0 & 5 & 7 & 8 \\ 0 & 5 & 1 & 14 \end{array}\right] \begin{array}{l} \\ \\ R_3 - R_2 \to R_3 \end{array}$$

$$\left[\begin{array}{cccc} 1 & -2 & -3 & -1 \\ 0 & 5 & 7 & 8 \\ 0 & 0 & -6 & 6 \end{array}\right]$$

$R_3$: $-6z = 6 \Rightarrow z = -1$

$R_2$: $5y + 7z = 8 \Rightarrow 5y + 7(-1) = 8 \Rightarrow 5y = 15 \Rightarrow y = 3$ ★ (2, 3, −1)

$R_1$: $x - 2y - 3z = -1 \Rightarrow x - 2(3) - 3(-1) = -1 \Rightarrow x = 2$

[3] $$\left[\begin{array}{cccc} 1 & -2 & 2 & 0 \\ 5 & 2 & -1 & -7 \\ 0 & 3 & 1 & 17 \end{array}\right] \begin{array}{l} \\ R_2 - 5R_1 \to R_2 \\ \\ \end{array}$$

$$\left[\begin{array}{cccc} 1 & -2 & 2 & 0 \\ 0 & 12 & -11 & -7 \\ 0 & 3 & 1 & 17 \end{array}\right] \begin{array}{l} \\ \\ 4R_3 - R_2 \to R_3 \end{array}$$

$$\left[\begin{array}{cccc} 1 & -2 & 2 & 0 \\ 0 & 12 & -11 & -7 \\ 0 & 0 & 15 & 75 \end{array}\right]$$

$R_3$: $15z = 75 \Rightarrow z = 5$

$R_2$: $12y - 11z = -7 \Rightarrow 12y - 11(5) = -7 \Rightarrow 12y = 48 \Rightarrow y = 4$ ★ (−2, 4, 5)

$R_1$: $x - 2y + 2z = 0 \Rightarrow x - 2(4) + 2(5) = 0 \Rightarrow x = -2$

[5] $$\left[\begin{array}{cccc} 2 & 6 & -4 & 1 \\ 1 & 3 & -2 & 4 \\ 2 & 1 & -3 & -7 \end{array}\right] \begin{array}{l} R_1 \leftrightarrow R_2 \\ \\ \\ \end{array}$$

$$\left[\begin{array}{cccc} 1 & 3 & -2 & 4 \\ 2 & 6 & -4 & 1 \\ 2 & 1 & -3 & -7 \end{array}\right] \begin{array}{l} \\ R_2 - 2R_1 \to R_2 \\ R_3 - 2R_1 \to R_3 \end{array}$$

$$\left[\begin{array}{cccc} 1 & 3 & -2 & 4 \\ 0 & 0 & 0 & -7 \\ 0 & -5 & 1 & -15 \end{array}\right]$$

The second row, $0x + 0y + 0z = -7$, has *no solution.*

[7] $\left[\begin{array}{rrrr} 2 & -3 & 2 & -3 \\ -3 & 2 & 1 & 1 \\ 4 & 1 & -3 & 4 \end{array}\right]$ $R_1 \leftrightarrow R_2$ and then $R_2 \leftrightarrow R_3$

$\left[\begin{array}{rrrr} -3 & 2 & 1 & 1 \\ 4 & 1 & -3 & 4 \\ 2 & -3 & 2 & -3 \end{array}\right]$ $R_1 + R_2 \rightarrow R_1$

$\left[\begin{array}{rrrr} 1 & 3 & -2 & 5 \\ 4 & 1 & -3 & 4 \\ 2 & -3 & 2 & -3 \end{array}\right]$ $R_2 - 4R_1 \rightarrow R_2$, $R_3 - 2R_1 \rightarrow R_3$

$\left[\begin{array}{rrrr} 1 & 3 & -2 & 5 \\ 0 & -11 & 5 & -16 \\ 0 & -9 & 6 & -13 \end{array}\right]$ $4R_2 - 5R_3 \rightarrow R_2$ (* see note)

$\left[\begin{array}{rrrr} 1 & 3 & -2 & 5 \\ 0 & 1 & -10 & 1 \\ 0 & -9 & 6 & -13 \end{array}\right]$ $R_3 + 9R_2 \rightarrow R_3$

$\left[\begin{array}{rrrr} 1 & 3 & -2 & 5 \\ 0 & 1 & -10 & 1 \\ 0 & 0 & -84 & -4 \end{array}\right]$

$R_3$: $-84z = -4 \Rightarrow z = \frac{1}{21}$

$R_2$: $y - 10z = 1 \Rightarrow y = \frac{31}{21}$ ★ $\left(\frac{2}{3}, \frac{31}{21}, \frac{1}{21}\right)$

$R_1$: $x + 3y - 2z = 5 \Rightarrow x = \frac{14}{21} = \frac{2}{3}$

(*) We could have just used $-\frac{1}{11}R_2 \rightarrow R_2$, but we would then have to work with many fractions and increase our chance of making a mistake. We will solve this system again, but this time we will obtain the reduced echelon form. The first 4 matrices are exactly the same. Starting with the fourth matrix we have:

$\left[\begin{array}{rrrr} 1 & 3 & -2 & 5 \\ 0 & 1 & -10 & 1 \\ 0 & -9 & 6 & -13 \end{array}\right]$ $R_1 - 3R_2 \rightarrow R_1$, $R_3 + 9R_2 \rightarrow R_3$

$\left[\begin{array}{rrrr} 1 & 0 & 28 & 2 \\ 0 & 1 & -10 & 1 \\ 0 & 0 & -84 & -4 \end{array}\right]$ $-\frac{1}{84}R_3 \rightarrow R_3$

$\left[\begin{array}{rrrr} 1 & 0 & 28 & 2 \\ 0 & 1 & -10 & 1 \\ 0 & 0 & 1 & \frac{1}{21} \end{array}\right]$ $R_1 - 28R_3 \rightarrow R_1$, $R_2 + 10R_3 \rightarrow R_2$

$\left[\begin{array}{rrrr} 1 & 0 & 0 & \frac{14}{21} \\ 0 & 1 & 0 & \frac{31}{21} \\ 0 & 0 & 1 & \frac{1}{21} \end{array}\right]$

$R_1$: $x = \frac{14}{21} = \frac{2}{3}$; $R_2$: $y = \frac{31}{21}$; $R_3$: $z = \frac{1}{21}$

*Note:* Exer. 9–16: There are other forms for the answers; $c$ is any real number.

**9** $\left[\begin{array}{cccc} 1 & 3 & 1 & 0 \\ 1 & 1 & -1 & 0 \\ 1 & -2 & -4 & 0 \end{array}\right]$ $\begin{array}{l} \\ R_2 - R_1 \to R_2 \\ R_3 - R_1 \to R_3 \end{array}$

$\left[\begin{array}{cccc} 1 & 3 & 1 & 0 \\ 0 & -2 & -2 & 0 \\ 0 & -5 & -5 & 0 \end{array}\right]$ $\begin{array}{l} -\frac{1}{2}R_2 \to R_2 \\ \\ -\frac{1}{5}R_3 \to R_3 \end{array}$

$\left[\begin{array}{cccc} 1 & 3 & 1 & 0 \\ 0 & 1 & 1 & 0 \\ 0 & 1 & 1 & 0 \end{array}\right]$ $\begin{array}{l} R_1 - 3R_2 \to R_1 \\ \\ R_3 - R_2 \to R_3 \end{array}$

$\left[\begin{array}{cccc} 1 & 0 & -2 & 0 \\ 0 & 1 & 1 & 0 \\ 0 & 0 & 0 & 0 \end{array}\right]$

$R_1$: $x - 2z = 0 \Rightarrow x = 2z$
$R_2$: $y + z = 0 \Rightarrow y = -z$

★ $(2c, -c, c)$

**11** $\left[\begin{array}{cccc} 2 & 1 & 1 & 0 \\ 1 & -2 & -2 & 0 \\ 1 & 1 & 1 & 0 \end{array}\right]$ $R_1 \leftrightarrow R_2$

$\left[\begin{array}{cccc} 1 & -2 & -2 & 0 \\ 2 & 1 & 1 & 0 \\ 1 & 1 & 1 & 0 \end{array}\right]$ $\begin{array}{l} \\ R_2 - 2R_1 \to R_2 \\ R_3 - R_1 \to R_3 \end{array}$

$\left[\begin{array}{cccc} 1 & -2 & -2 & 0 \\ 0 & 5 & 5 & 0 \\ 0 & 3 & 3 & 0 \end{array}\right]$ $\begin{array}{l} \frac{1}{5}R_2 \to R_2 \\ \\ \frac{1}{3}R_3 \to R_3 \end{array}$

$\left[\begin{array}{cccc} 1 & -2 & -2 & 0 \\ 0 & 1 & 1 & 0 \\ 0 & 1 & 1 & 0 \end{array}\right]$ $\begin{array}{l} R_1 + 2R_2 \to R_1 \\ \\ R_3 - R_2 \to R_3 \end{array}$

$\left[\begin{array}{cccc} 1 & 0 & 0 & 0 \\ 0 & 1 & 1 & 0 \\ 0 & 0 & 0 & 0 \end{array}\right]$

$R_1$: $x = 0$

$R_2$: $y + z = 0 \Rightarrow y = -z$

★ $(0, -c, c)$

**13** $\left[\begin{array}{cccc} 1 & 4 & -1 & -2 \\ 3 & -2 & 5 & 7 \end{array}\right]$ $R_2 - 3R_1 \to R_2$

$\left[\begin{array}{cccc} 1 & 4 & -1 & -2 \\ 0 & -14 & 8 & 13 \end{array}\right]$

$R_2$: $-14y + 8z = 13 \Rightarrow y = \frac{4}{7}z - \frac{13}{14}$

$R_1$: $x + 4y - z = -2 \Rightarrow x = -4(\frac{4}{7}z - \frac{13}{14}) + z - 2 = -\frac{9}{7}z + \frac{12}{7}$

★ $(\frac{12}{7} - \frac{9}{7}c, \frac{4}{7}c - \frac{13}{14}, c)$

**15** $\left[\begin{array}{cccc} 4 & -2 & 1 & 5 \\ 3 & 1 & -4 & 0 \end{array}\right]$ $R_1 - R_2 \to R_1$

$\left[\begin{array}{cccc} 1 & -3 & 5 & 5 \\ 3 & 1 & -4 & 0 \end{array}\right]$ $R_2 - 3R_1 \to R_2$

$\left[\begin{array}{cccc} 1 & -3 & 5 & 5 \\ 0 & 10 & -19 & -15 \end{array}\right]$

$R_2$: $10y - 19z = -15 \Rightarrow y = \frac{19}{10}z - \frac{3}{2}$

$R_1$: $x - 3y + 5z = 5 \Rightarrow x = 3(\frac{19}{10}z - \frac{3}{2}) - 5z + 5 = \frac{7}{10}z + \frac{1}{2}$ ★ $(\frac{7}{10}c + \frac{1}{2}, \frac{19}{10}c - \frac{3}{2}, c)$

**17** $\left[\begin{array}{cccc} 5 & 0 & 2 & 1 \\ 0 & 1 & -3 & 2 \\ 2 & 1 & 0 & 3 \end{array}\right]$ $R_1 - 2R_3 \to R_1$

$\left[\begin{array}{cccc} 1 & -2 & 2 & -5 \\ 0 & 1 & -3 & 2 \\ 2 & 1 & 0 & 3 \end{array}\right]$ $R_3 - 2R_1 \to R_3$

$\left[\begin{array}{cccc} 1 & -2 & 2 & -5 \\ 0 & 1 & -3 & 2 \\ 0 & 5 & -4 & 13 \end{array}\right]$ $R_3 - 5R_2 \to R_3$

$\left[\begin{array}{cccc} 1 & -2 & 2 & -5 \\ 0 & 1 & -3 & 2 \\ 0 & 0 & 11 & 3 \end{array}\right]$

$R_3$: $11z = 3 \Rightarrow z = \frac{3}{11}$

$R_2$: $y - 3z = 2 \Rightarrow y = \frac{31}{11}$

$R_1$: $x - 2y + 2z = -5 \Rightarrow x = \frac{1}{11}$ ★ $(\frac{1}{11}, \frac{31}{11}, \frac{3}{11})$

**19** $\left[\begin{array}{ccc} 4 & -3 & 1 \\ 2 & 1 & -7 \\ -1 & 1 & -1 \end{array}\right]$ $R_1 + 3R_3 \to R_1$

$\left[\begin{array}{ccc} 1 & 0 & -2 \\ 2 & 1 & -7 \\ -1 & 1 & -1 \end{array}\right]$ $R_2 - 2R_1 \to R_2$, $R_3 + R_1 \to R_3$

$\left[\begin{array}{ccc} 1 & 0 & -2 \\ 0 & 1 & -3 \\ 0 & 1 & -3 \end{array}\right]$ $R_3 - R_2 \to R_3$

$\left[\begin{array}{ccc} 1 & 0 & -2 \\ 0 & 1 & -3 \\ 0 & 0 & 0 \end{array}\right]$

$R_1$: $x = -2$

$R_2$: $y = -3$ ★ $(-2, -3)$

**21** $\left[\begin{array}{rrr} 2 & 3 & 5 \\ 1 & -3 & 4 \\ 1 & 1 & -2 \end{array}\right]$ $R_1 \leftrightarrow R_2$

$\left[\begin{array}{rrr} 1 & -3 & 4 \\ 2 & 3 & 5 \\ 1 & 1 & -2 \end{array}\right]$ $R_2 - 2R_1 \rightarrow R_2$, $R_3 - R_1 \rightarrow R_3$

$\left[\begin{array}{rrr} 1 & -3 & 4 \\ 0 & 9 & -3 \\ 0 & 4 & -6 \end{array}\right]$ $R_2 - 2R_3 \rightarrow R_2$

$\left[\begin{array}{rrr} 1 & -3 & 4 \\ 0 & 1 & 9 \\ 0 & 4 & -6 \end{array}\right]$ $R_1 + 3R_2 \rightarrow R_1$, $R_3 - 4R_2 \rightarrow R_3$

$\left[\begin{array}{rrr} 1 & 0 & 31 \\ 0 & 1 & 9 \\ 0 & 0 & -42 \end{array}\right]$

There is a contradiction in row 3, therefore there is *no solution.*

**23** Let $x$, $y$, and $z$ denote the number of liters of the 10% acid, 30% acid, and 50% acid.

$$\begin{cases} x + y + z = 50 & \textit{quantity} \quad (E_1) \\ 0.10x + 0.30y + 0.50z = (0.32)(50) & \textit{quality} \quad (E_2) \\ z = 2y & \textit{constraint} \quad (E_3) \end{cases}$$

Substitute $z = 2y$ into $E_1$ and $100\,E_2$ to obtain

$$\begin{cases} x + 3y = 50 & (E_4) \\ 10x + 130y = 1600 & (E_5) \end{cases}$$

$E_5 - 10\,E_4 \;\Rightarrow\; 100y = 1100 \;\Rightarrow\; y = 11;\ x = 17;\ z = 22$

**25** Let $x$, $y$, and $z$ denote the number of hours needed for A, B, and C, respectively, to produce 1000 items. In one hour, A, B, and C produce $\frac{1000}{x}$, $\frac{1000}{y}$, and $\frac{1000}{z}$ items, respectively. In 6 hours, A and B produce $\frac{6000}{x}$ and $\frac{6000}{y}$ items. From the table, this sum must equal 4500. The system of equations is then:

$$\begin{cases} \dfrac{6000}{x} + \dfrac{6000}{y} = 4500 \\ \dfrac{8000}{x} + \dfrac{8000}{z} = 3600 \\ \dfrac{7000}{y} + \dfrac{7000}{z} = 4900 \end{cases}$$

To simplify, let $a = 1/x$, $b = 1/y$, $c = 1/z$ and divide each equation by its greatest common factor { 1500, 400, and 700 }.

$$\begin{cases} 4a + 4b = 3 & (E_1) \\ 20a + 20c = 9 & (E_2) \\ 10b + 10c = 7 & (E_3) \end{cases} \qquad E_2 - 5\,E_1 \;\Rightarrow\; 20c - 20b = -6 \quad (E_4)$$

$E_4 + 2\,E_3 \;\Rightarrow\; 40c = 8 \;\Rightarrow\; c = \frac{1}{5};\ b = \frac{1}{2};\ a = \frac{1}{4}$.

Resubstituting, $x = 4$, $y = 2$, and $z = 5$.

[27] Let $x$, $y$, and $z$ denote the amounts of $G_1$, $G_2$, and $G_3$, respectively.

$$\begin{cases} x+y+z=600 & \textit{quantity} & (E_1) \\ 0.30x+0.20y+0.15z=(0.25)(600) & \textit{quality} & (E_2) \\ z=100+y & \textit{constraint} & (E_3) \end{cases}$$

Substitute $z=100+y$ into $E_1$ and $100\,E_2$ to obtain

$$\begin{cases} x + 2y = 500 & (E_4) \\ 30x + 35y = 13{,}500 & (E_5) \end{cases}$$

$$E_5 - 30\,E_4 \;\Rightarrow\; -25y=-1500 \;\Rightarrow\; y=60;\ z=160;\ x=380.$$

[29] (a) Let $R_1=R_2=R_3=3$ in the given system of equations.

$$\begin{cases} I_1 - I_2 + I_3 = 0 \\ 3I_1 + 3I_2 = 6 \\ 3I_2 + 3I_3 = 12 \end{cases} \Rightarrow \begin{cases} I_1 - I_2 + I_3 = 0 & (E_1) \\ I_1 + I_2 = 2 & (E_2) \\ I_2 + I_3 = 4 & (E_3) \end{cases}$$

Substitute $I_1=I_2-I_3$ from $E_1$ into $E_2$ to obtain $2I_2-I_3=2$ $(E_4)$.

Now $E_4+E_3 \;\Rightarrow\; 3I_2=6 \;\Rightarrow\; I_2=2;\ I_3=2;\ I_1=0.$

(b) Let $R_1=4$, $R_2=1$, and $R_3=4$ in the given system of equations.

$$\begin{cases} I_1 - I_2 + I_3 = 0 & (E_1) \\ 4I_1 + 1I_2 = 6 & (E_2) \\ 1I_2 + 4I_3 = 12 & (E_3) \end{cases}$$

Substitute $I_1=I_2-I_3$ from $E_1$ into $E_2$ to obtain $4(I_2-I_3)+I_2=6 \;\Rightarrow$

$5I_2-4I_3=6$ $(E_4)$. Now $E_4+E_3 \;\Rightarrow\; 6I_2=18 \;\Rightarrow\; I_2=3;\ I_3=\frac{9}{4};\ I_1=\frac{3}{4}.$

[31] Let $x$, $y$, and $z$ denote the amount of Colombian, Brazilian, and Kenyan coffee used, respectively.

$$\begin{cases} x+y+z=1 & \textit{quantity} & (E_1) \\ 10x+6y+8z=(8.50)(1) & \textit{quality} & (E_2) \\ x=3y & \textit{constraint} & (E_3) \end{cases}$$

Substitute $x=3y$ into $E_1$ and $E_2$ to obtain

$$\begin{cases} 4y + z = 1 & (E_4) \\ 36y + 8z = 8.5 & (E_5) \end{cases} \quad E_5-8\,E_4 \;\Rightarrow\; 4y=\tfrac{1}{2} \;\Rightarrow\; y=\tfrac{1}{8};\ z=\tfrac{1}{2};\ x=\tfrac{3}{8}.$$

[33] (a) A: $x_1+x_4=50+25=75$, B: $x_1+x_2=100+50=150$,

C: $x_2+x_3=150+75=225$, D: $x_3+x_4=100+50=150$

(b) From C, $x_3=100 \;\Rightarrow\; x_2=225-100=125$.

From D, $x_3=100 \;\Rightarrow\; x_4=150-100=50$.

From A, $x_4=50 \;\Rightarrow\; x_1=75-50=25$.

(c) From D in part (a), $x_3=150-x_4 \;\Rightarrow\; x_3 \le 150$ since $x_4 \ge 0$. From C,

$x_3=225-x_2=225-(150-x_1)$ {from B} $=75+x_1 \;\Rightarrow\; x_3 \ge 75$ since $x_1 \ge 0$.

**35** $t = 2070 - 1990 = 80$. $rt = (0.025)(80) = 2$ and $A = 800$ for $E_1$,
$(0.015)(80) = 1.2$ and $A = 560$ for $E_2$, and $(0.01)(0) = 0$ and $A = 340$ for $E_3$.
Substituting into $A = a + ct + ke^{rt}$ and summarizing as a system, we have:

$$\begin{cases} a + 80c + e^2k = 800 & (E_1) \\ a + 80c + e^{1.2}k = 560 & (E_2) \\ a \qquad\quad + \quad k = 340 & (E_3) \end{cases}$$

We want to find $t$ when $A$ has doubled—that is, $A = 2(340)$. First we find $c$ and $k$.

$E_1 - E_2 \Rightarrow e^2k - e^{1.2}k = 240 \Rightarrow k = \dfrac{240}{e^2 - e^{1.2}} \approx 58.98.$

Substituting into $E_3$ gives $a = 340 - k \approx 281.02$.

Substituting into $E_1$ gives $c = \dfrac{800 - a - e^2k}{80} \approx \dfrac{800 - 281.02 - (58.98)e^2}{80} \approx 1.04.$

Thus, $A = 281.02 + 1.04t + 58.98e^{rt}$. If $A = 680$ and $r = 0.01$, then
$680 = 281.02 + 1.04t + 58.98e^{0.01t} \Rightarrow 1.04t + 58.98e^{0.01t} - 398.98 = 0$. Graphing
$y = 1.04t + 58.98e^{0.01t} - 398.98$, we see there is an $x$-intercept at $x \approx 144.08$.

$1990 + 144.08 = 2134.08$, or during the year 2134.

[0, 1050, 100] by [−350, 350, 100]

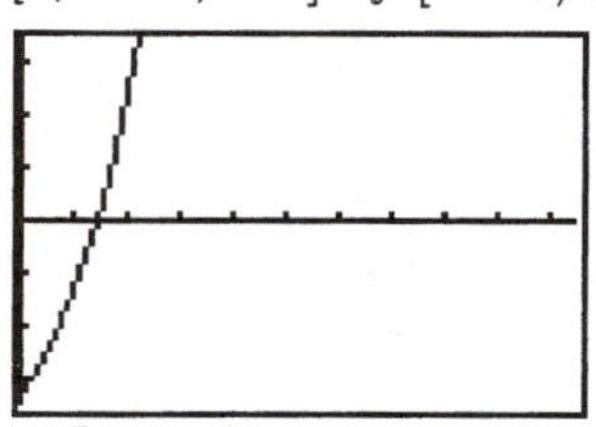

*Figure 35*

**37** The circle has an equation of the form $x^2 + y^2 + ax + by + c = 0$.
Substituting the $x$ and $y$ values of $P$, $Q$, and $R$ into this equation yields:

$$\begin{cases} 2a + b + c = -5 & P\ (E_1) \\ -a - 4b + c = -17 & Q\ (E_2) \\ 3a \qquad + c = -9 & R\ (E_3) \end{cases}$$

Solving $E_3$ for $c$ $\{c = -9 - 3a\}$ and substituting into $E_1$ and $E_2$ yields:

$$\begin{cases} -a + b = 4 & (E_4) \\ -4a - 4b = -8 & (E_5) \end{cases} \Rightarrow \begin{cases} -a + b = 4 & (E_6) \\ a + b = 2 & (E_7) \end{cases}$$

$E_6 + E_7 \Rightarrow 2b = 6 \Rightarrow b = 3;\ a = -1;\ c = -6.$

The equation is $x^2 + y^2 - x + 3y - 6 = 0$.

**39** Using $y = f(x) = ax^3 + bx^2 + cx + d$ and the point $P(0, -6)$, we should recognize that $d = -6$ since substituting 0 for $x$ leaves only the constant term. Now use the points $Q(1, -11)$, $R(-1, -5)$, and $S(2, -14)$ to obtain the system: (continued)

$$\begin{cases} a + b + c - 6 = -11 & Q \; (E_1) \\ -a + b - c - 6 = -5 & R \; (E_2) \\ 8a + 4b + 2c - 6 = -14 & S \; (E_3) \end{cases}$$

Add 6 to each equation and divide the resulting third equation by 2 to obtain:

$$\left[\begin{array}{cccc} 1 & 1 & 1 & -5 \\ -1 & 1 & -1 & 1 \\ 4 & 2 & 1 & -4 \end{array}\right] \begin{array}{l} \\ R_2 + R_1 \to R_2 \\ R_3 - 4R_1 \to R_3 \end{array}$$

$$\left[\begin{array}{cccc} 1 & 1 & 1 & -5 \\ 0 & 2 & 0 & -4 \\ 0 & -2 & -3 & 16 \end{array}\right] \; \tfrac{1}{2}R_2 \to R_2$$

$$\left[\begin{array}{cccc} 1 & 1 & 1 & -5 \\ 0 & 1 & 0 & -2 \\ 0 & -2 & -3 & 16 \end{array}\right] \begin{array}{l} R_1 - R_2 \to R_1 \\ \\ R_3 + 2R_2 \to R_3 \end{array}$$

$$\left[\begin{array}{cccc} 1 & 0 & 1 & -3 \\ 0 & 1 & 0 & -2 \\ 0 & 0 & -3 & 12 \end{array}\right] \begin{array}{l} \\ \\ -\tfrac{1}{3}R_3 \to R_3 \end{array}$$

$$\left[\begin{array}{cccc} 1 & 0 & 1 & -3 \\ 0 & 1 & 0 & -2 \\ 0 & 0 & 1 & -4 \end{array}\right] \begin{array}{l} R_1 - R_3 \to R_1 \\ \\ \\ \end{array}$$

$$\left[\begin{array}{cccc} 1 & 0 & 0 & 1 \\ 0 & 1 & 0 & -2 \\ 0 & 0 & 1 & -4 \end{array}\right]$$

Thus, $a = 1$, $b = -2$, and $c = -4$, so $f(x) = x^3 - 2x^2 - 4x - 6$.

**41** For $y = f(x) = ax^3 + bx^2 + cx + d$ and the points $(-1, 2)$, $(0.5, 2)$, $(1, 3)$, and $(2, 4.5)$, we obtain the system:

$$\begin{cases} -a + b - c + d = 2 & (-1, 2) \\ 0.125a + 0.25b + 0.5c + d = 2 & (0.5, 2) \\ a + b + c + d = 3 & (1, 3) \\ 8a + 4b + 2c + d = 4.5 & (2, 4.5) \end{cases}$$

Solving the system yields $a = -\frac{4}{9}$, $b = \frac{11}{9}$, $c = \frac{17}{18}$, and $d = \frac{23}{18}$.

## 6.6 Exercises

*Note:* For Exercises 1–8, $(A + B)$ and $(A - B)$ are possible only if $A$ and $B$ have the same number of rows and columns. {#7 & #8 are not possible} $(2A)$ and $(-3B)$ are always possible by simply multiplying the elements of $A$ and $B$ by 2 and $-3$, respectively.

**1** $A + B = \begin{bmatrix} 5 & -2 \\ 1 & 3 \end{bmatrix} + \begin{bmatrix} 4 & 1 \\ -3 & 2 \end{bmatrix} = \begin{bmatrix} 5+4 & -2+1 \\ 1+(-3) & 3+2 \end{bmatrix} = \begin{bmatrix} 9 & -1 \\ -2 & 5 \end{bmatrix}$

$A - B = \begin{bmatrix} 5 & -2 \\ 1 & 3 \end{bmatrix} - \begin{bmatrix} 4 & 1 \\ -3 & 2 \end{bmatrix} = \begin{bmatrix} 5-4 & -2-1 \\ 1-(-3) & 3-2 \end{bmatrix} = \begin{bmatrix} 1 & -3 \\ 4 & 1 \end{bmatrix}$

$$2A = 2\begin{bmatrix} 5 & -2 \\ 1 & 3 \end{bmatrix} = \begin{bmatrix} 2(5) & 2(-2) \\ 2(1) & 2(3) \end{bmatrix} = \begin{bmatrix} 10 & -4 \\ 2 & 6 \end{bmatrix}$$

$$-3B = -3\begin{bmatrix} 4 & 1 \\ -3 & 2 \end{bmatrix} = \begin{bmatrix} -3(4) & -3(1) \\ -3(-3) & -3(2) \end{bmatrix} = \begin{bmatrix} -12 & -3 \\ 9 & -6 \end{bmatrix}$$

**3** $A + B = \begin{bmatrix} 6 & -1 \\ 2 & 0 \\ -3 & 4 \end{bmatrix} + \begin{bmatrix} 3 & 1 \\ -1 & 5 \\ 6 & 0 \end{bmatrix} = \begin{bmatrix} 6+3 & -1+1 \\ 2+(-1) & 0+5 \\ -3+6 & 4+0 \end{bmatrix} = \begin{bmatrix} 9 & 0 \\ 1 & 5 \\ 3 & 4 \end{bmatrix}$

$$A - B = \begin{bmatrix} 6 & -1 \\ 2 & 0 \\ -3 & 4 \end{bmatrix} - \begin{bmatrix} 3 & 1 \\ -1 & 5 \\ 6 & 0 \end{bmatrix} = \begin{bmatrix} 6-3 & -1-1 \\ 2-(-1) & 0-5 \\ -3-6 & 4-0 \end{bmatrix} = \begin{bmatrix} 3 & -2 \\ 3 & -5 \\ -9 & 4 \end{bmatrix}$$

$$2A = 2\begin{bmatrix} 6 & -1 \\ 2 & 0 \\ -3 & 4 \end{bmatrix} = \begin{bmatrix} 2(6) & 2(-1) \\ 2(2) & 2(0) \\ 2(-3) & 2(4) \end{bmatrix} = \begin{bmatrix} 12 & -2 \\ 4 & 0 \\ -6 & 8 \end{bmatrix}$$

$$-3B = -3\begin{bmatrix} 3 & 1 \\ -1 & 5 \\ 6 & 0 \end{bmatrix} = \begin{bmatrix} -3(3) & -3(1) \\ -3(-1) & -3(5) \\ -3(6) & -3(0) \end{bmatrix} = \begin{bmatrix} -9 & -3 \\ 3 & -15 \\ -18 & 0 \end{bmatrix}$$

**5** $A + B = \begin{bmatrix} 4 & -3 & 2 \end{bmatrix} + \begin{bmatrix} 7 & 0 & -5 \end{bmatrix} = \begin{bmatrix} 11 & -3 & -3 \end{bmatrix}$,

$A - B = \begin{bmatrix} -3 & -3 & 7 \end{bmatrix}$, $2A = \begin{bmatrix} 8 & -6 & 4 \end{bmatrix}$, $-3B = \begin{bmatrix} -21 & 0 & 15 \end{bmatrix}$

**7** $A + B$ and $A - B$ are not possible since $A$ and $B$ are different sizes.

$$2A = 2\begin{bmatrix} 3 & -2 & 2 \\ 0 & 1 & -4 \\ -3 & 2 & -1 \end{bmatrix} = \begin{bmatrix} 6 & -4 & 4 \\ 0 & 2 & -8 \\ -6 & 4 & -2 \end{bmatrix}, \; -3B = -3\begin{bmatrix} 4 & 0 \\ 2 & -1 \\ -1 & 3 \end{bmatrix} = \begin{bmatrix} -12 & 0 \\ -6 & 3 \\ 3 & -9 \end{bmatrix}$$

**9** To find $c_{21}$ in Exercise 15, use the second row of $A$ and the first column of $B$.

$$c_{21} = (-5)(2) + (2)(0) + (2)(-4) = -10 + 0 - 8 = -18.$$

*Note:* For Exercises 11–22, $AB$ is possible only if the number of columns of $A$ equal the number of rows of $B$. $\{BA$ is not possible in #21, $AB$ is not possible in #22$\}$ Multiplying an $\boxed{m} \times n$ matrix with an $n \times \boxed{k}$ matrix will result in an $m \times k$ matrix.

**11** $AB = \begin{bmatrix} 2 & 6 \\ 3 & -4 \end{bmatrix}\begin{bmatrix} 5 & -2 \\ 1 & 7 \end{bmatrix}$

$$= \begin{bmatrix} 2(5)+6(1) & 2(-2)+6(7) \\ 3(5)+(-4)(1) & 3(-2)+(-4)(7) \end{bmatrix} = \begin{bmatrix} 10+6 & -4+42 \\ 15-4 & -6-28 \end{bmatrix} = \begin{bmatrix} 16 & 38 \\ 11 & -34 \end{bmatrix}$$

$BA = \begin{bmatrix} 5 & -2 \\ 1 & 7 \end{bmatrix}\begin{bmatrix} 2 & 6 \\ 3 & -4 \end{bmatrix}$

$$= \begin{bmatrix} 5(2)+(-2)(3) & 5(6)+(-2)(-4) \\ 1(2)+7(3) & 1(6)+7(-4) \end{bmatrix} = \begin{bmatrix} 4 & 38 \\ 23 & -22 \end{bmatrix}$$

**13** $AB = \begin{bmatrix} 3 & 0 & -1 \\ 0 & 4 & 2 \\ 5 & -3 & 1 \end{bmatrix}\begin{bmatrix} 1 & -5 & 0 \\ 4 & 1 & -2 \\ 0 & -1 & 3 \end{bmatrix}$

$$= \begin{bmatrix} 3(1)+0(4)+(-1)(0) & 3(-5)+0(1)+(-1)(-1) & 3(0)+0(-2)+(-1)(3) \\ 0(1)+4(4)+2(0) & 0(-5)+4(1)+2(-1) & 0(0)+4(-2)+2(3) \\ 5(1)+(-3)(4)+1(0) & 5(-5)+(-3)(1)+1(-1) & 5(0)+(-3)(-2)+1(3) \end{bmatrix}$$

$$= \begin{bmatrix} 3 & -14 & -3 \\ 16 & 2 & -2 \\ -7 & -29 & 9 \end{bmatrix}$$

$BA = \begin{bmatrix} 1 & -5 & 0 \\ 4 & 1 & -2 \\ 0 & -1 & 3 \end{bmatrix}\begin{bmatrix} 3 & 0 & -1 \\ 0 & 4 & 2 \\ 5 & -3 & 1 \end{bmatrix}$

$$= \begin{bmatrix} 1(3)+(-5)(0)+0(5) & 1(0)+(-5)(4)+0(-3) & 1(-1)+(-5)(2)+0(1) \\ 4(3)+1(0)+(-2)(5) & 4(0)+1(4)+(-2)(-3) & 4(-1)+1(2)+(-2)(1) \\ 0(3)+(-1)(0)+3(5) & 0(0)+(-1)(4)+3(-3) & 0(-1)+(-1)(2)+3(1) \end{bmatrix}$$

$$= \begin{bmatrix} 3 & -20 & -11 \\ 2 & 10 & -4 \\ 15 & -13 & 1 \end{bmatrix}$$

**15** $AB = \begin{bmatrix} 4 & -3 & 1 \\ -5 & 2 & 2 \end{bmatrix}\begin{bmatrix} 2 & 1 \\ 0 & 1 \\ -4 & 7 \end{bmatrix}$

$$= \begin{bmatrix} 4(2)+(-3)(0)+1(-4) & 4(1)+(-3)(1)+1(7) \\ -5(2)+2(0)+2(-4) & -5(1)+2(1)+2(7) \end{bmatrix} = \begin{bmatrix} 4 & 8 \\ -18 & 11 \end{bmatrix}$$

$BA = \begin{bmatrix} 2 & 1 \\ 0 & 1 \\ -4 & 7 \end{bmatrix}\begin{bmatrix} 4 & -3 & 1 \\ -5 & 2 & 2 \end{bmatrix}$

$$= \begin{bmatrix} 2(4)+1(-5) & 2(-3)+1(2) & 2(1)+1(2) \\ 0(4)+1(-5) & 0(-3)+1(2) & 0(1)+1(2) \\ -4(4)+7(-5) & -4(-3)+7(2) & -4(1)+7(2) \end{bmatrix} = \begin{bmatrix} 3 & -4 & 4 \\ -5 & 2 & 2 \\ -51 & 26 & 10 \end{bmatrix}$$

**17** $AB = \begin{bmatrix} 1 & 2 & 3 \\ 4 & 5 & 6 \\ 7 & 8 & 9 \end{bmatrix}\begin{bmatrix} 1 & 0 & 0 \\ 0 & 1 & 0 \\ 0 & 0 & 1 \end{bmatrix} = \begin{bmatrix} 1 & 2 & 3 \\ 4 & 5 & 6 \\ 7 & 8 & 9 \end{bmatrix}$, $BA = \begin{bmatrix} 1 & 2 & 3 \\ 4 & 5 & 6 \\ 7 & 8 & 9 \end{bmatrix}$

**19** $AB = \begin{bmatrix} -3 & 7 & 2 \end{bmatrix}\begin{bmatrix} 1 \\ 4 \\ -5 \end{bmatrix} = \begin{bmatrix} -3(1)+7(4)+2(-5) \end{bmatrix} = \begin{bmatrix} 15 \end{bmatrix}$

(continued)

$$BA = \begin{bmatrix} 1 \\ 4 \\ -5 \end{bmatrix} \begin{bmatrix} -3 & 7 & 2 \end{bmatrix} = \begin{bmatrix} 1(-3) & 1(7) & 1(2) \\ 4(-3) & 4(7) & 4(2) \\ -5(-3) & -5(7) & -5(2) \end{bmatrix} = \begin{bmatrix} -3 & 7 & 2 \\ -12 & 28 & 8 \\ 15 & -35 & -10 \end{bmatrix}$$

**21** $AB = \begin{bmatrix} 2 & 0 & 1 \\ -1 & 2 & 0 \end{bmatrix} \begin{bmatrix} 1 & -1 & 2 \\ 3 & 1 & 0 \\ 0 & 2 & 1 \end{bmatrix}$

$$= \begin{bmatrix} 2(1)+0(3)+1(0) & 2(-1)+0(1)+1(2) & 2(2)+0(0)+1(1) \\ -1(1)+2(3)+0(0) & -1(-1)+2(1)+0(2) & -1(2)+2(0)+0(1) \end{bmatrix}$$

$= \begin{bmatrix} 2 & 0 & 5 \\ 5 & 3 & -2 \end{bmatrix}$. $BA$ is a $3 \times 3$ matrix times a $2 \times 3$ matrix. Since the number of columns of $B$, 3, is not equal to the number of rows of $A$, 2, $BA$ is not possible.

**23** $AB = \begin{bmatrix} 4 & -2 \\ 0 & 3 \\ -7 & 5 \end{bmatrix} \begin{bmatrix} 3 \\ 4 \end{bmatrix} = \begin{bmatrix} 4(3)+(-2)(4) \\ 0(3)+3(4) \\ (-7)(3)+5(4) \end{bmatrix} = \begin{bmatrix} 4 \\ 12 \\ -1 \end{bmatrix}$

**25** $AB = \begin{bmatrix} 2 & 1 & 0 & -3 \\ -7 & 0 & -2 & 4 \end{bmatrix} \begin{bmatrix} 4 & -2 & 0 \\ 1 & 1 & -2 \\ 0 & 0 & 5 \\ -3 & -1 & 0 \end{bmatrix} = \begin{bmatrix} c_{11} & c_{12} & c_{13} \\ c_{21} & c_{22} & c_{23} \end{bmatrix}$, where

$c_{11} = 2(4)+1(1)+0(0)+(-3)(-3)$ $\quad$ $c_{12} = 2(-2)+1(1)+0(0)+(-3)(-1)$

$c_{13} = 2(0)+1(-2)+0(5)+(-3)(0)$ $\quad$ $c_{21} = -7(4)+0(1)+(-2)(0)+4(-3)$

$c_{22} = -7(-2)+0(1)+(-2)(0)+4(-1)$ $\quad$ $c_{23} = -7(0)+0(-2)+(-2)(5)+4(0)$

Hence, $AB = \begin{bmatrix} 18 & 0 & -2 \\ -40 & 10 & -10 \end{bmatrix}$.

**27**

$$A + B = \begin{bmatrix} 1 & 2 \\ 0 & -3 \end{bmatrix} + \begin{bmatrix} 2 & -1 \\ 3 & 1 \end{bmatrix} = \begin{bmatrix} 3 & 1 \\ 3 & -2 \end{bmatrix}$$

$$A - B = \begin{bmatrix} 1 & 2 \\ 0 & -3 \end{bmatrix} - \begin{bmatrix} 2 & -1 \\ 3 & 1 \end{bmatrix} = \begin{bmatrix} -1 & 3 \\ -3 & -4 \end{bmatrix}$$

So $(A+B)(A-B) = \begin{bmatrix} 3 & 1 \\ 3 & -2 \end{bmatrix} \begin{bmatrix} -1 & 3 \\ -3 & -4 \end{bmatrix} = \begin{bmatrix} -6 & 5 \\ 3 & 17 \end{bmatrix}$.

$$A^2 = AA = \begin{bmatrix} 1 & 2 \\ 0 & -3 \end{bmatrix} \begin{bmatrix} 1 & 2 \\ 0 & -3 \end{bmatrix} = \begin{bmatrix} 1 & -4 \\ 0 & 9 \end{bmatrix}$$

$$B^2 = BB = \begin{bmatrix} 2 & -1 \\ 3 & 1 \end{bmatrix} \begin{bmatrix} 2 & -1 \\ 3 & 1 \end{bmatrix} = \begin{bmatrix} 1 & -3 \\ 9 & -2 \end{bmatrix}$$

So $A^2 - B^2 = \begin{bmatrix} 1 & -4 \\ 0 & 9 \end{bmatrix} - \begin{bmatrix} 1 & -3 \\ 9 & -2 \end{bmatrix} = \begin{bmatrix} 0 & -1 \\ -9 & 11 \end{bmatrix}$.

Thus, $(A+B)(A-B) \neq A^2 - B^2$ since $\begin{bmatrix} -6 & 5 \\ 3 & 17 \end{bmatrix} \neq \begin{bmatrix} 0 & -1 \\ -9 & 11 \end{bmatrix}$.

**29** $A(B+C)=\begin{bmatrix} 1 & 2 \\ 0 & -3 \end{bmatrix}\begin{bmatrix} 5 & 0 \\ 1 & 1 \end{bmatrix}=\begin{bmatrix} 7 & 2 \\ -3 & -3 \end{bmatrix}$;

$$AB+AC=\begin{bmatrix} 8 & 1 \\ -9 & -3 \end{bmatrix}+\begin{bmatrix} -1 & 1 \\ 6 & 0 \end{bmatrix}=\begin{bmatrix} 7 & 2 \\ -3 & -3 \end{bmatrix}$$

**31** $m(A+B) = m\begin{bmatrix} a+p & b+q \\ c+r & d+s \end{bmatrix}$

$$= \begin{bmatrix} m(a+p) & m(b+q) \\ m(c+r) & m(d+s) \end{bmatrix}$$

$$= \begin{bmatrix} ma+mp & mb+mq \\ mc+mr & md+ms \end{bmatrix}$$

$$= \begin{bmatrix} ma & mb \\ mc & md \end{bmatrix}+\begin{bmatrix} mp & mq \\ mr & ms \end{bmatrix}$$

$$= m\begin{bmatrix} a & b \\ c & d \end{bmatrix}+m\begin{bmatrix} p & q \\ r & s \end{bmatrix}$$

$$= mA+mB$$

**33** $A(B+C) = \begin{bmatrix} a & b \\ c & d \end{bmatrix}\begin{bmatrix} p+w & q+x \\ r+y & s+z \end{bmatrix}$

$$= \begin{bmatrix} a(p+w)+b(r+y) & a(q+x)+b(s+z) \\ c(p+w)+d(r+y) & c(q+x)+d(s+z) \end{bmatrix}$$

$$= \begin{bmatrix} ap+aw+br+by & aq+ax+bs+bz \\ cp+cw+dr+dy & cq+cx+ds+dz \end{bmatrix}$$

$$= \begin{bmatrix} ap+br & aq+bs \\ cp+dr & cq+ds \end{bmatrix}+\begin{bmatrix} aw+by & ax+bz \\ cw+dy & cx+dz \end{bmatrix}$$

$$= \begin{bmatrix} a & b \\ c & d \end{bmatrix}\begin{bmatrix} p & q \\ r & s \end{bmatrix}+\begin{bmatrix} a & b \\ c & d \end{bmatrix}\begin{bmatrix} w & x \\ y & z \end{bmatrix}= AB+AC$$

*Note:* For Exercises 35–38, $A=\begin{bmatrix} 3 & -3 & 7 \\ 2 & 6 & -2 \\ 4 & 2 & 5 \end{bmatrix}$ and $B=\begin{bmatrix} -9 & 5 & -8 \\ 3 & -7 & 1 \\ -1 & 2 & 6 \end{bmatrix}$.

**35** $A^2 = AA = \begin{bmatrix} 3 & -3 & 7 \\ 2 & 6 & -2 \\ 4 & 2 & 5 \end{bmatrix}\begin{bmatrix} 3 & -3 & 7 \\ 2 & 6 & -2 \\ 4 & 2 & 5 \end{bmatrix}=\begin{bmatrix} 31 & -13 & 62 \\ 10 & 26 & -8 \\ 36 & 10 & 49 \end{bmatrix}$

$$B^2 = BB = \begin{bmatrix} -9 & 5 & -8 \\ 3 & -7 & 1 \\ -1 & 2 & 6 \end{bmatrix}\begin{bmatrix} -9 & 5 & -8 \\ 3 & -7 & 1 \\ -1 & 2 & 6 \end{bmatrix}=\begin{bmatrix} 104 & -96 & 29 \\ -49 & 66 & -25 \\ 9 & -7 & 46 \end{bmatrix}$$

$$A^2+B^2 = \begin{bmatrix} 31 & -13 & 62 \\ 10 & 26 & -8 \\ 36 & 10 & 49 \end{bmatrix}+\begin{bmatrix} 104 & -96 & 29 \\ -49 & 66 & -25 \\ 9 & -7 & 46 \end{bmatrix}=\begin{bmatrix} 135 & -109 & 91 \\ -39 & 92 & -33 \\ 45 & 3 & 95 \end{bmatrix}$$

37 $A^2 - 5B = \begin{bmatrix} 31 & -13 & 62 \\ 10 & 26 & -8 \\ 36 & 10 & 49 \end{bmatrix} - \begin{bmatrix} -45 & 25 & -40 \\ 15 & -35 & 5 \\ -5 & 10 & 30 \end{bmatrix} = \begin{bmatrix} 76 & -38 & 102 \\ -5 & 61 & -13 \\ 41 & 0 & 19 \end{bmatrix}$

39 (a) For the inventory matrix, we have 5 colors and 3 sizes of towels, and for each size of towel, we have 1 price. We choose a $5 \times 3$ matrix $A$ and a $3 \times 1$ matrix $B$.

$$\text{inventory matrix } A = \begin{bmatrix} 400 & 550 & 500 \\ 400 & 450 & 500 \\ 300 & 500 & 600 \\ 250 & 200 & 300 \\ 100 & 100 & 200 \end{bmatrix}, \text{ price matrix } B = \begin{bmatrix} \$1.39 \\ \$2.99 \\ \$4.99 \end{bmatrix}$$

(b) $C = AB = \begin{bmatrix} 400 & 550 & 500 \\ 400 & 450 & 500 \\ 300 & 500 & 600 \\ 250 & 200 & 300 \\ 100 & 100 & 200 \end{bmatrix} \begin{bmatrix} \$1.39 \\ \$2.99 \\ \$4.99 \end{bmatrix} = \begin{bmatrix} \$4695.50 \\ \$4396.50 \\ \$4906.00 \\ \$2442.50 \\ \$1436.00 \end{bmatrix}$

(c) The \$1436.00 represents the amount the store would receive if all the yellow towels were sold.

## 6.7 Exercises

*Note:* Exer. 1–12: Let $A$ denote the given matrix.

1 $AB = \begin{bmatrix} 5 & 7 \\ 2 & 3 \end{bmatrix} \begin{bmatrix} 3 & -7 \\ -2 & 5 \end{bmatrix} = \begin{bmatrix} 1 & 0 \\ 0 & 1 \end{bmatrix} = I_2$ and

$BA = \begin{bmatrix} 3 & -7 \\ -2 & 5 \end{bmatrix} \begin{bmatrix} 5 & 7 \\ 2 & 3 \end{bmatrix} = \begin{bmatrix} 1 & 0 \\ 0 & 1 \end{bmatrix} = I_2$.

Since $AB = I_2$ and $BA = I_2$, $B$ is the inverse of $A$.

3 $\left[\begin{array}{cc|cc} 2 & -4 & 1 & 0 \\ 1 & 3 & 0 & 1 \end{array}\right] R_1 - R_2 \rightarrow R_1 \Rightarrow \left[\begin{array}{cc|cc} 1 & -7 & 1 & -1 \\ 1 & 3 & 0 & 1 \end{array}\right] R_2 - R_1 \rightarrow R_2$

$\left[\begin{array}{cc|cc} 1 & -7 & 1 & -1 \\ 0 & 10 & -1 & 2 \end{array}\right] \frac{1}{10} R_2 \rightarrow R_2 \Rightarrow \left[\begin{array}{cc|cc} 1 & -7 & 1 & -1 \\ 0 & 1 & -\frac{1}{10} & \frac{2}{10} \end{array}\right] R_1 + 7R_2 \rightarrow R_1$

$\left[\begin{array}{cc|cc} 1 & 0 & \frac{3}{10} & \frac{4}{10} \\ 0 & 1 & -\frac{1}{10} & \frac{2}{10} \end{array}\right] \Rightarrow A^{-1} = \frac{1}{10}\begin{bmatrix} 3 & 4 \\ -1 & 2 \end{bmatrix}$

5 $\left[\begin{array}{cc|cc} 2 & 4 & 1 & 0 \\ 4 & 8 & 0 & 1 \end{array}\right] \frac{1}{2} R_1 \rightarrow R_1 \Rightarrow \left[\begin{array}{cc|cc} 1 & 2 & \frac{1}{2} & 0 \\ 4 & 8 & 0 & 1 \end{array}\right] R_2 - 4R_1 \rightarrow R_2$

$\left[\begin{array}{cc|cc} 1 & 2 & \frac{1}{2} & 0 \\ 0 & 0 & -2 & 1 \end{array}\right]$

Since the identity matrix cannot be obtained on the left, *no inverse exists.*

[7]
$$\left[\begin{array}{ccc|ccc} 3 & -1 & 0 & 1 & 0 & 0 \\ 2 & 2 & 0 & 0 & 1 & 0 \\ 0 & 0 & 4 & 0 & 0 & 1 \end{array}\right] \begin{array}{l} \mathrm{R}_1 - \mathrm{R}_2 \rightarrow \mathrm{R}_1 \\ \\ \\ \end{array}$$

$$\left[\begin{array}{ccc|ccc} 1 & -3 & 0 & 1 & -1 & 0 \\ 2 & 2 & 0 & 0 & 1 & 0 \\ 0 & 0 & 4 & 0 & 0 & 1 \end{array}\right] \mathrm{R}_2 - 2\,\mathrm{R}_1 \rightarrow \mathrm{R}_2$$

$$\left[\begin{array}{ccc|ccc} 1 & -3 & 0 & 1 & -1 & 0 \\ 0 & 8 & 0 & -2 & 3 & 0 \\ 0 & 0 & 4 & 0 & 0 & 1 \end{array}\right] \begin{array}{l} (1/8)\,\mathrm{R}_2 \rightarrow \mathrm{R}_2 \\ (1/4)\,\mathrm{R}_3 \rightarrow \mathrm{R}_3 \end{array}$$

$$\left[\begin{array}{ccc|ccc} 1 & -3 & 0 & 1 & -1 & 0 \\ 0 & 1 & 0 & -\frac{2}{8} & \frac{3}{8} & 0 \\ 0 & 0 & 1 & 0 & 0 & \frac{1}{4} \end{array}\right] \begin{array}{l} \mathrm{R}_1 + 3\,\mathrm{R}_2 \rightarrow \mathrm{R}_1 \\ \\ \\ \end{array}$$

$$\left[\begin{array}{ccc|ccc} 1 & 0 & 0 & \frac{2}{8} & \frac{1}{8} & 0 \\ 0 & 1 & 0 & -\frac{2}{8} & \frac{3}{8} & 0 \\ 0 & 0 & 1 & 0 & 0 & \frac{2}{8} \end{array}\right] \qquad A^{-1} = \frac{1}{8}\begin{bmatrix} 2 & 1 & 0 \\ -2 & 3 & 0 \\ 0 & 0 & 2 \end{bmatrix}$$

[9]
$$\left[\begin{array}{ccc|ccc} -2 & 2 & 3 & 1 & 0 & 0 \\ 1 & -1 & 0 & 0 & 1 & 0 \\ 0 & 1 & 4 & 0 & 0 & 1 \end{array}\right] \begin{array}{l} \mathrm{R}_1 + 2\,\mathrm{R}_2 \leftrightarrow \mathrm{R}_2 \\ \\ \\ \end{array}$$

$$\left[\begin{array}{ccc|ccc} 1 & -1 & 0 & 0 & 1 & 0 \\ 0 & 0 & 3 & 1 & 2 & 0 \\ 0 & 1 & 4 & 0 & 0 & 1 \end{array}\right] \begin{array}{l} \mathrm{R}_1 + \mathrm{R}_3 \rightarrow \mathrm{R}_1 \\ \\ \\ \end{array}$$

$$\left[\begin{array}{ccc|ccc} 1 & 0 & 4 & 0 & 1 & 1 \\ 0 & 0 & 3 & 1 & 2 & 0 \\ 0 & 1 & 4 & 0 & 0 & 1 \end{array}\right] \tfrac{1}{3}\mathrm{R}_2 \leftrightarrow \mathrm{R}_3$$

$$\left[\begin{array}{ccc|ccc} 1 & 0 & 4 & 0 & 1 & 1 \\ 0 & 1 & 4 & 0 & 0 & 1 \\ 0 & 0 & 1 & \frac{1}{3} & \frac{2}{3} & 0 \end{array}\right] \begin{array}{l} \mathrm{R}_1 - 4\,\mathrm{R}_3 \rightarrow \mathrm{R}_1 \\ \mathrm{R}_2 - 4\,\mathrm{R}_3 \rightarrow \mathrm{R}_2 \\ \\ \end{array}$$

$$\left[\begin{array}{ccc|ccc} 1 & 0 & 0 & -\frac{4}{3} & -\frac{5}{3} & 1 \\ 0 & 1 & 0 & -\frac{4}{3} & -\frac{8}{3} & 1 \\ 0 & 0 & 1 & \frac{1}{3} & \frac{2}{3} & 0 \end{array}\right] \qquad A^{-1} = \frac{1}{3}\begin{bmatrix} -4 & -5 & 3 \\ -4 & -8 & 3 \\ 1 & 2 & 0 \end{bmatrix}$$

[11] $\left[\begin{array}{ccc|ccc} 2 & 0 & 0 & 1 & 0 & 0 \\ 0 & 4 & 0 & 0 & 1 & 0 \\ 0 & 0 & 6 & 0 & 0 & 1 \end{array}\right]$ $\begin{array}{l} (1/2)\,R_1 \rightarrow R_1 \\ (1/4)\,R_2 \rightarrow R_2 \\ (1/6)\,R_3 \rightarrow R_3 \end{array}$

$\left[\begin{array}{ccc|ccc} 1 & 0 & 0 & \frac{1}{2} & 0 & 0 \\ 0 & 1 & 0 & 0 & \frac{1}{4} & 0 \\ 0 & 0 & 1 & 0 & 0 & \frac{1}{6} \end{array}\right]$ $\qquad A^{-1} = \frac{1}{12}\begin{bmatrix} 6 & 0 & 0 \\ 0 & 3 & 0 \\ 0 & 0 & 2 \end{bmatrix}$

[13] $\left[\begin{array}{cc|cc} a & 0 & 1 & 0 \\ 0 & b & 0 & 1 \end{array}\right]$ $\begin{array}{l} (1/a)\,R_1 \rightarrow R_1 \\ (1/b)\,R_2 \rightarrow R_2 \end{array}$ $\Rightarrow \left[\begin{array}{cc|cc} 1 & 0 & 1/a & 0 \\ 0 & 1 & 0 & 1/b \end{array}\right]$

The inverse is the matrix with main diagonal elements $(1/a)$ and $(1/b)$.

The required conditions are that $a$ and $b$ are nonzero to avoid division by zero.

[15] $AI_3 = \begin{bmatrix} a_{11} & a_{12} & a_{13} \\ a_{21} & a_{22} & a_{23} \\ a_{31} & a_{32} & a_{33} \end{bmatrix}\begin{bmatrix} 1 & 0 & 0 \\ 0 & 1 & 0 \\ 0 & 0 & 1 \end{bmatrix} = \begin{bmatrix} a_{11} & a_{12} & a_{13} \\ a_{21} & a_{22} & a_{23} \\ a_{31} & a_{32} & a_{33} \end{bmatrix} = A$

$I_3A = \begin{bmatrix} 1 & 0 & 0 \\ 0 & 1 & 0 \\ 0 & 0 & 1 \end{bmatrix}\begin{bmatrix} a_{11} & a_{12} & a_{13} \\ a_{21} & a_{22} & a_{23} \\ a_{31} & a_{32} & a_{33} \end{bmatrix} = \begin{bmatrix} a_{11} & a_{12} & a_{13} \\ a_{21} & a_{22} & a_{23} \\ a_{31} & a_{32} & a_{33} \end{bmatrix} = A$

[17] (a) $X = A^{-1}B = \frac{1}{10}\begin{bmatrix} 3 & 4 \\ -1 & 2 \end{bmatrix}\begin{bmatrix} 3 \\ 1 \end{bmatrix} = \frac{1}{10}\begin{bmatrix} 13 \\ -1 \end{bmatrix}$; $(\frac{13}{10}, -\frac{1}{10})$

(b) $X = A^{-1}B = \frac{1}{10}\begin{bmatrix} 3 & 4 \\ -1 & 2 \end{bmatrix}\begin{bmatrix} -2 \\ 5 \end{bmatrix} = \frac{1}{10}\begin{bmatrix} 14 \\ 12 \end{bmatrix}$; $(\frac{7}{5}, \frac{6}{5})$

[19] (a) $X = A^{-1}B = \frac{1}{3}\begin{bmatrix} -4 & -5 & 3 \\ -4 & -8 & 3 \\ 1 & 2 & 0 \end{bmatrix}\begin{bmatrix} 1 \\ 3 \\ -2 \end{bmatrix} = \frac{1}{3}\begin{bmatrix} -25 \\ -34 \\ 7 \end{bmatrix}$; $(-\frac{25}{3}, -\frac{34}{3}, \frac{7}{3})$

(b) $X = A^{-1}B = \frac{1}{3}\begin{bmatrix} -4 & -5 & 3 \\ -4 & -8 & 3 \\ 1 & 2 & 0 \end{bmatrix}\begin{bmatrix} -1 \\ 0 \\ 4 \end{bmatrix} = \frac{1}{3}\begin{bmatrix} 16 \\ 16 \\ -1 \end{bmatrix}$; $(\frac{16}{3}, \frac{16}{3}, -\frac{1}{3})$

[21] The inverse should be found using some type of computational device. If you are using a TI-83/86, enter the 9 values into the matrix [A]. Change Float to 5 via MODE. Now find $[A]^{-1}$ { be sure to use the $\boxed{x^{-1}}$ key }. Use the right arrow key to see the rightmost elements in [A].

$A = \begin{bmatrix} 2 & -5 & 8 \\ 3 & 7 & -1 \\ 0 & 2 & 1 \end{bmatrix} \Rightarrow A^{-1} \approx \begin{bmatrix} 0.11111 & 0.25926 & -0.62963 \\ -0.03704 & 0.02469 & 0.32099 \\ 0.07407 & -0.04938 & 0.35802 \end{bmatrix}$

[23] $A = \begin{bmatrix} 2 & -1 & 1 & 4 \\ 7 & 1.2 & -8 & 0 \\ 2.5 & 0 & 1.9 & 7.9 \\ 1 & -1 & 3 & 1 \end{bmatrix} \Rightarrow A^{-1} \approx \begin{bmatrix} -0.22278 & 0.12932 & 0.06496 & 0.37796 \\ -1.17767 & 0.09503 & 0.55936 & 0.29171 \\ -0.37159 & 0.00241 & 0.14074 & 0.37447 \\ 0.15987 & -0.04150 & 0.07218 & -0.20967 \end{bmatrix}$

**25** (a) $AX = B \Leftrightarrow \begin{bmatrix} 4.0 & 7.1 \\ 2.2 & -4.9 \end{bmatrix}\begin{bmatrix} x \\ y \end{bmatrix} = \begin{bmatrix} 6.2 \\ 2.9 \end{bmatrix}$

(b) On your calculator, Find $[A]^{-1}$ and STOre this matrix into matrix [B] to use in part (c).

$$A^{-1} \approx \begin{bmatrix} 0.1391 & 0.2016 \\ 0.0625 & -0.1136 \end{bmatrix}$$

(c) Following the instructions in part (b), enter the 3 values into [C] {a $3 \times 1$ matrix}, and then evaluate [B]*[C].

$$X = A^{-1}B \approx \begin{bmatrix} 0.1391 & 0.2016 \\ 0.0625 & -0.1136 \end{bmatrix}\begin{bmatrix} 6.2 \\ 2.9 \end{bmatrix} \approx \begin{bmatrix} 1.4472 \\ 0.0579 \end{bmatrix}.$$

**27** (a) $AX = B \Leftrightarrow \begin{bmatrix} 3.1 & 6.7 & -8.7 \\ 4.1 & -5.1 & 0.2 \\ 0.6 & 1.1 & -7.4 \end{bmatrix}\begin{bmatrix} x \\ y \\ z \end{bmatrix} = \begin{bmatrix} 1.5 \\ 2.1 \\ 3.9 \end{bmatrix}$

(b) $A^{-1} \approx \begin{bmatrix} 0.1474 & 0.1572 & -0.1691 \\ 0.1197 & -0.0696 & -0.1426 \\ 0.0297 & 0.0024 & -0.1700 \end{bmatrix}$

(c) $X = A^{-1}B \approx \begin{bmatrix} 0.1474 & 0.1572 & -0.1691 \\ 0.1197 & -0.0696 & -0.1426 \\ 0.0297 & 0.0024 & -0.1700 \end{bmatrix}\begin{bmatrix} 1.5 \\ 2.1 \\ 3.9 \end{bmatrix} \approx \begin{bmatrix} -0.1081 \\ -0.5227 \\ -0.6135 \end{bmatrix}$

**29** (a) $f(2) = 4a + 2b + c = 19$; $f(8) = 64a + 8b + c = 59$; $f(11) = 121a + 11b + c = 26$

Solve the $3 \times 3$ linear system using the inverse method.

$$\begin{bmatrix} 4 & 2 & 1 \\ 64 & 8 & 1 \\ 121 & 11 & 1 \end{bmatrix}\begin{bmatrix} a \\ b \\ c \end{bmatrix} = \begin{bmatrix} 19 \\ 59 \\ 26 \end{bmatrix} \Rightarrow \begin{bmatrix} a \\ b \\ c \end{bmatrix} \approx \begin{bmatrix} -1.9630 \\ 26.2963 \\ -25.7407 \end{bmatrix}$$

Thus, let $f(x) = -1.9630x^2 + 26.2963x - 25.7407$.

(b) [1, 12] by [−15, 70, 5]

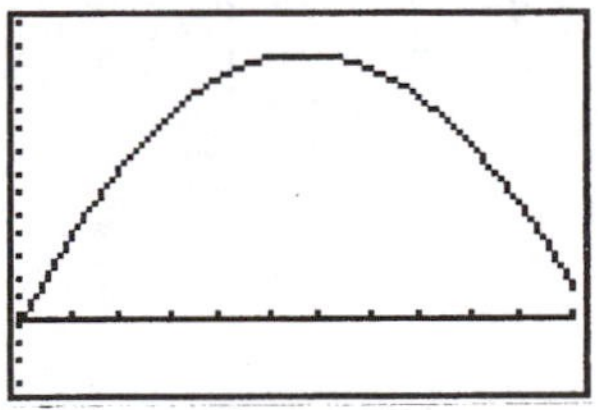

*Figure 29*

(c) For June, $f(6) \approx 61$°F and for October, $f(10) \approx 41$°F.

## 6.8 Exercises

*Note:* The minor $M_{ij}$ and the cofactor $A_{ij}$ are equal if $(i+j)$ is even and of opposite sign if $(i+j)$ is odd.

[1] The minor $M_{11}$ is obtained by deleting the first row and first column from $A = \begin{bmatrix} 7 & -1 \\ 5 & 0 \end{bmatrix}$. Thus, $M_{11} = 0 = A_{11}$. Similarly, $M_{12} = 5$ and $A_{12} = -5$; $M_{21} = -1$ and $A_{21} = 1$; and $M_{22} = 7 = A_{22}$.

[3] Let $A = \begin{bmatrix} 2 & 4 & -1 \\ 0 & 3 & 2 \\ -5 & 7 & 0 \end{bmatrix}$.

Be sure you understand the note preceding the solution for Exercise 1.

$$M_{11} = \begin{vmatrix} 3 & 2 \\ 7 & 0 \end{vmatrix} = (3)(0) - (7)(2) = 0 - 14 = -14 = A_{11};$$

$$M_{12} = \begin{vmatrix} 0 & 2 \\ -5 & 0 \end{vmatrix} = (0)(0) - (-5)(2) = 0 - (-10) = 10;\ A_{12} = -10;$$

$$M_{13} = \begin{vmatrix} 0 & 3 \\ -5 & 7 \end{vmatrix} = 15 = A_{13}; \qquad M_{21} = \begin{vmatrix} 4 & -1 \\ 7 & 0 \end{vmatrix} = 7;\ A_{21} = -7;$$

$$M_{22} = \begin{vmatrix} 2 & -1 \\ -5 & 0 \end{vmatrix} = -5 = A_{22}; \qquad M_{23} = \begin{vmatrix} 2 & 4 \\ -5 & 7 \end{vmatrix} = 34;\ A_{23} = -34; \quad \text{(cont.)}$$

$$M_{31} = \begin{vmatrix} 4 & -1 \\ 3 & 2 \end{vmatrix} = 11 = A_{31}; \qquad M_{32} = \begin{vmatrix} 2 & -1 \\ 0 & 2 \end{vmatrix} = 4;\ A_{32} = -4;$$

$$M_{33} = \begin{vmatrix} 2 & 4 \\ 0 & 3 \end{vmatrix} = 6 = A_{33}.$$

*Note:* Exercises 5–20: Let $A$ denote the given matrix.

[5] $\begin{vmatrix} 7 & -1 \\ 5 & 0 \end{vmatrix} = (7)(0) - (-1)(5) = 0 + 5 = 5$

[7] Expanding $|A|$ by the first column and using the cofactor values from Exercise 3, we obtain $|A| = a_{11}A_{11} + a_{21}A_{21} + a_{31}A_{31} = 2(-14) + 0(A_{21}) - 5(11) = -83$.

[9] By the definition of the determinant of a $2 \times 2$ matrix $A$ (on page 468),

$$|A| = \begin{vmatrix} -5 & 4 \\ -3 & 2 \end{vmatrix} = (-5)(2) - (4)(-3) = -10 + 12 = 2.$$

[11] $|A| = (a)(-b) - (-a)(b) = -ab + ab = 0$

[13] Expand by the first row.

$$|A| = a_{11}A_{11} + a_{12}A_{12} + a_{13}A_{13}$$

$$= (3)(-1)^{1+1}\begin{vmatrix} 2 & 5 \\ 3 & -1 \end{vmatrix} + (1)(-1)^{1+2}\begin{vmatrix} 4 & 5 \\ -6 & -1 \end{vmatrix} + (-2)(-1)^{1+3}\begin{vmatrix} 4 & 2 \\ -6 & 3 \end{vmatrix}$$

$$= (3)(1)(-17) + (1)(-1)(26) + (-2)(1)(24)$$

$$= -51 - 26 - 48 = -125$$

[15] Expand by the third row.

$$|A| = a_{31}A_{31} + a_{32}A_{32} + a_{33}A_{33} = 2(30) + 0(A_{32}) + 6(-2) = 48$$

[17] Expand $|A|$ by the third row. $|A| = 6\,A_{32} = -6\,M_{32} = -6\begin{vmatrix} 3 & 2 & 0 \\ 4 & -3 & 5 \\ 1 & -4 & 2 \end{vmatrix}$.

Expand $M_{32}$ by the first row.

$M_{32} = 3(14) + 2(-3) + 0(-13) = 36 \;\Rightarrow\; |A| = -6(36) = -216.$

[19] Expand by the first row. $|A| = -b\begin{vmatrix} 0 & c & 0 \\ a & 0 & 0 \\ 0 & 0 & d \end{vmatrix}$.

Expand again by the first row. $|A| = (-b)(-c)\begin{vmatrix} a & 0 \\ 0 & d \end{vmatrix} = bc(ad - 0) = abcd.$

[21] LS $= ad - bc$; RS $= -(bc - ad) = ad - bc$

[23] LS $= adk - bck$; RS $= k(ad - bc) = adk - bck$

[25] LS $= ad - bc$; RS $= abk + ad - abk - bc = ad - bc$

[27] LS $= ad - bc + af - ce$; RS $= ad + af - bc - ce$

[29] Consider the matrix in Exercise 20. If we expanded along the first column, we would obtain $a$ times its cofactor. Expanding along the first column again, we obtain $ab$ times another cofactor. This exercise is similar since all elements in $A$ *above* {rather than below} the main diagonal are zero. We can evaluate the determinant using $n$ expansions by the first row, and obtain $|A| = a_{11}a_{22}\cdots a_{nn}$.

[31] (a) $A - xI = \begin{bmatrix} 1 & 2 \\ 3 & 2 \end{bmatrix} - x\begin{bmatrix} 1 & 0 \\ 0 & 1 \end{bmatrix} = \begin{bmatrix} 1-x & 2 \\ 3 & 2-x \end{bmatrix}$.

$$f(x) = |A - xI| = \begin{vmatrix} 1-x & 2 \\ 3 & 2-x \end{vmatrix}$$

$$= (1-x)(2-x) - (3)(2) = (2 - 3x + x^2) - 6 = x^2 - 3x - 4.$$

(b) $x^2 - 3x - 4 = 0 \;\Rightarrow\; (x-4)(x+1) = 0 \;\Rightarrow\; x = -1, 4$

**33** (a) $f(x) = |A - xI| = \begin{vmatrix} -3-x & -2 \\ 2 & 2-x \end{vmatrix} = x^2 + x - 2$

(b) $x^2 + x - 2 = 0 \Rightarrow (x+2)(x-1) = 0 \Rightarrow x = -2,\ 1$

**35** (a) $f(x) = \begin{vmatrix} 1-x & 0 & 0 \\ 1 & 0-x & -2 \\ -1 & 1 & -3-x \end{vmatrix}$ { Expand by the first row. }

$= (1-x)[(-x)(-3-x) - (-2)]$

$= (1-x)(x^2 + 3x + 2)$

$= (1-x)(x+1)(x+2)$ or

$(-x^3 - 2x^2 + x + 2)$

(b) $(1-x)(x+1)(x+2) = 0 \Rightarrow x = -2,\ -1,\ 1$

**37** (a) $f(x) = \begin{vmatrix} 0-x & 2 & -2 \\ -1 & 3-x & 1 \\ -3 & 3 & 1-x \end{vmatrix}$ { Expand by the first row. }

$= (-x)[(3-x)(1-x) - 3] - 2[(x-1) + 3] - 2[-3 + 3(3-x)]$

$= (-x)[(3 - 4x + x^2) - 3] - 2(x+2) - 2(-3x+6)$

$= (-x^3 + 4x^2) - 2x - 4 + 6x - 12$

$= -x^3 + 4x^2 + 4x - 16$

(b) By trying possible rational roots, we determine that 2 is a zero of $f$.

Thus, $-x^3 + 4x^2 + 4x - 16 = 0 \Rightarrow (x-2)(-x^2 + 2x + 8) = 0 \Rightarrow$

$(x+2)(x-2)(-x+4) = 0 \Rightarrow x = -2,\ 2,\ 4.$

**39** Expand the determinant by the first row.

$$\begin{vmatrix} i & j & k \\ 2 & -1 & 6 \\ -3 & 5 & 1 \end{vmatrix} = i\begin{vmatrix} -1 & 6 \\ 5 & 1 \end{vmatrix} - j\begin{vmatrix} 2 & 6 \\ -3 & 1 \end{vmatrix} + k\begin{vmatrix} 2 & -1 \\ -3 & 5 \end{vmatrix} = -31i - 20j + 7k$$

**41** $\begin{vmatrix} i & j & k \\ 5 & -6 & -1 \\ 3 & 0 & 1 \end{vmatrix} = i\begin{vmatrix} -6 & -1 \\ 0 & 1 \end{vmatrix} - j\begin{vmatrix} 5 & -1 \\ 3 & 1 \end{vmatrix} + k\begin{vmatrix} 5 & -6 \\ 3 & 0 \end{vmatrix} = -6i - 8j + 18k$

**43** $A = \begin{bmatrix} 29 & -17 & 90 \\ -34 & 91 & -34 \\ 48 & 7 & 10 \end{bmatrix} \Rightarrow |A| = -359{,}284.$

**45** $A = \begin{bmatrix} 4 & -7 & -3 & 13 \\ -17 & -0.8 & 5 & 0.9 \\ 1.1 & 0.2 & 10 & -4 \\ 3 & -6 & 2 & 1 \end{bmatrix} \Rightarrow |A| = 10{,}739.92.$

**47** (a) $f(x) = |A - xI| = \begin{vmatrix} 1-x & 0 & 1 \\ 0 & 2-x & 1 \\ 1 & 1 & -2-x \end{vmatrix} = -x^3 + x^2 + 6x - 7$

(b) The characteristic values of $A$ are equal to the zeros of $f$. From the graph,

we see that the zeros are approximately $-2.51$, $1.22$, and $2.29$.

$[-10, 11]$ by $[-12, 2]$

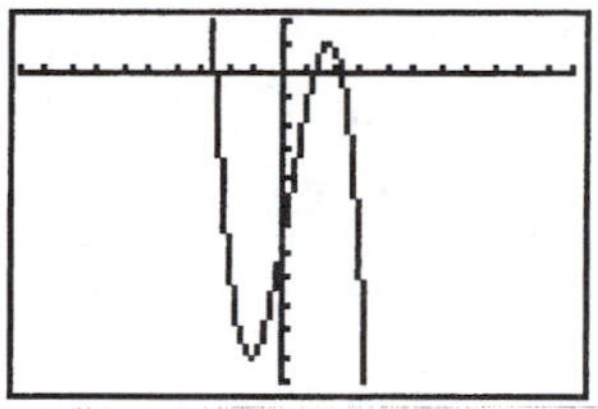

*Figure 47*

## *6.9 Exercises*

1 $R_2$ and $R_3$ are interchanged. The determinant value is multiplied by $-1$.

3 $R_3$ is replaced by $(R_3 - R_1)$. There is no change in the determinant value.

5 The number 2 can be factored out of $R_1$, yielding $2\begin{vmatrix} 1 & 2 & 1 \\ 1 & 2 & 4 \\ 2 & 6 & 4 \end{vmatrix}$.

Next, the number 2 can be factored out of $R_3$, yielding $4\begin{vmatrix} 1 & 2 & 1 \\ 1 & 2 & 4 \\ 1 & 3 & 2 \end{vmatrix}$.

7 $R_1$ and $R_3$ are identical. The determinant is 0.

9 The number $-1$ can be factored out of $R_2$, yielding the determinant on the right side.

11 Every number in $C_2$ is 0. The determinant is zero.

13 $C_3$ is replaced by $(2C_1 + C_3)$. There is no change in the determinant value.

*Note:* The notation $\{R_i\ (C_i)\}$ means expand the determinant by the $i$th row (column).

15 There are many possibilities for introducing zeros. In this case, obtaining a zero in the third row, second column would lead to an easy evaluation using the second column.

$$\begin{vmatrix} 3 & 1 & 0 \\ -2 & 0 & 1 \\ 1 & 3 & -1 \end{vmatrix} R_3 - 3R_1 \to R_3 = \begin{vmatrix} 3 & 1 & 0 \\ -2 & 0 & 1 \\ -8 & 0 & -1 \end{vmatrix} \{C_2\}$$

$$= (-1)\begin{vmatrix} -2 & 1 \\ -8 & -1 \end{vmatrix} = (-1)(2+8) = -10$$

**17** $\begin{vmatrix} 5 & 4 & 3 \\ -3 & 2 & 1 \\ 0 & 7 & -2 \end{vmatrix} \begin{matrix} R_1 - 3R_2 \to R_1 \\ \\ R_3 + 2R_2 \to R_3 \end{matrix} = \begin{vmatrix} 14 & -2 & 0 \\ -3 & 2 & 1 \\ -6 & 11 & 0 \end{vmatrix} \{C_3\}$

$= (-1)\begin{vmatrix} 14 & -2 \\ -6 & 11 \end{vmatrix} = (-1)(154 - 12) = -142$

**19** $\begin{vmatrix} 2 & 2 & -3 \\ 3 & 6 & 9 \\ -2 & 5 & 4 \end{vmatrix} \{\text{3 is a common factor of } R_2\} = (3)\begin{vmatrix} 2 & 2 & -3 \\ 1 & 2 & 3 \\ -2 & 5 & 4 \end{vmatrix}$

Since there is a "1" in the second row, first column, we will obtain zeros in the other two locations of the first column.

$(3)\begin{vmatrix} 2 & 2 & -3 \\ 1 & 2 & 3 \\ -2 & 5 & 4 \end{vmatrix} \begin{matrix} R_1 - 2R_2 \to R_1 \\ \\ R_3 + 2R_2 \to R_3 \end{matrix} = (3)\begin{vmatrix} 0 & -2 & -9 \\ 1 & 2 & 3 \\ 0 & 9 & 10 \end{vmatrix} \{C_1\}$

$= (3)(-1)\begin{vmatrix} -2 & -9 \\ 9 & 10 \end{vmatrix} = (-3)(-20 + 81) = -183$

**21** $\begin{vmatrix} 3 & 1 & -2 & 2 \\ 2 & 0 & 1 & 4 \\ 0 & 1 & 3 & 5 \\ -1 & 2 & 0 & -3 \end{vmatrix} \begin{matrix} \\ \\ R_3 - R_1 \to R_3 \\ R_4 - 2R_1 \to R_4 \end{matrix} = \begin{vmatrix} 3 & 1 & -2 & 2 \\ 2 & 0 & 1 & 4 \\ -3 & 0 & 5 & 3 \\ -7 & 0 & 4 & -7 \end{vmatrix} \{C_2\}$

$= (-1)\begin{vmatrix} 2 & 1 & 4 \\ -3 & 5 & 3 \\ -7 & 4 & -7 \end{vmatrix} \begin{matrix} \\ R_2 - 5R_1 \to R_2 \\ R_3 - 4R_1 \to R_3 \end{matrix} = (-1)\begin{vmatrix} 2 & 1 & 4 \\ -13 & 0 & -17 \\ -15 & 0 & -23 \end{vmatrix} \{C_2\}$

$= (-1)(-1)\begin{vmatrix} -13 & -17 \\ -15 & -23 \end{vmatrix} = (1)(299 - 255) = 44$

**23** $\begin{vmatrix} 2 & -2 & 0 & 0 & -3 \\ 3 & 0 & 3 & 2 & -1 \\ 0 & 1 & -2 & 0 & 2 \\ -1 & 2 & 0 & 3 & 0 \\ 0 & 4 & 1 & 0 & 0 \end{vmatrix} C_2 - 4C_3 \to C_2 = \begin{vmatrix} 2 & -2 & 0 & 0 & -3 \\ 3 & -12 & 3 & 2 & -1 \\ 0 & 9 & -2 & 0 & 2 \\ -1 & 2 & 0 & 3 & 0 \\ 0 & 0 & 1 & 0 & 0 \end{vmatrix} \{R_5\}$

$= (1)\begin{vmatrix} 2 & -2 & 0 & -3 \\ 3 & -12 & 2 & -1 \\ 0 & 9 & 0 & 2 \\ -1 & 2 & 3 & 0 \end{vmatrix} \begin{matrix} R_1 + 2R_4 \to R_1 \\ R_2 + 3R_4 \to R_2 \\ \\ \\ \end{matrix} = \begin{vmatrix} 0 & 2 & 6 & -3 \\ 0 & -6 & 11 & -1 \\ 0 & 9 & 0 & 2 \\ -1 & 2 & 3 & 0 \end{vmatrix} \{C_1\}$

$= (1)\begin{vmatrix} 2 & 6 & -3 \\ -6 & 11 & -1 \\ 9 & 0 & 2 \end{vmatrix} \begin{matrix} R_1 - 3R_2 \to R_1 \\ \\ R_3 + 2R_2 \to R_3 \end{matrix} = \begin{vmatrix} 20 & -27 & 0 \\ -6 & 11 & -1 \\ -3 & 22 & 0 \end{vmatrix} \{C_3\}$

$= (1)\begin{vmatrix} 20 & -27 \\ -3 & 22 \end{vmatrix} = (1)(440 - 81) = 359$

**25** $\begin{vmatrix} 1 & 1 & 1 \\ a & b & c \\ a^2 & b^2 & c^2 \end{vmatrix}$ $\begin{matrix} C_1 - C_2 \rightarrow C_1 \\ C_3 - C_2 \rightarrow C_3 \end{matrix}$

$= \begin{vmatrix} 0 & 1 & 0 \\ a-b & b & c-b \\ a^2-b^2 & b^2 & c^2-b^2 \end{vmatrix}$ $\begin{matrix} a-b \text{ is a common factor of } C_1 \\ c-b \text{ is a common factor of } C_3 \end{matrix}$

$= (a-b)(c-b)\begin{vmatrix} 0 & 1 & 0 \\ 1 & b & 1 \\ a+b & b^2 & c+b \end{vmatrix} \{R_1\}$

$= (a-b)(c-b)(-1)\begin{vmatrix} 1 & 1 \\ a+b & c+b \end{vmatrix}$

$= (a-b)(b-c)(c+b-a-b) = \underline{(a-b)(b-c)(c-a)}$

**27** $\begin{vmatrix} a_{11} & a_{12} & a_{13} & a_{14} \\ 0 & a_{22} & a_{23} & a_{24} \\ 0 & 0 & a_{33} & a_{34} \\ 0 & 0 & 0 & a_{44} \end{vmatrix} \{C_1\} = (a_{11})\begin{vmatrix} a_{22} & a_{23} & a_{24} \\ 0 & a_{33} & a_{34} \\ 0 & 0 & a_{44} \end{vmatrix} \{C_1\}$

$= (a_{11})(a_{22})\begin{vmatrix} a_{33} & a_{34} \\ 0 & a_{44} \end{vmatrix} = (a_{11}\,a_{22})(a_{33}\,a_{44} - 0) = a_{11}\,a_{22}\,a_{33}\,a_{44}$

**29** $|\,AB\,| = \begin{vmatrix} a_{11}b_{11} + a_{12}b_{21} & a_{11}b_{12} + a_{12}b_{22} \\ a_{21}b_{11} + a_{22}b_{21} & a_{21}b_{12} + a_{22}b_{22} \end{vmatrix}$

$= (a_{11}b_{11} + a_{12}b_{21})(a_{21}b_{12} + a_{22}b_{22}) - (a_{11}b_{12} + a_{12}b_{22})(a_{21}b_{11} + a_{22}b_{21})$

$= a_{11}b_{11}a_{21}b_{12} + a_{11}b_{11}a_{22}b_{22} + a_{12}b_{21}a_{21}b_{12} + a_{12}b_{21}a_{22}b_{22}$

$- a_{11}b_{12}a_{21}b_{11} - a_{11}b_{12}a_{22}b_{21} - a_{12}b_{22}a_{21}b_{11} - a_{12}b_{22}a_{22}b_{21}$

$= a_{11}a_{22}b_{11}b_{22} - a_{11}a_{22}b_{21}b_{12} - a_{21}a_{12}b_{11}b_{22} + a_{21}a_{12}b_{21}b_{12}$

$= (a_{11}a_{22} - a_{21}a_{12})(b_{11}b_{22} - b_{21}b_{12}) = |\,A\,|\,|\,B\,|$

**31** Expanding by the first row yields $Ax + By + C = 0$ {an equation of a line} where $A$, $B$, and $C$ are constants. To show that the line contains $(x_1, y_1)$ and $(x_2, y_2)$, we must show that these points are solutions of the equation. Substituting $x_1$ for $x$ and $y_1$ for $y$, we obtain two identical rows and the determinant is zero. Hence, $(x_1, y_1)$ is a solution of the equation and a similar argument can be made for $(x_2, y_2)$.

**33** For the system $\begin{cases} 2x + 3y = 2 \\ x - 2y = 8 \end{cases}$, $|\,D\,| = \begin{vmatrix} 2 & 3 \\ 1 & -2 \end{vmatrix} = -4 - 3 = -7.$

Since $|\,D\,| = -7 \neq 0$, we may solve the system using Cramer's rule.

$|\,D_x\,| = \begin{vmatrix} 2 & 3 \\ 8 & -2 \end{vmatrix} = -4 - 24 = -28.$ $|\,D_y\,| = \begin{vmatrix} 2 & 2 \\ 1 & 8 \end{vmatrix} = 16 - 2 = 14.$

$x = \dfrac{|\,D_x\,|}{|\,D\,|} = \dfrac{-28}{-7} = 4$ and $y = \dfrac{|\,D_y\,|}{|\,D\,|} = \dfrac{14}{-7} = -2.$ ★ (4, −2)

**35** For the system $\begin{cases} 2x + 5y = 16 \\ 3x - 7y = 24 \end{cases}$, $|D| = \begin{vmatrix} 2 & 5 \\ 3 & -7 \end{vmatrix} = -14 - 15 = -29.$

Since $|D| = -29 \neq 0$, we may solve the system using Cramer's rule.

$|D_x| = \begin{vmatrix} 16 & 5 \\ 24 & -7 \end{vmatrix} = -112 - 120 = -232.$ $|D_y| = \begin{vmatrix} 2 & 16 \\ 3 & 24 \end{vmatrix} = 48 - 48 = 0.$

$x = \dfrac{|D_x|}{|D|} = \dfrac{-232}{-29} = 8$ and $y = \dfrac{|D_y|}{|D|} = \dfrac{0}{-29} = 0.$ ★ (8, 0)

**37** $|D| = \begin{vmatrix} 2 & -3 \\ -6 & 9 \end{vmatrix} = 18 - 18 = 0$, so Cramer's rule cannot be used.

**39** $|D| = \begin{vmatrix} 1 & -2 & -3 \\ 2 & 1 & 1 \\ 1 & 3 & -2 \end{vmatrix} \{R_1\} = 1(-5) - (-2)(-5) + (-3)(5) = -30$

$|D_x| = \begin{vmatrix} -1 & -2 & -3 \\ 6 & 1 & 1 \\ 13 & 3 & -2 \end{vmatrix} \{R_1\} = (-1)(-5) - (-2)(-25) + (-3)(5) = -60$

$|D_y| = \begin{vmatrix} 1 & -1 & -3 \\ 2 & 6 & 1 \\ 1 & 13 & -2 \end{vmatrix} \{R_1\} = 1(-25) - (-1)(-5) + (-3)(20) = -90$

$|D_z| = \begin{vmatrix} 1 & -2 & -1 \\ 2 & 1 & 6 \\ 1 & 3 & 13 \end{vmatrix} \{R_1\} = 1(-5) - (-2)(20) + (-1)(5) = 30$

$x = \dfrac{|D_x|}{|D|} = \dfrac{-60}{-30} = 2;\ y = \dfrac{|D_y|}{|D|} = \dfrac{-90}{-30} = 3;\ z = \dfrac{|D_z|}{|D|} = \dfrac{30}{-30} = -1.$

★ (2, 3, −1)

**41** $|D| = \begin{vmatrix} 5 & 2 & -1 \\ 1 & -2 & 2 \\ 0 & 3 & 1 \end{vmatrix} \{R_3\} = -3(11) + 1(-12) = -45$

$|D_x| = \begin{vmatrix} -7 & 2 & -1 \\ 0 & -2 & 2 \\ 17 & 3 & 1 \end{vmatrix} \{C_1\} = -7(-8) + 17(2) = 90$

$|D_y| = \begin{vmatrix} 5 & -7 & -1 \\ 1 & 0 & 2 \\ 0 & 17 & 1 \end{vmatrix} \{R_2\} = -1(10) - 2(85) = -180$

$|D_z| = \begin{vmatrix} 5 & 2 & -7 \\ 1 & -2 & 0 \\ 0 & 3 & 17 \end{vmatrix} \{C_1\} = 5(-34) - 1(55) = -225$

$x = \dfrac{|D_x|}{|D|} = \dfrac{90}{-45} = -2;\ y = \dfrac{|D_y|}{|D|} = \dfrac{-180}{-45} = 4;\ z = \dfrac{|D_z|}{|D|} = \dfrac{-225}{-45} = 5.$

★ (−2, 4, 5)

**43** To solve for $x$, find $|D|$ and $|D_x|$, and then use Cramer's Rule.

$$|D| = \begin{vmatrix} a & b & c \\ e & 0 & f \\ h & i & 0 \end{vmatrix} \quad (1/c)\mathrm{R}_1 \rightarrow \mathrm{R}_1$$

$$= \begin{vmatrix} a/c & b/c & 1 \\ e & 0 & f \\ h & i & 0 \end{vmatrix} \quad \mathrm{R}_2 - f\,\mathrm{R}_1 \rightarrow \mathrm{R}_2$$

$$= \begin{vmatrix} a/c & b/c & 1 \\ e - af/c & -bf/c & 0 \\ h & i & 0 \end{vmatrix} \quad \{\mathrm{C}_3\}$$

$$= (1)\begin{vmatrix} e - af/c & -bf/c \\ h & i \end{vmatrix}$$

$$= \frac{ce - af}{c} \cdot i - \left(-\frac{bf}{c} \cdot h\right) = \frac{cei - afi + bfh}{c}$$

$$|D_x| = \begin{vmatrix} d & b & c \\ g & 0 & f \\ j & i & 0 \end{vmatrix} \quad (1/c)\mathrm{R}_1 \rightarrow \mathrm{R}_1$$

$$= \begin{vmatrix} d/c & b/c & 1 \\ g & 0 & f \\ j & i & 0 \end{vmatrix} \quad \mathrm{R}_2 - f\,\mathrm{R}_1 \rightarrow \mathrm{R}_2$$

$$= \begin{vmatrix} d/c & b/c & 1 \\ g - df/c & -bf/c & 0 \\ j & i & 0 \end{vmatrix} \quad \{\mathrm{C}_3\}$$

$$= (1)\begin{vmatrix} g - df/c & -bf/c \\ j & i \end{vmatrix}$$

$$= \frac{cg - df}{c} \cdot i - \left(-\frac{bf}{c} \cdot j\right) = \frac{cgi - dfi + bfj}{c}$$

Thus, $x = \dfrac{|D_x|}{|D|} = \dfrac{cgi - dfi + bfj}{cei - afi + bfh}$.

## 6.10 Exercises

*Note:* The general outline for the solutions in this section is as follows:

1st line) The expression is shown on the left side of the equation and its decomposition is on the right side.

2nd line) The equation in the first line is multiplied by its least common denominator and left in factored form.

3rd line and beyond) Values are substituted into the equation in the second line and the coefficients are found by solving the resulting equations. It will be stated when the method of equating coefficients is used.

[1] $\dfrac{8x-1}{(x-2)(x+3)}=\dfrac{A}{x-2}+\dfrac{B}{x+3}$

$8x-1=A(x+3)+B(x-2)$

$x=-3$: $-25=-5B \Rightarrow B=5$ ★ $\dfrac{3}{x-2}+\dfrac{5}{x+3}$

$x=2$: $15=5A \Rightarrow A=3$

[3] $\dfrac{x+34}{(x-6)(x+2)}=\dfrac{A}{x-6}+\dfrac{B}{x+2}$

$x+34=A(x+2)+B(x-6)$

$x=-2$: $32=-8B \Rightarrow B=-4$ ★ $\dfrac{5}{x-6}-\dfrac{4}{x+2}$

$x=6$: $40=8A \Rightarrow A=5$

[5] $\dfrac{4x^2-15x-1}{(x-1)(x+2)(x-3)}=\dfrac{A}{x-1}+\dfrac{B}{x+2}+\dfrac{C}{x-3}$

$4x^2-15x-1=A(x+2)(x-3)+B(x-1)(x-3)+C(x-1)(x+2)$

$x=-2$: $45=15B \Rightarrow B=3$

$x=1$: $-12=-6A \Rightarrow A=2$ ★ $\dfrac{2}{x-1}+\dfrac{3}{x+2}-\dfrac{1}{x-3}$

$x=3$: $-10=10C \Rightarrow C=-1$

[7] $\dfrac{4x^2-5x-15}{x(x-5)(x+1)}=\dfrac{A}{x}+\dfrac{B}{x-5}+\dfrac{C}{x+1}$

$4x^2-5x-15=A(x-5)(x+1)+Bx(x+1)+Cx(x-5)$

$x=-1$: $-6=6C \Rightarrow C=-1$

$x=0$: $-15=-5A \Rightarrow A=3$ ★ $\dfrac{3}{x}+\dfrac{2}{x-5}-\dfrac{1}{x+1}$

$x=5$: $60=30B \Rightarrow B=2$

[9] $\dfrac{2x+3}{(x-1)^2}=\dfrac{A}{x-1}+\dfrac{B}{(x-1)^2}$

$2x+3=A(x-1)+B$

$x=1$: $5=B$ ★ $\dfrac{2}{x-1}+\dfrac{5}{(x-1)^2}$

$x=0$: $3=-A+B \Rightarrow A=2$

[11] $\dfrac{19x^2+50x-25}{x^2(3x-5)}=\dfrac{A}{x}+\dfrac{B}{x^2}+\dfrac{C}{3x-5}$

$19x^2+50x-25=Ax(3x-5)+B(3x-5)+Cx^2$

$x=\frac{5}{3}$: $\frac{1000}{9}=\frac{25}{9}C \Rightarrow C=40$

$x=0$: $-25=-5B \Rightarrow B=5$ ★ $-\frac{7}{x}+\frac{5}{x^2}+\dfrac{40}{3x-5}$

$x=1$: $44=-2A-2B+C \Rightarrow A=-7$

[13] $\dfrac{x^2-6}{(x+2)^2(2x-1)} = \dfrac{A}{x+2} + \dfrac{B}{(x+2)^2} + \dfrac{C}{2x-1}$

$x^2 - 6 = A(x+2)(2x-1) + B(2x-1) + C(x+2)^2$

$x = -2$: $-2 = -5B \;\Rightarrow\; B = \frac{2}{5}$

$x = \frac{1}{2}$: $-\frac{23}{4} = \frac{25}{4}C \;\Rightarrow\; C = -\frac{23}{25}$

$x = 1$: $-5 = 3A + B + 9C \;\Rightarrow\; A = \frac{24}{25}$

★ $\dfrac{\frac{24}{25}}{x+2} + \dfrac{\frac{2}{5}}{(x+2)^2} - \dfrac{\frac{23}{25}}{2x-1}$

[15] $\dfrac{3x^3 + 11x^2 + 16x + 5}{x(x+1)^3} = \dfrac{A}{x} + \dfrac{B}{x+1} + \dfrac{C}{(x+1)^2} + \dfrac{D}{(x+1)^3}$

$3x^3 + 11x^2 + 16x + 5 = A(x+1)^3 + Bx(x+1)^2 + Cx(x+1) + Dx$

$x = -1$: $-3 = -D \;\Rightarrow\; D = 3$

$x = 0$: $5 = A$

$x = 1$: $35 = 8A + 4B + 2C + D \quad (E_1)$

$x = -2$: $-7 = -A - 2B + 2C - 2D \quad (E_2)$

Substituting the values for $A$ and $D$ into $E_1$ and $E_2$ yields

$$\begin{cases} 4B + 2C = -8 \\ -2B + 2C = 4 \end{cases} \;\Rightarrow\; \begin{cases} 2B + C = -4 & (E_3) \\ -B + C = 2 & (E_4) \end{cases}$$

$E_3 + 2E_4 \;\Rightarrow\; 3C = 0 \;\Rightarrow\; C = 0;\; B = -2$

★ $\dfrac{5}{x} - \dfrac{2}{x+1} + \dfrac{3}{(x+1)^3}$

[17] $\dfrac{x^2 + x - 6}{(x^2+1)(x-1)} = \dfrac{Ax+B}{x^2+1} + \dfrac{C}{x-1}$

$x^2 + x - 6 = (Ax+B)(x-1) + C(x^2+1)$

$x = 1$: $-4 = 2C \;\Rightarrow\; C = -2$

$x = 0$: $-6 = -B + C \;\Rightarrow\; B = 4$

$x = 2$: $0 = 2A + B + 5C \;\Rightarrow\; A = 3$

★ $-\dfrac{2}{x-1} + \dfrac{3x+4}{x^2+1}$

[19] $\dfrac{9x^2 - 3x + 8}{x(x^2+2)} = \dfrac{A}{x} + \dfrac{Bx+C}{x^2+2}$

$9x^2 - 3x + 8 = A(x^2+2) + (Bx+C)x$

$x = 0$: $8 = 2A \;\Rightarrow\; A = 4$

$x = 1$: $14 = 3A + B + C \quad (E_1)$

$x = -1$: $20 = 3A + B - C \quad (E_2)$

$E_1 - E_2 \;\Rightarrow\; -6 = 2C \;\Rightarrow\; C = -3;\; B = 5$

★ $\dfrac{4}{x} + \dfrac{5x-3}{x^2+2}$

**21** $\dfrac{4x^3 - x^2 + 4x + 2}{(x^2+1)^2} = \dfrac{Ax+B}{x^2+1} + \dfrac{Cx+D}{(x^2+1)^2}$

$$\begin{aligned} 4x^3 - x^2 + 4x + 2 &= (Ax+B)(x^2+1) + Cx + D \\ &= Ax^3 + Bx^2 + (A+C)x + (B+D) \end{aligned}$$

Equating coefficients, that is, the coefficient of $x^3$ on the left side must equal the coefficient of $x^3$ on the right side, we have the following:

$x^3$ : $A = 4$

$x^2$ : $B = -1$

$x$ : $A + C = 4 \Rightarrow C = 0$

constant : $B + D = 2 \Rightarrow D = 3$

★ $\dfrac{4x-1}{x^2+1} + \dfrac{3}{(x^2+1)^2}$

**23** The degree of the numerator is not lower than the degree of the denominator. Thus, we must first use long division and then decompose the remaining expression. Hence, by first dividing and then factoring, we have the following:

$$2x + \frac{4x^2 - 3x + 1}{(x^2+1)(x-1)} = 2x + \frac{Ax+B}{x^2+1} + \frac{C}{x-1}$$

$4x^2 - 3x + 1 = (Ax+B)(x-1) + C(x^2+1)$

$x = 1$: $2 = 2C \Rightarrow C = 1$

$x = 0$: $1 = -B + C \Rightarrow B = 0$

$x = -1$: $8 = 2A - 2B + 2C \Rightarrow A = 3$

★ $2x + \dfrac{1}{x-1} + \dfrac{3x}{x^2+1}$

**25** By first dividing and then factoring, we have the following:

$$3 + \frac{12x - 16}{x(x-4)} = 3 + \frac{A}{x} + \frac{B}{x-4}$$

$12x - 16 = A(x-4) + Bx$

$x = 0$: $-16 = -4A \Rightarrow A = 4$

$x = 4$: $32 = 4B \Rightarrow B = 8$

★ $3 + \frac{4}{x} + \dfrac{8}{x-4}$

**27** By first dividing and then factoring, we have the following:

$$2x + 3 + \frac{x+5}{(2x+1)(x-1)} = 2x + 3 + \frac{A}{2x+1} + \frac{B}{x-1}$$

$x + 5 = A(x-1) + B(2x+1)$

$x = 1$: $6 = 3B \Rightarrow B = 2$

$x = -\frac{1}{2}$: $\frac{9}{2} = -\frac{3}{2}A \Rightarrow A = -3$

★ $2x + 3 + \dfrac{2}{x-1} - \dfrac{3}{2x+1}$

## *Chapter 6 Review Exercises*

[1] $4\,E_1 + 3\,E_2 \;\Rightarrow\; 8x + 15x = 16 + 3 \;\Rightarrow\; 23x = 19 \;\Rightarrow\; x = \frac{19}{23};$

$-5\,E_1 + 2\,E_2 \;\Rightarrow\; 15y + 8y = -20 + 2 \;\Rightarrow\; 23y = -18 \;\Rightarrow\; y = -\frac{18}{23}$

★ $(\frac{19}{23}, -\frac{18}{23})$

[3] Solve $E_2$ for $y$, $y = -2x - 1$, and substitute into $E_1$ to yield $-2x - 1 + 4 = x^2 \;\Rightarrow$

$x^2 + 2x - 3 = 0 \;\Rightarrow\; (x+3)(x-1) = 0 \;\Rightarrow\; x = -3,\ 1$ and $y = 5,\ -3$.

★ $(-3, 5), (1, -3)$

[5] $4\,E_2 + E_1 \;\Rightarrow\; 4x^2 + 9x^2 = 16 + 140 \;\Rightarrow\; 13x^2 = 156 \;\Rightarrow\; x^2 = 12 \;\Rightarrow$

$x = \pm\sqrt{12} = \pm 2\sqrt{3}$. Substituting 12 for $x^2$ in $E_2$ gives us $12 - 4y^2 = 4 \;\Rightarrow$

$8 = 4y^2 \;\Rightarrow\; 2 = y^2 \;\Rightarrow\; y = \pm\sqrt{2}$.

There are four solutions: $(2\sqrt{3}, \pm\sqrt{2})$, $(-2\sqrt{3}, \pm\sqrt{2})$.

[7] $-4\,E_1 + E_2 \;\Rightarrow\; -12/y - 2/y = -28 + 1 \;\Rightarrow\; -14/y = -27 \;\Rightarrow\; y = \frac{14}{27};$

$2\,E_1 + 3\,E_2 \;\Rightarrow\; 2/x + 12/x = 14 + 3 \;\Rightarrow\; 14/x = 17 \;\Rightarrow\; x = \frac{14}{17}$ ★ $(\frac{14}{17}, \frac{14}{27})$

[8] Treat $3^{y+1}$ as $3 \cdot 3^y$ and $2^{x+1}$ as $2 \cdot 2^x$.

Thus, $E_1$ is $1(2^x) + 3(3^y) = 10$ and $E_2$ is $2(2^x) - 1(3^y) = 5$.

Now $E_1 + 3\,E_2 \;\Rightarrow\; 7 \cdot 2^x = 25 \;\Rightarrow\; 2^x = \frac{25}{7} \;\Rightarrow\; x = \log_2 \frac{25}{7} = \dfrac{\log \frac{25}{7}}{\log 2} \approx 1.84.$

Resolving the original system for $y$, we have

$$-2\,E_1 + E_2 \;\Rightarrow\; -7 \cdot 3^y = -15 \;\Rightarrow\; 3^y = \tfrac{15}{7} \;\Rightarrow\; y = \log_3 \tfrac{15}{7} = \frac{\log \frac{15}{7}}{\log 3} \approx 0.69.$$

[9] Solve $E_3$ for $z$, $z = 4x + 5y + 2$, and substitute into $E_1$ and $E_2$ to yield

$$\begin{cases} 3x + y - 2(4x + 5y + 2) = -1 \\ 2x - 3y + (4x + 5y + 2) = 4 \end{cases} \;\Rightarrow\; \begin{cases} -5x - 9y = 3 & (E_4) \\ 6x + 2y = 2 & (E_5) \end{cases}$$

$6\,E_4 + 5\,E_5 \;\Rightarrow\; -44y = 28 \;\Rightarrow\; y = -\frac{7}{11};$

$2\,E_4 + 9\,E_5 \;\Rightarrow\; 44x = 24 \;\Rightarrow\; x = \frac{6}{11};\; z = 1$ ★ $(\frac{6}{11}, -\frac{7}{11}, 1)$

[11] Solve $E_2$ for $x$, $x = y + z$, and substitute into $E_1$ and $E_3$ to yield

$$\begin{cases} 4(y + z) - 3y - z = 0 \\ 3(y + z) - y + 3z = 0 \end{cases} \;\Rightarrow\; \begin{cases} y + 3z = 0 & (E_4) \\ 2y + 6z = 0 & (E_5) \end{cases}$$

Now $E_5$ is $2\,E_4$, hence $y = -3z$ and $x = y + z = -3z + z = -2z$.

The general solution is $(-2c, -3c, c)$ for any real number $c$.

[13] $E_1 - E_2 \;\Rightarrow\; x - 5z = -1 \;\Rightarrow\; x = 5z - 1;$

Substitute this value into $E_1$ to obtain $y = \dfrac{-19z + 5}{2}$; $\left(5c - 1, \dfrac{-19c + 5}{2}, c\right)$ is

the general solution, where $c$ is any real number.

[15] Let $a = 1/x$, $b = 1/y$, and $c = 1/z$ to obtain the system

$$\begin{cases} 4a + b + 2c = 4 & (E_1) \\ 2a + 3b - c = 1 & (E_2) \\ a + b + c = 4 & (E_3) \end{cases}$$

Solving $E_2$ for $c$ and substituting into $E_1$ and $E_3$ yields

$$\begin{cases} 4a + b + 2(2a + 3b - 1) = 4 \\ a + b + (2a + 3b - 1) = 4 \end{cases} \Rightarrow \begin{cases} 8a + 7b = 6 & (E_4) \\ 3a + 4b = 5 & (E_5) \end{cases}$$

$3\,E_4 - 8\,E_5 \Rightarrow -11b = -22 \Rightarrow b = 2$

$4\,E_4 - 7\,E_5 \Rightarrow 11a = -11 \Rightarrow a = -1,\ c = 3;\ (x, y, z) = (-1, \frac{1}{2}, \frac{1}{3})$

[17] $\begin{cases} x^2 + y^2 < 16 \\ y - x^2 > 0 \end{cases} \Leftrightarrow \begin{cases} x^2 + y^2 < 4^2 \\ y > x^2 \end{cases}$

$V$ @ $(\pm\sqrt{-\frac{1}{2} + \frac{1}{2}\sqrt{65}},\ -\frac{1}{2} + \frac{1}{2}\sqrt{65}) \approx (\pm 1.88,\ 3.53)$

Shade inside the dashed circle $x^2 + y^2 = 4^2$ and inside the dashed parabola $y = x^2$.

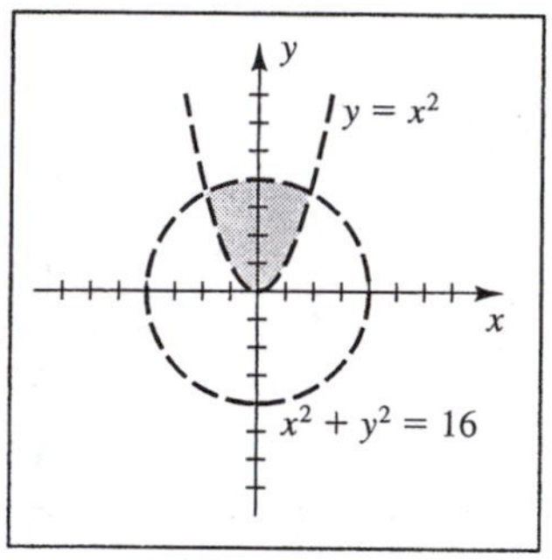


*Figure 17*

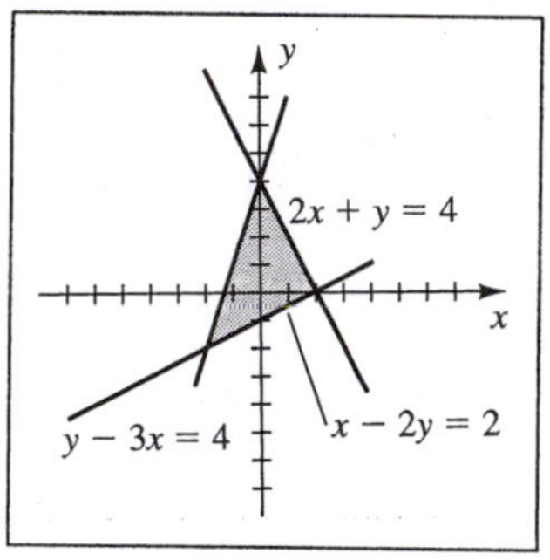


*Figure 19*

[19] $\begin{cases} x - 2y \le 2 \\ y - 3x \le 4 \\ 2x + y \le 4 \end{cases} \Leftrightarrow \begin{cases} y \ge \frac{1}{2}x - 1 \\ y \le 3x + 4 \\ y \le -2x + 4 \end{cases}$ $\quad V$ @ $(-2, -2), (0, 4), (2, 0)$

Shade above the solid line $y = \frac{1}{2}x - 1$, under the solid line $y = 3x + 4$, and under the solid line $y = -2x + 4$.

[21] $$\begin{bmatrix} 2 & -1 & 0 \\ 3 & 0 & -2 \end{bmatrix} \begin{bmatrix} 2 & -1 & 3 \\ 0 & 3 & 0 \\ 1 & 4 & 2 \end{bmatrix} =$$

$$= \begin{bmatrix} 2(2) + (-1)(0) + 0(1) & 2(-1) + (-1)(3) + 0(4) & 2(3) + (-1)(0) + 0(2) \\ 3(2) + 0(0) + (-2)(1) & 3(-1) + 0(3) + (-2)(4) & 3(3) + 0(0) + (-2)(2) \end{bmatrix}$$

$$= \begin{bmatrix} 4 & -5 & 6 \\ 4 & -11 & 5 \end{bmatrix}$$

**23** $\begin{bmatrix} 2 & 0 \\ 1 & 4 \\ -2 & 3 \end{bmatrix}\begin{bmatrix} 0 & 2 & -3 \\ 4 & 5 & 1 \end{bmatrix} =$

$$\begin{bmatrix} 2(0)+0(4) & 2(2)+0(5) & 2(-3)+0(1) \\ 1(0)+4(4) & 1(2)+4(5) & 1(-3)+4(1) \\ -2(0)+3(4) & -2(2)+3(5) & -2(-3)+3(1) \end{bmatrix} = \begin{bmatrix} 0 & 4 & -6 \\ 16 & 22 & 1 \\ 12 & 11 & 9 \end{bmatrix}$$

**25** $2\begin{bmatrix} 0 & -1 & -4 \\ 3 & 2 & 1 \end{bmatrix} - 3\begin{bmatrix} 4 & -2 & 1 \\ 0 & 5 & -1 \end{bmatrix} =$

$$\begin{bmatrix} 0 & -2 & -8 \\ 6 & 4 & 2 \end{bmatrix} + \begin{bmatrix} -12 & 6 & -3 \\ 0 & -15 & 3 \end{bmatrix} = \begin{bmatrix} -12 & 4 & -11 \\ 6 & -11 & 5 \end{bmatrix}$$

**27** $\begin{bmatrix} a & 0 \\ 0 & b \end{bmatrix}\begin{bmatrix} 1 & 3 \\ 2 & 4 \end{bmatrix} = \begin{bmatrix} a & 3a \\ 2b & 4b \end{bmatrix}$

**29** $\begin{bmatrix} 1 & 2 \\ 3 & 4 \end{bmatrix}\left\{\begin{bmatrix} 2 & -4 \\ 3 & 7 \end{bmatrix} + \begin{bmatrix} 1 & 5 \\ -2 & -3 \end{bmatrix}\right\} = \begin{bmatrix} 1 & 2 \\ 3 & 4 \end{bmatrix}\begin{bmatrix} 3 & 1 \\ 1 & 4 \end{bmatrix} = \begin{bmatrix} 5 & 9 \\ 13 & 19 \end{bmatrix}$

**30** Don't multiply, just remember that $AA^{-1} = I_3$.

*Note:* Let $A$ denote each of the matrices in Exercises 31–46.

**31** $\left[\begin{array}{cc|cc} 5 & -4 & 1 & 0 \\ -3 & 2 & 0 & 1 \end{array}\right] 2R_1 + 3R_2 \to R_1 \Rightarrow \left[\begin{array}{cc|cc} 1 & -2 & 2 & 3 \\ -3 & 2 & 0 & 1 \end{array}\right] R_2 + 3R_1 \to R_2$

$\left[\begin{array}{cc|cc} 1 & -2 & 2 & 3 \\ 0 & -4 & 6 & 10 \end{array}\right] -\frac{1}{4}R_2 \to R_2 \quad \Rightarrow \left[\begin{array}{cc|cc} 1 & -2 & 2 & 3 \\ 0 & 1 & -\frac{3}{2} & -\frac{5}{2} \end{array}\right] R_1 + 2R_2 \to R_1$

$\left[\begin{array}{cc|cc} 1 & 0 & -1 & -2 \\ 0 & 1 & -\frac{3}{2} & -\frac{5}{2} \end{array}\right] \qquad A^{-1} = -\frac{1}{2}\begin{bmatrix} 2 & 4 \\ 3 & 5 \end{bmatrix}$

**33** $\left[\begin{array}{ccc|ccc} 1 & 0 & 0 & 1 & 0 & 0 \\ 0 & 4 & 7 & 0 & 1 & 0 \\ 0 & 1 & 2 & 0 & 0 & 1 \end{array}\right] R_3 \leftrightarrow R_2$

$\left[\begin{array}{ccc|ccc} 1 & 0 & 0 & 1 & 0 & 0 \\ 0 & 1 & 2 & 0 & 0 & 1 \\ 0 & 4 & 7 & 0 & 1 & 0 \end{array}\right] R_3 - 4R_2 \to R_3$

$\left[\begin{array}{ccc|ccc} 1 & 0 & 0 & 1 & 0 & 0 \\ 0 & 1 & 2 & 0 & 0 & 1 \\ 0 & 0 & -1 & 0 & 1 & -4 \end{array}\right] -R_3 \to R_3$

$\left[\begin{array}{ccc|ccc} 1 & 0 & 0 & 1 & 0 & 0 \\ 0 & 1 & 2 & 0 & 0 & 1 \\ 0 & 0 & 1 & 0 & -1 & 4 \end{array}\right] R_2 - 2R_3 \to R_2$

$\left[\begin{array}{ccc|ccc} 1 & 0 & 0 & 1 & 0 & 0 \\ 0 & 1 & 0 & 0 & 2 & -7 \\ 0 & 0 & 1 & 0 & -1 & 4 \end{array}\right] \qquad A^{-1} = \begin{bmatrix} 1 & 0 & 0 \\ 0 & 2 & -7 \\ 0 & -1 & 4 \end{bmatrix}$

**35** $X = A^{-1}B = -\frac{1}{2}\begin{bmatrix} 2 & 4 \\ 3 & 5 \end{bmatrix}\begin{bmatrix} 30 \\ -16 \end{bmatrix} = -\frac{1}{2}\begin{bmatrix} -4 \\ 10 \end{bmatrix} = \begin{bmatrix} 2 \\ -5 \end{bmatrix}$; $(x, y) = (2, -5)$

**37** $A = [\,-6\,] \Rightarrow |A| = -6.$

**39** $A = \begin{bmatrix} 3 & -4 \\ 6 & 8 \end{bmatrix} \Rightarrow |A| = \begin{vmatrix} 3 & -4 \\ 6 & 8 \end{vmatrix} = 3(8) - 6(-4) = 24 + 24 = 48$

**41** $|A| = \begin{vmatrix} 2 & -3 & 5 \\ -4 & 1 & 3 \\ 3 & 2 & -1 \end{vmatrix}\ \{R_1\} = 2(-7) + 3(-5) + 5(-11) = -84$

**43** From Exercise 29 of §6.8, the determinant of $A$ is the product of the main diagonal elements of $A$—that is, $|A| = (5)(-3)(-4)(2) = 120.$

**44** $$\begin{vmatrix} 1 & 2 & 0 & 3 & 1 \\ -2 & -1 & 4 & 1 & 2 \\ 3 & 0 & -1 & 0 & -1 \\ 2 & -3 & 2 & -4 & 2 \\ -1 & 1 & 0 & 1 & 3 \end{vmatrix} \begin{matrix} R_2 + 4R_3 \to R_2 \\ \\ R_4 + 2R_3 \to R_4 \end{matrix} = \begin{vmatrix} 1 & 2 & 0 & 3 & 1 \\ 10 & -1 & 0 & 1 & -2 \\ 3 & 0 & -1 & 0 & -1 \\ 8 & -3 & 0 & -4 & 0 \\ -1 & 1 & 0 & 1 & 3 \end{vmatrix} \{C_3\}$$

$$= (-1)\begin{vmatrix} 1 & 2 & 3 & 1 \\ 10 & -1 & 1 & -2 \\ 8 & -3 & -4 & 0 \\ -1 & 1 & 1 & 3 \end{vmatrix} \begin{matrix} R_2 + 2R_1 \to R_2 \\ \\ R_4 - 3R_1 \to R_4 \end{matrix} = (-1)\begin{vmatrix} 1 & 2 & 3 & 1 \\ 12 & 3 & 7 & 0 \\ 8 & -3 & -4 & 0 \\ -4 & -5 & -8 & 0 \end{vmatrix} \{C_4\}$$

$$= (-1)(-1)\begin{vmatrix} 12 & 3 & 7 \\ 8 & -3 & -4 \\ -4 & -5 & -8 \end{vmatrix} \{4 \text{ is a common factor of } C_1 \text{ and } -1 \text{ of } R_3\}$$

$$= (-4)\begin{vmatrix} 3 & 3 & 7 \\ 2 & -3 & -4 \\ 1 & 5 & 8 \end{vmatrix} \begin{matrix} R_1 - 3R_3 \to R_1 \\ R_2 - 2R_3 \to R_2 \end{matrix} = (-4)\begin{vmatrix} 0 & -12 & -17 \\ 0 & -13 & -20 \\ 1 & 5 & 8 \end{vmatrix} \{C_1\}$$

$$= (-4)\begin{vmatrix} -12 & -17 \\ -13 & -20 \end{vmatrix} = (-4)(240 - 221) = -76$$

**45** $C_2$ and $C_4$ are identical, so $|A| = 0.$

**47** $\begin{vmatrix} 2-x & 3 \\ 1 & -4-x \end{vmatrix} = 0 \Rightarrow (2-x)(-4-x) - 3 = 0 \Rightarrow$

$x^2 + 2x - 11 = 0 \Rightarrow x = \dfrac{-2 \pm \sqrt{4+44}}{2} = \dfrac{-2 \pm 4\sqrt{3}}{2} = -1 \pm 2\sqrt{3}$

**49** 2 is a common factor of $R_1$, 2 is a common factor of $C_2$, and 3 is a common factor of $C_3$.

**51** This is an extension of Exercise 29 of §6.8. Expanding by $C_1$, only $a_{11}$ is not 0. Expanding by the new $C_1$ again, only $a_{22}$ is not 0. Repeating this process yields $|A| = a_{11}a_{22}a_{33}\cdots a_{nn}$, the product of the main diagonal elements.

[52] $\begin{vmatrix} 1 & a & b+c \\ 1 & b & a+c \\ 1 & c & a+b \end{vmatrix}$ $C_3 + C_2 \rightarrow C_2$

$= \begin{vmatrix} 1 & a & a+b+c \\ 1 & b & a+b+c \\ 1 & c & a+b+c \end{vmatrix}$ $C_3 - (a+b+c)\,C_1 \rightarrow C_3$

$= \begin{vmatrix} 1 & a & 0 \\ 1 & b & 0 \\ 1 & c & 0 \end{vmatrix} = 0$, since $C_3$ consists of all zeros.

[53] For the system $\begin{cases} 5x - 6y = 4 \\ 3x + 7y = 8 \end{cases}$, $|D| = \begin{vmatrix} 5 & -6 \\ 3 & 7 \end{vmatrix} = 35 + 18 = 53.$

Since $|D| = 53 \neq 0$, we may solve the system using Cramer's rule.

$|D_x| = \begin{vmatrix} 4 & -6 \\ 8 & 7 \end{vmatrix} = 28 + 48 = 76.$ $|D_y| = \begin{vmatrix} 5 & 4 \\ 3 & 8 \end{vmatrix} = 40 - 12 = 28.$

$x = \dfrac{|D_x|}{|D|} = \dfrac{76}{53}$ and $y = \dfrac{|D_y|}{|D|} = \dfrac{28}{53}.$ ★ $(\frac{76}{53}, \frac{28}{53})$

[55] $\dfrac{4x^2 + 54x + 134}{(x+3)(x^2+4x-5)} = \dfrac{A}{x+3} + \dfrac{B}{x+5} + \dfrac{C}{x-1}$

$4x^2 + 54x + 134 = A(x+5)(x-1) + B(x+3)(x-1) + C(x+3)(x+5)$

$x = 1$: $192 = 24C \Rightarrow C = 8$

$x = -3$: $8 = -8A \Rightarrow A = -1$ ★ $\dfrac{8}{x-1} - \dfrac{3}{x+5} - \dfrac{1}{x+3}$

$x = -5$: $-36 = 12B \Rightarrow B = -3$

[57] $\dfrac{x^2 + 14x - 13}{x^3 + 5x^2 + 4x + 20} = \dfrac{A}{x+5} + \dfrac{Bx + C}{x^2 + 4}$

$x^2 + 14x - 13 = A(x^2 + 4) + (Bx + C)(x + 5)$

$x = -5$: $-58 = 29A \Rightarrow A = -2$

$x = 0$: $-13 = 4A + 5C \Rightarrow C = -1$ ★ $-\dfrac{2}{x+5} + \dfrac{3x-1}{x^2+4}$

$x = 1$: $2 = 5A + 6B + 6C \Rightarrow B = 3$

[59] Let $x$ and $y$ denote the length and width, respectively, of the rectangle. A diagonal of the field is 100 ft. We can use the Pythagorean theorem to formulate an equation that relates the sides and a diagonal.

$\begin{cases} xy = 4000 & \textit{area} & (E_1) \\ x^2 + y^2 = 100^2 & \textit{diagonal} & (E_2) \end{cases}$

(continued)

Solve $E_1$ for $y$, $y = 4000/x$, and substitute into $E_2$.

$x^2 + \dfrac{4000^2}{x^2} = 100^2 \Rightarrow x^4 - 10{,}000x^2 + 16{,}000{,}000 = 0 \Rightarrow$

$(x^2 - 2000)(x^2 - 8000) = 0 \Rightarrow x = 20\sqrt{5}, 40\sqrt{5}$ and $y = 40\sqrt{5}, 20\sqrt{5}$.

The dimensions are $20\sqrt{5}$ ft $\times$ $40\sqrt{5}$ ft.

**61** Let $x$ and $y$ denote the total amount of taxes paid and bonus money, respectively.

$$\begin{cases} x = 0.40(50{,}000 - y) & \textit{taxes} \\ y = 0.10(50{,}000 - x) & \textit{bonuses} \end{cases} \Rightarrow \begin{cases} 10x + 4y = 200{,}000 & (E_1) \\ x + 10y = 50{,}000 & (E_2) \end{cases}$$

$E_1 - 10\,E_2 \Rightarrow -96y = -300{,}000 \Rightarrow y = \$3{,}125;\ x = \$18{,}750$

**63** Let $x$, $y$, and $z$ denote the number of $\text{ft}^3/\text{hr}$ flowing through pipes A, B, and C, respectively.

$$\begin{cases} 10x + 10y + 10z = 1000 & \textit{all 3 working} \\ 20x + 20y = 1000 & \text{A } \textit{and} \text{ B } \textit{only} \\ 12.5x + 12.5z = 1000 & \text{A } \textit{and} \text{ C } \textit{only} \end{cases} \Rightarrow \begin{cases} x + y + z = 100 & (E_1) \\ x + y = 50 & (E_2) \\ x + z = 80 & (E_3) \end{cases}$$

$E_1 - E_2 \Rightarrow z = 50;\ E_1 - E_3 \Rightarrow y = 20$; from $E_2$, $x = 30$.

**65** If $x$ and $y$ denote the length and the width, respectively, then a system is $x \le 12$, $y \le 8$, $y \ge \frac{1}{2}x$. The graph is the region bounded by the quadrilateral with vertices (0, 0), (0, 8), (12, 8), and (12, 6).

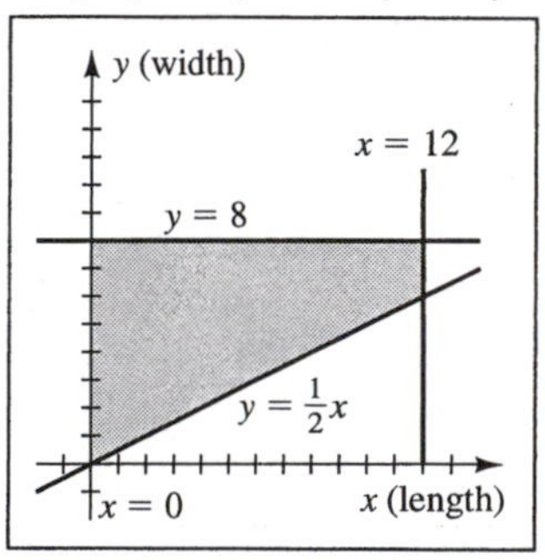


Figure 65

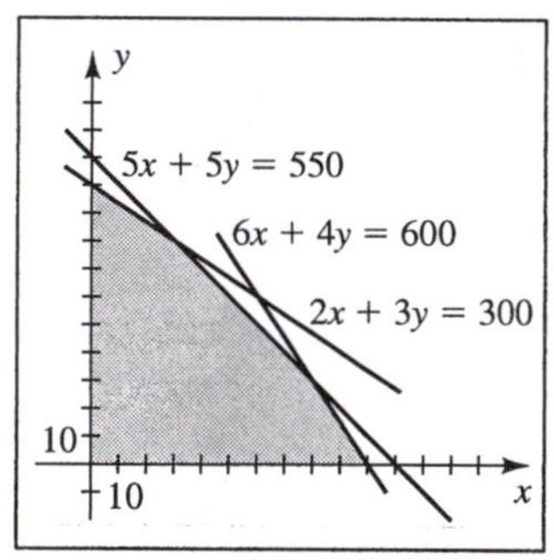


Figure 67

**67** Let $x$ and $y$ denote the number of lawn mowers and edgers produced, respectively.

Profit function: $P = 100x + 80y$

$$\begin{cases} 6x + 4y \le 600 & \textit{machining} \\ 2x + 3y \le 300 & \textit{welding} \\ 5x + 5y \le 550 & \textit{assembly} \\ x,\ y \ge 0 \end{cases}$$

| $(x, y)$ | (100, 0) | (80, 30) | (30, 80) | (0, 100) | (0, 0) |
|---|---|---|---|---|---|
| $P$ | 10,000 | 10,400 ■ | 9400 | 8000 | 0 |

The maximum weekly profit of \$10,400 occurs when 80 lawn mowers and 30 edgers are produced.

## Chapter 6 Discussion Exercises

[1] (a) For $b = 1.99$, we get $x = 204$ and $y = -100$. For $b = 1.999$, we get $x = 2004$ and $y = -1000$.

(b) Solving $\left\{\begin{array}{rl} x + 2y = & 4 \\ x + by = & 5 \end{array}\right.$ for $y$ gives us $y = \dfrac{1}{b-2}$ and then solving for $x$ we obtain $x = \dfrac{4b-10}{b-2}$. Both $x$ and $y$ are rational functions of $b$ and as $b \to 2^-$, $x \to \infty$ and $y \to -\infty$.

(c) If $b$ gets very large, $(x, y)$ gets close to $(4, 0)$.

[3] If we let $A$ be an $m \times n$ matrix ($m \neq n$), then $B$ would have to be an $n \times m$ matrix so that $AB$ and $BA$ are both defined. But then $AB = I_m$ and $BA = I_n$, different orders of the identity matrix, which we can't have (they must be the same).

[5] Synthetically dividing $x^4 + ax^2 + bx + c$ by $x + 1$, $x - 2$, and $x - 3$ yields the remainders $a - b + c + 1$, $4a + 2b + c + 16$, and $9a + 3b + c + 81$, respectively. Setting each of the remainders equal to 0 gives us the following system of equations in matrix form.

$$AX = B \quad \Leftrightarrow \quad \begin{bmatrix} 1 & -1 & 1 \\ 4 & 2 & 1 \\ 9 & 3 & 1 \end{bmatrix} \begin{bmatrix} a \\ b \\ c \end{bmatrix} = \begin{bmatrix} -1 \\ -16 \\ -81 \end{bmatrix} \quad \Rightarrow \quad X = A^{-1}B = \begin{bmatrix} -15 \\ 10 \\ 24 \end{bmatrix}.$$

Hence, $a = -15$, $b = 10$, $c = 24$, and we graph $Y_1 = x^4 - 15x^2 + 10x + 24$.

The roots of $Y_1$ are $-1$, 2, 3, and $-4$.

7 Using $a_{11}x + a_{12}y = k_1$ and the value of $y$ that was found, we substitute for $y$ and solve for $x$. Let $|A| = \begin{vmatrix} a_{11} & a_{12} \\ a_{21} & a_{22} \end{vmatrix}$ and $|K| = \begin{vmatrix} a_{11} & k_1 \\ a_{21} & k_2 \end{vmatrix}$.

$$a_{11}x + a_{12}y = k_1 \quad \Rightarrow \quad a_{11}x + a_{12} \cdot \frac{|K|}{|A|} = k_1 \quad \Rightarrow$$

$$x = \frac{k_1 - a_{12} \cdot \dfrac{|K|}{|A|}}{a_{11}}$$

$$= \frac{|A| \cdot k_1 - a_{12} \cdot |K|}{a_{11} \cdot |A|}$$

$$= \frac{(a_{11}a_{22} - a_{12}a_{21})k_1 - a_{12}(a_{11}k_2 - k_1a_{21})}{a_{11} \cdot |A|}$$

$$= \frac{a_{11}a_{22}k_1 - a_{12}a_{21}k_1 - a_{12}a_{11}k_2 - a_{12}a_{21}k_1}{a_{11} \cdot |A|}$$

$$= \frac{a_{11}a_{22}k_1 - a_{12}a_{11}k_2}{a_{11} \cdot |A|}$$

$$= \frac{a_{11}(a_{22}k_1 - a_{12}k_2)}{a_{11} \cdot |A|}$$

$$= \frac{\begin{vmatrix} k_1 & a_{12} \\ k_2 & a_{22} \end{vmatrix}}{|A|},$$ which is what we wanted to show.

# Chapter 7: Sequences, Series, and Probability

## 7.1 Exercises

*Note:* For Exercises 1–16, the answers are listed in the order $a_1$, $a_2$, $a_3$, $a_4$; and $a_8$. Simply substitute 1, 2, 3, 4, and 8 for $n$ in the formula for $a_n$ to obtain the results.

[1] $a_n = 12 - 3n$ • $a_1 = 12 - 3(1) = 9$,
$a_2 = 12 - 3(2) = 6$, $a_3 = 12 - 3(3) = 3$,
$a_4 = 12 - 3(4) = 0$; $a_8 = 12 - 3(8) = -12$ ★ 9, 6, 3, 0; −12

[3] $a_n = \dfrac{3n-2}{n^2+1}$ • $a_1 = \dfrac{3(1)-2}{(1)^2+1} = \frac{1}{2}$, $a_2 = \dfrac{3(2)-2}{(2)^2+1} = \frac{4}{5}$,
$a_3 = \dfrac{3(3)-2}{(3)^2+1} = \frac{7}{10}$, $a_4 = \dfrac{3(4)-2}{(4)^2+1} = \frac{10}{17}$; $a_8 = \dfrac{3(8)-2}{(8)^2+1} = \frac{22}{65}$ ★ $\frac{1}{2}, \frac{4}{5}, \frac{7}{10}, \frac{10}{17}; \frac{22}{65}$

[5] $a_n = 9$ • $a_{(\text{any allowable value})} = 9$ ★ 9, 9, 9, 9; 9

[7] $a_n = 2 + (-0.1)^n$ • $a_1 = 2 + (-0.1)^1 = 2 - 0.1 = 1.9$,
$a_2 = 2 + (-0.1)^2 = 2 + 0.01 = 2.01$,
$a_3 = 2 + (-0.1)^3 = 2 - 0.001 = 1.999$,
$a_4 = 2 + (-0.1)^4 = 2 + 0.0001 = 2.0001$;
$a_8 = 2 + (-0.1)^8 = 2 + 0.00000001 = 2.00000001$
★ 1.9, 2.01, 1.999, 2.0001; 2.00000001

[9] $a_n = (-1)^{n-1}\left(\dfrac{n+7}{2n}\right)$ •

$a_1 = (-1)^{1-1}\left(\dfrac{1+7}{2(1)}\right) = (-1)^0\left(\frac{8}{2}\right) = (1)(4) = 4$,

$a_2 = (-1)^{2-1}\left(\dfrac{2+7}{2(2)}\right) = (-1)^1\left(\frac{9}{4}\right) = -\frac{9}{4}$,

$a_3 = (-1)^{3-1}\left(\dfrac{3+7}{2(3)}\right) = (-1)^2\left(\frac{10}{6}\right) = \frac{5}{3}$,

$a_4 = (-1)^{4-1}\left(\dfrac{4+7}{2(4)}\right) = (-1)^3\left(\frac{11}{8}\right) = -\frac{11}{8}$;

$a_8 = (-1)^{8-1}\left(\dfrac{8+7}{2(8)}\right) = (-1)^7\left(\frac{15}{16}\right) = -\frac{15}{16}$ ★ $4, -\frac{9}{4}, \frac{5}{3}, -\frac{11}{8}; -\frac{15}{16}$

[11] $a_n = 1 + (-1)^{n+1}$ • $a_1 = 1 + (-1)^{1+1} = 1 + (-1)^2 = 1 + 1 = 2$,
$a_2 = 1 + (-1)^{2+1} = 1 + (-1)^3 = 1 - 1 = 0$,
$a_3 = 1 + (-1)^{3+1} = 1 + (-1)^4 = 1 + 1 = 2$,
$a_4 = 1 + (-1)^{4+1} = 1 + (-1)^5 = 1 - 1 = 0$;
$a_8 = 1 + (-1)^{8+1} = 1 + (-1)^9 = 1 - 1 = 0$
★ 2, 0, 2, 0; 0

[13] $a_n = \dfrac{2^n}{n^2+2}$ • $a_1 = \dfrac{2^1}{1^2+2} = \dfrac{2}{3}$, $a_2 = \dfrac{2^2}{2^2+2} = \dfrac{4}{6} = \dfrac{2}{3}$, $a_3 = \dfrac{2^3}{3^2+2} = \dfrac{8}{11}$,

$a_4 = \dfrac{2^4}{4^2+2} = \dfrac{16}{18} = \dfrac{8}{9}$; $a_8 = \dfrac{2^8}{8^2+2} = \dfrac{256}{66} = \dfrac{128}{33}$ ★ $\dfrac{2}{3}, \dfrac{2}{3}, \dfrac{8}{11}, \dfrac{8}{9}; \dfrac{128}{33}$

[15] $a_n$ is the number of decimal places in $(0.1)^n$. •

$(0.1)^1 = 0.1$ {1 decimal place} $\Rightarrow$ $a_1 = 1$,

$(0.1)^2 = 0.01$ {2 decimal places} $\Rightarrow$ $a_2 = 2$,

$(0.1)^3 = 0.001$ {3 decimal places} $\Rightarrow$ $a_3 = 3$,

$(0.1)^4 = 0.0001$ {4 decimal places} $\Rightarrow$ $a_4 = 4$;

$(0.1)^8 = 0.00000001$ {8 decimal places} $\Rightarrow$ $a_8 = 8$ ★ 1, 2, 3, 4; 8

[17] $\left\{\dfrac{1}{\sqrt{n}}\right\} = \dfrac{1}{\sqrt{1}}, \dfrac{1}{\sqrt{2}}, \dfrac{1}{\sqrt{3}}, \dfrac{1}{\sqrt{4}}, \dfrac{1}{\sqrt{5}}, \ldots \approx 1, 0.71, 0.58, 0.5, 0.45, \ldots$

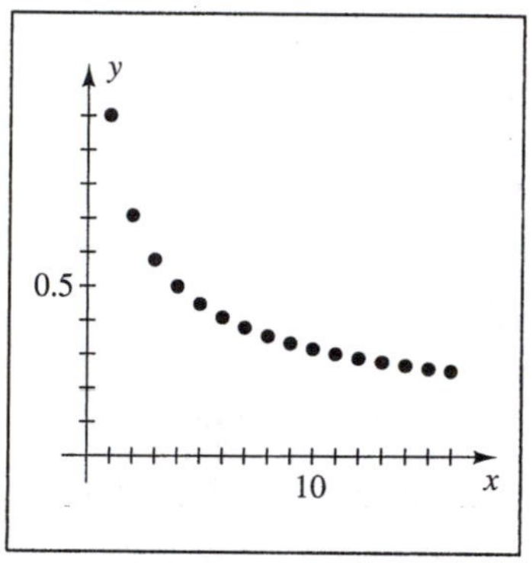


*Figure 17*

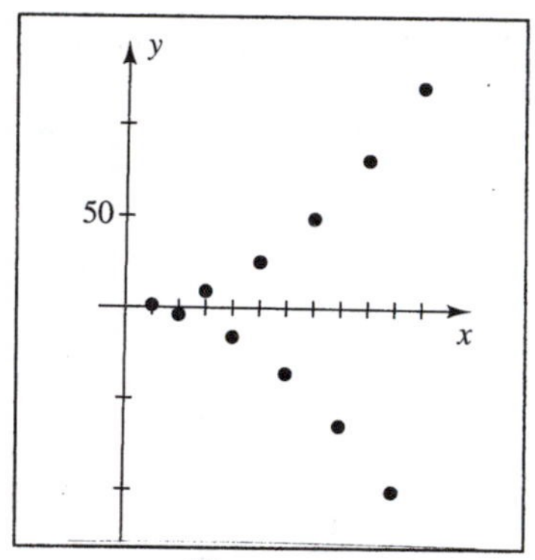


*Figure 19*

[19] $\{(-1)^{n+1}n^2\} = 1\cdot 1^2, -1\cdot 2^2, 1\cdot 3^2, -1\cdot 4^2, \ldots = 1, -4, 9, -16, \ldots$

[21] $a_1 = 2$, $a_{k+1} = 3a_k - 5$ •

$a_2 = a_{1+1}$ $\{k=1\}$ $= 3a_1 - 5 = 3(2) - 5 = 1$, $a_3 = 3a_2 - 5 = 3(1) - 5 = -2$,

$a_4 = 3a_3 - 5 = 3(-2) - 5 = -11$, $a_5 = 3a_4 - 5 = 3(-11) - 5 = -38$

[23] $a_1 = -3$, $a_{k+1} = a_k^2$ •

$a_2 = a_{1+1}$ $\{k=1\}$ $= a_1^2 = (-3)^2 = 3^2 = 9$, $a_3 = a_2^2 = (3^2)^2 = 3^4$,

$a_4 = a_3^2 = (3^4)^2 = 3^8$, $a_5 = a_4^2 = (3^8)^2 = 3^{16}$

[25] $a_1 = 5$, $a_{k+1} = ka_k$ •

$a_2 = a_{1+1}$ $\{k=1\}$ $= 1\,a_1 = 1(5) = 5$, $a_3 = 2\,a_2 = 2(5) = 10$,

$a_4 = 3\,a_3 = 3(10) = 30$, $a_5 = 4\,a_4 = 4(30) = 120$

[27] $a_1 = 2$, $a_{k+1} = (a_k)^k$ •

$a_2 = a_{1+1}$ $\{k=1\}$ $= (a_1)^1 = (2)^1 = 2$, $a_3 = (a_2)^2 = (2)^2 = 4$,

$a_4 = (a_3)^3 = (4)^3 = 4^3$ or 64, $a_5 = (a_4)^4 = (4^3)^4 = 4^{12}$ or 16,777,216

[29] $a_n = 3 + \frac{1}{2}n$ • $S_1 = a_1 = 3 + \frac{1}{2} = \frac{7}{2}$. $S_2 = S_1 + a_2 = \frac{7}{2} + 4 = \frac{15}{2}$. Alternatively, to find $S_2$ we could use $a_1 + a_2$. However, in most cases, it is easier to compute $S_{n+1}$ by using $S_n + a_{n+1}$. $S_3 = S_2 + a_3 = \frac{15}{2} + \frac{9}{2} = 12$. $S_4 = S_3 + a_4 = 12 + 5 = 17$.

[31] $\{(-1)^n n^{-1/2}\} = \left\{(-1)^n \frac{1}{\sqrt{n}}\right\}$ • $S_1 = a_1 = -1$.

$S_2 = S_1 + a_2 = -1 + \frac{1}{\sqrt{2}}$. $S_3 = S_2 + a_3 = -1 + \frac{1}{\sqrt{2}} - \frac{1}{\sqrt{3}}$.

$S_4 = S_3 + a_4 = -1 + \frac{1}{\sqrt{2}} - \frac{1}{\sqrt{3}} + \frac{1}{2} = -\frac{1}{2} + \frac{1}{\sqrt{2}} - \frac{1}{\sqrt{3}}$.

[33] Substitute $k = 1$, 2, 3, 4, and 5 into $2k - 7$ and then add the results.

$$\sum_{k=1}^{5} (2k-7) = (-5) + (-3) + (-1) + 1 + 3 = -5$$

[35] $\sum_{k=1}^{4} (k^2 - 5) = (1^2 - 5) + (2^2 - 5) + (3^2 - 5) + (4^2 - 5) = (-4) + (-1) + 4 + 11 = 10$

[37] $$\begin{aligned}\sum_{k=0}^{5} k(k-2) &= 0(-2) + 1(-1) + 2(0) + 3(1) + 4(2) + 5(3) \\ &= 0 + (-1) + 0 + 3 + 8 + 15 = 25\end{aligned}$$

[39] $\sum_{k=3}^{6} \frac{k-5}{k-1} = \frac{-2}{2} + \frac{-1}{3} + \frac{0}{4} + \frac{1}{5} = -\frac{15}{15} - \frac{5}{15} + \frac{3}{15} = -\frac{17}{15}$

[41] $$\begin{aligned}\sum_{k=1}^{5} (-3)^{k-1} &= (-3)^0 + (-3)^1 + (-3)^2 + (-3)^3 + (-3)^4 \\ &= 1 + (-3) + 9 + (-27) + 81 = 61\end{aligned}$$

[43] Using part (1) of the theorem on the sum of a constant, we have

$$\sum_{k=1}^{100} 100 = 100(100) = 10{,}000.$$

[45] Using part (2) of the theorem on the sum of a constant, we have

$$\sum_{k=253}^{571} \tfrac{1}{3} = (571 - 253 + 1)(\tfrac{1}{3}) = 319(\tfrac{1}{3}) = \tfrac{319}{3}.$$

[47] Note that $j$, not $k$, is the summation variable. Hence, we again use part (1) of the theorem on the sum of a constant, obtaining $\sum_{j=1}^{7} \frac{1}{2}k^2 = 7(\frac{1}{2}k^2) = \frac{7}{2}k^2$.

[49] $$\begin{aligned}\sum_{k=1}^{n} (a_k - b_k) &= (a_1 - b_1) + (a_2 - b_2) + \cdots + (a_n - b_n) \\ &= (a_1 + a_2 + \cdots + a_n) + (-b_1 - b_2 - \cdots - b_n) \\ &= (a_1 + a_2 + \cdots + a_n) - (b_1 + b_2 + \cdots + b_n) \\ &= \sum_{k=1}^{n} a_k - \sum_{k=1}^{n} b_k\end{aligned}$$

[51] $a_1 = 5$, $a_2 = 5^{1/2}$, $a_3 = 5^{1/4}$, $a_4 = 5^{1/8}$, ....

As $k$ increases, the exponent gets closer to zero and the terms approach 1.

53 $a_1 = 0.4,\ a_k = 0.1(3 \cdot 2^{k-2} + 4) \ \Rightarrow$

$a_2 = 0.1(3 \cdot 2^{2-2} + 4) = 0.1(3 \cdot 2^0 + 4) = 0.1(3 \cdot 1 + 4) = 0.1(7) = 0.7$

$a_3 = 0.1(3 \cdot 2^{3-2} + 4) = 0.1(3 \cdot 2^1 + 4) = 0.1(3 \cdot 2 + 4) = 0.1(10) = 1$

$a_4 = 0.1(3 \cdot 2^{4-2} + 4) = 0.1(3 \cdot 2^2 + 4) = 0.1(3 \cdot 4 + 4) = 0.1(16) = 1.6$

$a_5 = 0.1(3 \cdot 2^{5-2} + 4) = 0.1(3 \cdot 2^3 + 4) = 0.1(3 \cdot 8 + 4) = 0.1(28) = 2.8$

55 (a) $a_1 = 1,\ a_2 = 1,\ a_{k+1} = a_k + a_{k-1}$ for $k \geq 2$. So

$a_3 = a_2 + a_1 = 1 + 1 = 2,$

$a_4 = a_3 + a_2 = 2 + 1 = 3,$

$a_5 = a_4 + a_3 = 3 + 2 = 5,$

$a_6 = a_5 + a_4 = 5 + 3 = 8,$

$a_7 = a_6 + a_5 = 8 + 5 = 13,$

$a_8 = a_7 + a_6 = 13 + 8 = 21,$

$a_9 = a_8 + a_7 = 21 + 13 = 34$, and

$a_{10} = a_9 + a_8 = 34 + 21 = 55.$

The following screens show the assignment needed to generate the Fibonacci sequence on the TI-83 Plus and a listing of the first 14 Fibonacci numbers.

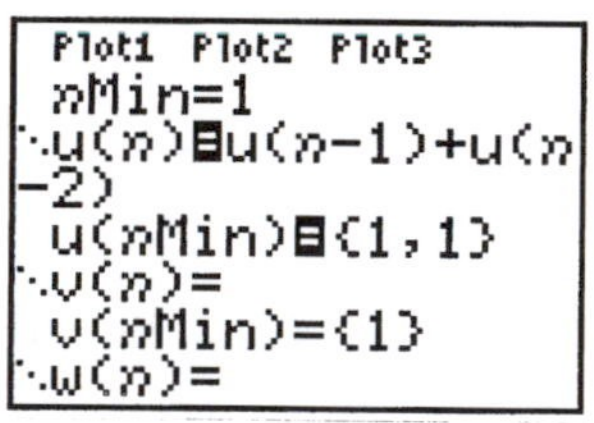


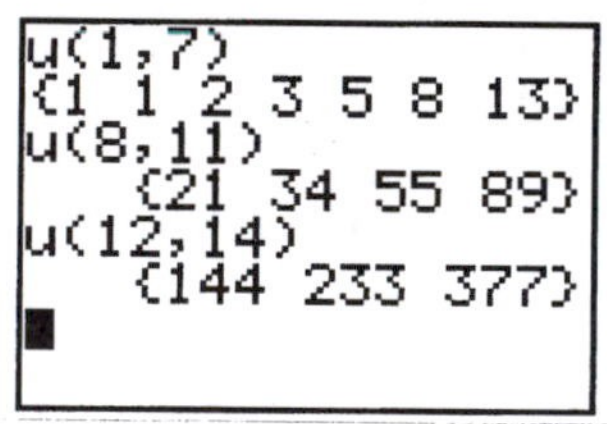


(b) $r_1 = \frac{1}{1} = 1,$ $r_2 = \frac{2}{1} = 2,$ $r_3 = \frac{3}{2} = 1.5,$

$r_4 = \frac{5}{3} = 1.\overline{6},$ $r_5 = \frac{8}{5} = 1.6,$ $r_6 = \frac{13}{8} = 1.625,$

$r_7 = \frac{21}{13} \approx 1.6153846,$ $r_8 = \frac{34}{21} \approx 1.6190476,$ $r_9 = \frac{55}{34} \approx 1.6176471,$

and $r_{10} = \frac{89}{55} \approx 1.6181818.$

57 (a) Since the amount of chlorine decreases by a factor of 0.20 each day {retain 80%}, $a_n = 0.8a_{n-1}$, where $a_0$ is the initial amount of chlorine in the pool.

(b) Let $a_0 = 7$ and $a_n = 0.8a_{n-1}$.

| Day ($n$) | 0 | 1 | 2 | 3 | 4 | 5 |
|---|---|---|---|---|---|---|
| Chlorine ($a_n$) | 7.00 | 5.60 | 4.48 | 3.58 | 2.87 | 2.29 |

The chlorine level will drop below 3 ppm during the fourth day.

**59** If $1 \leq n \leq 4$, then $C(n) = 89.95n$. If $5 \leq n \leq 9$, then $C(n) = 87.95n$—there are no changes in prices, that is, you don't pay \$89.95 for the first 4 and then \$87.95 for the next one. If $n \geq 10$, then $C(n) = 85.95n$. Summarizing in a piecewise-defined function gives us

$$C(n) = \begin{cases} 89.95n & \text{if } 1 \leq n \leq 4 \\ 87.95n & \text{if } 5 \leq n \leq 9 \\ 85.95n & \text{if } n \geq 10 \end{cases}$$

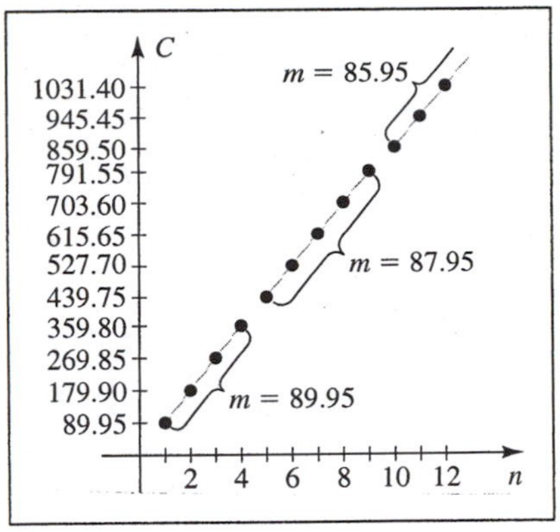


*Figure 59*

**61** $N = 5$ and $x_1 = \frac{5}{2} \Rightarrow x_2 = \frac{1}{2}\left(x_1 + \frac{N}{x_1}\right) = \frac{1}{2}\left(2.5 + \frac{5}{2.5}\right) = 2.25 \Rightarrow$

$x_3 = \frac{1}{2}\left(2.25 + \frac{5}{2.25}\right) \approx 2.236111 \Rightarrow x_4 \approx 2.236068 \Rightarrow x_5 \approx 2.236068.$

Thus, $\sqrt{5} \approx 2.236068$.

*Note:* TI-users—see the key sequence at the end of the solution of #65.

**63** $x_1 = 2$ and $x_2 = \frac{1}{3}\sqrt[3]{x_1} + 2 \Rightarrow x_2 \approx 2.419974 \Rightarrow x_3 \approx 2.447523 \Rightarrow$

$x_4 \approx 2.449215 \Rightarrow x_5 \approx 2.449319 \Rightarrow x_6 \approx 2.449325$. The root is approximately 2.4493. *Note:* TI-users—see the key sequence at the end of the solution of #65.

**65** (a) $f(1) = (1 - 2 + \log 1) = (1 - 2 + 0) = -1 < 0$ and

$f(2) = (2 - 2 + \log 2) = \log 2 \approx 0.30 > 0$.

Thus, $f$ assumes both positive and negative values on [1, 2].

(b) Solving $\underline{x - 2 + \log x = 0}$ for $x$ in terms of $x$ gives us $\underline{x = 2 - \log x}$.

$x_1 = 1.5 \Rightarrow x_2 = 2 - \log x_1 = 2 - \log 1.5 \approx 1.823909 \Rightarrow x_3 \approx 1.738997 \Rightarrow$

$x_4 \approx 1.759701 \Rightarrow x_5 \approx 1.754561 \Rightarrow x_6 \approx 1.755832 \Rightarrow x_7 \approx 1.755517.$

The zero is approximately 1.76. *Note:* If you are using a TI-8x, type 1.5 and press the ENTER key to store 1.5 in the memory location ANS. Now type 2 − LOG ANS and then successively press the ENTER key to obtain the approximations for $x_1$, $x_2$, ... (see the display below).

1.5 [ENTER] 2 [−] [LOG] [ANS] [ENTER] [ENTER] ...

Similar key sequences for **Exercises 61 and 63**, respectively, are:

2.5 [ENTER] .5 [(] [ANS] [+] 5 [÷] [ANS] [)] [ENTER] [ENTER] ...

2 [ENTER] [MATH] [4] [ANS] [÷] 3 [+] 2 [ENTER] [ENTER] ...

(continued)

Alternatively, we can use the memory location X instead of ANS by using the STO ▷ key as shown in *Figure 65.*

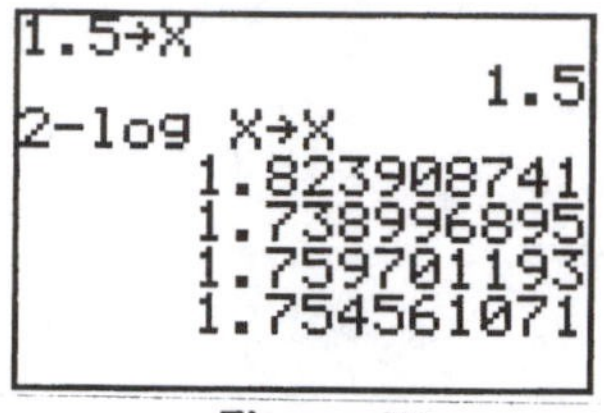


*Figure 65*

67 Graph $y = \left(1 + \frac{1}{x} + \frac{1}{2x^2}\right)^x$ on the interval [1, 100].

The graph approaches the horizontal asymptote $y \approx 2.718 \approx e$.

For increasing values of $n$, the terms of the sequence appear to approximate $e$.

[1, 100, 10] by [0, 3]

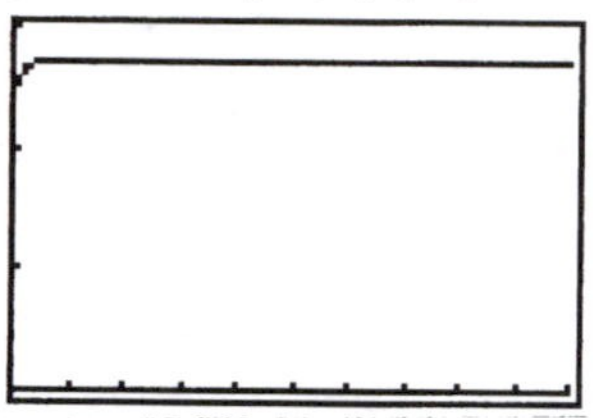

*Figure 67*

[1, 100, 10] by [0, 3]

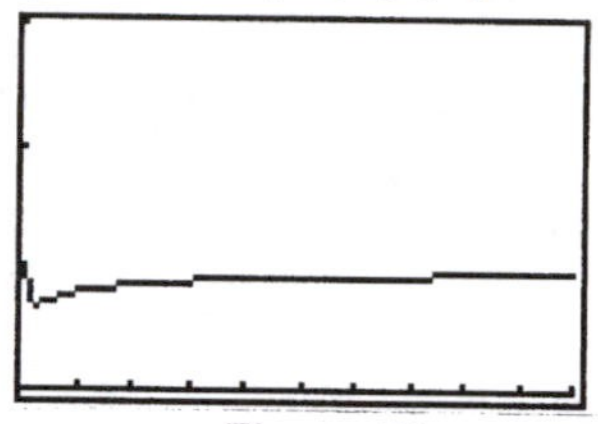

*Figure 69*

69 Graph $y = \left(\frac{1}{x}\right)^{1/x}$ on the interval [1, 100].

The graph approaches the horizontal asymptote $y \approx 1$ from below.

For increasing values of $n$, the terms of the sequence appear to approximate 1.

71 *Note:* **TI-82 users:** Under MODE, make sure you are in Seq and Dot mode.

Under Y = , assign 1.7U$n$ − 1 + .5 to U$n$ {U$n$ − 1 is 2nd 7}.

Under WINDOW, let U$n$Start = .25, $n$Start = 1, $n$Min = 1, $n$Max = 20, and the rest of the window parameters as shown in *Figure 71.*

**TI-83 users:** Under MODE, make sure you are in Seq mode.

Under Y = , let $n$Min = 1, u($n$Min) = .25, and u($n$) = 1.7u($n$ − 1) + .5 {u is 2nd 7 and $n$ is X,T,$\theta$,$n$}.

Under WINDOW, let $n$Min = 1, $n$Max = 20, PlotStart = 1, PlotStep = 1, and the rest of the window parameters as shown in the figure.

By tracing the graph we see that $a_9 \approx 66.55$ and $a_{10} \approx 113.64$. Thus, $k = 10$.

See *Figure 71* on the next page.

[0, 20, 5] by [0, 125, 25]

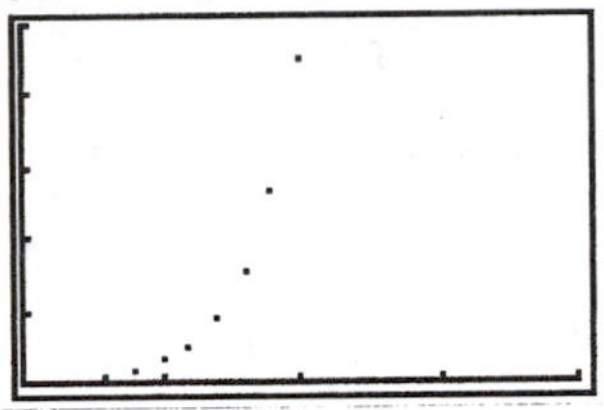

*Figure 71*

[0, 20, 5] by [0, 300, 50]

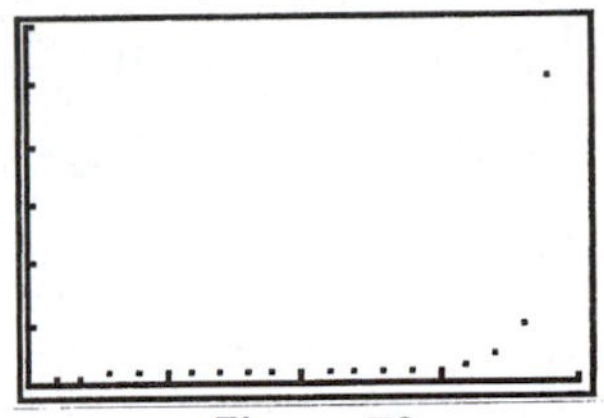

*Figure 73*

[73] By tracing the graph we see that $a_{18} \approx 50.39$ and $a_{19} \approx 255.96$. Thus, $k = 19$.

[75] (a) Since $c = 0.5$, let the sequence be defined by $a_{k+1} = 0.5a_k(1 - a_k)$. Then, $a_1 = 0.25$, $a_2 = 0.5a_1(1 - a_1) = 0.5(0.25)(1 - 0.25) = 0.09375$. In a similar manner, $a_3 \approx 0.04248$, $a_4 \approx 0.02034$, ..., $a_{10} \approx 0.00031$, $a_{11} \approx 0.00015$, $a_{12} \approx 0.00008$. The insect population is initially $1000a_1 = 1000(0.25) = 250$. It then becomes approximately 94, 42, 20, and so on, until the population decreases to zero.

(b) The sequence determined is $a_1 = 0.25$, $a_2 \approx 0.28125$, $a_3 \approx 0.30322$, $a_4 \approx 0.31692$, ..., $a_{18} \approx 0.33333$, $a_{19} \approx 0.33333$, $a_{20} \approx 0.33333$. The insect population is initially 250. It then becomes approximately 281, 303, 317, and so on, until the population stabilizes at 333.

(c) The sequence determined is $a_1 = 0.25$, $a_2 \approx 0.51563$, $a_3 \approx 0.68683$, $a_4 \approx 0.59151$, ..., $a_{40} \approx 0.63636$, $a_{41} \approx 0.63636$, $a_{42} \approx 0.63636$. The insect population is initially 250. It then becomes approximately 516, 687, 592, and so on, until the population stabilizes at 636.

## *7.2 Exercises*

[1] To show that the given sequence, $-6, -2, 2, \ldots, 4n - 10, \ldots$, is arithmetic, we must show that $a_{k+1} - a_k$ is equal to some constant, which is the common difference.

$a_n = 4n - 10 \Rightarrow$

$a_{k+1} - a_k = [4(k+1) - 10] - [4(k) - 10] = 4k + 4 - 10 - 4k + 10 = 4$

*Note:* For Exercises 3–10, we will find the $n$th term first, and then use that term to find $a_5$ and $a_{10}$.

[3] The common difference can be found by *subtracting* any term from its successor.

$d = 6 - 2 = 4$; $a_n = a_1 + (n - 1)d = 2 + (n - 1)(4) = 2 + 4n - 4 = 4n - 2$;

$a_5 = 4(5) - 2 = 18$; $a_{10} = 4(10) - 2 = 38$

[5] $d = 2.7 - 3 = -0.3$; $a_n = a_1 + (n - 1)d = 3 + (n - 1)(-0.3) = -0.3n + 3.3$;

$a_5 = -0.3(5) + 3.3 = 1.8$; $a_{10} = -0.3(10) + 3.3 = 0.3$

[7] $d = -3.9 - (-7) = 3.1$; $a_n = a_1 + (n-1)d = -7 + (n-1)(3.1) = 3.1n - 10.1$;

$a_5 = 3.1(5) - 10.1 = 5.4$; $a_{10} = 3.1(10) - 10.1 = 20.9$

[9] An equivalent sequence is $\ln 3$, $\ln 3^2$, $\ln 3^3$, $\ln 3^4$, ..., which is also equivalent to the sequence $\ln 3$, $2\ln 3$, $3\ln 3$, $4\ln 3$, ...; $d = 2\ln 3 - \ln 3 = \ln 3$;

$a_n = \ln 3 + (n-1)(\ln 3) = n\ln 3$ or $\ln 3^n$; $a_5 = 5\ln 3$ or $\ln 3^5$; $a_{10} = 10\ln 3$ or $\ln 3^{10}$

[11] $a_6 = a_1 + 5d$ and $a_2 = a_1 + d \Rightarrow a_6 - a_2 = (a_1 + 5d) - (a_1 + d) = 4d$.

But $a_6 - a_2 = -11 - 21 = -32$. Hence, $4d = -32$ and $d = -8$.

[13] Given $a_1$ and $a_2$, we can find the difference $d$. $d = a_2 - a_1 = 7.5 - 9.1 = -1.6$.

The twelfth term is equal to the first term plus eleven differences.

$a_{12} = 9.1 + (11)(-1.6) = -8.5$.

[15] $d = a_7 - a_6 = 5.2 - 2.7 = 2.5$; $a_6 = a_1 + 5d \Rightarrow 2.7 = a_1 + 5(2.5) \Rightarrow a_1 = -9.8$

[17] $a_3 = a_1 + 2d$ and $a_{20} = a_1 + 19d \Rightarrow a_{20} - a_3 = (a_1 + 19d) - (a_1 + 2d) = 17d$.

$a_3 = 7$ and $a_{20} = 43 \Rightarrow 17d = 43 - 7 \Rightarrow d = \frac{36}{17}$.

$a_{15} = a_3 + 12d = 7 + 12(\frac{36}{17}) = \frac{551}{17}$.

*Note:* To find the sums in Exercises 19–28, we use the sum formulas

$$\textbf{(1)}\ S_n = \tfrac{n}{2}[2a_1 + (n-1)d] \quad \text{and} \quad \textbf{(2)}\ S_n = \tfrac{n}{2}(a_1 + a_n).$$

[19] $a_1 = 40$, $d = -3$, $n = 30 \Rightarrow S_{30} = \frac{30}{2}[2(40) + (29)(-3)] = -105$.

[21] $a_1 = -9$, $a_{10} = 15$, $n = 10 \Rightarrow S_{10} = \frac{10}{2}(-9 + 15) = 30$.

[23] $\sum_{k=1}^{20} (3k - 5)$ • Using the second formula for $S_n$ with $n = 20$, $a_1 = 3(1) - 5 = -2$,

and $a_{20} = 3(20) - 5 = 55$, we obtain $S_{20} = \frac{20}{2}(-2 + 55) = 530$.

[25] $\sum_{k=1}^{18} (\frac{1}{2}k + 7)$ • $a_1 = \frac{15}{2}$, $a_{18} = 16$, $n = 18 \Rightarrow S_{18} = \frac{18}{2}(\frac{15}{2} + 16) = \frac{423}{2}$.

[27] $$\sum_{k=126}^{592} (5k + 2j) = \sum_{k=126}^{592} 5k + \sum_{k=126}^{592} 2j$$
$$= 5\sum_{k=126}^{592} k + 2\sum_{k=126}^{592} j = 5(\text{SUM1}) + 2(\text{SUM2}).$$

The number of terms is $n = 592 - 126 + 1 = 467$.

To evaluate SUM1, we use **(1)** with $a_1 = 126$, $d = 1$, and $n = 467$ to get:

$$S_{467} = \tfrac{467}{2}[2(126) + (466)(1)] = \tfrac{467}{2}[718] = 167{,}653.$$

SUM2 is just $467 \cdot j$ because $j$ is a constant ($k$ is the index).

Thus, $5(\text{SUM1}) + 2(\text{SUM2}) = 5(167{,}653) + 2(467j) = 838{,}265 + 934j$.

[29] $4 + 11 + 18 + 25 + 32$ • The number of terms is $n = 5$.

The difference in terms is $d = 11 - 4 = 7$.

The general term is $a_n = a_1 + (n-1)d = 4 + (n-1)7 = 4 + 7n - 7 = 7n - 3$.

We can represent the sum in terms of summation notation as $\sum_{n=1}^{5} a_n = \sum_{n=1}^{5} (7n-3)$.

Sometimes it is easier to use $\sum_{n=0}^{4} a_{n+1}$ than $\sum_{n=1}^{5} a_n$.

In this case, $a_{n+1} = a_1 + nd = 4 + n \cdot 7$,

so an alternate form of the summation notation is $\sum_{n=0}^{4} (4 + 7n)$.

[31] $4 + 11 + 18 + \cdots + 466$ • From Exercise 29, the general term is $7n - 3$ with $n$ starting at 1. To find the largest value of $n$, set the general term equal to the largest value of the sum and solve for $n$. $7n - 3 = 466 \Rightarrow 7n = 469 \Rightarrow n = 67$.

$$\sum_{n=1}^{67} (7n-3) \quad \text{or} \quad \sum_{n=0}^{66} (4+7n)$$

[33] $\frac{3}{7} + \frac{6}{11} + \frac{9}{15} + \frac{12}{19} + \frac{15}{23} + \frac{18}{27}$. The numerators increase by 3, the denominators increase by 4. The general terms are $3 + (n-1)3 = 3n$ and $7 + (n-1)4 = 4n + 3$. $\sum_{n=1}^{6} \frac{3n}{4n+3}$

[35] $8 + 19 + 30 + \cdots + 16{,}805$ • $d = 19 - 8 = 11$ and $a_n = a_1 + (n-1)d = 8 + (n-1)11 = 11n - 3$. Find the largest value of $n$: $11n - 3 = 16{,}805 \Rightarrow 11n = 16{,}808 \Rightarrow n = 1528$. Thus, the summation notation is $\sum_{n=1}^{1528} (11n - 3)$.

Now find the sum: $S_{1528} = \frac{1528}{2}(8 + 16{,}805) = 12{,}845{,}132$.

[37] $a_1 = -2$, $d = \frac{1}{4}$, $S = 21$, and $S_n = \frac{n}{2}[2a_1 + (n-1)(d)] \Rightarrow$

$21 = \frac{n}{2}\left[2(-2) + (n-1)(\frac{1}{4})\right] \Rightarrow 42 = n(\frac{1}{4}n - \frac{17}{4}) \Rightarrow 168 = n^2 - 17n \Rightarrow$

$n^2 - 17n - 168 = 0 \Rightarrow (n-24)(n+7) = 0 \Rightarrow n = 24, -7$.

Since $n$ can not be negative, $n = 24$.

[39] $S_n = \frac{n}{2}[2a_1 + (n-1)(d)] \Rightarrow -36 = \frac{n}{2}\left[2(-\frac{29}{6}) + (n-1)(\frac{1}{3})\right] \Rightarrow$

$-72 = n(-\frac{29}{3} + \frac{1}{3}n - \frac{1}{3}) \Rightarrow -216 = n(n-30) \Rightarrow$

$n^2 - 30n + 216 = 0 \Rightarrow (n-12)(n-18) = 0 \Rightarrow n = 12, 18$.

There are two sequences that satisfy the given conditions.

[41] If we insert five arithmetic means between 2 and 10, there will be 6 differences that span the distance from 2 to 10. Hence, $6d = 10 - 2 \Rightarrow d = \frac{4}{3}$;

The terms are 2, $\frac{10}{3}$, $\frac{14}{3}$, 6, $\frac{22}{3}$, $\frac{26}{3}$, 10.

[43] (a) The first integer greater than 32 that is divisible by 6 is 36 $\{6 \cdot 6\}$ and the last integer less than 395 that is divisible by 6 is 390 $\{65 \cdot 6\}$.

The number of terms is then $65 - 6 + 1 = 60$.

(b) The sum is $S_{60} = \frac{60}{2}(36 + 390) = 12{,}780$.

[45] There are $(24 - 10 + 1) = 15$ layers. Model this problem as an arithmetic sequence with $a_1 = 10$ and $a_{15} = 24$. $S_{15} = \frac{15}{2}(10 + 24) = 255$.

[47] This is similar to inserting 9 arithmetic means between 4 and 24.

$10d = 20 \Rightarrow d = 2$. The circumference of each ring is $\pi D$ with $D = 4, 6, 8, \ldots, 24$.

$$a_1 = 4\pi \text{ and } a_{11} = 24\pi \Rightarrow S_{11} = \frac{11}{2}(4\pi + 24\pi) = 154\pi \text{ ft.}$$

[49] $n = 5$, $S_5 = 5000$, $d = -100 \Rightarrow$

$$5000 = \frac{5}{2}[2a_1 + 4(-100)] \Rightarrow 2000 = 2a_1 - 400 \Rightarrow a_1 = \$1200$$

[51] The sequence 16, 48, 80, 112, ..., is an arithmetic sequence with $a_1 = 16$ and $d = 48 - 16 = 32$. The total distance traveled in $n$ seconds is

$$a_1 + a_2 + \cdots + a_n = \frac{n}{2}[2a_1 + (n-1)d] = \frac{n}{2}[2(16) + (n-1)(32)] = \frac{n}{2}(32n) = 16n^2.$$

[53] If the $n$th term is $\frac{1}{x_n}$ and $x_{n+1} = \frac{x_n}{1 + x_n}$, then the $(n+1)$st term is

$$\frac{1}{x_{n+1}} = \frac{1}{\dfrac{x_n}{1 + x_n}} = \frac{1 + x_n}{x_n} = \frac{1}{x_n} + \frac{x_n}{x_n} = 1 + \frac{1}{x_n},$$

which is 1 greater than the $n$th term and therefore the sequence is arithmetic.

[55] (a) $T_8 = 1 + 2 + \cdots + 8 = 36$. $A_k = \dfrac{n - k + 1}{T_n} \Rightarrow A_1 = \dfrac{8 - 1 + 1}{36} = \dfrac{8}{36}$.

$$A_2 = \tfrac{7}{36},\ A_3 = \tfrac{6}{36},\ A_4 = \tfrac{5}{36},\ A_5 = \tfrac{4}{36},\ A_6 = \tfrac{3}{36},\ A_7 = \tfrac{2}{36},\ A_8 = \tfrac{1}{36}.$$

(b) $d = A_{k+1} - A_k = -\frac{1}{36}$ for $k = 1, 2, \ldots, 7$. $S_8 = \sum_{k=1}^{8} A_k = \frac{8}{36} + \frac{7}{36} + \cdots + \frac{1}{36} = 1$.

(c) $\$1000\left(\frac{8}{36} + \frac{7}{36} + \frac{6}{36} + \frac{5}{36}\right) \approx \$722.22$

## 7.3 Exercises

[1] To show that the given sequence, $5, -\frac{5}{4}, \frac{5}{16}, \ldots, 5(-\frac{1}{4})^{n-1}, \ldots$, is geometric, we must show that $\dfrac{a_{k+1}}{a_k}$ is equal to some constant, which is the common ratio.

$$a_n = a_1 r^{n-1} = 5(-\tfrac{1}{4})^{n-1} \Rightarrow \frac{a_{k+1}}{a_k} = \frac{5(-\frac{1}{4})^{(k+1)-1}}{5(-\frac{1}{4})^{k-1}} = -\frac{1}{4}$$

*Note:* For Exercises 3–14, we will find the $n$th term first and then use that term to find $a_5$ and $a_8$.

[3] The common ratio can be found by *dividing* any term by its successor. $r = \frac{4}{8} = \frac{1}{2}$;

$a_n = a_1 r^{n-1} = 8(\frac{1}{2})^{n-1} = 2^3(2^{-1})^{n-1} = 2^3 2^{1-n} = 2^{4-n}$;

$a_5 = 2^{4-5} = 2^{-1} = \frac{1}{2}$; $\qquad a_8 = 2^{4-8} = 2^{-4} = \frac{1}{16}$

[5] $r = \frac{-30}{300} = -0.1$; $a_n = a_1 r^{n-1} = 300(-0.1)^{n-1}$;

$a_5 = 300(-0.1)^4 = 0.03$; $\qquad a_8 = 300(-0.1)^7 = -0.00003$

[7] $r = \frac{25}{5} = 5$; $a_n = 5(5)^{n-1} = 5^n$; $a_5 = 5^5 = 3125$; $a_8 = 5^8 = 390{,}625$

[9] $r = \frac{-6}{4} = -1.5$; $a_n = 4(-1.5)^{n-1}$; $a_5 = 4(-1.5)^4 = 20.25$; $a_8 = 4(-1.5)^7 = -68.34375$

[11] $r = \dfrac{-x^2}{1} = -x^2$; $a_n = a_1 r^{n-1} = 1(-x^2)^{n-1} = (-1)^{n-1} x^{2n-2}$;

$a_5 = (-1)^4 x^{10-2} = x^8$; $\qquad a_8 = (-1)^7 x^{16-2} = -x^{14}$

[13] $r = \dfrac{2^{x+1}}{2} = 2^x$;

$$a_n = 2(2^x)^{n-1} = 2^{[x(n-1)]} \cdot 2^1 = 2^{(n-1)x+1};\ a_5 = 2^{4x+1};\ a_8 = 2^{7x+1}$$

[15] $\frac{a_6}{a_4} = \frac{9}{3} = 3$ and $\frac{a_6}{a_4} = \dfrac{a_1 r^5}{a_1 r^3} = r^2$. Hence, $r^2 = 3$ and $r = \pm\sqrt{3}$.

[17] $r = \frac{6}{4} = \frac{3}{2}$; $a_6 = a_1 r^{n-1} = 4(\frac{3}{2})^5 = \frac{243}{8}$

[19] $\frac{a_7}{a_4} = \dfrac{12}{4} = 3$ and $\frac{a_7}{a_4} = \dfrac{a_1 r^6}{a_1 r^3} = r^3$. Hence, $r^3 = 3$ and $r = \sqrt[3]{3}$. Since the tenth term can be obtained by multiplying the seventh term by the common ratio three times, $a_{10} = a_7 r^3 = 12(3) = 36$.

[21] Using the formula for $S_n$ with $n = 10$, $r = 3$, and $a_1 = 3^1 = 3$, we get

$$\sum_{k=1}^{10} 3^k = 3 \cdot \frac{1 - 3^{10}}{1 - 3} = 3 \cdot \frac{-59{,}048}{-2} = 88{,}572.$$

[23] $$\sum_{k=0}^{9} (-\tfrac{1}{2})^{k+1} = \sum_{k=1}^{10} (-\tfrac{1}{2})^k = -\frac{1}{2} \cdot \frac{1 - (-\frac{1}{2})^{10}}{1 - (-\frac{1}{2})} = -\frac{1}{2} \cdot \frac{\frac{1023}{1024}}{\frac{3}{2}} = -\frac{1023}{3072} = -\frac{341}{1024}$$

[25] $$\sum_{k=16}^{26} (2^{k-14} + 5j) = \sum_{k=16}^{26} 2^{k-14} + 5 \sum_{k=16}^{26} j = \text{SUM1} + 5(\text{SUM2}).$$

To evaluate SUM1, we use $a_1 = 2^{16-14} = 4$, $r = 2$, and $n = 26 - 16 + 1 = 11$ to get:

$$S_{11} = 4 \cdot \frac{1 - 2^{11}}{1 - 2} = 4 \cdot \frac{-2047}{-1} = 8188.$$

SUM2 is just $11 \cdot j$ because $j$ is a constant ($k$ is the index).

Thus, $\text{SUM1} + 5(\text{SUM2}) = 8188 + 5(11j) = 8188 + 55j$.

[27] $2 + 4 + 8 + 16 + 32 + 64 + 128 = 2^1 + 2^2 + 2^3 + 2^4 + 2^5 + 2^6 + 2^7 = \sum_{n=1}^{7} 2^n$

[29] $\frac{1}{4} - \frac{1}{12} + \frac{1}{36} - \frac{1}{108} = \frac{1}{4} - \frac{1}{4}\cdot\frac{1}{3^1} + \frac{1}{4}\cdot\frac{1}{3^2} - \frac{1}{4}\cdot\frac{1}{3^3} = \sum_{n=1}^{4} (-1)^{n+1}\frac{1}{4}\left(\frac{1}{3}\right)^{n-1}$

[31] $1 - \frac{1}{2} + \frac{1}{4} - \frac{1}{8} + \cdots$ • $a_1 = 1$, $r = -\frac{1}{2}$, $S = \frac{a}{1-r} = \frac{1}{1-(-\frac{1}{2})} = \frac{1}{\frac{3}{2}} = \frac{2}{3}$

[33] $1.5 + 0.015 + 0.00015 + \cdots$ •

$$a_1 = 1.5,\ r = \frac{0.015}{1.5} = 0.01,\ S = \frac{1.5}{1 - 0.01} = \frac{1.5}{0.99} = \frac{150}{99} = \frac{50}{33}$$

[35] $\sqrt{2} - 2 + \sqrt{8} - 4 + \cdots$ •

The ratio $r = \frac{-2}{\sqrt{2}} = -\sqrt{2}$, but since $|r| = \sqrt{2} > 1$, the sum does not exist.

[37] $256 + 192 + 144 + 108 + \cdots$ • $a_1 = 256$, $r = \frac{192}{256} = \frac{3}{4}$, $S = \frac{a}{1-r} = \frac{256}{1-\frac{3}{4}} = 1024$

[39] $0.\overline{23}$ • If we write $0.\overline{23}$ as $\frac{23}{100} + \frac{23}{10{,}000} + \cdots$, we see that the first term is $\frac{23}{100}$ and the common ratio is $\frac{1}{100}$. $a_1 = 0.23$, $r = 0.01$, $S = \frac{0.23}{1 - 0.01} = \frac{23}{99}$

[41] $2.4\overline{17} = 2.4171717\ldots = 2.4 + \frac{17}{1000} + \frac{17}{1000}\cdot\frac{1}{100} + \frac{17}{1000}\cdot\left(\frac{1}{100}\right)^2 + \cdots$.

$a_1 = \frac{17}{1000}$ and $r = \frac{1}{100}$ $\Rightarrow$ $S = \dfrac{\frac{17}{1000}}{1 - \frac{1}{100}} = \dfrac{\frac{17}{1000}}{\frac{99}{100}} = \frac{17}{990}$, so

$$2.4\overline{17} = 2.4 + \frac{17}{990} = \frac{2376}{990} + \frac{17}{990} = \frac{2393}{990}.$$

[43] $5.\overline{146}$ • In this problem, we work only with $0.\overline{146}$ and add its rational number representation to 5. $a_1 = 0.146$, $r = 0.001$,

$$S = \frac{0.146}{1 - 0.001} = \frac{146}{999};\ 5.\overline{146} = 5 + \frac{146}{999} = \frac{5141}{999}$$

[45] $1.\overline{6124}$ • $a_1 = 0.6124$, $r = 0.0001$, $S = \frac{0.6124}{1 - 0.0001} = \frac{6124}{9999}$;

$$1.\overline{6124} = 1 + \frac{6124}{9999} = \frac{16{,}123}{9999}$$

[47] The geometric mean of 12 and 48 is $\sqrt{12 \cdot 48} = \sqrt{576} = 24$.

[49] Inserting 2 geometric means results in a sequence that looks like 4, $x$, $y$, 500.
Now $4r = x$, $4r^2 = y$, and $4r^3 = 500$. Thus, $4r^3 = 500 \Rightarrow r^3 = \frac{500}{4} = 125 \Rightarrow$ $r = 5$. The terms are 4, 20, 100, and 500.

[51] Let $a_1 = x$ {the original amount of air in the container} and $r = \frac{1}{2}$.

$a_{11} = x(\frac{1}{2})^{10} = \frac{1}{1024}x$. This is $(\frac{1}{1024}\cdot 100)\%$ or $\frac{25}{256}\%$ or approximately 0.1% of $x$.

[53] Let $a_1 = 10{,}000$ and $r = 1.2$, i.e., 120% every hour.

(a) $N(1) = a_2 = 10{,}000(1.2)^1$, $N(2) = a_3 = 10{,}000(1.2)^2$, ...,

$$N(t) = a_{t+1} = 10{,}000(1.2)^t$$

(b) $N(10) = a_{11} = 10{,}000(1.2)^{10} \approx 61{,}917$.

[55] $\text{Distance}_{\text{total}} = \text{Distance}_{\text{down}} + \text{Distance}_{\text{up}} = 60 + 2 \cdot \text{Distance}_{\text{up}}$

$$= 60 + 2\left[60(\tfrac{2}{3}) + 60(\tfrac{2}{3})^2 + \cdots\right] = 60 + 2\left(\frac{60(\frac{2}{3})}{1 - \frac{2}{3}}\right) = 60 + 2(120) = 300 \text{ ft.}$$

[57] Total amount of local spending $= 2{,}000{,}000(0.60) + 2{,}000{,}000(0.60)^2 + \cdots$. We have $a_1 = 2{,}000{,}000(0.60) = 1{,}200{,}000$ and $r = 0.60$, so $S = = \dfrac{1{,}200{,}000}{1 - 0.60} = \$3{,}000{,}000$.

[59] (a) A half-life of 2 hours means there will be $(\frac{1}{2})(\frac{1}{2}D) = \frac{1}{4}D$ after 4 hours. The amount remaining after $n$ doses {not hours} for a given dose is $a_n = D(\frac{1}{4})^{n-1}$. Since $D$ mg are administered every 4 hours, the amount of the drug in the bloodstream after $n$ doses is $\sum_{k=1}^{n} a_k = \sum_{k=1}^{n} D(\frac{1}{4})^{k-1} = D + \frac{1}{4}D + \cdots + (\frac{1}{4})^{n-1}D$. Since $r = \frac{1}{4} < 1$, $S_n$ may be approximated by $S = \dfrac{a_1}{1 - r}$ for large $n$. $S = \dfrac{D}{1 - \frac{1}{4}} = \frac{4}{3}D$.

(b) Since the amount of the drug in the bloodstream is given by $\frac{4}{3}D$, and this amount must be less than or equal to 500 mg, we have $\frac{4}{3}D \le 500$ mg, or, equivalently, $D \le 375$ mg.

[61] (a) From the figure on the right, we see that

$$(\tfrac{1}{4}a_k)^2 + (\tfrac{3}{4}a_k)^2 = (a_{k+1})^2 \Rightarrow \tfrac{10}{16}a_k^2 = a_{k+1}^2 \Rightarrow a_{k+1} = \tfrac{1}{4}\sqrt{10}\,a_k.$$

(b) From part (a), the common ratio $r = \dfrac{a_{k+1}}{a_k} = \frac{1}{4}\sqrt{10}$, so $a_n = a_1 r^{n-1} = a_1(\frac{1}{4}\sqrt{10})^{n-1}$. For the square $S_{k+1}$, the area $A_{k+1} = a_{k+1}^2$. But from part (a), $a_{k+1}^2 = \frac{10}{16}a_k^2$. Since $a_k^2$ is the area $A_k$ of square $S_k$, we have $A_{k+1} = \frac{5}{8}A_k$, and hence $A_n = (\frac{5}{8})^{n-1}A_1$. Similarly, for the perimeter of the square $S_k$, $P_{k+1} = 4a_{k+1} = 4 \cdot \frac{1}{4}\sqrt{10}\,a_k = \sqrt{10}\,a_k = \sqrt{10}(\frac{1}{4}P_k)$, and hence $P_n = (\frac{1}{4}\sqrt{10})^{n-1}P_1$.

(c) $\sum_{n=1}^{\infty} P_n$ is an infinite geometric series with first term $P_1$ and $r = \frac{1}{4}\sqrt{10}$.

$$S = \frac{P_1}{1 - \frac{1}{4}\sqrt{10}} \cdot \frac{4}{4} = \frac{4P_1}{4 - \sqrt{10}} = \frac{16a_1}{4 - \sqrt{10}}.$$

**63** (a) The sequence is 1, 3, 9, 27, 81, .... Thus, $a_k = 3^{k-1}$ for $k = 1, 2, 3, \ldots$.

(b) $a_{15} = 3^{14} = 4{,}782{,}969$

(c) The area of the triangle removed first is $\frac{1}{4}$. During the next step, 3 triangles with an area of $\frac{1}{16}$ are removed. Then, 9 triangles with an area of $\frac{1}{64}$ are removed. At each step the number of triangles increase by a factor of 3, while the area of each triangle decreases by a factor of 4. Thus, $b_k = \dfrac{3^{k-1}}{4^k} = \frac{1}{4}\left(\frac{3}{4}\right)^{k-1}$.

(d) $b_7 = \frac{1}{4}\left(\frac{3}{4}\right)^6 = \dfrac{729}{16{,}384} \approx 0.0445 = 4.45\%$.

**65** Let $a_k = 100\left(1 + \frac{0.06}{12}\right)^k = 100(1.005)^k$, where $k$ represents the number of compounding periods for each deposit. For the first deposit, $k = 18 \cdot 12 = 216$. For the last deposit, $k = 1$.

$$\begin{aligned} S_{216} &= a_1 + a_2 + \cdots + a_{216} \\ &= 100(1.005)^1 + 100(1.005)^2 + \cdots + 100(1.005)^{216} \\ &= 100(1.005)\left(\frac{1-(1.005)^{216}}{1-(1.005)}\right) \approx \$38{,}929.00 \end{aligned}$$

**67** Use $A = P\left(\frac{12}{r} + 1\right)\left[\left(1 + \frac{r}{12}\right)^n - 1\right]$ with $P = 100$, $r = 0.08$, and $n = 60$ to get

$$A = 100\left(\tfrac{12}{0.08} + 1\right)\left[\left(1 + \tfrac{0.08}{12}\right)^{60} - 1\right] \approx \$7396.67.$$

**69** (a) $A_1 = \frac{2}{5}\left(1 - \frac{2}{5}\right)^{1-1} = \frac{2}{5}\left(\frac{3}{5}\right)^0 = \frac{2}{5}$.

$A_2 = \frac{2}{5}\left(\frac{3}{5}\right)^1 = \frac{6}{25}$, $A_3 = \frac{2}{5}\left(\frac{3}{5}\right)^2 = \frac{18}{125}$, $A_4 = \frac{2}{5}\left(\frac{3}{5}\right)^3 = \frac{54}{625}$, $A_5 = \frac{2}{5}\left(\frac{3}{5}\right)^4 = \frac{162}{3125}$.

(b) $r = A_{k+1}/A_k = \frac{3}{5}$ for $k = 1, 2, 3, 4$.

$$S_5 = \sum_{k=1}^{5} A_k = A_1 + A_2 + A_3 + A_4 + A_5 = \frac{2}{5} \cdot \frac{1 - (\frac{3}{5})^5}{1 - \frac{3}{5}} = 1 - (\tfrac{3}{5})^5 = \tfrac{2882}{3125} = 0.92224.$$

(c) $\$25{,}000\left(\frac{2}{5} + \frac{6}{25}\right) = \$25{,}000(\frac{16}{25}) = \$16{,}000$

## 7.4 Exercises

**1** (1) $P_1$ is true, since $2(1) = 1(1+1) = 2$.

(2) Assume $P_k$ is true:

$2 + 4 + 6 + \cdots + 2k = k(k+1)$. Hence,

$$\begin{aligned} 2 + 4 + 6 + \cdots + 2k + 2(k+1) &= k(k+1) + 2(k+1) \\ &= (k+1)(k+2) \\ &= (k+1)(k+1+1). \end{aligned}$$

Thus, $P_{k+1}$ is true, and the proof is complete.

[3] (1) $P_1$ is true, since $2(1) - 1 = (1)^2 = 1$.

(2) Assume $P_k$ is true:

$1 + 3 + 5 + \cdots + (2k - 1) = k^2$. Hence,

$$\begin{aligned} 1 + 3 + 5 + \cdots + (2k - 1) + 2(k + 1) - 1 &= k^2 + 2(k + 1) - 1 \\ &= k^2 + 2k + 1 \\ &= (k + 1)^2. \end{aligned}$$

Thus, $P_{k+1}$ is true, and the proof is complete.

[5] (1) $P_1$ is true, since $5(1) - 3 = \frac{1}{2}(1)[5(1) - 1] = 2$.

(2) Assume $P_k$ is true:

$2 + 7 + 12 + \cdots + (5k - 3) = \frac{1}{2}k(5k - 1)$. Hence,

$$\begin{aligned} 2 + 7 + 12 + \cdots + (5k - 3) + 5(k + 1) - 3 &= \tfrac{1}{2}k(5k - 1) + 5(k + 1) - 3 \\ &= \tfrac{5}{2}k^2 + \tfrac{9}{2}k + 2 \\ &= \tfrac{1}{2}(5k^2 + 9k + 4) \\ &= \tfrac{1}{2}(k + 1)(5k + 4) \\ &= \tfrac{1}{2}(k + 1)[5(k + 1) - 1]. \end{aligned}$$

Thus, $P_{k+1}$ is true, and the proof is complete.

[7] (1) $P_1$ is true, since $1 \cdot 2^{1-1} = 1 + (1 - 1) \cdot 2^1 = 1$.

(2) Assume $P_k$ is true:

$1 + 2 \cdot 2 + 3 \cdot 2^2 + \cdots + k \cdot 2^{k-1} = 1 + (k - 1) \cdot 2^k$. Hence,

$$\begin{aligned} 1 + 2 \cdot 2 + 3 \cdot 2^2 + \cdots + k \cdot 2^{k-1} + (k + 1) \cdot 2^k &= 1 + (k - 1) \cdot 2^k + (k + 1) \cdot 2^k \\ &= 1 + k \cdot 2^k - 2^k + k \cdot 2^k + 2^k \\ &= 1 + k \cdot 2^1 \cdot 2^k \\ &= 1 + [(k + 1) - 1] \cdot 2^{k+1}. \end{aligned}$$

Thus, $P_{k+1}$ is true, and the proof is complete.

[9] (1) $P_1$ is true, since $(1)^1 = \dfrac{1(1 + 1)[2(1) + 1]}{6} = 1$.

(2) Assume $P_k$ is true:

$1^2 + 2^2 + 3^2 + \cdots + k^2 = \dfrac{k(k + 1)(2k + 1)}{6}$. Hence,

$$\begin{aligned} 1^2 + 2^2 + 3^2 + \cdots + k^2 + (k + 1)^2 &= \frac{k(k + 1)(2k + 1)}{6} + (k + 1)^2 \\ &= (k + 1)\left[\frac{k(2k + 1)}{6} + \frac{6(k + 1)}{6}\right] \\ &= \frac{(k + 1)(2k^2 + 7k + 6)}{6} \\ &= \frac{(k + 1)(k + 2)(2k + 3)}{6}. \end{aligned}$$

Thus, $P_{k+1}$ is true, and the proof is complete.

[11] (1) $P_1$ is true, since $\frac{1}{1(1+1)} = \frac{1}{1+1} = \frac{1}{2}$.

(2) Assume $P_k$ is true:

$\frac{1}{1\cdot 2} + \frac{1}{2\cdot 3} + \frac{1}{3\cdot 4} + \cdots + \frac{1}{k(k+1)} = \frac{k}{k+1}$. Hence,

$$\begin{aligned}\frac{1}{1\cdot 2} + \frac{1}{2\cdot 3} + \frac{1}{3\cdot 4} + \cdots + \frac{1}{k(k+1)} + \frac{1}{(k+1)(k+2)} &= \frac{k}{k+1} + \frac{1}{(k+1)(k+2)} \\ &= \frac{k}{k+1} + \frac{1}{(k+1)(k+2)} \\ &= \frac{k(k+2)+1}{(k+1)(k+2)} \\ &= \frac{k^2+2k+1}{(k+1)(k+2)} \\ &= \frac{k+1}{(k+1)+1}.\end{aligned}$$

Thus, $P_{k+1}$ is true, and the proof is complete.

[13] (1) $P_1$ is true, since $3^1 = \frac{3}{2}(3^1 - 1) = 3$.

(2) Assume $P_k$ is true:

$3 + 3^2 + 3^3 + \cdots + 3^k = \frac{3}{2}(3^k - 1)$. Hence,

$$\begin{aligned}3 + 3^2 + 3^3 + \cdots + 3^k + 3^{k+1} &= \tfrac{3}{2}(3^k - 1) + 3^{k+1} \\ &= \tfrac{3}{2}\cdot 3^k - \tfrac{3}{2} + 3\cdot 3^k \\ &= \tfrac{9}{2}\cdot 3^k - \tfrac{3}{2} \\ &= \tfrac{3}{2}(3\cdot 3^k - 1) \\ &= \tfrac{3}{2}(3^{k+1} - 1).\end{aligned}$$

Thus, $P_{k+1}$ is true, and the proof is complete.

[15] (1) $P_1$ is true, since $1 < 2^1$.

(2) Assume $P_k$ is true: $k < 2^k$. Now $k+1 < k+k = 2(k)$ for $k > 1$.

From $P_k$, we see that $2(k) < 2(2^k) = 2^{k+1}$ and conclude that $k+1 < 2^{k+1}$.

Thus, $P_{k+1}$ is true, and the proof is complete.

[17] (1) $P_1$ is true, since $1 < \frac{1}{8}[2(1)+1]^2 = \frac{9}{8}$.

(2) Assume $P_k$ is true: $1 + 2 + 3 + \cdots + k < \frac{1}{8}(2k+1)^2$. Hence,

$$\begin{aligned}1 + 2 + 3 + \cdots + k + (k+1) &< \tfrac{1}{8}(2k+1)^2 + (k+1) \\ &= \tfrac{1}{2}k^2 + \tfrac{3}{2}k + \tfrac{9}{8} \\ &= \tfrac{1}{8}(4k^2 + 12k + 9) \\ &= \tfrac{1}{8}(2k+3)^2 \\ &= \tfrac{1}{8}[2(k+1)+1]^2.\end{aligned}$$

Thus, $P_{k+1}$ is true, and the proof is complete.

[19] (1) For $n = 1$, $n^3 - n + 3 = 3$ and 3 is a factor of 3.

(2) Assume 3 is a factor of $k^3 - k + 3$. The $(k+1)$st term is

$$\begin{aligned}(k+1)^3 - (k+1) + 3 &= k^3 + 3k^2 + 2k + 3\\ &= (k^3 - k + 3) + 3k^2 + 3k\\ &= (k^3 - k + 3) + 3(k^2 + k).\end{aligned}$$

By the induction hypothesis, 3 is a factor of $k^3 - k + 3$ and 3 is a factor of $3(k^2 + k)$, so 3 is a factor of the $(k+1)$st term. Thus, $P_{k+1}$ is true, and the proof is complete.

[21] (1) For $n = 1$, $5^n - 1 = 4$ and 4 is a factor of 4.

(2) Assume 4 is a factor of $5^k - 1$. The $(k+1)$st term is

$$\begin{aligned}5^{k+1} - 1 &= 5 \cdot 5^k - 1\\ &= 5 \cdot 5^k - 5 + 4\\ &= 5(5^k - 1) + 4.\end{aligned}$$

By the induction hypothesis, 4 is a factor of $5^k - 1$ and 4 is a factor of 4, so 4 is a factor of the $(k+1)$st term. Thus, $P_{k+1}$ is true, and the proof is complete.

[23] (1) If $a > 1$, then $a^1 = a > 1$, so $P_1$ is true.

(2) Assume $P_k$ is true: $a^k > 1$.

Multiply both sides by $a$ to obtain $a^{k+1} > a$, but since $a > 1$, we have $a^{k+1} > 1$.

Thus, $P_{k+1}$ is true, and the proof is complete.

[25] (1) For $n = 1$, $a - b$ is a factor of $a^1 - b^1$.

(2) Assume $a - b$ is a factor of $a^k - b^k$. Following the hint for the $(k+1)$st term, $a^{k+1} - b^{k+1} = a^k \cdot a - b \cdot a^k + b \cdot a^k - b^k \cdot b = a^k(a - b) + (a^k - b^k)b$. Since $(a - b)$ is a factor of $a^k(a - b)$ and since by the induction hypothesis $a - b$ is a factor of $(a^k - b^k)$, it follows that $a - b$ is a factor of the $(k+1)$st term. Thus, $P_{k+1}$ is true, and the proof is complete.

*Note:* For Exercises 27–32 in this section and Exercises 49–50 in the Chapter Review, there are several ways to find $j$. Possibilities include: solve the inequality, sketch the graphs of functions representing each side, and trial and error. Trial and error may be the easiest to use.

[27] For $j$: $n^2 \geq n + 12 \;\Rightarrow\; n^2 - n - 12 \geq 0 \;\Rightarrow\; (n-4)(n+3) \geq 0 \;\Rightarrow\; n \geq 4 \; \{n > 0\}$

(1) $P_4$ is true, since $4 + 12 \leq 4^2$.

(2) Assume $P_k$ is true: $k + 12 \leq k^2$. Hence,

$(k+1) + 12 = (k + 12) + 1 \leq (k^2) + 1 < k^2 + 2k + 1 = (k+1)^2$.

Thus, $P_{k+1}$ is true, and the proof is complete.

**29** For $j$: By sketching $y = 5 + \log_2 x$ and $y = x$, we see that the solution for $x > 1$ must be larger than 5. By trial and error, $j = 8$.

(1) $P_8$ is true, since $5 + \log_2 8 \leq 8$.

(2) Assume $P_k$ is true: $5 + \log_2 k \leq k$. Hence,

$$\begin{aligned} 5 + \log_2 (k+1) &< 5 + \log_2 (k+k) \\ &= 5 + \log_2 2k \\ &= 5 + \log_2 2 + \log_2 k \\ &= (5 + \log_2 k) + 1 \\ &\leq k + 1. \end{aligned}$$

Thus, $P_{k+1}$ is true, and the proof is complete.

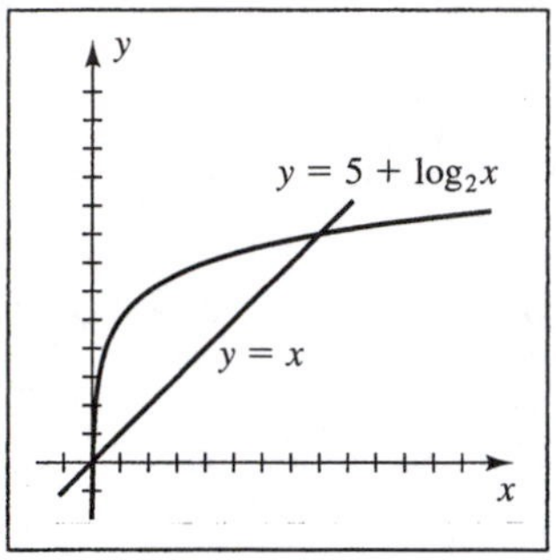


Figure 29

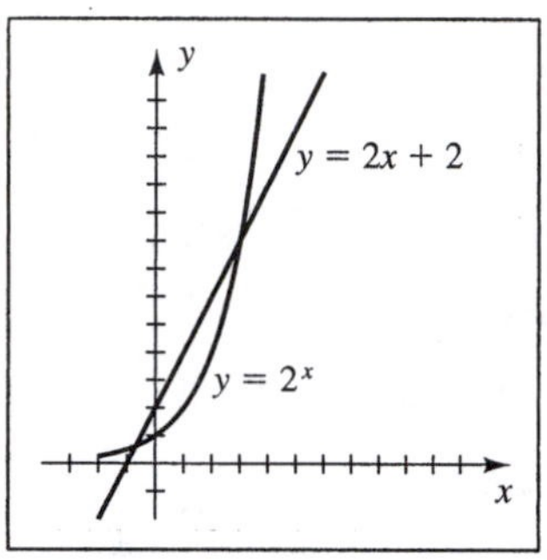


Figure 31

**31** For $j$: By sketching $y = 2x + 2$ and $y = 2^x$, we see there is one positive solution. By trial and error, $j = 3$.

(1) $P_3$ is true, since $2(3) + 2 \leq 2^3$.

(2) Assume $P_k$ is true: $2k + 2 \leq 2^k$. Hence,

$$2(k+1) + 2 = (2k+2) + 2 \leq 2^k + 2^k = 2 \cdot 2^k = 2^{k+1}.$$

Thus, $P_{k+1}$ is true, and the proof is complete.

**33** Following the hint in the text:

$$\begin{aligned} \sum_{k=1}^{n} (k^2 + 3k + 5) &= \sum_{k=1}^{n} k^2 + 3\sum_{k=1}^{n} k + \sum_{k=1}^{n} 5 \\ &= \frac{n(n+1)(2n+1)}{6} + 3\left[\frac{n(n+1)}{2}\right] + 5n \\ &= \frac{n(n+1)(2n+1) + 9n(n+1) + 30n}{6} \\ &= \frac{2n^3 + 12n^2 + 40n}{6} \\ &= \frac{n^3 + 6n^2 + 20n}{3} \end{aligned}$$

35 $\sum_{k=1}^{n}(2k-3)^2 = \sum_{k=1}^{n}(4k^2-12k+9) = 4\sum_{k=1}^{n}k^2 - 12\sum_{k=1}^{n}k + \sum_{k=1}^{n}9$

$$= 4\left[\frac{n(n+1)(2n+1)}{6}\right] - 12\left[\frac{n(n+1)}{2}\right] + 9n = \frac{4n^3-12n^2+11n}{3}$$

37 (a) $n=1 \Rightarrow a(1)^3+b(1)^2+c(1)=1^2 \Rightarrow a+b+c=1$

$n=2 \Rightarrow a(2)^3+b(2)^2+c(2)=1^2+2^2 \Rightarrow 8a+4b+2c=5$

$n=3 \Rightarrow a(3)^3+b(3)^2+c(3)=1^2+2^2+3^2 \Rightarrow 27a+9b+3c=14$

$$AX=B \Rightarrow \begin{bmatrix} 1 & 1 & 1 \\ 8 & 4 & 2 \\ 27 & 9 & 3 \end{bmatrix}\begin{bmatrix} a \\ b \\ c \end{bmatrix} = \begin{bmatrix} 1 \\ 5 \\ 14 \end{bmatrix} \Rightarrow X = A^{-1}B = \begin{bmatrix} 1/3 \\ 1/2 \\ 1/6 \end{bmatrix}.$$

(b) $a=\frac{1}{3},\ b=\frac{1}{2},\ c=\frac{1}{6} \Rightarrow$

$$1^2+2^2+3^2+\cdots+n^2 = \tfrac{1}{3}n^3+\tfrac{1}{2}n^2+\tfrac{1}{6}n = \frac{n(n+1)(2n+1)}{6},$$

which is the formula found in Exercise 9. This method does not verify the formula for all $n$ but only for $n = 1, 2, 3$. Mathematical induction should be used to verify the formula for all $n$ as in Exercise 9.

## *7.5 Exercises*

1 $2!\,6! = (2\cdot1)\cdot(6\cdot5\cdot4\cdot3\cdot2\cdot1) = 2\cdot720 = 1440$

3 $7!\,0! = (7\cdot6\cdot5\cdot4\cdot3\cdot2\cdot1)\cdot(1)$ {Remember that $0! = 1$.} $= 5040$

5 $\frac{8!}{5!} = \frac{8\cdot7\cdot6\cdot5!}{5!}$ {cancel $5!$} $= 8\cdot7\cdot6 = 336$

7 $\binom{5}{5} = \frac{5!}{5!\,0!}$ {cancel $5!$} $= \frac{1}{0!} = \frac{1}{1} = 1$

9 $\binom{7}{5} = \frac{7!}{5!\,2!}$ {cancel $5!$} $= \frac{7\cdot6}{2} = 21$

11 $\binom{13}{4} = \frac{13!}{4!\,9!}$ {cancel $9!$} $= \frac{13\cdot12\cdot11\cdot10}{4\cdot3\cdot2} = 715$

13 $\frac{n!}{(n-2)!} = \frac{n(n-1)(n-2)!}{(n-2)!}$ {cancel $(n-2)!$} $= n(n-1)$

15 $\frac{(2n+2)!}{(2n)!} = \frac{(2n+2)(2n+1)(2n)!}{(2n)!}$ {cancel $(2n)!$} $= (2n+2)(2n+1)$

17 We use the binomial theorem formula with $a = 4x$, $b = -y$, and $n = 3$.

$$\begin{aligned}(4x-y)^3 &= \binom{3}{0}(4x)^3(-y)^0 + \binom{3}{1}(4x)^2(-y)^1 + \binom{3}{2}(4x)^1(-y)^2 + \binom{3}{3}(4x)^0(-y)^3 \\ &= (1)(64x^3)(1) + (3)(16x^2)(-y) + (3)(4x)(y^2) + (1)(1)(-y^3) \\ &= 64x^3 - 48x^2y + 12xy^2 - y^3\end{aligned}$$

[19] We use the binomial theorem formula with $a = x$, $b = y$, and $n = 6$.

$$(x+y)^6 = x^6 + \binom{6}{1}x^5y^1 + \binom{6}{2}x^4y^2 + \binom{6}{3}x^3y^3 + \binom{6}{4}x^2y^4 + \binom{6}{5}x^1y^5 + y^6$$
$$= x^6 + 6x^5y + 15x^4y^2 + 20x^3y^3 + 15x^2y^4 + 6xy^5 + y^6$$

[21] We use the binomial theorem formula with $a = x$, $b = -y$, and $n = 7$.

$$(x-y)^7 = x^7 + \binom{7}{1}x^6(-y)^1 + \binom{7}{2}x^5(-y)^2 + \binom{7}{3}x^4(-y)^3 + \binom{7}{4}x^3(-y)^4 + \binom{7}{5}x^2(-y)^5 + \binom{7}{6}x^1(-y)^6 + (-y)^7$$
$$= x^7 - 7x^6y + 21x^5y^2 - 35x^4y^3 + 35x^3y^4 - 21x^2y^5 + 7xy^6 - y^7$$

[23] We use the binomial theorem formula with $a = 3t$, $b = -5s$, and $n = 4$.

$$(3t-5s)^4 = \binom{4}{0}(3t)^4(-5s)^0 + \binom{4}{1}(3t)^3(-5s)^1 + \binom{4}{2}(3t)^2(-5s)^2 + \binom{4}{3}(3t)^1(-5s)^3 + \binom{4}{4}(3t)^0(-5s)^4$$
$$= (1)(81t^4)(1) + (4)(27t^3)(-5s) + (6)(9t^2)(25s^2) + (4)(3t)(-125s^3) + (1)(1)(625s^4)$$
$$= 81t^4 - 540t^3s + 1350t^2s^2 - 1500ts^3 + 625s^4$$

[25] We use the binomial theorem formula with $a = \frac{1}{3}x$, $b = y^2$, and $n = 5$.

$$(\tfrac{1}{3}x + y^2)^5 = \binom{5}{0}(\tfrac{1}{3}x)^5(y^2)^0 + \binom{5}{1}(\tfrac{1}{3}x)^4(y^2)^1 + \binom{5}{2}(\tfrac{1}{3}x)^3(y^2)^2 + \binom{5}{3}(\tfrac{1}{3}x)^2(y^2)^3 + \binom{5}{4}(\tfrac{1}{3}x)^1(y^2)^4 + \binom{5}{5}(\tfrac{1}{3}x)^0(y^2)^5$$
$$= (1)(\tfrac{1}{243}x^5)(1) + (5)(\tfrac{1}{81}x^4)(y^2) + (10)(\tfrac{1}{27}x^3)(y^4) + (10)(\tfrac{1}{9}x^2)(y^6) + (5)(\tfrac{1}{3}x)(y^8) + (1)(1)(y^{10})$$
$$= \tfrac{1}{243}x^5 + \tfrac{5}{81}x^4y^2 + \tfrac{10}{27}x^3y^4 + \tfrac{10}{9}x^2y^6 + \tfrac{5}{3}xy^8 + y^{10}$$

[27] We use the binomial theorem formula with $a = x^{-2}$, $b = 3x$, and $n = 6$.

$$\left(\frac{1}{x^2} + 3x\right)^6 = (x^{-2} + 3x)^6$$
$$= \binom{6}{0}(x^{-2})^6(3x)^0 + \binom{6}{1}(x^{-2})^5(3x)^1 + \binom{6}{2}(x^{-2})^4(3x)^2 + \binom{6}{3}(x^{-2})^3(3x)^3 + \binom{6}{4}(x^{-2})^2(3x)^4 + \binom{6}{5}(x^{-2})^1(3x)^5 + \binom{6}{6}(x^{-2})^0(3x)^6$$
$$= (1)(x^{-12})(1) + (6)(x^{-10})(3x^1) + (15)(x^{-8})(9x^2) + (20)(x^{-6})(27x^3) + (15)(x^{-4})(81x^4) + (6)(x^{-2})(243x^5) + (1)(1)(729x^6)$$
$$= x^{-12} + 18x^{-9} + 135x^{-6} + 540x^{-3} + 1215 + 1458x^3 + 729x^6$$

**29** We use the binomial theorem formula with $a = x^{1/2}$, $b = -x^{-1/2}$, and $n = 5$.

$$\left(\sqrt{x} - \frac{1}{\sqrt{x}}\right)^5 = (x^{1/2} - x^{-1/2})^5$$

$$= \binom{5}{0}(x^{1/2})^5(-x^{-1/2})^0 + \binom{5}{1}(x^{1/2})^4(-x^{-1/2})^1 + \binom{5}{2}(x^{1/2})^3(-x^{-1/2})^2 +$$

$$\binom{5}{3}(x^{1/2})^2(-x^{-1/2})^3 + \binom{5}{4}(x^{1/2})^1(-x^{-1/2})^4 + \binom{5}{5}(x^{1/2})^0(-x^{-1/2})^5$$

$$= (1)(x^{5/2})(1) + (5)(x^2)(-x^{-1/2}) + (10)(x^{3/2})(x^{-1}) +$$

$$(10)(x^1)(-x^{-3/2}) + (5)(x^{1/2})(x^{-2}) + (1)(1)(-x^{-5/2})$$

$$= x^{5/2} - 5x^{3/2} + 10x^{1/2} - 10x^{-1/2} + 5x^{-3/2} - x^{-5/2}$$

**31** For the binomial expression $(3c^{2/5} + c^{4/5})^{25}$, the first three terms are

$$\sum_{k=0}^{2}\binom{25}{k}(3c^{2/5})^{25-k}(c^{4/5})^k$$

$$= \binom{25}{0}(3c^{2/5})^{25}(c^{4/5})^0 + \binom{25}{1}(3c^{2/5})^{24}(c^{4/5})^1 + \binom{25}{2}(3c^{2/5})^{23}(c^{4/5})^2$$

$$= (1)(3^{25}c^{10})(1) + (25)(3^{24}c^{48/5})(c^{4/5}) + (300)(3^{23}c^{46/5})(c^{8/5})$$

$$= 3^{25}c^{10} + 25 \cdot 3^{24}c^{52/5} + 300 \cdot 3^{23}c^{54/5}$$

**33** For the binomial expression $(4z^{-1} - 3z)^{15}$, the last three terms are

$$\sum_{k=13}^{15}\binom{15}{k}(4z^{-1})^{15-k}(-3z)^k$$

$$= \binom{15}{13}(4z^{-1})^2(-3z)^{13} + \binom{15}{14}(4z^{-1})^1(-3z)^{14} + \binom{15}{15}(4z^{-1})^0(-3z)^{15}$$

$$= (105)(16z^{-2})(-3^{13}z^{13}) + (15)(4z^{-1})(3^{14}z^{14}) + (1)(1)(-3^{15}z^{15})$$

$$= -1680 \cdot 3^{13}z^{11} + 60 \cdot 3^{14}z^{13} - 3^{15}z^{15}$$

*Note:* For the following exercises, the general formula for the

<u>$(k+1)$st term of the expansion of $(a+b)^n$</u> is $\boxed{\binom{n}{k}(a)^{n-k}(b)^k}$.

**35** $\left(\frac{3}{c} + \frac{c^2}{4}\right)^7$; sixth term $\{k = 5\} = \binom{7}{5}\left(\frac{3}{c}\right)^2\left(\frac{c^2}{4}\right)^5 = 21\left(\frac{9}{c^2}\right)\left(\frac{c^{10}}{1024}\right) = \frac{189}{1024}c^8$

**37** $(\frac{1}{3}u + 4v)^8$; seventh term $\{k = 6\} = \binom{8}{6}(\frac{1}{3}u)^2(4v)^6 = 28\left(\frac{u^2}{9}\right)(4096v^6) = \frac{114{,}688}{9}u^2v^6$

**39** Since there are 9 terms, the middle term is the fifth term $\left\{\frac{9+1}{2} = 5\right\}$.

Using the formula in the *Note* above, we obtain the fifth term of $(x^{1/2} + y^{1/2})^8$,

$$\binom{8}{4}(x^{1/2})^4(y^{1/2})^4 = 70x^2y^2.$$

**41** $(2y + x^2)^8$; term that contains $x^{10}$ •

Consider only the variable $x$ in the expansion: $(x^2)^k = x^{10} \Rightarrow 2k = 10 \Rightarrow k = 5$;

sixth term $= \binom{8}{5}(2y)^3(x^2)^5 = 448y^3x^{10}$

43 $(3y^3 - 2x^2)^4$; term that contains $y^9$ •

Consider only the variable $y$ in the expansion:

$(y^3)^{4-k} = y^9 \Rightarrow 12 - 3k = 9 \Rightarrow k = 1$; 2nd term $= \binom{4}{1}(3y^3)^3(-2x^2)^1 = -216y^9x^2$

45 $\left(3x - \frac{1}{4x}\right)^6$; term that does not contain $x$ •

Consider only the variable $x$ in the expansion:

$x^{6-k}(x^{-1})^k = x^0 \Rightarrow x^{6-2k} = x^0 \Rightarrow k = 3$; 4th term $= \binom{6}{3}(3x)^3\left(-\frac{1}{4x}\right)^3 = -\frac{135}{16}$

47 The first three terms in the expansion of $(1 + 0.2)^{10}$ are

$$\sum_{k=0}^{2}\binom{10}{k}(1)^{10-k}(0.2)^k = \binom{10}{0}(1)^{10}(0.2)^0 + \binom{10}{1}(1)^9(0.2)^1 + \binom{10}{2}(1)^8(0.2)^2$$

$$= (1)(1)(1) + (10)(1)(0.2) + (45)(1)(0.04) = 1 + 2 + 1.8 = 4.8.$$

The calculator result for $(1.2)^{10}$ is approximately 6.19.

49 $$\frac{(x+h)^4 - x^4}{h} = \frac{(x^4 + 4x^3h + 6x^2h^2 + 4xh^3 + h^4) - x^4}{h} = \frac{h(4x^3 + 6x^2h + 4xh^2 + h^3)}{h} = 4x^3 + 6x^2h + 4xh^2 + h^3$$

51 $\binom{n}{1} = \frac{n!}{(n-1)!\,1!} = n$ and $\binom{n}{n-1} = \frac{n!}{[n-(n-1)]!\,(n-1)!} = \frac{n!}{1!\,(n-1)!} = n$

## 7.6 Exercises

1 $P(7, 3) = \frac{7!}{(7-3)!} = \frac{7!}{4!} = \frac{7 \cdot 6 \cdot 5 \cdot 4!}{4!}$ {cancel 4!} $= 7 \cdot 6 \cdot 5 = 210$

3 $P(9, 6) = \frac{9!}{(9-6)!} = \frac{9!}{3!} = \frac{9 \cdot 8 \cdot 7 \cdot 6 \cdot 5 \cdot 4 \cdot 3!}{3!}$ {cancel 3!} $= 9 \cdot 8 \cdot 7 \cdot 6 \cdot 5 \cdot 4 = 60{,}480$

5 $P(5, 5) = \frac{5!}{(5-5)!} = \frac{5!}{0!} = \frac{5 \cdot 4 \cdot 3 \cdot 2 \cdot 1}{1} = 120$

7 $P(6, 1) = \frac{6!}{(6-1)!} = \frac{6!}{5!} = \frac{6 \cdot 5!}{5!} = 6$

9 $P(n, 0) = \frac{n!}{(n-0)!} = \frac{n!}{n!} = 1$

11 $P(n, n-1) = \frac{n!}{[n-(n-1)]!} = \frac{n!}{1!} = \frac{n!}{1} = n!$

13 (a) We can think of the three-digit numbers as filling 3 slots. There are 5 digits to pick from for filling the first slot. There are 4 remaining digits to pick from to fill the second slot. There are 3 remaining digits to pick from to fill the third slot. By the fundamental counting principle, there are a total of $5 \cdot 4 \cdot 3 = 60$ three-digit numbers.

(b) The difference between parts (a) and (b) is that we can use any of the 5 digits for all 3 slots. Hence, there are a total of $5 \cdot 5 \cdot 5 = 125$ three-digit numbers if repetitions are allowed.

[15] There are 4 one digit numbers; $4 \cdot 3 = 12$ two digit numbers;
$4 \cdot 3 \cdot 2 = 24$ three digit numbers; $4 \cdot 3 \cdot 2 \cdot 1 = 24$ four digit numbers.
The total number of numbers is $4 + 12 + 24 + 24 = 64$.

[17] We want eight teams taken three at a time. $P(8, 3) = \frac{8!}{(8-3)!} = \frac{8!}{5!} = 8 \cdot 7 \cdot 6 = 336$

[19] By the fundamental counting principle, $4 \cdot 6 = 24$.

[21] (a) By the fundamental counting principle, $26 \cdot 9 \cdot 10^4 = 2{,}340{,}000$.
(b) By the fundamental counting principle, $24 \cdot 9 \cdot 10^4 = 2{,}160{,}000$.

[23] (a) $P(10, 6) = \frac{10!}{(10-6)!} = \frac{10!}{4!} = 10 \cdot 9 \cdot 8 \cdot 7 \cdot 6 \cdot 5 = 151{,}200$
(b) $\boxed{\text{Boy}}$-$\underline{\text{girl}}$ seatings: $\boxed{6} \cdot \underline{4} \cdot \boxed{5} \cdot \underline{3} \cdot \boxed{4} \cdot \underline{2} = 2880$.
$\underline{\text{Girl}}$-$\boxed{\text{boy}}$ seatings: $\underline{4} \cdot \boxed{6} \cdot \underline{3} \cdot \boxed{5} \cdot \underline{2} \cdot \boxed{4} = 2880$.
The total number of seatings is $2880 + 2880 = 5760$.

[25] There are 2 choices for each of the 10 questions. 2 times itself 10 times $= 2^{10} = 1024$

[27] We want eight people taken eight at a time. $P(8, 8) = \frac{8!}{(8-8)!} = \frac{8!}{0!} = 8! = 40{,}320$

[29] We want six flags taken three at a time. $P(6, 3) = \frac{6!}{(6-3)!} = \frac{6!}{3!} = 6 \cdot 5 \cdot 4 = 120$

[31] (a) The number of choices for each letter are: $\underline{2} \cdot \underline{25} \cdot \underline{24} \cdot \underline{23} = 27{,}600$
(b) The number of choices for each letter are: $\underline{2} \cdot \underline{26} \cdot \underline{26} \cdot \underline{26} = 35{,}152$

[33] We have a choice of 9 digits for the first number and a choice of 10 digits for the other numbers. By the fundamental counting principle, $9 \cdot 10^6 = 9{,}000{,}000$.

[35] $P(4, 4) = \frac{4!}{(4-4)!} = \frac{4!}{0!} = 4! = 24$

[37] There are 3! ways to choose the ordering of the couples and 2 ways for each couple to sit. By the fundamental counting principle, $3! \cdot 2^3 = 48$.

[39] There are 16 "fixins" and each one can be "on" or "off" your sandwich (or hot dog or salad). With just the "fixins," we have $\underbrace{2 \cdot 2 \cdot 2 \cdot \cdots \cdot 2}_{16\ 2\text{'s}} = 2^{16}$, but we subtract 1 because that would be the one choice that would have no "fixins." Now we have 17 sandwiches, hot dogs, or salads that can be combined with any of the choices of "fixins," so we use the fundamental counting principle to obtain $(2^{16} - 1) \cdot 17 = 1{,}114{,}095$ different lunches.

[41] (a) There are 9 choices for the first digit {can't have 0}, 10 for the second, 10 for the third, and 1 for the fourth and fifth. $9 \cdot 10 \cdot 10 \cdot 1 \cdot 1 = 900$

(b) If $n$ is even, we need to select the first $\frac{n}{2}$ digits. There are 9 choices for the first digit and 10 choices for each of the next $\left(\frac{n}{2}-1\right)$ digits, so the total number of even $n$-digit palindromes is $9 \cdot 10^{(n/2)-1}$.

If $n$ is odd, we need to select the first $\left(\frac{n}{2}+\frac{1}{2}\right)$ digits. There are 9 choices for the first digit and 10 choices for each of the next $\left(\frac{n}{2}+\frac{1}{2}-1\right)=\left(\frac{n}{2}-\frac{1}{2}\right)=\frac{n-1}{2}$ digits, so the total number of odd $n$-digit palindromes is $9 \cdot 10^{(n-1)/2}$.

43 (a) There is a horizontal asymptote of $y = 1$.

(b) $\dfrac{n!\,e^n}{n^n\sqrt{2\pi n}} \approx 1 \Rightarrow n! \approx \dfrac{n^n\sqrt{2\pi n}}{e^n}$.

Example: $50! \approx \dfrac{50^{50}\sqrt{2\pi(50)}}{e^{50}} \approx 3.0363 \times 10^{64}$.

The actual value is closer to $3.0414 \times 10^{64}$.

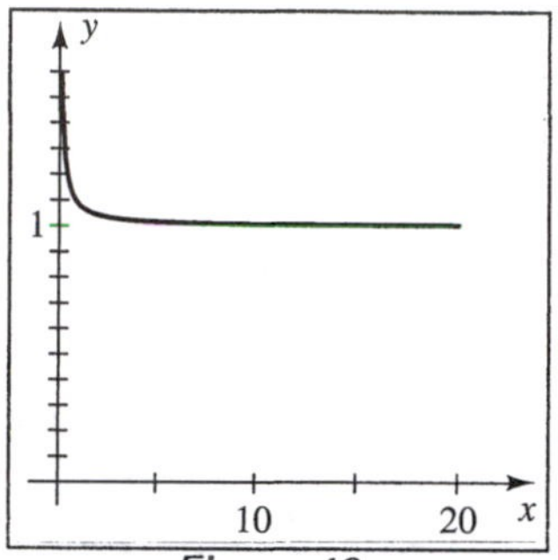

Figure 43

## 7.7 Exercises

1 $C(7,\,3)=\dfrac{7!}{(7-3)!\,3!}=\dfrac{7!}{4!\,3!}=\dfrac{7\cdot6\cdot5\cdot4!}{4!\,3!}=\dfrac{7\cdot6\cdot5}{3!}=\dfrac{7\cdot6\cdot5}{6}=7\cdot5=35$

3 $C(9,\,8)=\dfrac{9!}{1!\,8!}=\dfrac{9\cdot8!}{1\cdot8!}=9$

5 $C(n,\,n-1)=\dfrac{n!}{\left[n-(n-1)\right]!\,(n-1)!}=\dfrac{n!}{1!\,(n-1)!}=n$

7 $C(7,\,0)=\dfrac{7!}{7!\,0!}=\dfrac{1}{0!}=\dfrac{1}{1}=1$

9 There are 12! total permutations. We want the number of *distinguishable* permutations of the 12 disks. Using the second theorem on distinguishable permutations, we have $\dfrac{(5+3+2+2)!}{5!\,3!\,2!\,2!}=\dfrac{12!}{5!\,3!\,2!\,2!}=166{,}320$.

11 In the 10-letter word *bookkeeper*, there are 3 *e*'s, 2 *o*'s, and 2 *k*'s. Using the second theorem on distinguishable permutations, we have $\dfrac{10!}{3!\,2!\,2!\,1!\,1!\,1!}=151{,}200$.

13 There are $C(10,\,5)$ ways to pick the first team.
The second team is determined once the first team is selected. $C(10,\,5)=252$

15 Two points determine a unique line. $C(8,\,2)=28$

17 There are 3! ways to order the categories. $(5!\cdot4!\cdot8!)\cdot3!=696{,}729{,}600$

**19** Pick the center, $C(3, 1)$; two guards, $C(10, 2)$;

two tackles from the 8 remaining linemen, $C(8, 2)$; two ends, $C(4, 2)$;

two halfbacks, $C(6, 2)$; the quarterback, $C(3, 1)$; and the fullback, $C(4, 1)$.

$$3 \cdot C(10, 2) \cdot C(8, 2) \cdot C(4, 2) \cdot C(6, 2) \cdot 3 \cdot 4 = 4{,}082{,}400$$

**21** There are C(12, 3) = 220 ways to pick the men and C(8, 2) = 28 ways to pick the women. By the fundamental counting principle, the total number of ways to pick the committee is $220 \cdot 28 = 6160$.

**23** We need 3 U's out of 8 moves. $C(8, 3) = 56$

**25** (a) We want 6 of the 49 numbers and order is not important. $C(49, 6) = 13{,}983{,}816$

(b) There are 24 even numbers from 1 to 49. $C(24, 6) = 134{,}596$

**27** Let $n$ denote the number of players. $C(n, 2) = 45 \Rightarrow \dfrac{n!}{(n-2)!\,2!} = 45 \Rightarrow$

$$n(n-1) = 90 \Rightarrow n^2 - n - 90 = 0 \Rightarrow (n-10)(n+9) = 0 \Rightarrow \{n > 0\}\ n = 10.$$

**29** Each team must win 3 of the first 6 games for the series to be extended to a 7th game. $C(6, 3) = 20$

**31** They may have computed $C(31, 3)$, which is 4495.

**33** (a) The amounts received are the same, so the order of selection *is not* important, and we use a combination. $C(1000, 30) = \dfrac{1000!}{970!\,30!} \approx 2.43 \times 10^{57}$

(b) The amounts received are different, so the order of selection *is* important, and we use a permutation. $P(1000, 30) = \dfrac{1000!}{970!} \approx 6.44 \times 10^{89}$

**35** We want to select 3 of the 4 kings and 2 of the remaining 48 cards. The order of selection is not important, so we use combinations. Any of the groups of 3 kings can be selected with any pair of the remaining 48 cards, so by the fundamental counting principle the total number of hands is $C(4, 3) \cdot C(48, 2) = 4 \cdot 1128 = 4512$.

**37** (a) $S_1 = \binom{1}{1} + \binom{1}{3} + \binom{1}{5} + \cdots = 1 + 0 + 0 + \cdots = 1.$

$S_2 = \binom{2}{1} + \binom{2}{3} + \binom{2}{5} + \cdots = 2 + 0 + 0 + \cdots = 2.$

$S_3 = 3 + 1 + 0 + \cdots = 4.$ $S_4 = 4 + 4 + 0 + \cdots = 8.$

$S_5 = 16,\ S_6 = 32,\ S_7 = 64,\ S_8 = 128,\ S_9 = 256,\ S_{10} = 512.$

(b) It appears that $S_n = 2^{n-1}$.

[39] (a) Graph the values of $C(10, 1)$, $C(10, 2)$, $C(10, 3)$, ... , $C(10, 10)$.

(b) The maximum value of $C(10, r)$ is 252 and occurs at $r = 5$.

[0, 10] by [0, 300, 50]

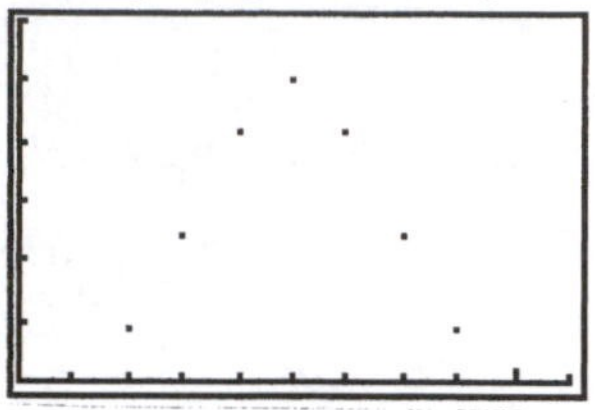

Figure 39

[0, 19] by [0, $10^5$, $10^4$]

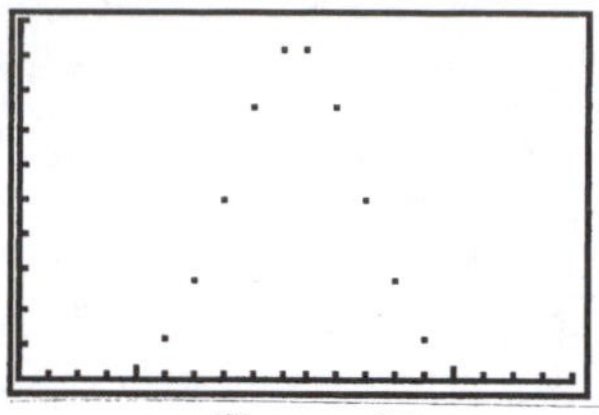

Figure 41

[41] (a) Graph the values of $C(19, 1)$, $C(19, 2)$, $C(19, 3)$, ... , $C(19, 19)$.

(b) The maximum value of $C(19, r)$ is 92,378 and occurs at $r = 9, 10$.

## 7.8 Exercises

[1] (a) The sample space is the 52-card deck, thus, $n(S) = 52$.

There are 4 kings, so the probability is $P(E) = \dfrac{n(E)}{n(S)} = \dfrac{4}{52} = \dfrac{1}{13}$.

The odds are $n(E)$ to $n(E')$, which are 4 to 48, or 1 to 12.

(b) $\frac{4}{52} + \frac{4}{52} = \frac{8}{52} = \frac{2}{13}$; $O(E)$ are 8 to 44, or 2 to 11

(c) $\frac{4}{52} + \frac{4}{52} + \frac{4}{52} = \frac{12}{52} = \frac{3}{13}$; $O(E)$ are 12 to 40, or 3 to 10

[3] $S = \{1, 2, 3, 4, 5, 6\}$ and $n(S) = 6$.

(a) $P(\text{rolling a } 4) = \frac{1}{6}$. There's 1 way to get a 4 and 5 ways to not get a 4,

so $O(E)$ are 1 to 5.

(b) $P(\text{rolling a } 6) = \frac{1}{6}$; $O(E)$ are 1 to 5.

(c) $\frac{1}{6} + \frac{1}{6} = \frac{2}{6} = \frac{1}{3}$; $O(E)$ are 2 to 4, or 1 to 2

[5] $n(S) = 5 + 6 + 4 = 15$

(a) $P(\text{drawing a red ball}) = \frac{5}{15} = \frac{1}{3}$; $O(E)$ are 5 to 10, or 1 to 2.

(b) $P(\text{drawing a green ball}) = \frac{6}{15} = \frac{2}{5}$; $O(E)$ are 6 to 9, or 2 to 3.

(c) $\frac{5}{15} + \frac{4}{15} = \frac{9}{15} = \frac{3}{5}$; $O(E)$ are 9 to 6, or 3 to 2

[7] *Note:* The following table summarizes the results for the sum of two dice being tossed. Notice the symmetry about the sum of 7 in the # of ways to obtain.

| Sum of two dice | 2 | 3 | 4 | 5 | 6 | 7 | 8 | 9 | 10 | 11 | 12 |
|---|---|---|---|---|---|---|---|---|---|---|---|
| # of ways to obtain | 1 | 2 | 3 | 4 | 5 | 6 | 5 | 4 | 3 | 2 | 1 |

(a) $\frac{2}{36} = \frac{1}{18}$; $O(E)$ are 2 to 34, or 1 to 17

(b) $\frac{5}{36}$; $O(E)$ are 5 to 31

(c) $\frac{2}{36} + \frac{5}{36} = \frac{7}{36}$; $O(E)$ are 7 to 29

[9] $n(S) = 6 \times 6 \times 6 = 216$.

There are 6 ways to make a sum of 5 (3 with 1, 1, 3 and 3 with 1, 2, 2). $\frac{6}{216} = \frac{1}{36}$

[11] There are 2 choices for each of the 3 coins, so $n(S) = 2 \cdot 2 \cdot 2 = 8$. Getting two heads to turn up is the same as getting one tail to turn up. There are 3 ways to get one tail: THH, HTH, HHT. So the probability is $\frac{3}{2^3} = \frac{3}{8}$.

[13] If $P(E) = \frac{5}{7}$, then $O(E)$ are 5 to 2 and $O(E')$ are 2 to 5.

[15] If $O(E)$ are 9 to 5, then $O(E')$ are 5 to 9 and $P(E) = \frac{9}{5+9} = \frac{9}{14}$.

[17] $P(E) \approx 0.659 = \frac{659}{1000}$, so $n(E) = 659$ and $n(E') = 1000 - 659 = 341$ and $O(E)$ are 659 to 341. Divide both numbers by 341 to get odds of 1.93 to 1.

*Note:* For Exercises 19–24, there are $C(52, 5) = 2{,}598{,}960$ ways to draw 5 cards.

[19] There are 13 denominations to pick from and any one of them could be combined with any one of the remaining 48 cards. $\frac{48 \cdot 13}{C(52, 5)} = \frac{1}{4165} \approx 0.00024$

[21] Pick 4 of the 13 diamonds and 1 of the 13 spades.

$$\frac{C(13, 4) \cdot C(13, 1)}{C(52, 5)} = \frac{143}{39{,}984} \approx 0.00358$$

[23] Pick 5 of the 13 cards in one suit. There are 4 suits. $\frac{C(13, 5) \cdot 4}{C(52, 5)} = \frac{33}{16{,}660} \approx 0.00198$

[25] Let $E_1$ be the event that the number is odd, $E_2$ that the number is prime.

$E_1 = \{1, 3, 5\}$ and $E_2 = \{2, 3, 5\}$.

$$P(E_1 \cup E_2) = P(E_1) + P(E_2) - P(E_1 \cap E_2) = \tfrac{3}{6} + \tfrac{3}{6} - \tfrac{2}{6} = \tfrac{4}{6} = \tfrac{2}{3}.$$

[27] The probability that the player does *not* get a hit is $1 - 0.326 = 0.674$. The probability that the player does not get a hit four times in a row is $(0.674) \cdot (0.674) \cdot (0.674) \cdot (0.674) = (0.674)^4 \approx 0.2064$.

[29] (a) $P(E_2) = P(2) + P(3) + P(4) = 0.10 + 0.15 + 0.20 = 0.45$

(b) $P(E_1 \cap E_2) = P(2) = 0.10$

(c) $P(E_1 \cup E_2) = P(E_1) + P(E_2) - P(E_1 \cap E_2) = 0.35 + 0.45 - 0.10 = 0.70$

(d) $P(E_2 \cup E_3') = P(E_2) + P(E_3') - P(E_2 \cap E_3')$. $E_3' = \{1, 2, 3, 5\}$ and

$$E_2 \cap E_3' = \{2, 3\} \Rightarrow P(E_2 \cup E_3') = 0.45 + 0.75 - 0.25 = 0.95.$$

*Note:* For Exercises 31–32, there are $C(60, 5) = 5{,}461{,}512$ ways to draw 5 chips.

[31] (a) We want 5 blue and 0 non-blue. There are 20 blue and 40 non-blue chips in the box. $\frac{C(20, 5) \cdot C(40, 0)}{C(60, 5)} = \frac{34}{11{,}977} \approx 0.0028$.

(b) $P(\text{at least 1 green}) = 1 - P(\text{no green}) =$

$$1 - \frac{C(30,\,0)\cdot C(30,\,5)}{C(60,\,5)} = 1 - \frac{117}{4484} = \frac{4367}{4484} \approx 0.9739.$$

(c) $P(\text{at most 1 red}) = P(0\text{ red}) + P(1\text{ red}) =$

$$\frac{C(10,\,0)\cdot C(50,\,5)}{C(60,\,5)} + \frac{C(10,\,1)\cdot C(50,\,4)}{C(60,\,5)} = \frac{26{,}320}{32{,}509} \approx 0.8096.$$

**33** (a) $\frac{C(8,\,8)}{2^8} = \frac{1}{256} \approx 0.00391$ (b) $\frac{C(8,\,7)}{2^8} = \frac{1}{32} = 0.03125$

(c) $\frac{C(8,\,6)}{2^8} = \frac{7}{64} = 0.109375$ (d) $\frac{C(8,\,6) + C(8,\,7) + C(8,\,8)}{2^8} = \frac{37}{256} \approx 0.14453$

**35** $P(\text{obtaining at least one ace}) = 1 - P(\text{no aces}) =$

$$1 - \frac{C(48,\,5)}{C(52,\,5)} = 1 - \frac{35{,}673}{54{,}145} = \frac{18{,}472}{54{,}145} \approx 0.34116$$

**37** (a) We may use ordered pairs to represent the outcomes of the sample space $S$ of the experiment. A representative outcome is (nine of clubs, 3). The number of outcomes in the sample space $S$ is $n(S) = 52\cdot 6 = 312$.

(b) For each integer $k$, where $2 \le k \le 6$, there are 4 ways to obtain an outcome of the form $(k,\ k)$ since there are 4 suits. Because there are 5 values of $k$, $n(E_1) = 5\cdot 4 = 20$. $n(E_1') = n(S) - n(E_1) = 312 - 20 = 292$.

$$P(E_1) = \frac{n(E_1)}{n(S)} = \frac{20}{312} = \frac{5}{78}.$$

(c) No, if $E_2$ or $E_3$ occurs, then the other event may occur.

Yes, the occurrence of either $E_2$ or $E_3$ has no effect on the other event.

$P(E_2) = \frac{n(E_2)}{n(S)} = \frac{12\cdot 6}{312} = \frac{72}{312} = \frac{3}{13}.$ $P(E_3) = \frac{n(E_3)}{n(S)} = \frac{52\cdot 3}{312} = \frac{156}{312} = \frac{1}{2}.$

Since $E_2$ and $E_3$ are indep., $P(E_2 \cap E_3) = P(E_2)\cdot P(E_3) = \frac{3}{13}\cdot\frac{1}{2} = \frac{3}{26} = \frac{36}{312}.$

$P(E_2 \cup E_3) = P(E_2) + P(E_3) - P(E_2 \cap E_3) = \frac{72}{312} + \frac{156}{312} - \frac{36}{312} = \frac{192}{312} = \frac{8}{13}.$

(d) Yes, if $E_1$ or $E_2$ occurs, then the other event cannot occur. No, the occurrence of either $E_1$ or $E_2$ influences the occurrence of the other event.

Remember, (non-empty) *mutually exclusive events* ***cannot*** *be independent events.*

Since $E_1$ and $E_2$ are mutually exclusive, $P(E_1 \cap E_2) = 0$ and

$$P(E_1 \cup E_2) = P(E_1) + P(E_2) = \frac{20}{312} + \frac{72}{312} = \frac{92}{312} = \frac{23}{78}.$$

**39** Let $k$ denote the sum.

$$P(k > 5) = 1 - P(k \le 5) = 1 - \left(\frac{1}{36} + \frac{2}{36} + \frac{3}{36} + \frac{4}{36}\right) = 1 - \frac{10}{36} = \frac{26}{36} = \frac{13}{18}.$$

**41** (a) Each birth is independent of the others and the probability of having a boy is $\frac{1}{2}$.

$P(\text{all boys}) = \frac{1}{2}\cdot\frac{1}{2}\cdot\frac{1}{2}\cdot\frac{1}{2}\cdot\frac{1}{2} = \frac{1}{32} = 0.03125.$

(b) $P(\text{at least one girl}) = 1 - P(\text{all boys}) = 1 - \frac{1}{32} = \frac{31}{32} = 0.96875.$

**43** There are 4! ways to place the cards, so $n(S) = 4!$.

(a) $\dfrac{C(4,\,4)}{4!} = \frac{1}{24} \approx 0.04167$ (b) $\dfrac{C(4,\,2)}{4!} = \frac{1}{4} = 0.25$

**45** (a) The 3, 4, or 5 on the left die would need to combine with a 4, 3, or 2 on the right die to sum to 7, but the right die only has 1, 5, or 6. The probability is 0.

(b) To obtain 8, we would need a 3 on the left die and a 5 on the right. $\frac{1}{3}\cdot\frac{1}{3} = \frac{1}{9} = 0.\overline{1}$

**47** There are 3273 smoking-related deaths that are not in any of the 3 categories.

(a) $P(\text{cardiovascular disease or cancer}) = \dfrac{179{,}820 + 151{,}322}{418{,}890} = \dfrac{331{,}142}{418{,}890} \approx 0.791$

(b) $P(\textit{not}\text{ respiratory}) = 1 - P(\text{respiratory}) = \dfrac{418{,}890 - 84{,}475}{418{,}890} = \dfrac{334{,}415}{418{,}890} \approx 0.798$

**49** Start the tree diagram by making branches for those with cancer ($C$, 2%) and those without cancer ($C'$, 98%). On the $C$ branch, make a branch for those exposed to high levels of arsenic ($E$, 70%) and a branch for those not exposed ($E'$, 30%). On the $C'$ branch, make a branch for those exposed to high levels of arsenic ($E$, 10%) and a branch for those not exposed ($E'$, 90%). Find the products: $0.02 \cdot 0.70 = \mathbf{0.014}$, $0.02 \cdot 0.30 = \mathbf{0.006}$, $0.98 \cdot 0.10 = \mathbf{0.098}$, and $0.98 \cdot 0.90 = \mathbf{0.882}$. Note that the sum of the boldfaced numbers equals 1. The fraction of people that have been exposed to high levels of arsenic and have cancer is

$$\frac{\text{exposed and have cancer}}{\text{all those exposed}} = \frac{C \text{ and } E}{C \text{ and } E + C' \text{ and } E} = \frac{0.014}{0.014 + 0.098} = \frac{0.014}{0.112} = 0.125, \text{ so}$$

the percentage is 12.5%.

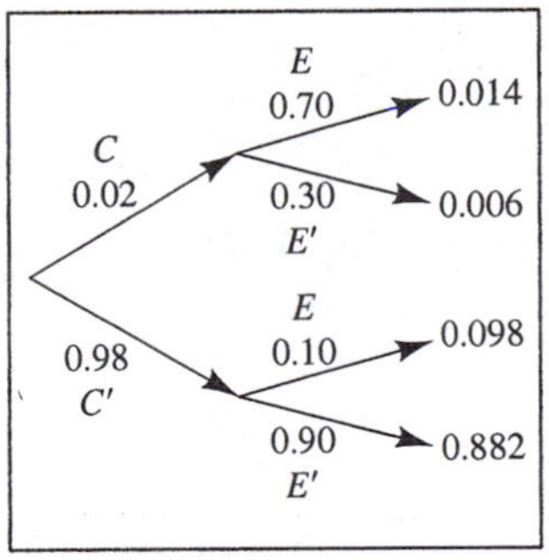


*Figure 49*

**51** (a) The ball must take 4 "lefts". $\frac{1}{2}\cdot\frac{1}{2}\cdot\frac{1}{2}\cdot\frac{1}{2} = \frac{1}{16} = 0.0625$

(b) We need two "lefts". $\dfrac{C(4,\,2)}{2^4} = \dfrac{6}{16} = \dfrac{3}{8} = 0.375$

**53** For one ticket, $P(E) = \dfrac{n(E)}{n(S)} = \dfrac{C(6,\,6)}{C(54,\,6)} = \dfrac{1}{25{,}827{,}165}$.

For two tickets, $P(E) = \dfrac{2 \times 1}{25{,}827{,}165}$, or about 1 chance in 13 million.

[55] The probability that the first bulb is not defective is $\frac{195}{200}$ since 195 of the 200 bulbs are not defective. If the first bulb is not replaced and not defective, then there are 199 bulbs left, and 194 of them are not defective. The probability that the second bulb is not defective is then $\frac{194}{199}$. Thus, the probability that both bulbs are not defective is $\frac{195}{200} \times \frac{194}{199} = \frac{37{,}830}{39{,}800} \approx 0.9505$. The event that either light bulb is defective is the complement of the event that neither bulb is defective. The probability that the sample will be rejected is $1 - \frac{37{,}830}{39{,}800} = \frac{1970}{39{,}800} \approx 0.0495$.

[57] (a) $P(7 \text{ or } 11) = P(7) + P(11) = \frac{6}{36} + \frac{2}{36} = \frac{8}{36}$

(b) To win with a 4 on the first roll, we must first get a 4, and then get another 4 before a 7. The probability of getting a 4 is $\frac{3}{36}$. The probability of getting another 4 before a 7 is $\frac{3}{3+6}$ since there are 3 ways to get a 4, 6 ways to get a 7, and numbers other than 4 and 7 are immaterial.

Thus, $P(\text{winning with } 4) = \frac{3}{36} \cdot \frac{3}{3+6} = \frac{1}{36}$.

(c) Let $P(k)$ denote the probability of winning a pass line bet with the number $k$. We first note that $P(4) = P(10)$, $P(5) = P(9)$, and $P(6) = P(8)$.

$$P(\text{winning}) = 2 \cdot P(4) + 2 \cdot P(5) + 2 \cdot P(6) + P(7) + P(11)$$
$$= 2 \cdot \frac{3}{36} \cdot \frac{3}{3+6} + 2 \cdot \frac{4}{36} \cdot \frac{4}{4+6} + 2 \cdot \frac{5}{36} \cdot \frac{5}{5+6} + \frac{6}{36} + \frac{2}{36}$$
$$= 2 \cdot \frac{1}{36} + 2 \cdot \frac{2}{45} + 2 \cdot \frac{25}{396} + \frac{1}{6} + \frac{1}{18} = \frac{488}{990} = \frac{244}{495} \approx 0.4929$$

[59] (a) $p = P((S_1 \cap S_2) \cup (S_3 \cap S_4))$
$$= P(S_1 \cap S_2) + P(S_3 \cap S_4) - P((S_1 \cap S_2) \cap (S_3 \cap S_4))$$
$$= P(S_1) \cdot P(S_2) + P(S_3) \cdot P(S_4) - P(S_1 \cap S_2) \cdot P(S_3 \cap S_4)$$
$$= P(S_1) \cdot P(S_2) + P(S_3) \cdot P(S_4) - P(S_1) \cdot P(S_2) \cdot P(S_3) \cdot P(S_4)$$

Let $P(S_k) = x$. Then $p = x \cdot x + x \cdot x - x \cdot x \cdot x \cdot x = -x^4 + 2x^2$.

$x = 0.9 \Rightarrow p = 0.9639$.

[−2.25, 2.25, 0.5] by [−2, 1, 0.5]

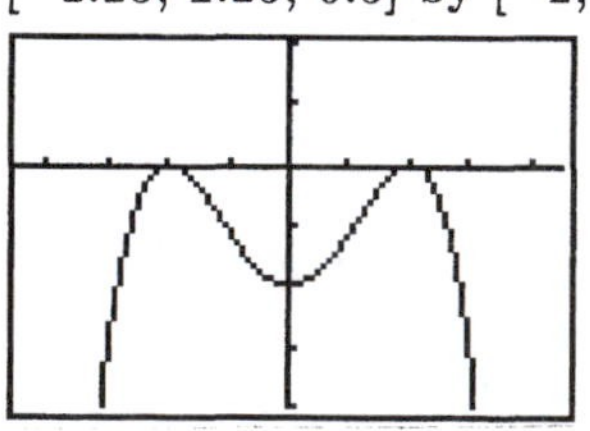

*Figure 59(a)*

[0.96, 1.05, 0.5] by [−0.03, 0.04, 0.5]

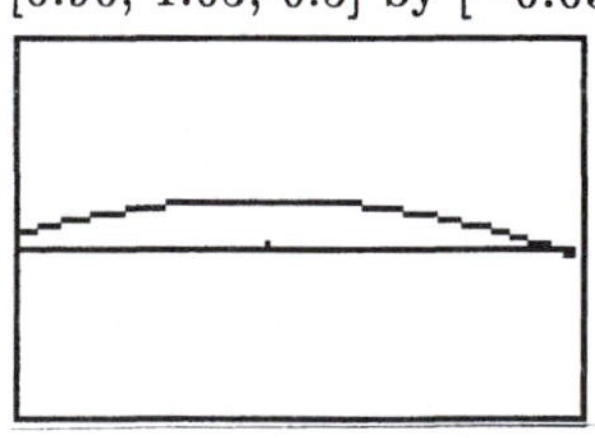

*Figure 59(b)*

(b) $p = 0.99 \Rightarrow -x^4 + 2x^2 = 0.99$. The graph of $y = -x^4 + 2x^2 - 0.99$ is shown in *Figure 59(a)*. The region near $x = 1$ is enlarged in *Figure 59(b)* to show that the graph is above the $x$-axis for some values of $x$. The approximate $x$-intercepts are $\pm 0.95$, $\pm 1.05$. Since $0 \le P(S_k) \le 1$, $P(S_k) = 0.95$.

**61** (a) The number of ways that $n$ people can all have a different birthday is $P(365, n)$. The number of outcomes in the sample space is $365^n$.

$$\text{Thus, } p = \frac{P(365,\, n)}{365^n} = \frac{365!}{365^n(365-n)!}.$$

(b) $n = 32 \Rightarrow p = \dfrac{365!}{365^{32}\,333!} \Rightarrow \ln p = \ln \dfrac{365!}{365^{32}\,333!} =$

$\ln 365! - \ln 365^{32} - \ln 333! \approx (365 \ln 365 - 365) - (32 \ln 365) - (333 \ln 333 - 333) \approx -1.45$. Thus, $p \approx e^{-1.45} \approx 0.24$.

The probability that two or more people have the same birthday is $1 - p \approx 0.76$.

**63** From Exercise 57(c), the payoff amount of \$2 has a probability of $\frac{244}{495}$. Hence, $EV = 2 \cdot \frac{244}{495} = \frac{488}{495} \approx \$0.986$, or about \$0.99. Of course, after paying \$1 to play, you can expect to lose about \$7 for every 495 one-dollar pass line bets.

**65** $EV = 1{,}000{,}000 \cdot \frac{1}{20{,}000{,}000} + 100{,}000 \cdot \frac{10}{20{,}000{,}000} + 10{,}000 \cdot \frac{100}{20{,}000{,}000} + 1000 \cdot \frac{1000}{20{,}000{,}000}$

$= \$0.20$ {less than the cost of a first class stamp}

## *Chapter 7 Review Exercises*

**1** $a_n = \dfrac{5n}{3 - 2n^2}$ •

$a_1 = \dfrac{5(1)}{3 - 2(1)^2} = \dfrac{5}{1} = 5,$ $\quad a_2 = \dfrac{5(2)}{3 - 2(2)^2} = \dfrac{10}{-5} = -2,$

$a_3 = \dfrac{5(3)}{3 - 2(3)^2} = \dfrac{15}{-15} = -1,$ $\quad a_4 = \dfrac{5(4)}{3 - 2(4)^2} = \dfrac{20}{-29} = -\dfrac{20}{29},$

$a_7 = \dfrac{5(7)}{3 - 2(7)^2} = \dfrac{35}{-95} = -\dfrac{7}{19}.$ $\quad$ ★ $5, -2, -1, -\frac{20}{29}; -\frac{7}{19}$

**3** $a_n = 1 + (-\frac{1}{2})^{n-1}$ •

$a_1 = 1 + (-\frac{1}{2})^0 = 1 + 1 = 2,$ $\quad a_2 = 1 + (-\frac{1}{2})^1 = 1 - \frac{1}{2} = \frac{1}{2},$

$a_3 = 1 + (-\frac{1}{2})^2 = 1 + \frac{1}{4} = \frac{5}{4},$ $\quad a_4 = 1 + (-\frac{1}{2})^3 = 1 - \frac{1}{8} = \frac{7}{8},$

$a_7 = 1 + (-\frac{1}{2})^6 = 1 + \frac{1}{64} = \frac{65}{64}$ $\quad$ ★ $2, \frac{1}{2}, \frac{5}{4}, \frac{7}{8}; \frac{65}{64}$

[5] $a_1 = 10,\ a_{k+1} = 1 + (1/a_k)$ •

$a_2 = a_{1+1}\ \{k = 1\} = 1 + 1/a_1 = 1 + 1/10 = \frac{10}{10} + \frac{1}{10} = \frac{11}{10}$

$a_3 = a_{2+1} = 1 + 1/a_2 = 1 + 1/\frac{11}{10} = 1 + \frac{10}{11} = \frac{11}{11} + \frac{10}{11} = \frac{21}{11}$

$a_4 = a_{3+1} = 1 + 1/a_3 = 1 + 1/\frac{21}{11} = 1 + \frac{11}{21} = \frac{21}{21} + \frac{11}{21} = \frac{32}{21}$

$a_5 = a_{4+1} = 1 + 1/a_4 = 1 + 1/\frac{32}{21} = 1 + \frac{21}{32} = \frac{32}{32} + \frac{21}{32} = \frac{53}{32}$ ★ $10, \frac{11}{10}, \frac{21}{11}, \frac{32}{21}, \frac{53}{32}$

[7] $a_1 = 9,\ a_{k+1} = \sqrt{a_k}$ • $a_2 = a_{1+1}\ \{k = 1\} = \sqrt{a_1} = \sqrt{9} = 3$

$a_3 = a_{2+1} = \sqrt{a_2} = \sqrt{3}$ $a_4 = a_{3+1} = \sqrt{a_3} = \sqrt{\sqrt{3}} = \sqrt[4]{3}$

$a_5 = a_{4+1} = \sqrt{a_4} = \sqrt{\sqrt[4]{3}} = \sqrt[8]{3}$ ★ $9, 3, \sqrt{3}, \sqrt[4]{3}, \sqrt[8]{3}$

[9] $\displaystyle\sum_{k=1}^{5} (k^2 + 4) = (1^2 + 4) + (2^2 + 4) + (3^2 + 4) + (4^2 + 4) + (5^2 + 4)$
$= 5 + 8 + 13 + 20 + 29 = 75$

[11] $\displaystyle\sum_{k=7}^{100} 10 = (100 - 7 + 1)(10) = 94(10) = 940$

[13] $3 + 6 + 9 + 12 + 15 = 3(1) + 3(2) + 3(3) + 3(4) + 3(5) = \displaystyle\sum_{n=1}^{5} (3n)$

[14] We first note that $4 + 2 + 1 + \frac{1}{2} + \frac{1}{4} + \frac{1}{8}$ is the sum of six terms of a geometric sequence with first term 4 and ratio $\frac{1}{2}$—that is, $a_n = a_1 r^{n-1} = 4(\frac{1}{2})^{n-1}$.

$4 + 2 + 1 + \frac{1}{2} + \frac{1}{4} + \frac{1}{8} = \displaystyle\sum_{n=1}^{6} 4(\tfrac{1}{2})^{n-1} = \sum_{n=1}^{6} 2^2(2^{-1})^{n-1} = \sum_{n=1}^{6} 2^2 2^{1-n} = \sum_{n=1}^{6} 2^{3-n}$

[15] $\dfrac{1}{1 \cdot 2} + \dfrac{1}{2 \cdot 3} + \dfrac{1}{3 \cdot 4} + \cdots + \dfrac{1}{99 \cdot 100} = \displaystyle\sum_{n=1}^{99} \frac{1}{n(n+1)}$

[17] $\frac{1}{2} + \frac{2}{5} + \frac{3}{8} + \frac{4}{11}$ • The numerators are just the positive integers, so we'll use the summation notation with $n$ having values from 1 to 4. The denominators increase by 3, so it must be of the form $3n + k$. When $n = 1$, $3n + k = 3(1) + k = 3 + k$ and $3 + k$ must equal 2 (the first denominator). Thus, $3 + k = 2 \Rightarrow k = -1$, so the denominator is $3n - 1$ and the summation notation is $\displaystyle\sum_{n=1}^{4} \frac{n}{3n-1}$.

[19] $100 - 95 + 90 - 85 + 80$ • The terms have alternating signs and decrease by 5 in magnitude. For $n = 1$, we subtract 5 from 105 to get 100. $\displaystyle\sum_{n=1}^{5} (-1)^{n+1}(105 - 5n)$

[20] $1 - \frac{1}{2} + \frac{1}{3} - \frac{1}{4} + \frac{1}{5} - \frac{1}{6} + \frac{1}{7}$ • The terms have alternating signs,

the numerator is 1, and the denominators increase by 1. $\displaystyle\sum_{n=1}^{7} (-1)^{n-1}\frac{1}{n}$

[21] $a_0 + a_1x^4 + a_2x^8 + \cdots + a_{25}x^{100}$ { the exponents are multiples of 4 } $= \sum_{n=0}^{25} a_n x^{4n}$

[23] $1 - \frac{x^2}{2} + \frac{x^4}{4} - \frac{x^6}{6} + \cdots + (-1)^n \frac{x^{2n}}{2n}$ • The pattern begins with the second term and the general term is listed. $1 + \sum_{k=1}^{n} (-1)^k \frac{x^{2k}}{2k}$

[25] $d = 3 - (4 + \sqrt{3}) = -1 - \sqrt{3}$; $a_{10} = (4 + \sqrt{3}) + (9)(-1 - \sqrt{3}) = -5 - 8\sqrt{3}$;

$$S_{10} = \tfrac{10}{2}\Big[(4 + \sqrt{3}) + (-5 - 8\sqrt{3})\Big] = -5 - 35\sqrt{3}.$$

[27] $a_5 = a_1 + 4d = 5$ and $a_{13} = a_1 + 12d = 77$. $a_{13} - a_5 = (a_1 + 12d) - (a_1 + 4d) = 8d$ and

$a_{13} - a_5 = 77 - 5 = 72 \Rightarrow 8d = 72 \Rightarrow d = 9$.

$a_5 = a_1 + 4d \Rightarrow 5 = a_1 + 36 \Rightarrow a_1 = -31$. $a_{10} = -31 + 9(9) = 50$.

[29] Four arithmetic means $\Rightarrow 5d = -10 - 20 \Rightarrow d = -6$.

The terms are 20, 14, 8, 2, $-4$, and $-10$.

[31] We can divide the fourth term by the third term to obtain the common ratio,

$r = \frac{-0.3}{3} = -0.1$. If we multiply the third term by $r$ five times,

we will get the eighth term. $a_8 = a_3r^5 = 3(-0.1)^5 = -0.00003$.

[33] The geometric mean of 4 and 8 is $\sqrt{4 \cdot 8} = \sqrt{32} = 4\sqrt{2}$.

[35] $402 = \frac{12}{2}(a_1 + 50) \Rightarrow a_1 = 17$. $a_{12} = a_1 + 11d \Rightarrow d = (50 - 17)/11 = 3$.

[36] $a_5 = a_1r^4 \Rightarrow \frac{1}{16} = a_1\left(\frac{3}{2}\right)^4 \Rightarrow a_1 = \frac{1}{16} \cdot \frac{16}{81} = \frac{1}{81}$.

$$S_5 = \frac{1}{81} \cdot \frac{1 - \left(\frac{3}{2}\right)^5}{1 - \frac{3}{2}} = \frac{1}{81} \cdot \frac{1 - \frac{243}{32}}{-\frac{1}{2}} = \frac{1}{81} \cdot \frac{-\frac{211}{32}}{-\frac{1}{2}} = \frac{211}{1296}.$$

[37] The sequence of terms is arithmetic.

$a_1 = 5(1) - 2 = 3$ and $a_{15} = 5(15) - 2 = 73$, so $S_{15} = \frac{15}{2}(3 + 73) = 570$.

[39] This sum can be written as a sum of a geometric sequence and an arithmetic

sequence. $\sum_{k=1}^{10} (2^k - \frac{1}{2}) = \sum_{k=1}^{10} 2^k - \sum_{k=1}^{10} \frac{1}{2} = 2 \cdot \frac{1 - 2^{10}}{1 - 2} - 10(\frac{1}{2}) = 2046 - 5 = 2041$

[41] $a_1 = 1$, $r = -\frac{2}{5} \Rightarrow S = \frac{1}{1 - (-\frac{2}{5})} = \frac{1}{\frac{7}{5}} = \frac{5}{7}$.

**43** (1) $P_1$ is true, since $3(1)-1=\frac{1[3(1)+1]}{2}=2$.

(2) Assume $P_k$ is true: $2+5+8+\cdots+(3k-1)=\frac{k(3k+1)}{2}$. Hence,

$$\begin{aligned}2+5+8+\cdots+(3k-1)+3(k+1)-1 &= \frac{k(3k+1)}{2}+3(k+1)-1\\ &= \frac{3k^2+k+6k+4}{2}\\ &= \frac{3k^2+7k+4}{2}\\ &= \frac{(k+1)(3k+4)}{2}\\ &= \frac{(k+1)[3(k+1)+1]}{2}.\end{aligned}$$

Thus, $P_{k+1}$ is true, and the proof is complete.

**45** (1) $P_1$ is true, since $\frac{1}{[2(1)-1][2(1)+1]}=\frac{1}{2(1)+1}=\frac{1}{3}$.

(2) Assume $P_k$ is true:

$$\frac{1}{1\cdot 3}+\frac{1}{3\cdot 5}+\frac{1}{5\cdot 7}+\cdots+\frac{1}{(2k-1)(2k+1)}=\frac{k}{2k+1}. \text{ Hence,}$$

$$\begin{aligned}\frac{1}{1\cdot 3}+\frac{1}{3\cdot 5}+\frac{1}{5\cdot 7}+\cdots+\frac{1}{(2k-1)(2k+1)}+\frac{1}{(2k+1)(2k+3)}\\ &= \frac{k}{2k+1}+\frac{1}{(2k+1)(2k+3)}\\ &= \frac{k(2k+3)+1}{(2k+1)(2k+3)}\\ &= \frac{2k^2+3k+1}{(2k+1)(2k+3)}\\ &= \frac{(2k+1)(k+1)}{(2k+1)(2k+3)}\\ &= \frac{k+1}{2(k+1)+1}.\end{aligned}$$

Thus, $P_{k+1}$ is true, and the proof is complete.

**47** (1) For $n=1$, $n^3+2n=3$ and 3 is a factor of 3.

(2) Assume 3 is a factor of $k^3+2k$. The $(k+1)$st term is

$$\begin{aligned}(k+1)^3+2(k+1) &= k^3+3k^2+5k+3\\ &= (k^3+2k)+(3k^2+3k+3)\\ &= (k^3+2k)+3(k^2+k+1).\end{aligned}$$

By the induction hypothesis, 3 is a factor of $k^3+2k$ and 3 is a factor of $3(k^2+k+1)$, so 3 is a factor of the $(k+1)$st term. Thus, $P_{k+1}$ is true, and the proof is complete.

[49] For $j$: Examining the pattern formed by letting $n = 1, 2, 3, 4$ leads us to the conclusion that $j = 4$.

(1) $P_4$ is true, since $2^4 \le 4!$.

(2) Assume $P_k$ is true: $2^k \le k!$. Hence,

$$2^{k+1} = 2 \cdot 2^k \le 2 \cdot k! < (k+1) \cdot k! = (k+1)!.$$

Thus, $P_{k+1}$ is true, and the proof is complete.

[51] We use the binomial theorem formula with $a = x^2$, $b = -3y$, and $n = 6$.

$$\begin{aligned}(x^2 - 3y)^6 &= (x^2)^6 + \binom{6}{1}(x^2)^5(-3y)^1 + \binom{6}{2}(x^2)^4(-3y)^2 + \binom{6}{3}(x^2)^3(-3y)^3 \\ &\quad + \binom{6}{4}(x^2)^2(-3y)^4 + \binom{6}{5}(x^2)^1(-3y)^5 + (-3y)^6 \\ &= x^{12} - 6 \cdot 3x^{10}y + 15 \cdot 9x^8y^2 - 20 \cdot 27x^6y^3 \\ &\quad + 15 \cdot 81x^4y^4 - 6 \cdot 243x^2y^5 + 729y^6 \\ &= x^{12} - 18x^{10}y + 135x^8y^2 - 540x^6y^3 + 1215x^4y^4 - 1458x^2y^5 + 729y^6\end{aligned}$$

[53] $(x^{2/5} + 2x^{-3/5})^{20}$; first three terms •

$$\begin{aligned}&= \binom{20}{0}(x^{2/5})^{20}(2x^{-3/5})^0 + \binom{20}{1}(x^{2/5})^{19}(2x^{-3/5})^1 + \binom{20}{2}(x^{2/5})^{18}(2x^{-3/5})^2 \\ &= 1 \cdot x^8 \cdot 1 + 20 \cdot x^{38/5} \cdot 2x^{-3/5} + 190 \cdot x^{36/5} \cdot 4x^{-6/5} \\ &= x^8 + 40 \cdot x^{35/5} + 760 \cdot x^{30/5} = x^8 + 40x^7 + 760x^6\end{aligned}$$

[55] $(4x^2 - y)^7$; term that contains $x^{10}$ •

Consider only the variable $x$ in the expansion:

$(x^2)^{7-k} = x^{10} \Rightarrow 14 - 2k = 10 \Rightarrow 2k = 4 \Rightarrow k = 2;$

$$(k+1)\text{st term} = 3\text{rd term} = \binom{7}{2}(4x^2)^5(-y)^2 = 21{,}504x^{10}y^2$$

[57] (a) $S_5 = 10 \Rightarrow 10 = \frac{5}{2}(2a_1 + 4d) \Rightarrow 4 = 2a_1 + 4d \Rightarrow 4d = 4 - 2a_1 \Rightarrow$

$d = 1 - \frac{1}{2}a_1$. Since $a_1$ is positive, $1 - \frac{1}{2}a_1$ is less than 1 ft.

(b) $a_1 = \frac{1}{2} \Rightarrow d = 1 - \frac{1}{2}(\frac{1}{2}) = \frac{3}{4}$.

The lengths of the other four pieces are $1\frac{1}{4}$, 2, $2\frac{3}{4}$, and $3\frac{1}{2}$ ft.

[59] If $s_1 = 1$, then $s_2 = f$, $s_3 = f^2$, .... There are two of each of the $s_k$'s.

Since $0 < f < 1$, and $f$ is the common ratio, we can sum the infinite sequence.

The sum of the $s_k$'s is $2(1 + f + f^2 + \cdots) = 2\left(\dfrac{1}{1-f}\right) = \dfrac{2}{1-f}$.

[60] Two ways to sum these integers are

$1 + 2 + 3 + 4 + 5 + 6 + 7 + 8 + 9 + 10$ and $2 + 1 + 4 + 3 + 6 + 5 + 8 + 7 + 10 + 9$.

$$P(10, 10) = \frac{10!}{(10-10)!} = \frac{10!}{0!} = 10! = 3{,}628{,}800$$

**62** (a) We want 6 digits taken 4 at a time. Order is important, so we use a permutation. $P(6,\ 4) = \dfrac{6!}{(6-4)!} = \dfrac{6\cdot5\cdot4\cdot3\cdot2!}{2!} = 6\cdot5\cdot4\cdot3 = 360$

(b) Since repetitions are allowed, we can use any of the 6 digits for each choice:

$$6\cdot6\cdot6\cdot6 = 6^4 = 1296$$

**64** $\dfrac{(6+5+4+2)!}{6!\,5!\,4!\,2!} = \dfrac{17!}{6!\,5!\,4!\,2!} = 85{,}765{,}680$

**66** There is one way to get all heads and one way to get all tails, so there are only two ways to get the coins to match.

(a) With two boys, $n(S) = 2\cdot2 = 4$, so the probability of a match is $\frac{2}{4} = \frac{1}{2}$.

(b) With three boys, $n(S) = 2\cdot2\cdot2 = 8$, so the probability of a match is $\frac{2}{8} = \frac{1}{4}$.

**68** (a) $\frac{1}{1000}$ (b) $\frac{10}{1000} = \frac{1}{100}$ (c) $\frac{50}{1000} = \frac{1}{20}$

**69** $P(1 \text{ head}) = \dfrac{C(4,\ 1)}{2^4} = \dfrac{4}{16} = \dfrac{1}{4} = 0.25$. Odds are 4 to 12, or 1 to 3.

**70** (a) $P(\text{passing}) = P(4,\ 5, \text{ or } 6 \text{ correct}) =$

$$\frac{C(6,\ 4) + C(6,\ 5) + C(6,\ 6)}{2^6} = \frac{15+6+1}{64} = \frac{22}{64} = \frac{11}{32}$$

(b) $P(\text{failing}) = 1 - P(\text{passing}) = 1 - \frac{22}{64} = \frac{42}{64}$

**72** Let $O$ denote the event that the individual is over 60 and $F$ denote the event that the individual is female.

$$P(O \cup F) = P(O) + P(F) - P(O \cap F) = \tfrac{1000}{5000} + \tfrac{2000}{5000} - \tfrac{0.40(2000)}{5000} = \tfrac{2200}{5000} = 0.44$$

**74** The two teams, A and B, can play as few as 4 games or as many as 7 games. Since the two teams are equally matched, the probability that either team wins any particular game is $\frac{1}{2}$.

$P(\text{team A wins in 4 games}) = \frac{1}{2}\cdot\frac{1}{2}\cdot\frac{1}{2}\cdot\frac{1}{2} = (\frac{1}{2})^4 = \frac{1}{16} = 0.0625$.

$P(\text{team A wins in 5 games})$

$= P(\text{team A wins 3 of the first 4 games \{losing 1 game\} } \textit{and} \text{ then wins game 5})$

$= \binom{4}{3}(\frac{1}{2})^3\cdot(\frac{1}{2})^1\cdot\frac{1}{2} = \binom{4}{3}(\frac{1}{2})^5 = \frac{4}{32} = 0.125$. In a similar fashion,

$$P(\text{team A wins in 6 games}) = \tbinom{5}{3}(\tfrac{1}{2})^6 = \tfrac{10}{64} = \tfrac{5}{32} = 0.15625$$

and $$P(\text{team A wins in 7 games}) = \tbinom{6}{3}(\tfrac{1}{2})^7 = \tfrac{20}{128} = \tfrac{5}{32} = 0.15625.$$

Thus, the probability that team A wins the series is

$$0.0625 + 0.125 + 0.15625 + 0.15625 = 0.5,$$

(continued)

which agrees with our common sense. Since the probabilities for team B winning the series are the same, the expected number of games is

$$4(2\cdot\tfrac{1}{16})+5(2\cdot\tfrac{4}{32})+6(2\cdot\tfrac{5}{32})+7(2\cdot\tfrac{5}{32})=5\tfrac{13}{16}=5.8125.$$

## Chapter 7 Discussion Exercises

[1] Probably the easiest way to describe this sequence is by using a piecewise-defined function. If $1 \le n \le 4$, then $2n$ describes the sequence 2, 4, 6, 8. For $n=5$, we need to obtain $a$ as the general term—one way to symbolize this is $(n-4)a$. Hence, one possibility is $a_n = \begin{cases} 2n & \text{if } 1 \le n \le 4 \\ (n-4)a & \text{if } n \ge 5 \end{cases}$

A more sophisticated (and complicated) approach is to add a term to $2n$—call this term $k$—so that $k$ is 0 for $n = 1, 2, 3, 4$, but equal to $(a-10)$ for $n=5$ {the *minus* 10 is needed to compensate for the *plus* 10 from the $2n$ term}. We can do this by starting with the expression $(n-1)(n-2)(n-3)(n-4)$, which is zero for $n = 1$, 2, 3, 4. For $n=5$, this expression is $4\cdot3\cdot2\cdot1=24$, so we will divide it by 24 and multiply it by $(a-10)$ to obtain $(a-10)$. Thus, another possibility is

$$a_n = 2n + \frac{(n-1)(n-2)(n-3)(n-4)(a-10)}{24}.$$

[3] (a) Following the pattern from Exercises 37–38 in Section 7.4, write $1^4+2^4+3^4+\cdots+n^4 = an^5+bn^4+cn^3+dn^2+en$. Then, it follows that:

$n=1 \;\Rightarrow\; a(1)^5+b(1)^4+c(1)^3+d(1)^2+e(1)=1^4 \;\Rightarrow\; a+b+c+d+e=1,$

$n=2 \;\Rightarrow\; a(2)^5+b(2)^4+c(2)^3+d(2)^2+e(2)=1^4+2^4 \;\Rightarrow$

$32a+16b+8c+4d+2e=17,$

$n=3 \;\Rightarrow\; a(3)^5+b(3)^4+c(3)^3+d(3)^2+e(3)=1^4+2^4+3^4 \;\Rightarrow$

$243a+81b+27c+9d+3e=98,$

$n=4 \;\Rightarrow\; a(4)^5+b(4)^4+c(4)^3+d(4)^2+e(4)=1^4+2^4+3^4+4^4 \;\Rightarrow$

$1024a+256b+64c+16d+4e=354,$

$n=5 \;\Rightarrow\; a(5)^5+b(5)^4+c(5)^3+d(5)^2+e(5)=1^4+2^4+3^4+4^4+5^4 \;\Rightarrow$

$3125a+625b+125c+25d+5e=979.$

(continued)

$$AX = B \Rightarrow \begin{bmatrix} 1 & 1 & 1 & 1 & 1 \\ 32 & 16 & 8 & 4 & 2 \\ 243 & 81 & 27 & 9 & 3 \\ 1024 & 256 & 64 & 16 & 4 \\ 3125 & 625 & 125 & 25 & 5 \end{bmatrix} \begin{bmatrix} a \\ b \\ c \\ d \\ e \end{bmatrix} = \begin{bmatrix} 1 \\ 17 \\ 98 \\ 354 \\ 979 \end{bmatrix} \Rightarrow X = A^{-1}B = \begin{bmatrix} 1/5 \\ 1/2 \\ 1/3 \\ 0 \\ -1/30 \end{bmatrix}$$

Thus, $1^4 + 2^4 + 3^4 + \cdots + n^4 = \frac{1}{5}n^5 + \frac{1}{2}n^4 + \frac{1}{3}n^3 - \frac{1}{30}n$.

This formula must be verified.

(b) Let $P_n$ be the statement that $1^4 + 2^4 + 3^4 + \cdots + n^4 = \frac{1}{5}n^5 + \frac{1}{2}n^4 + \frac{1}{3}n^3 - \frac{1}{30}n$.

(1) $1^4 = \frac{1}{5} + \frac{1}{2} + \frac{1}{3} - \frac{1}{30} = 1$ and $P_1$ is true.

(2) Assume that $P_k$ is true. We must show that $P_{k+1}$ is true.

$P_k \Rightarrow 1^4 + 2^4 + 3^4 + \cdots + k^4 = \frac{1}{5}k^5 + \frac{1}{2}k^4 + \frac{1}{3}k^3 - \frac{1}{30}k \Rightarrow$

$1^4 + 2^4 + 3^4 + \cdots + k^4 + (k+1)^4 = \frac{1}{5}k^5 + \frac{1}{2}k^4 + \frac{1}{3}k^3 - \frac{1}{30}k + (k+1)^4 \Rightarrow$

$1^4 + 2^4 + 3^4 + \cdots + k^4 + (k+1)^4 =$

$\frac{1}{5}k^5 + \frac{1}{2}k^4 + \frac{1}{3}k^3 - \frac{1}{30}k + (k^4 + 4k^3 + 6k^2 + 4k + 1) \Rightarrow$

$1^4 + 2^4 + 3^4 + \cdots + k^4 + (k+1)^4 = \frac{1}{5}k^5 + \frac{3}{2}k^4 + \frac{13}{3}k^3 + 6k^2 + \frac{119}{30}k + 1$.

$P_{k+1} \Rightarrow 1^4 + 2^4 + 3^4 + \cdots + k^4 + (k+1)^4$

$= \frac{1}{5}(k+1)^5 + \frac{1}{2}(k+1)^4 + \frac{1}{3}(k+1)^3 - \frac{1}{30}(k+1)$

$= \frac{1}{5}(k^5 + 5k^4 + 10k^3 + 10k^2 + 5k + 1) + \frac{1}{2}(k^4 + 4k^3 + 6k^2 + 4k + 1) +$

$\frac{1}{3}(k^3 + 3k^2 + 3k + 1) - \frac{1}{30}(k+1)$

$= \frac{1}{5}k^5 + \frac{3}{2}k^4 + \frac{13}{3}k^3 + 6k^2 + \frac{119}{30}k + 1$. Thus, the formula is true for all $n$.

[5] Examine the number of digits in the exponent of the value in scientific notation. The TI-82/83 can compute 69!, but not 70!, since 70! is larger than a 2-digit exponent. The TI-85/86 can compute 449!, but not 450!, since 450! is larger than a 3-digit exponent.

[7] $\text{Time}_{\text{total}} = \text{Time}_{\text{down}} + \text{Time}_{\text{up}}$ {note $h = 10$ and $d = 10 \cdot \frac{1}{2}$ for the first rebound}

$$= \left[\frac{\sqrt{10}}{4} + \frac{\sqrt{10 \cdot \frac{1}{2}}}{4} + \frac{\sqrt{10 \cdot (\frac{1}{2})^2}}{4} + \cdots\right] + \left[\frac{\sqrt{10 \cdot \frac{1}{2}}}{4} + \frac{\sqrt{10 \cdot (\frac{1}{2})^2}}{4} + \cdots\right]$$

$= \frac{\sqrt{10}}{4} + 2\left[\frac{\sqrt{10 \cdot \frac{1}{2}}}{4} + \frac{\sqrt{10 \cdot (\frac{1}{2})^2}}{4} + \cdots\right]$. The ratio of this geometric sequence is $\sqrt{\frac{1}{2}}$ and its absolute value is less than 1, so we can find its infinite sum using $S = a_1/(1-r)$. Thus, the total time is $\frac{\sqrt{10}}{4} + 2 \cdot \frac{\sqrt{5}/4}{1 - \sqrt{\frac{1}{2}}} \approx 4.61$ seconds.

[9] To the nearest penny with $r = 1.1008163$ (found by trial and error), we have the places 1st–10th:

| | | | | | | | | | | |
|---|---|---|---|---|---|---|---|---|---|---|
| \$0.01: | 237.37 | 215.63 | 195.89 | 177.95 | 161.65 | 146.85 | 133.40 | 121.18 | 110.08 | 100 |
| \$1.00: | 237.00 | 216.00 | 196.00 | 178.00 | 162.00 | 147.00 | 133.00 | 121.00 | 110.00 | 100 |
| \$5.00: | 240.00 | 215.00 | 195.00 | 180.00 | 160.00 | 145.00 | 135.00 | 120.00 | 110.00 | 100 |
| \$10.00: | 240.00 | 220.00 | 200.00 | 180.00 | 160.00 | 140.00 | 130.00 | 120.00 | 110.00 | 100 |

If the amounts are to be realistic, the amounts may not be rounded to the *nearest* amount—i.e, \$134 may be rounded to \$140 rather than \$130.

[11] First, calculate the probabilities for all prizes: $n(S) = C(49, 5) \cdot 42 = 80{,}089{,}128$

$$n(5\,\text{W},\,1\,\text{R}) = C(5,\,5) \cdot C(44,\,0) \cdot 1 \;\Rightarrow\; P(5\,\text{W},\,1\,\text{R}) = 1/80{,}089{,}128\ (p_1)$$
$$n(5\,\text{W},\,0\,\text{R}) = C(5,\,5) \cdot C(44,\,0) \cdot 41 \;\Rightarrow\; P(5\,\text{W},\,0\,\text{R}) = 41/80{,}089{,}128\ (p_2)$$
$$n(4\,\text{W},\,1\,\text{R}) = C(5,\,4) \cdot C(44,\,1) \cdot 1 \;\Rightarrow\; P(4\,\text{W},\,1\,\text{R}) = 220/80{,}089{,}128\ (p_3)$$
$$n(4\,\text{W},\,0\,\text{R}) = C(5,\,4) \cdot C(44,\,1) \cdot 41 \;\Rightarrow\; P(4\,\text{W},\,0\,\text{R}) = 9020/80{,}089{,}128\ (p_4)$$
$$n(3\,\text{W},\,1\,\text{R}) = C(5,\,3) \cdot C(44,\,2) \cdot 1 \;\Rightarrow\; P(3\,\text{W},\,1\,\text{R}) = 9460/80{,}089{,}128\ (p_5)$$
$$n(3\,\text{W},\,0\,\text{R}) = C(5,\,3) \cdot C(44,\,2) \cdot 41 \;\Rightarrow\; P(3\,\text{W},\,0\,\text{R}) = 387{,}860/80{,}089{,}128\ (p_6)$$
$$n(2\,\text{W},\,1\,\text{R}) = C(5,\,2) \cdot C(44,\,3) \cdot 1 \;\Rightarrow\; P(2\,\text{W},\,1\,\text{R}) = 132{,}440/80{,}089{,}128\ (p_7)$$
$$n(1\,\text{W},\,1\,\text{R}) = C(5,\,1) \cdot C(44,\,4) \cdot 1 \;\Rightarrow\; P(1\,\text{W},\,1\,\text{R}) = 678{,}755/80{,}089{,}128\ (p_8)$$
$$n(0\,\text{W},\,1\,\text{R}) = C(5,\,0) \cdot C(44,\,5) \cdot 1 \;\Rightarrow\; P(0\,\text{W},\,1\,\text{R}) = 1{,}086{,}008/80{,}089{,}128\ (p_9)$$

(a) The probability of winning the jackpot is $\frac{1}{80{,}089{,}128}$.

(b) The probability of winning any prize is the sum of all the probabilities, that is, $\frac{2{,}303{,}805}{80{,}089{,}128}$. This gives us odds of 77,785,323 to 2,303,805 (or about 34 to 1) for *not* winning any prize.

(c) The expected value of the game without the jackpot is

$$\sum_{k=2}^{9} a_k p_k = (\$100{,}000)p_2 + (\$5000)p_3 + (\$100)p_4 + (\$100)p_5 + (\$7)p_6 + (\$7)p_7 + (\$4)p_8 + (\$3)p_9 = \frac{16{,}663{,}144}{80{,}089{,}128} \approx 0.21$$

(d) Assuming that one ticket costs \$1, we must have an expected value of 1 to be a fair game. Since the probability of winning the jackpot is $\frac{1}{80{,}089{,}128}$, we must multiply that probability by \$63,425,984 {80,089,128 − 16,663,144} so that the numerator of the expected value in part (c) is equal to the denominator.

13 $(a+b)^n = \sum_{k=0}^{n} \binom{n}{k} (a)^{n-k} (b)^k$ given

$(0+1)^n = \sum_{k=0}^{n} \binom{n}{k} (0)^{n-k} (1)^k$ let $a = 0$ and $b = 1$

$1^n = 0^n \cdot 1^0 + 0^{n-1} \cdot 1^1 + \cdots + 0^0 \cdot 1^n$ expand

$1 = 0 + 0 + \cdots + 0^0$ simplify

$1 = 0^0$ simplify

So we can use the result $0^0 = 1$, which is consistent with results in mathematics courses at a higher level.

# Chapter 8: Topics from Analytic Geometry

## *8.1 Exercises*

*Note:* For Exercises 1–12, we will put each parabola equation in one of the forms listed on page 584—either

$$(x-h)^2 = 4p(y-k) \quad \text{or} \quad (y-k)^2 = 4p(x-h).$$

Once in one of those forms, the information concerning the vertex, focus, and directrix is easily obtainable and illustrated in the chart in the text. Let $V$, $F$, and $l$ denote the vertex, focus, and directrix, respectively.

[1] $8y = x^2 \Rightarrow y = \frac{1}{8}x^2 \Rightarrow p = \frac{1}{4a} = \frac{1}{4(\frac{1}{8})} = \frac{1}{\frac{1}{2}} = 2.$ We know that the parabola opens either upward or downward since the variable "$x$" is squared. Since $p$ is positive, we know that the parabola opens upward and that the focus is 2 units above the vertex. The directrix is 2 units below the vertex. $V(0,\ 0)$; $F(0,\ 2)$; $l$: $y = -2$

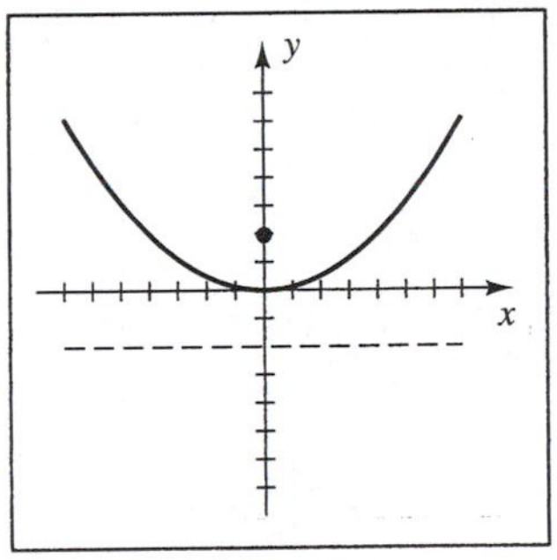


*Figure 1*

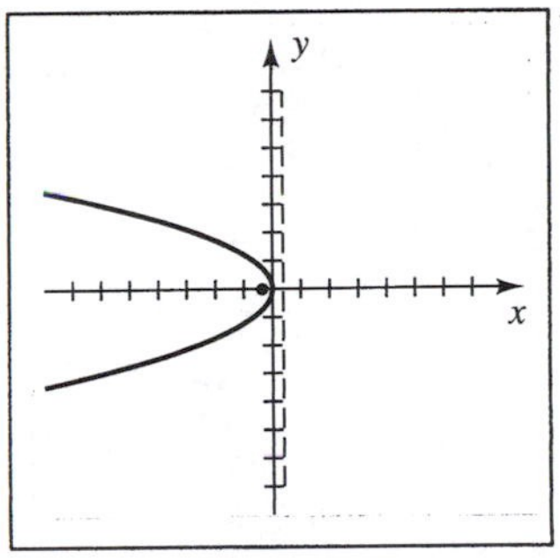


*Figure 3*

[3] $2y^2 = -3x \Rightarrow (y-0)^2 = -\frac{3}{2}(x-0) \Rightarrow 4p = -\frac{3}{2} \Rightarrow p = -\frac{3}{8}$. We know that the parabola opens either right or left since the variable "$y$" is squared. Since $p$ is negative, we know that the parabola opens left and that the focus is $\frac{3}{8}$ unit to the left of the vertex. The directrix is $\frac{3}{8}$ unit to the right of the vertex.

$V(0,\ 0)$; $F(-\frac{3}{8},\ 0)$; $l$: $x = \frac{3}{8}$

[5] $(x+2)^2 = -8(y-1) \Rightarrow 4p = -8 \Rightarrow p = -2.$ The $(x+2)$ and $(y-1)$ factors indicate that we need to shift the vertex, of the parabola having equation $x^2 = -8y$, 2 units left and 1 unit up—that is, move it from $(0,\ 0)$ to $(-2,\ 1)$.

$V(-2,\ 1)$; $F(-2,\ -1)$; $l$: $y = 3$

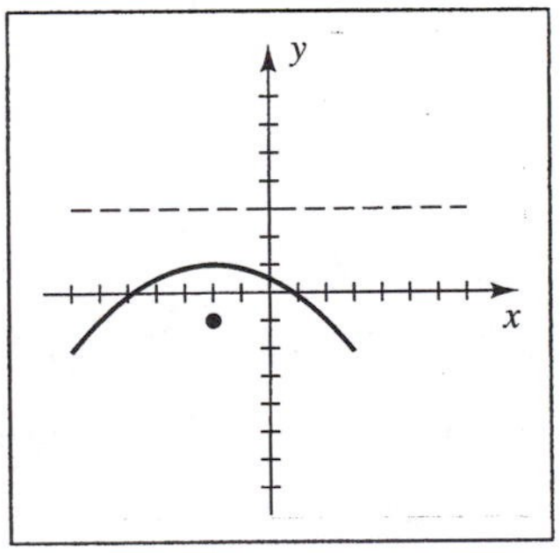


*Figure 5*

[7] $(y-2)^2 = \frac{1}{4}(x-3) \Rightarrow 4p = \frac{1}{4} \Rightarrow p = \frac{1}{16}$. "$y$" squared and $p$ positive imply that the parabola opens to the right and the focus is to the right of the vertex. The $(x-3)$ and $(y-2)$ factors indicate that we need to shift the vertex 3 units right and 2 units up from (0, 0) to (3, 2). $V(3,\ 2);\ F(\frac{49}{16},\ 2);\ l\colon x = \frac{47}{16}$

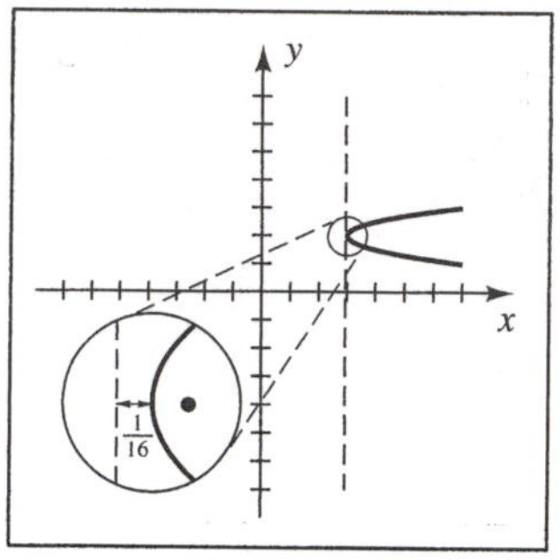


Figure 7

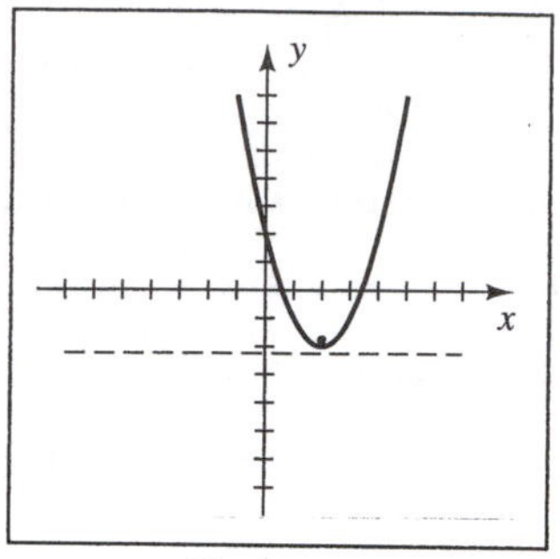


Figure 9

[9] For this exercise, we need to "complete the square" in order to get the equation in proper form. The term we need to add is $\left[\frac{1}{2}(\text{coefficient of } x)\right]^2$. In this case, that value is $\left[\frac{1}{2}(-4)\right]^2 = 4$. Notice that we add and subtract the value 4 from the same side of the equation as opposed to adding 4 to both sides of the equation.
$y = x^2 - 4x + 2 \Rightarrow y = (x^2 - 4x + \underline{\ 4\ }) + 2 - \underline{\ 4\ } \Rightarrow y = (x-2)^2 - 2 \Rightarrow$
$(y+2) = 1(x-2)^2 \Rightarrow 4p = 1 \Rightarrow p = \frac{1}{4}$. $V(2,\ -2);\ F(2,\ -\frac{7}{4});\ l\colon y = -\frac{9}{4}$

[11] $x^2 + 20y = 10 \Rightarrow x^2 = -20y + 10 \Rightarrow$
$(x-0)^2 = -20(y - \frac{1}{2}) \Rightarrow 4p = -20 \Rightarrow p = -5.$

$V(0,\ \frac{1}{2});\ F(0,\ -\frac{9}{2});\ l\colon y = \frac{11}{2}$

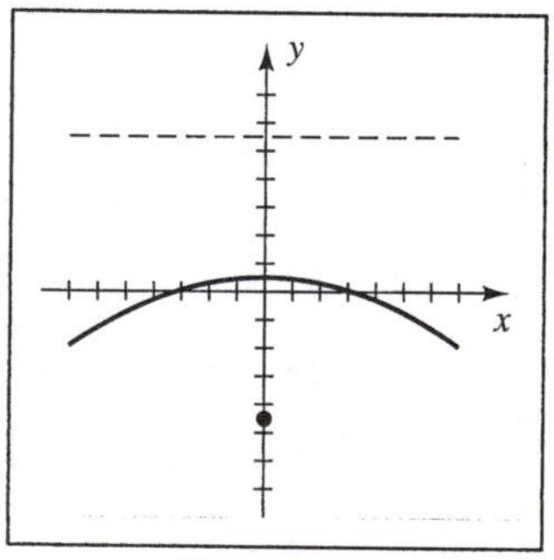


Figure 11

[13] Since the vertex is at (1, 0) and the parabola has a horizontal axis, the standard equation has the form $(y-0)^2 = 4p(x-1)$. The distance from the focus $F(6,\ 0)$ to the vertex $V(1,\ 0)$ is $6 - 1 = 5$, which is the value of $p$. Hence, an equation of the parabola is $y^2 = 20(x-1)$.

[15] An equation of a parabola with vertex $V(-2,\ 3)$ is $(x+2)^2 = 4p(y-3)$. Since the point $P(2,\ 2)$ is on the graph, we'll substitute 2 for $x$ and 2 for $y$ and then find $p$.
$x = 2,\ y = 2 \Rightarrow (2+2)^2 = 4p(2-3) \Rightarrow 16 = 4p(-1) \Rightarrow p = -4,$

so an equation is $(x+2)^2 = -16(y-3)$.

17 The distance from the focus $F(3,\ 2)$ to the directrix $l$: $y = -1$ is $2 - (-1) = 3$ units.
The vertex $V$ is the point $(3,\ \frac{1}{2})$ {halfway between $F$ and $l$},
so the form of an equation of the parabola is $(x-3)^2 = 4p(y - \frac{1}{2})$.
The value of $p$ is the difference in the $y$-values; that is, $p = 2 - \frac{1}{2} = \frac{3}{2}$.
Thus, the equation is $(x-3)^2 = 4(\frac{3}{2})(y+2)$, or, equivalently, $x^2 = 6(y+2)$.

19 The distance from the directrix to the focus is $2 - (-2) = 4$ units. The vertex $V(0,\ 0)$ is halfway between the directrix and the focus—that is, 2 units from either one. Since the focus is 2 units to the right of the vertex, $p$ is 2. Using one of the forms of an equation of a parabola with vertex at $(h,\ k)$, we have $(y-0)^2 = 4p(x-0)$, or, equivalently, $y^2 = 8x$.

21 $F(6,\ 4)$ and $l$: $y = -2 \Rightarrow 2p = 4 - (-2) \Rightarrow 2p = 6 \Rightarrow p = 3$ and $V(6,\ 1)$.

$$(x-6)^2 = 4p(y-1) \Rightarrow (x-6)^2 = 12(y-1).$$

23 $V(3,\ -5)$ and $l$: $x = 2 \Rightarrow p = 3 - 2 = 1$.

$$(y+5)^2 = 4p(x-3) \Rightarrow (y+5)^2 = 4(x-3).$$

25 $V(-1,\ 0)$ and $F(-4,\ 0) \Rightarrow p = -4 - (-1) = -3$.

$$(y-0)^2 = 4p(x+1) \Rightarrow y^2 = -12(x+1).$$

27 The vertex at the origin and symmetric to the $y$-axis imply that the equation is of the form $y = ax^2$. Substituting $x = 2$ and $y = -3$ into that equation yields

$$-3 = a \cdot 4 \Rightarrow a = -\tfrac{3}{4}. \text{ Thus, an equation is } y = -\tfrac{3}{4}x^2, \text{ or } 3x^2 = -4y.$$

29 The vertex at $(-3,\ 5)$ and axis parallel to the $x$-axis imply that the equation is of the form $(y-5)^2 = 4p(x+3)$. Substituting $x = 5$ and $y = 9$ into that equation yields $16 = 4p \cdot 8 \Rightarrow p = \frac{1}{2}$. Thus, an equation is $(y-5)^2 = 2(x+3)$.

31 Refer to the definition of a parabola. The point $P(0,\ 5)$ is the fixed point (focus) and the line $l$: $y = -3$ is the fixed line (directrix). The vertex is halfway between the focus and the directrix—that is, at $V(0,\ 1)$. An equation is of the form $(x-h)^2 = 4p(y-k) \Rightarrow (x-0)^2 = 4p(y-1)$. The distance from the vertex to the focus is $p = 5 - 1 = 4$. Thus, an equation is

$$(x-0)^2 = 4(4)(y-1) \Rightarrow x^2 = 16(y-1).$$

33 The point $P(-6,\ 3)$ is the fixed point (focus) and the line $l$: $x = -2$ is the fixed line (directrix). The vertex is halfway between the focus and the directrix—that is, at $V(-4,\ 3)$. An equation is of the form $(y-3)^2 = 4p(x+4)$. The distance from the vertex to the focus is $p = -6 - (-4) = -2$. Thus, an equation is $(y-3)^2 = -8(x+4)$.

*Note:* To find an equation for a lower or upper half, we need to solve for $y$ (use $-$ or $+$ respectively). For the left or right half, solve for $x$ (use $-$ or $+$ respectively).

[35] $(y+1)^2 = x+3 \Rightarrow y+1 = \pm\sqrt{x+3} \Rightarrow y = -\sqrt{x+3}-1$

[37] $(x+1)^2 = y-4 \Rightarrow x+1 = \pm\sqrt{y-4} \Rightarrow x = \sqrt{y-4}-1$

[39] The parabola has an equation of the form $y = ax^2+bx+c$. Substituting the $x$ and $y$ values of $P(2, 5)$, $Q(-2, -3)$, and $R(1, 6)$ into this equation yields:

$$\begin{cases} 4a + 2b + c = 5 & P \ (\mathrm{E}_1) \\ 4a - 2b + c = -3 & Q \ (\mathrm{E}_2) \\ a + b + c = 6 & R \ (\mathrm{E}_3) \end{cases}$$

Solving $\mathrm{E}_3$ for $c$ $\{c = 6-a-b\}$ and substituting into $\mathrm{E}_1$ and $\mathrm{E}_2$ yields:

$$\begin{cases} 3a + b = -1 & (\mathrm{E}_4) \\ 3a - 3b = -9 & (\mathrm{E}_5) \end{cases}$$

$\mathrm{E}_4 - \mathrm{E}_5 \Rightarrow 4b = 8 \Rightarrow b = 2;\ a = -1;\ c = 5$. The equation is $y = -x^2+2x+5$.

[41] The parabola has an equation of the form $x = ay^2+by+c$. Substituting the $x$ and $y$ values of $P(-1, 1)$, $Q(11, -2)$, and $R(5, -1)$ into this equation yields:

$$\begin{cases} a + b + c = -1 & P \ (\mathrm{E}_1) \\ 4a - 2b + c = 11 & Q \ (\mathrm{E}_2) \\ a - b + c = 5 & R \ (\mathrm{E}_3) \end{cases}$$

Solving $\mathrm{E}_3$ for $c$ $\{c = 5-a+b\}$ and substituting into $\mathrm{E}_1$ and $\mathrm{E}_2$ yields:

$$\begin{cases} 2b = -6 & (\mathrm{E}_4) \\ 3a - b = 6 & (\mathrm{E}_5) \end{cases}$$

$\mathrm{E}_4 \Rightarrow b = -3;\ a = 1;\ c = 1$. The equation is $x = y^2-3y+1$.

[43] A cross section of the mirror is a parabola with $V(0, 0)$ and passing through $P(4, 1)$. The incoming light will collect at the focus $F$. A general equation of this form of a parabola is $y = ax^2$. Substituting $x = 4$ and $y = 1$ gives us $1 = a(4)^2 \Rightarrow a = \frac{1}{16}$. $p = 1/(4a) = 1/(\frac{1}{4}) = 4$. The light will collect 4 inches from the center of the mirror.

[45] If we set up a coordinate system with a parabola opening upward and the vertex at the origin, then the phrase "3 feet across at the opening and 1 foot deep" implies that the points $(\pm\frac{3}{2}, 1)$ are on the parabola.

$y = ax^2 \Rightarrow 1 = a(\frac{3}{2})^2 \Rightarrow a = \frac{4}{9}$. $p = 1/(4a) = 1/(\frac{16}{9}) = \frac{9}{16}$ ft.

[47] $p = 5 \Rightarrow a = 1/(4p) = 1/(4\cdot 5) = \frac{1}{20}$.

$y = ax^2$ $\{y = 2\ \underline{\text{ft}} = 24 \text{ inches}\} \Rightarrow 24 = \frac{1}{20}x^2 \Rightarrow x^2 = 480 \Rightarrow x = \sqrt{480}$.

The width is twice the value of $x$. Width $= 2\sqrt{480} \approx 43.82$ in.

49 (a) Let the parabola have the equation $x^2 = 4py$. Since the point $(r, h)$ is on the parabola, we can substitute $r$ for $x$ and $h$ for $y$, giving us $r^2 = 4ph$. Solving for $p$ we have $p = \frac{r^2}{4h}$.

(b) $p = 10$ and $h = 5 \Rightarrow r^2 = 4(10)(5) \Rightarrow r = \sqrt{200} = 10\sqrt{2}$.

51 With $a = 125$ and $p = 50$, $S = \frac{8\pi p^2}{3}\left[\left(1 + \frac{a^2}{4p^2}\right)^{3/2} - 1\right] \approx 64{,}968 \text{ ft}^2$.

53 Depending on the type of calculator or software used, we may need to solve for $y$ in terms of $x$. $x = -y^2 + 2y + 5 \Rightarrow y^2 - 2y + (x - 5) = 0$. This is a quadratic equation in $y$. Using the quadratic formula to solve for $y$ yields

$$y = \frac{-(-2) \pm \sqrt{(-2)^2 - 4(1)(x-5)}}{2(1)} = \frac{2 \pm \sqrt{24 - 4x}}{2} = \frac{2 \pm \sqrt{4}\sqrt{6-x}}{2} = 1 \pm \sqrt{6-x}.$$

[−11, 10, 2] by [−7, 7]

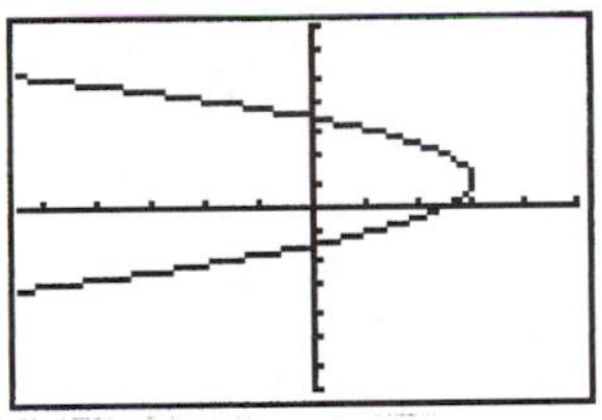

Figure 53

[−2, 4] by [−3, 3]

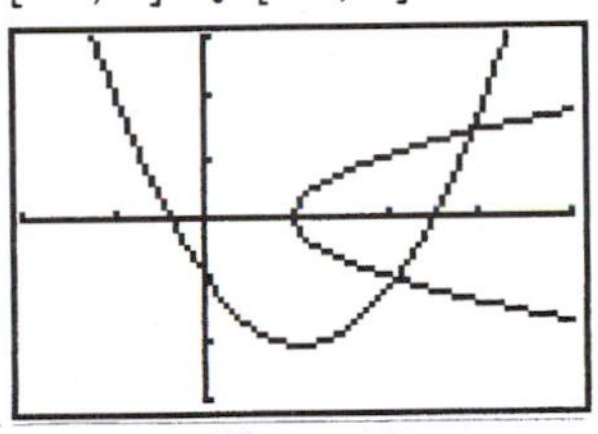

Figure 55

55 $x = y^2 + 1 \Rightarrow y = \pm\sqrt{x-1}$. From the graph, we can see that there are 2 points of intersection. Their coordinates are approximately (2.08, −1.04) and (2.92, 1.38).

## 8.2 Exercises

*Note:* Let $C$, $V$, $F$, and $M$ denote the center, the vertices, the foci, and the endpoints of the minor axis, respectively. Let $c$ denote the distance from the center of the ellipse to a focus.

[1] $\frac{x^2}{9}+\frac{y^2}{4}=1$ • The $x$-intercepts are $\pm\sqrt{9}=\pm 3$. The $y$-intercepts are $\pm\sqrt{4}=\pm 2$. The major axis {the longer of the two axes} is the horizontal axis and has length $2(3)=6$. The minor axis {the shorter} is the vertical axis and has length $2(2)=4$. To find the foci, it is helpful to remember the relationship

$\left[\frac{1}{2}(\text{minor axis})\right]^2+\left[c\right]^2=\left[\frac{1}{2}(\text{major axis})\right]^2.$

Using the values from above we have

$\left[\frac{1}{2}(4)\right]^2+c^2=\left[\frac{1}{2}(6)\right]^2 \Rightarrow 4+c^2=9 \Rightarrow$

$c^2=5 \Rightarrow c=\pm\sqrt{5}$. Simply put, $c^2$ is equal to difference of the denominators.

$V(\pm 3, 0)$; $F(\pm\sqrt{5}, 0)$; $M(0, \pm 2)$

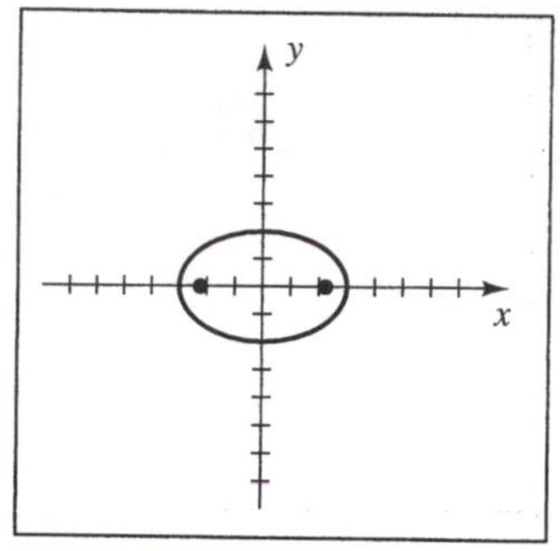

Figure 1

[3] $\frac{x^2}{15}+\frac{y^2}{16}=1$ • Since the 16 in the denominator of the term with the variable $y$ is larger than the 15 in the denominator of the term with the variable $x$, the vertices and foci are on the $y$-axis and the major axis is the vertical axis. $c^2=16-15 \Rightarrow c=\pm 1$.

$V(0, \pm 4)$; $F(0, \pm 1)$; $M(\pm\sqrt{15}, 0)$

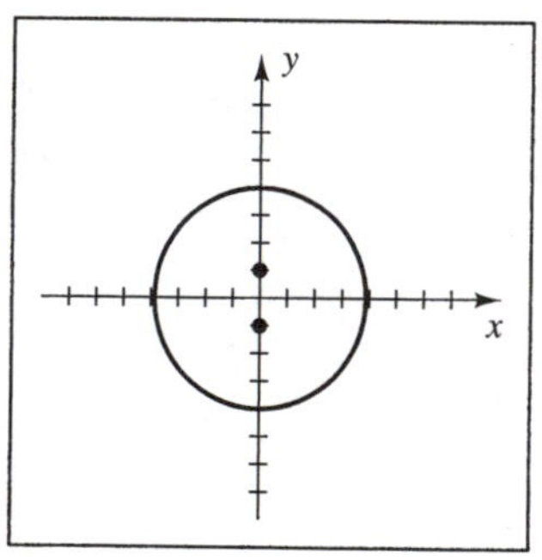

Figure 3

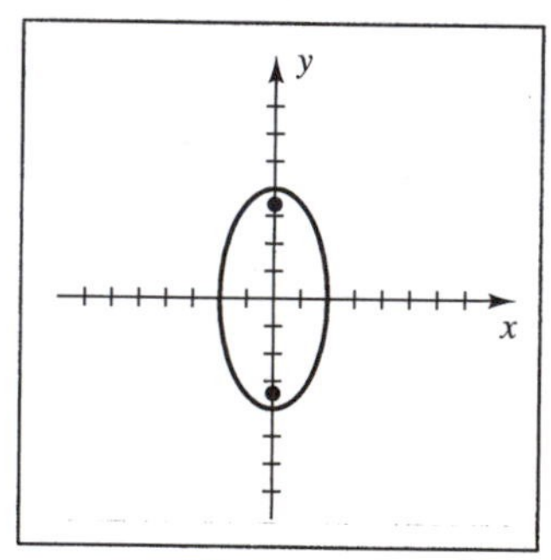

Figure 5

[5] We first divide by 16 to obtain the "1" on the right side of the equation. $4x^2+y^2=16 \Rightarrow \frac{x^2}{4}+\frac{y^2}{16}=1$. Since the 16 in the denominator of the term with the variable $y$ is larger than the 4 in the denominator of the term with the variable $x$, the vertices and foci are on the $y$-axis and the major axis is the vertical axis. $4+c^2=16 \Rightarrow c^2=16-4 \Rightarrow c=\pm\sqrt{12}=\pm 2\sqrt{3}$.

$V(0, \pm 4)$; $F(0, \pm 2\sqrt{3})$; $M(\pm 2, 0)$

[7] First write $4x^2 + 25y^2 = 1$ in standard form: $\frac{x^2}{\frac{1}{4}} + \frac{y^2}{\frac{1}{25}} = 1$.

$\frac{1}{25} + c^2 = \frac{1}{4} \Rightarrow c^2 = \frac{1}{4} - \frac{1}{25} = \frac{21}{100} \Rightarrow c = \pm\frac{1}{10}\sqrt{21}$.

$V(\pm\frac{1}{2}, 0)$; $F(\pm\frac{1}{10}\sqrt{21}, 0)$; $M(0, \pm\frac{1}{5})$ (See *Figure 7* below.)

[9] $\frac{(x-3)^2}{16} + \frac{(y+4)^2}{9} = 1$ • The effect of the factors $(x-3)$ and $(y+4)$ is to shift the center of the ellipse from $(0, 0)$ to $(3, -4)$. Since the larger denominator, 16, is in the term with $x$, the major axis will be horizontal. The endpoints are 4 units in either direction of the center. Their coordinates are the points $(3 \pm 4, -4)$, or, equivalently, $(7, -4)$ and $(-1, -4)$. The minor axis will be vertical with endpoints $(3, -4 \pm 3)$, or, equivalently, $(3, -1)$ and $(3, -7)$. $9 + c^2 = 16 \Rightarrow c^2 = 16 - 9 \Rightarrow c = \pm\sqrt{7}$. Remember that $c$ is the distance *from the center* to a focus. Hence, the coordinates of the foci are $(3 \pm \sqrt{7}, -4)$.

$C(3, -4)$; $V(3 \pm 4, -4)$; $F(3 \pm \sqrt{7}, -4)$; $M(3, -4 \pm 3)$

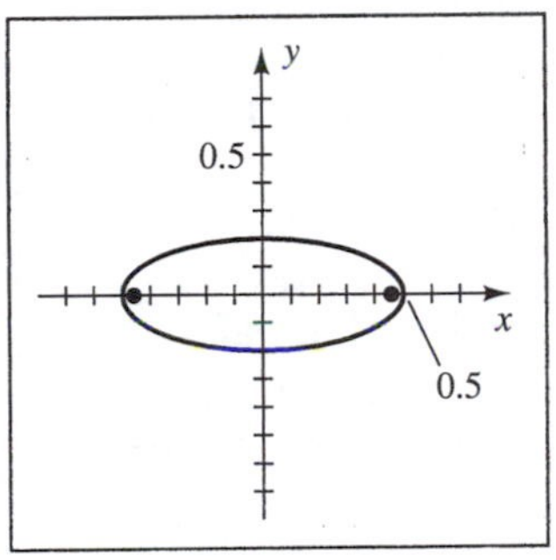

*Figure 7*

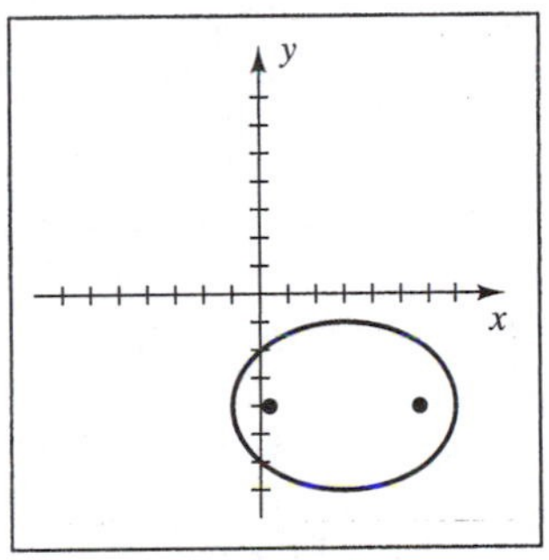

*Figure 9*

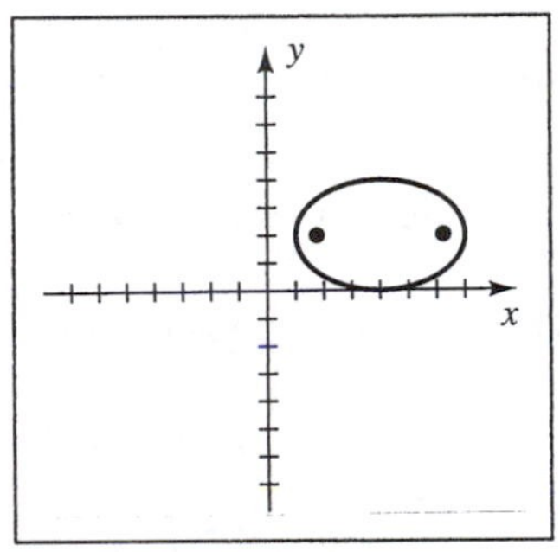

*Figure 11*

[11] $4x^2 + 9y^2 - 32x - 36y + 64 = 0 \Rightarrow (4x^2 - 32x) + (9y^2 - 36y) = -64 \Rightarrow$

{first group the $x$ terms and the $y$ terms} $4(x^2 - 8x) + 9(y^2 - 4y) = -64 \Rightarrow$

{factor out the coefficients of $x^2$ and $y^2$}

$4(x^2 - 8x + \underline{16}) + 9(y^2 - 4y + \underline{4}) = -64 + \underline{64} + \underline{36} \Rightarrow$

{Complete the squares—remember that we have added $\underline{4}(16)$ and $\underline{9}(4)$, <u>not</u> 16 and 4. Thus, we add 64 and 36 to the right side of the equation.}

$4(x-4)^2 + 9(y-2)^2 = 36 \Rightarrow \frac{(x-4)^2}{9} + \frac{(y-2)^2}{4} = 1$. $c^2 = 9 - 4 \Rightarrow c = \pm\sqrt{5}$.

$C(4, 2)$; $V(4 \pm 3, 2)$; $F(4 \pm \sqrt{5}, 2)$; $M(4, 2 \pm 2)$

[13] $25x^2 + 4y^2 - 250x - 16y + 541 = 0 \Rightarrow$

$25(x^2 - 10x + \underline{25}) + 4(y^2 - 4y + \underline{4}) = -541 + \underline{625} + \underline{16} \Rightarrow$

$25(x-5)^2 + 4(y-2)^2 = 100 \Rightarrow \frac{(x-5)^2}{4} + \frac{(y-2)^2}{25} = 1.$

$c^2 = 25 - 4 \Rightarrow c = \pm\sqrt{21}.$ $C(5, 2)$; $V(5, 2 \pm 5)$; $F(5, 2 \pm \sqrt{21})$; $M(5 \pm 2, 2)$

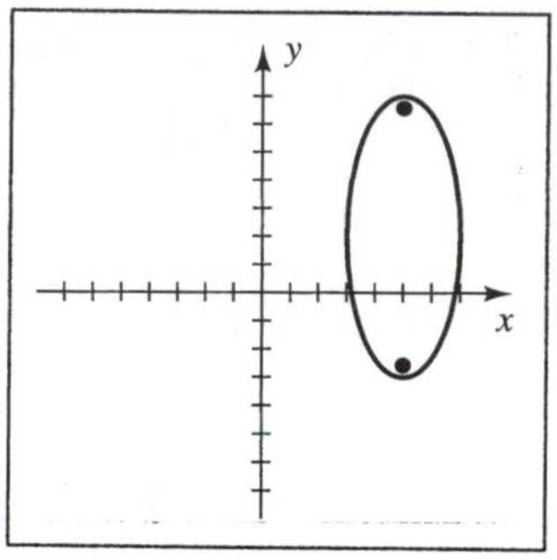


*Figure 13*

*Note:* Let $b$ denote the distance equal to $\frac{1}{2}$(minor axis length) and let $a$ denote the distance equal to $\frac{1}{2}$(major axis length).

[15] $a = 2$ and $b = 6 \Rightarrow \frac{x^2}{a^2} + \frac{y^2}{b^2} = 1$ is $\frac{x^2}{4} + \frac{y^2}{36} = 1.$

[17] The center of the ellipse is $(-2, 1)$. $a = 5$ and $b = 2$ give us $\frac{(x+2)^2}{25} + \frac{(y-1)^2}{4} = 1.$

[19] Since the vertices are at $(\pm 8, 0)$, $\frac{1}{2}$(major axis) $= 8$. Since the foci are at $(\pm 5, 0)$, $c = 5$. Using the relationship $\left[\frac{1}{2}(\text{minor axis})\right]^2 + \left[c\right]^2 = \left[\frac{1}{2}(\text{major axis})\right]^2$, we have $\left[\frac{1}{2}(\text{minor axis})\right]^2 + 5^2 = 8^2 \Rightarrow \left[\frac{1}{2}(\text{minor axis})\right]^2 = 64 - 25 = 39.$

An equation is $\frac{x^2}{64} + \frac{y^2}{39} = 1.$

[21] If the length of the minor axis is 3, then $b = \frac{1}{2}(3) = \frac{3}{2}$. The vertices are $V(0, \pm 5)$, so an equation is $\frac{x^2}{(\frac{3}{2})^2} + \frac{y^2}{5^2} = 1$, or, equivalently, $\frac{4x^2}{9} + \frac{y^2}{25} = 1.$

[23] With the vertices at $(0, \pm 6)$, an equation of the ellipse is $\frac{x^2}{b^2} + \frac{y^2}{6^2} = 1$. Substituting $x = 3$ and $y = 2$ and solving for $b^2$ yields $\frac{9}{b^2} + \frac{4}{36} = 1 \Rightarrow \frac{9}{b^2} = \frac{8}{9} \Rightarrow b^2 = \frac{81}{8}.$

An equation is $\frac{x^2}{\frac{81}{8}} + \frac{y^2}{36} = 1$, or, equivalently, $\frac{8x^2}{81} + \frac{y^2}{36} = 1.$

[25] With vertices $V(0,\ \pm 4)$, an equation of the ellipse is $\frac{x^2}{b^2}+\frac{y^2}{16}=1$.

Remember the formula for the eccentricity:

$$e=\frac{\text{distance from center to focus}}{\text{distance from center to vertex}}=\frac{c}{a}=\frac{\sqrt{a^2-b^2}}{a}$$

Hence, $e=\frac{c}{a}=\frac{3}{4}$ and $a=4 \Rightarrow c=3$. Thus, $b^2+c^2=a^2 \Rightarrow$

$$b^2=a^2-c^2=4^2-3^2=16-9=7. \text{ An equation is } \frac{x^2}{7}+\frac{y^2}{16}=1.$$

[27] $x$-intercepts $=\ \pm 2$ and $y$-intercepts $=\ \pm\frac{1}{3} \Rightarrow$

$$\frac{x^2}{2^2}+\frac{y^2}{(\frac{1}{3})^2}=1 \Rightarrow \frac{x^2}{4}+\frac{y^2}{\frac{1}{9}}=1 \Rightarrow \frac{x^2}{4}+9y^2=1.$$

[29] Remember to divide the lengths of the major and minor axes by 2.

$$\frac{x^2}{(\frac{1}{2}\cdot 8)^2}+\frac{y^2}{(\frac{1}{2}\cdot 5)^2}=1 \Rightarrow \frac{x^2}{16}+\frac{y^2}{\frac{25}{4}}=1 \Rightarrow \frac{x^2}{16}+\frac{4y^2}{25}=1.$$

[31] The graph of $x^2+4y^2=20$, or, equivalently, $\frac{x^2}{20}+\frac{y^2}{5}=1$, is that of an ellipse with $x$-intercepts at $\pm\sqrt{20}$ and $y$-intercepts $\pm\sqrt{5}$. The graph of $x+2y=6$, or, equivalently, $y=-\frac{1}{2}x+3$, is that of a line with $y$-intercept 3 and slope $-\frac{1}{2}$. Substituting $x=6-2y$ into $x^2+4y^2=20$ yields $(6-2y)^2+4y^2=20 \Rightarrow$ $8y^2-24y+16=0 \Rightarrow 8(y^2-3y+2)=0 \Rightarrow 8(y-1)(y-2)=0 \Rightarrow y=1,\ 2;$ $x=4,\ 2$. The two points of intersection are $(2,\ 2)$ and $(4,\ 1)$.

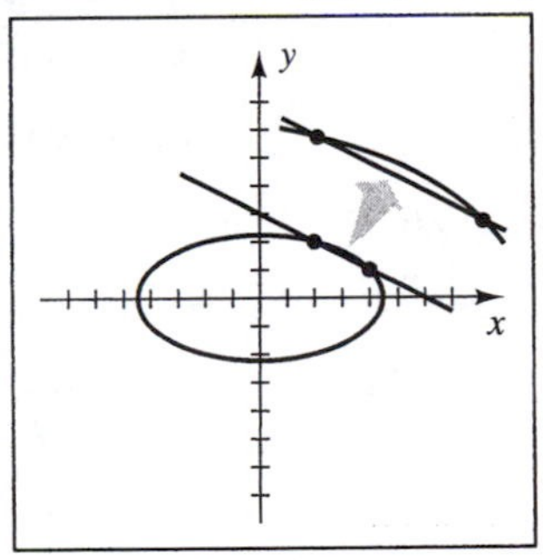


*Figure 31*

[33] Refer to the definition of an ellipse. As in the discussion on page 590, we will let the positive constant, $k$, equal $2a$. $k=10 \Rightarrow 2a=10 \Rightarrow a=5$.

$F(3,\ 0)$ and $F'(-3,\ 0) \Rightarrow c=3$.

$$b^2=a^2-c^2=25-9=16. \text{ An equation is } \frac{x^2}{25}+\frac{y^2}{16}=1.$$

[35] $k=34 \Rightarrow 2a=34 \Rightarrow a=17$. $F(0,\ 15)$ and $F'(0,\ -15) \Rightarrow c=15$.

$$b^2=a^2-c^2=289-225=64. \text{ An equation is } \frac{x^2}{64}+\frac{y^2}{289}=1.$$

[37] $k = 2a = 7 + 3 = 10 \Rightarrow a = 5.$

$F(4,\ 0)$ and $F'(-4,\ 0) \Rightarrow c = 4.$

$b^2 = a^2 - c^2 = 25 - 16 = 9.$

An equation is $\frac{x^2}{25} + \frac{y^2}{9} = 1.$

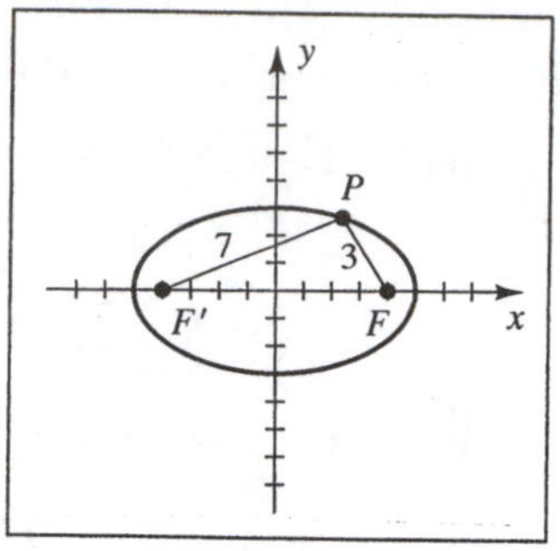


Figure 37

[39] $y = 11\sqrt{1 - \frac{x^2}{49}} \Rightarrow \frac{y}{11} = \sqrt{1 - \frac{x^2}{49}} \Rightarrow \frac{x^2}{49} + \frac{y^2}{121} = 1.$

Since $y \geq 0$ in the original equation, its graph is the upper half of the ellipse.

[41] $x = -\frac{1}{3}\sqrt{9 - y^2} \Rightarrow -3x = \sqrt{9 - y^2} \Rightarrow 9x^2 = 9 - y^2 \Rightarrow x^2 + \frac{y^2}{9} = 1.$

Since $x \leq 0$ in the original equation, its graph is the left half of the ellipse.

[43] $x = 1 + 2\sqrt{1 - \frac{(y+2)^2}{9}} \Rightarrow \frac{x-1}{2} = \sqrt{1 - \frac{(y+2)^2}{9}} \Rightarrow \frac{(x-1)^2}{4} + \frac{(y+2)^2}{9} = 1.$

Since $x \geq 1$ in the original equation, its graph is the right half of the ellipse.

[45] $y = 2 - 7\sqrt{1 - \frac{(x+1)^2}{9}} \Rightarrow \frac{y-2}{-7} = \sqrt{1 - \frac{(x+1)^2}{9}} \Rightarrow \frac{(x+1)^2}{9} + \frac{(y-2)^2}{49} = 1.$

Since $y \leq 2$ in the original equation, its graph is the lower half of the ellipse.

[47] Model this problem as an ellipse with $V(\pm 15,\ 0)$ and $M(0,\ \pm 10)$.

Substituting $x = 6$ into $\frac{x^2}{15^2} + \frac{y^2}{10^2} = 1$ yields $\frac{y^2}{100} = \frac{189}{225} \Rightarrow y^2 = 84.$

The desired height is $\sqrt{84} = 2\sqrt{21} \approx 9.165$ ft.

[49] $e = \frac{c}{a} = 0.017 \Rightarrow c = 0.017a = 0.017(93{,}000{,}000) = 1{,}581{,}000.$ The maximum and minimum distances are $a + c = 94{,}581{,}000$ miles and $a - c = 91{,}419{,}000$ miles.

[51] (a) Let $c$ denote the distance from the center of the hemi-ellipsoid to $F$.

Hence, $(\frac{1}{2}k)^2 + c^2 = h^2 \Rightarrow c^2 = h^2 - \frac{1}{4}k^2 \Rightarrow c = \sqrt{h^2 - \frac{1}{4}k^2}.$

$d = d(V,\ F) = h - c \Rightarrow d = h - \sqrt{h^2 - \frac{1}{4}k^2}$ and

$d' = d(V,\ F') = h + c \Rightarrow d' = h + \sqrt{h^2 - \frac{1}{4}k^2}.$

(b) From part (a), $d' = h + c \Rightarrow c = d' - h = 32 - 17 = 15.$ $c = \sqrt{h^2 - \frac{1}{4}k^2} \Rightarrow$

$15 = \sqrt{17^2 - \frac{1}{4}k^2} \Rightarrow 225 = 289 - \frac{1}{4}k^2 \Rightarrow \frac{1}{4}k^2 = 64 \Rightarrow k^2 = 256 \Rightarrow$

$k = 16$ cm. $d = h - c = 17 - 15 = 2 \Rightarrow F$ should be located 2 cm from $V$.

[53] $c^2 = (\frac{1}{2} \cdot 50)^2 - 15^2 = 625 - 225 = 400 \Rightarrow c = 20.$

Their feet should be $25 - 20 = 5$ ft from the vertices.

55 First determine an equation of the ellipse for the orbit of Earth.

$e = \frac{c}{a} \Rightarrow c = ae = 0.093 \times 149.6 = 13.9128.$

$b^2 = a^2 - c^2 = 149.6^2 - 13.9128^2 \Rightarrow b \approx 148.95 \approx 149.0.$

An approximate equation for the orbit of Earth is $\dfrac{x^2}{149.6^2} + \dfrac{y^2}{149.0^2} = 1.$

Solving for $y$ gives us $\dfrac{y^2}{149^2} = 1 - \dfrac{x^2}{149.6^2} \Rightarrow y^2 = 149^2\left(1 - \dfrac{x^2}{149.6^2}\right) \Rightarrow$

$y = \pm 149\sqrt{1 - \dfrac{x^2}{149.6^2}}$. Graph $Y_1 = 149\sqrt{1 - (x^2/149.6^2)}$ and $Y_2 = -Y_1$.

The sun is at $(\pm 13.9128, 0)$. Plot the point $(13.9128, 0)$ for the sun.

[−300, 300, 100] by [−200, 200, 100]

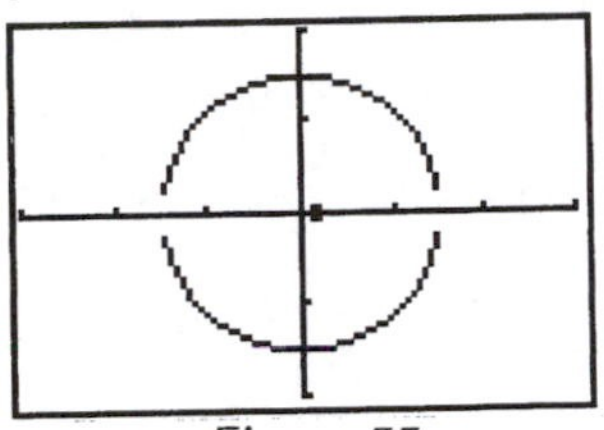

Figure 55

[−6, 6] by [−2, 6]

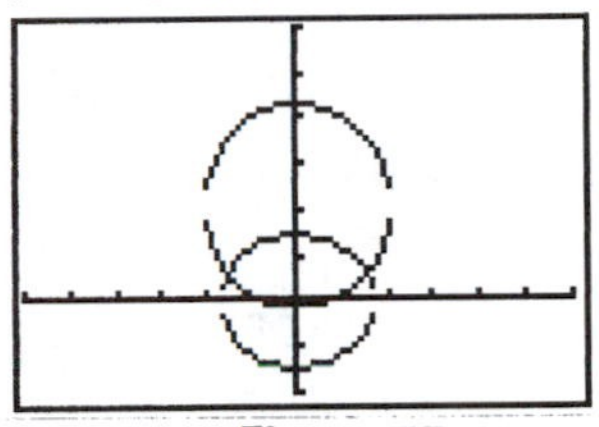

Figure 57

57 $\dfrac{x^2}{2.9} + \dfrac{y^2}{2.1} = 1 \Rightarrow \dfrac{y^2}{2.1} = 1 - \dfrac{x^2}{2.9} \Rightarrow y^2 = 2.1\left(1 - \dfrac{x^2}{2.9}\right) \Rightarrow y = \pm\sqrt{2.1(1 - x^2/2.9)}.$

$\dfrac{x^2}{4.3} + \dfrac{(y-2.1)^2}{4.9} = 1 \Rightarrow \dfrac{(y-2.1)^2}{4.9} = 1 - \dfrac{x^2}{4.3} \Rightarrow (y-2.1)^2 = 4.9\left(1 - \dfrac{x^2}{4.3}\right) \Rightarrow$

$y - 2.1 = \pm\sqrt{4.9(1 - x^2/4.3)} \Rightarrow y = 2.1 \pm \sqrt{4.9(1 - x^2/4.3)}.$

From the graph, the points of intersection are approximately $(\pm 1.540, 0.618)$.

59 $\dfrac{(x+0.1)^2}{1.7} + \dfrac{y^2}{0.9} = 1 \Rightarrow \dfrac{y^2}{0.9} = 1 - \dfrac{(x+0.1)^2}{1.7} \Rightarrow y^2 = 0.9[1 - (x+0.1)^2/1.7] \Rightarrow$

$y = \pm\sqrt{0.9[1 - (x+0.1)^2/1.7]}.$

$\dfrac{x^2}{0.9} + \dfrac{(y-0.25)^2}{1.8} = 1 \Rightarrow \dfrac{(y-0.25)^2}{1.8} = 1 - \dfrac{x^2}{0.9} \Rightarrow (y-0.25)^2 = 1.8\left(1 - \dfrac{x^2}{0.9}\right) \Rightarrow$

$y - 0.25 = \pm\sqrt{1.8(1 - x^2/0.9)} \Rightarrow y = 0.25 \pm \sqrt{1.8(1 - x^2/0.9)}.$

From the graph, the points of intersection are approximately (−0.88, 0.76), (−0.48,−0.91), (0.58,−0.81), and (0.92, 0.59).

[−3, 3] by [−2, 2]

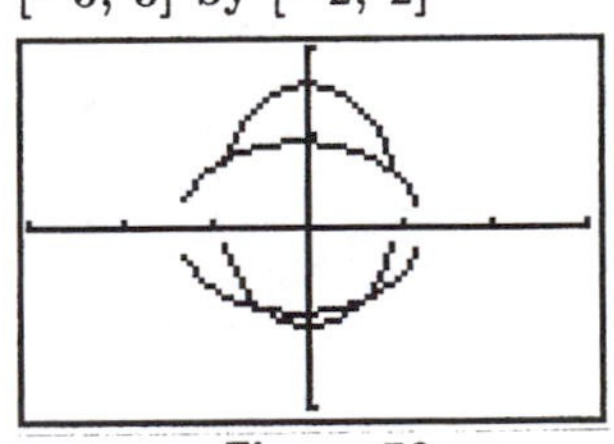

Figure 59

## 8.3 Exercises

*Note:* Let $C$, $V$, $F$, and $W$ denote the center, the vertices, the foci, and the endpoints of the conjugate axis, respectively. Let $c$ denote the distance from the center of the hyperbola to a focus.

[1] $\frac{x^2}{9} - \frac{y^2}{4} = 1$ • The hyperbola will have a right branch and a left branch since the term containing $x$ is positive. The vertices will be on the horizontal transverse axis, $\pm\sqrt{9} = \pm 3$ units from the center. The endpoints of the vertical conjugate axis are $(0, \pm\sqrt{4}) = (0, \pm 2)$. The asymptotes have equations

$$y = \pm\left[\frac{\sqrt{\text{constant under } y^2}}{\sqrt{\text{constant under } x^2}}\right](x) \quad \text{or} \quad y = \pm\left[\frac{\frac{1}{2}(\text{vertical axis length})}{\frac{1}{2}(\text{horizontal axis length})}\right](x).$$

Note that the terms are "vertical" and "horizontal" and not transverse and conjugate since the latter can be either vertical or horizontal. In this case, we have $y = \pm\frac{\sqrt{4}}{\sqrt{9}}(x) = \pm\frac{2}{3}x$. The positive sign corresponds to the asymptote with positive slope and the negative sign corresponds to the asymptote with negative slope. To find the foci, it is helpful to remember the relationship

$$\left[\tfrac{1}{2}(\text{transverse axis})\right]^2 + \left[\tfrac{1}{2}(\text{conjugate axis})\right]^2 = \left[c\right]^2,$$

which is the same as $c^2 = (\text{constant under } x^2) + (\text{constant under } y^2)$. Thus, $c^2 = 9 + 4 \Rightarrow c = \pm\sqrt{13}$. $V(\pm 3, 0)$; $F(\pm\sqrt{13}, 0)$; $W(0, \pm 2)$; $y = \pm\frac{2}{3}x$

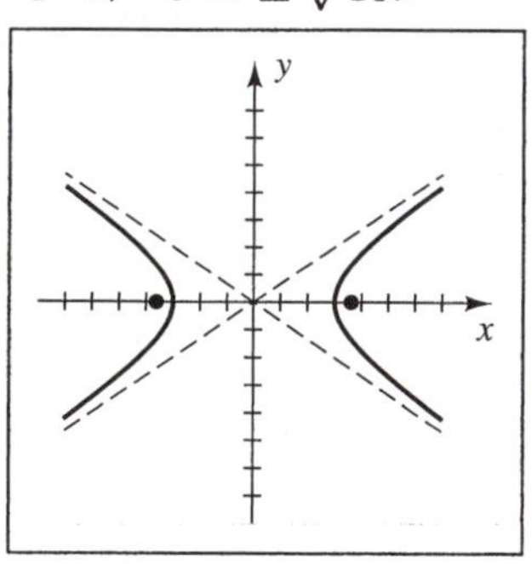


Figure 1

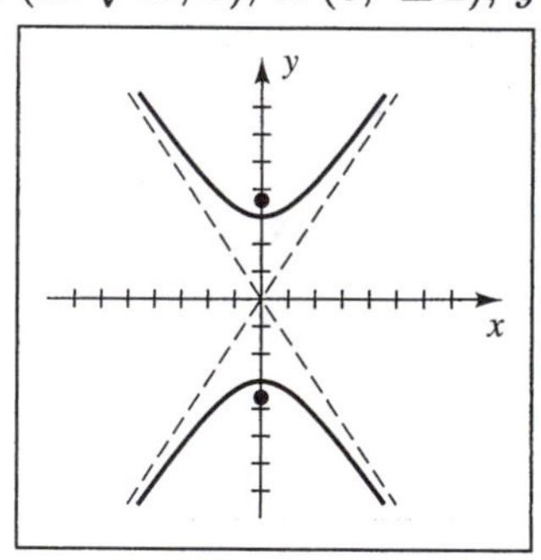


Figure 3

[3] $\frac{y^2}{9} - \frac{x^2}{4} = 1$ • The hyperbola will have an upper branch and a lower branch since the term containing $y$ is positive. The vertices will be on the vertical transverse axis, $\pm\sqrt{9} = \pm 3$ units from the center. The endpoints of the horizontal conjugate axis are $(\pm\sqrt{4}, 0) = (\pm 2, 0)$. The asymptotes have equations $y = \pm\frac{\sqrt{9}}{\sqrt{4}}(x) = \pm\frac{3}{2}x$. Using the foci relationship given in Exercise 1, we have $c^2 = 9 + 4 \Rightarrow c = \pm\sqrt{13}$.

$V(0, \pm 3)$; $F(0, \pm\sqrt{13})$; $W(\pm 2, 0)$; $y = \pm\frac{3}{2}x$

[5] $x^2 - \frac{y^2}{24} = 1 \;\Leftrightarrow\; \frac{x^2}{1} - \frac{y^2}{24} = 1$ •

The asymptotes have equations $y = \pm\frac{\sqrt{24}}{\sqrt{1}}(x) = \pm\sqrt{24}x$.

For the foci: $c^2 = 1 + 24 \;\Rightarrow\; c = \pm 5$. See *Figure 5* below.

$V(\pm 1, 0);\ F(\pm 5, 0);\ W(0, \pm\sqrt{24});\ y = \pm\sqrt{24}x$

[7] $y^2 - 4x^2 = 16$ {divide by 16} $\Rightarrow \frac{y^2}{16} - \frac{x^2}{4} = 1$ • The vertices will be $\pm\sqrt{16} = \pm 4$ units from the center on the vertical transverse axis. The endpoints of the horizontal conjugate axis are $(\pm\sqrt{4}, 0) = (\pm 2, 0)$. The asymptotes have equations $y = \pm\frac{\sqrt{16}}{\sqrt{4}}(x) = \pm\frac{4}{2}x = \pm 2x$. For the foci: $c^2 = 4 + 16 \;\Rightarrow$ $c^2 = 20 \;\Rightarrow\; c = \pm\sqrt{20} \;\Rightarrow\; c = \pm 2\sqrt{5}$.

$V(0, \pm 4);\ F(0, \pm 2\sqrt{5});\ W(\pm 2, 0);\ y = \pm 2x$

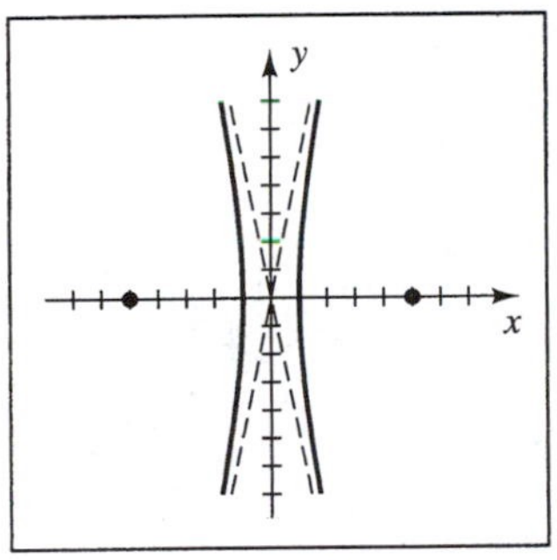

*Figure 5*

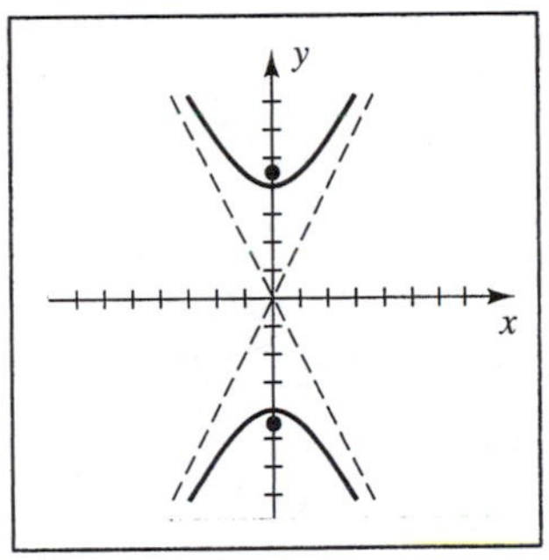

*Figure 7*

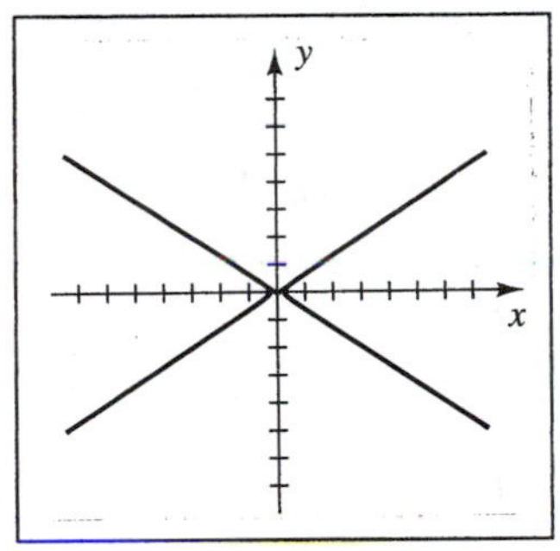

*Figure 9*

[9] $16x^2 - 36y^2 = 1$ {rewrite in standard form} $\Rightarrow \frac{x^2}{\frac{1}{16}} - \frac{y^2}{\frac{1}{36}} = 1$.

The asymptotes have equations $y = \pm\frac{\sqrt{\frac{1}{36}}}{\sqrt{\frac{1}{16}}}(x) = \pm\frac{\frac{1}{6}}{\frac{1}{4}}(x) = \pm\frac{4}{6}(x) = \pm\frac{2}{3}x$.

For the foci: $c^2 = \frac{1}{16} + \frac{1}{36} \;\Rightarrow\; c^2 = \frac{9}{144} + \frac{4}{144} \;\Rightarrow\; c^2 = \frac{13}{144} \;\Rightarrow\; c = \pm\frac{1}{12}\sqrt{13}$.

$V(\pm\frac{1}{4}, 0);\ F(\pm\frac{1}{12}\sqrt{13}, 0);\ W(0, \pm\frac{1}{6});\ y = \pm\frac{2}{3}x$

Note that the branches of the hyperbola almost coincide with the asymptotes.

[11] $\dfrac{(y+2)^2}{9} - \dfrac{(x+2)^2}{4} = 1$ • The effect of the factors $(x+2)$ and $(y+2)$ is to shift the center of the hyperbola from $(0,\ 0)$ to $(-2,\ -2)$. Since the term involving $y$ is positive, the transverse axis will be vertical. The vertices are 3 units in either direction of the center. Their coordinates are $(-2,\ -2\pm 3)$ or equivalently, $(-2,\ 1)$ and $(-2,\ -5)$. The conjugate axis will be horizontal with endpoints $(-2\pm 2,\ -2)$, or, equivalently, $(0,\ -2)$ and $(-4,\ -2)$. Remember that $c$ is the distance *from the center* to a focus. $c^2 = 9+4 \Rightarrow c = \pm\sqrt{13}$. Hence, the coordinates of the foci are $(-2,\ -2\pm\sqrt{13})$. If the center of the hyperbola was at the origin, we would have asymptote equations $y = \pm\frac{3}{2}x$. Since the center of the hyperbola has been shifted by the factors $(y+2)$ and $(x+2)$, we can shift the asymptote equations using the same factors. Hence, these equations are $(y+2) = \pm\frac{3}{2}(x+2)$.

$C(-2,\ -2);\ V(-2,\ -2\pm 3);\ F(-2,\ -2\pm\sqrt{13});\ W(-2\pm 2,\ -2)$

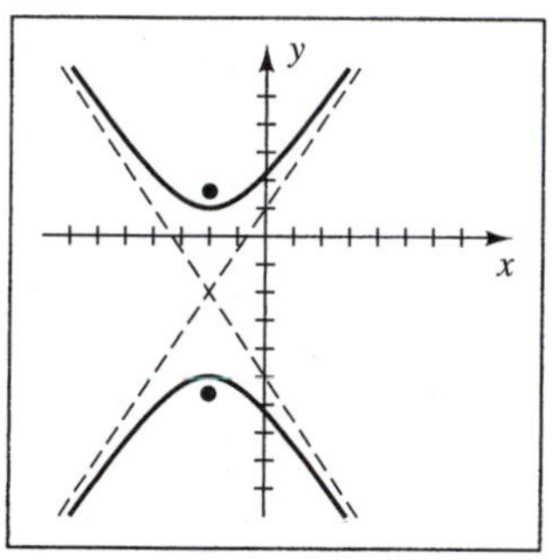

*Figure 11*

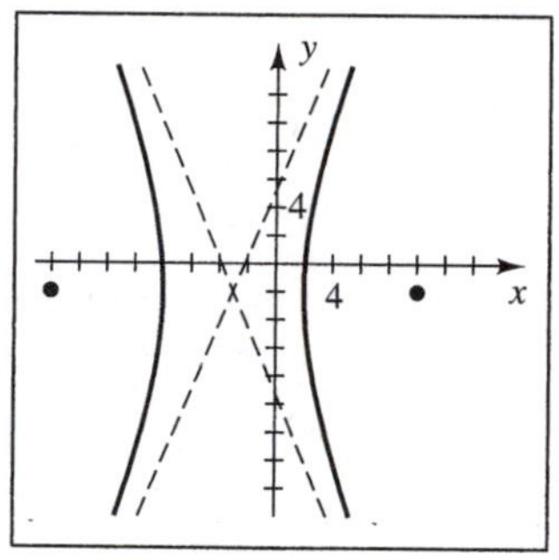

*Figure 13*

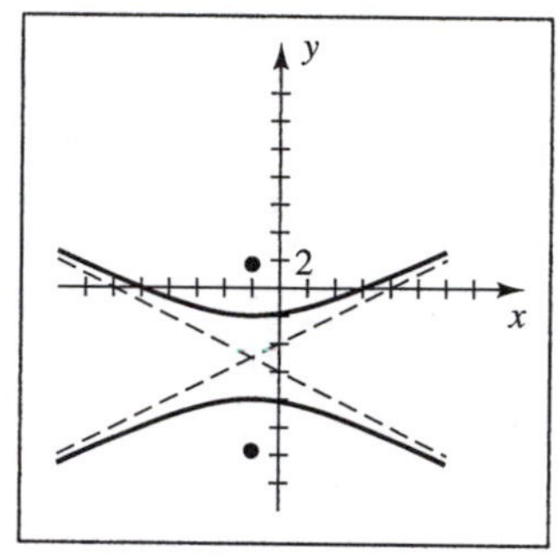

*Figure 15*

[13] $144x^2 - 25y^2 + 864x - 100y - 2404 = 0 \Rightarrow$

$144(x^2 + 6x + \underline{\ 9\ }) - 25(y^2 + 4y + \underline{\ 4\ }) = 2404 + \underline{\ 1296\ } - \underline{\ 100\ } \Rightarrow$

$144(x+3)^2 - 25(y+2)^2 = 3600 \Rightarrow \dfrac{(x+3)^2}{25} - \dfrac{(y+2)^2}{144} = 1.$

$c^2 = 25 + 144 \Rightarrow c = \pm 13.$

$C(-3,\ -2);\ V(-3\pm 5,\ -2);\ F(-3\pm 13,\ -2);\ W(-3,\ -2\pm 12);\ (y+2) = \pm\frac{12}{5}(x+3)$

[15] $4y^2 - x^2 + 40y - 4x + 60 = 0 \Rightarrow$

$4(y^2 + 10y + \underline{\ 25\ }) - 1(x^2 + 4x + \underline{\ 4\ }) = -60 + \underline{\ 100\ } - \underline{\ 4\ } \Rightarrow$

$4(y+5)^2 - (x+2)^2 = 36 \Rightarrow \dfrac{(y+5)^2}{9} - \dfrac{(x+2)^2}{36} = 1.$

$c^2 = 9 + 36 \Rightarrow c = \pm 3\sqrt{5}.$

$C(-2,\ -5);\ V(-2,\ -5\pm 3);\ F(-2,\ -5\pm 3\sqrt{5});\ W(-2\pm 6,\ -5);\ (y+5) = \pm\frac{1}{2}(x+2)$

See *Figure 15* above.

*Note:* Let $b$ denote the distance equal to $\frac{1}{2}$(conjugate axis length) and

let $a$ denote the distance equal to $\frac{1}{2}$(transverse axis length).

[17] $a = 3$ and $c = 5 \Rightarrow b^2 = c^2 - a^2 = 16$. $\frac{x^2}{a^2} - \frac{y^2}{b^2} = 1$ is then $\frac{x^2}{9} - \frac{y^2}{16} = 1$.

[19] The center of the hyperbola is $(-2, -3)$. $a = 1$ and $c = 2 \Rightarrow b^2 = 2^2 - 1^2 = 3$ and

an equation is $\frac{(y+3)^2}{1^2} - \frac{(x+2)^2}{3} = 1$, or, equivalently, $(y+3)^2 - \frac{(x+2)^2}{3} = 1$.

[21] $F(0, \pm 4) \Rightarrow c = 4$. $V(0, \pm 1) \Rightarrow a = 1$. $a^2 + b^2 = c^2 \Rightarrow b^2 = 4^2 - 1^2 = 15$.

Since the vertices are on the $y$-axis, the "$1^2$" is associated with the $y^2$ term.

An equation is $\frac{y^2}{1} - \frac{x^2}{15} = 1$.

[23] $F(\pm 5, 0)$ and $V(\pm 3, 0) \Rightarrow b^2 = c^2 - a^2 = 5^2 - 3^2 = 16 \Rightarrow W(0, \pm 4)$.

An equation is $\frac{x^2}{9} - \frac{y^2}{16} = 1$.

[25] Conjugate axis of length 4 implies that $b = 2$. $F(0, \pm 5) \Rightarrow c = 5$.

$a^2 + b^2 = c^2 \Rightarrow a^2 = 5^2 - 2^2 = 21$. An equation is $\frac{y^2}{21} - \frac{x^2}{4} = 1$.

[27] Since the asymptote equations are $y = \pm 2x$ and we know that the point $(3, 0)$ is on the hyperbola, we conclude that the upper right corner of the rectangle formed by the transverse and conjugate axes has coordinates $(3, 6)$ {substitute $x = 3$ in $y = 2x$ to obtain the 6}. Thus we have endpoints of the conjugate axis at $(0, \pm 6)$.

An equation is $\frac{x^2}{3^2} - \frac{y^2}{6^2} = 1$, or $\frac{x^2}{9} - \frac{y^2}{36} = 1$.

[29] $a = 5$, $b = 2(5) = 10$. $\frac{x^2}{5^2} - \frac{y^2}{10^2} = 1 \Rightarrow \frac{x^2}{25} - \frac{y^2}{100} = 1$.

[31] Since the transverse axis is vertical, the $y^2$ term will be positive.

An equation is $\frac{y^2}{(\frac{1}{2} \cdot 10)^2} - \frac{x^2}{(\frac{1}{2} \cdot 14)^2} = 1$, or $\frac{y^2}{25} - \frac{x^2}{49} = 1$.

[33] $\frac{1}{3}(x+2) = y^2$ • We have a first degree $x$-term and a second degree $y$-term, so this is a parabola with a horizontal axis.

[35] $x^2 + 6x - y^2 = 7$ • The coefficient of $x^2$ is positive and the coefficient of $y^2$ is negative, so this is a hyperbola.

[37] $-x^2 = y^2 - 25 \Leftrightarrow x^2 + y^2 = 25$, a circle with center at the origin and radius 5.

[39] $4x^2 - 16x + 9y^2 + 36y = -16$ • The coefficients of $x^2$ and $y^2$ are positive, so this is an ellipse.

[41] $x^2 + 3x = 3y - 6$ • We have a first degree $y$-term and a second degree $x$-term, so this is a parabola with a vertical axis.

[43] The graph of $y^2 - 4x^2 = 16$, or, equivalently, $\frac{y^2}{16} - \frac{x^2}{4} = 1$, is that of a hyperbola with $y$-intercepts $\pm 4$. The graph of $y - x = 4$, or, equivalently, $y = x + 4$, is that of a line with $y$-intercept 4 and slope 1. Substituting $y = x + 4$ into $y^2 - 4x^2 = 16$ yields $(x+4)^2 - 4x^2 = 16 \Rightarrow$ $3x^2 - 8x = 0 \Rightarrow x(3x - 8) = 0 \Rightarrow x = 0, \frac{8}{3}$; $y = 4, \frac{20}{3}$. The two points of intersection are $(0, 4)$ and $(\frac{8}{3}, \frac{20}{3})$.

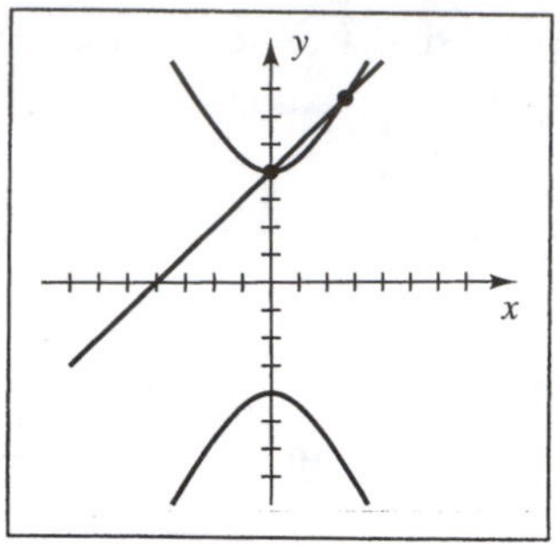

Figure 43

[45] Refer to the definition of a hyperbola. As in the discussion on page 604, we will let the positive constant, $k$, equal $2a$. $k = 24 \Rightarrow 2a = 24 \Rightarrow a = 12$.
$F(13, 0)$ and $F'(-13, 0) \Rightarrow c = 13$. $b^2 = c^2 - a^2 = 13^2 - 12^2 = 169 - 144 = 25$.

An equation is $\frac{x^2}{144} - \frac{y^2}{25} = 1$.

[47] $k = 16 \Rightarrow 2a = 16 \Rightarrow a = 8$. $F(0, 10)$ and $F'(0, -10) \Rightarrow c = 10$.
$b^2 = c^2 - a^2 = 10^2 - 8^2 = 100 - 64 = 36$. An equation is $\frac{y^2}{64} - \frac{x^2}{36} = 1$.

[49] $k = 2a = 11 - 3 = 8 \Rightarrow a = 4$.
$F(0, 5)$ and $F'(0, -5) \Rightarrow c = 5$.
$b^2 = c^2 - a^2 = 25 - 16 = 9$.
An equation is $\frac{y^2}{16} - \frac{x^2}{9} = 1$.

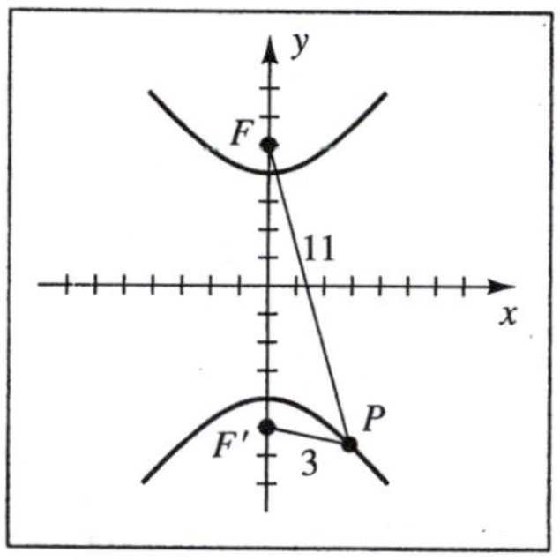

Figure 49

[51] $x = \frac{5}{4}\sqrt{y^2 + 16} \Rightarrow \frac{4}{5}x = \sqrt{y^2 + 16} \Rightarrow \frac{16}{25}x^2 = y^2 + 16 \Rightarrow \frac{16}{25}x^2 - y^2 = 16 \Rightarrow$
$\frac{x^2}{25} - \frac{y^2}{16} = 1$, which is an equation of a hyperbola with right and left branches.

Since $x > 0$ in the original equation, its graph is the right branch of the hyperbola.

[53] $y = \frac{3}{7}\sqrt{x^2 + 49} \Rightarrow \frac{7}{3}y = \sqrt{x^2 + 49} \Rightarrow \frac{49}{9}y^2 = x^2 + 49 \Rightarrow \frac{y^2}{9} - \frac{x^2}{49} = 1$.

Since $y > 0$ in the original equation, its graph is the upper branch of the hyperbola.

[55] $y = -\frac{9}{4}\sqrt{x^2 - 16} \Rightarrow -\frac{4}{9}y = \sqrt{x^2 - 16} \Rightarrow \frac{16}{81}y^2 = x^2 - 16 \Rightarrow \frac{x^2}{16} - \frac{y^2}{81} = 1$.
Since $y \le 0$ in the original equation,

its graph is the lower halves of the branches of the hyperbola.

[57] $x = -\frac{2}{3}\sqrt{y^2 - 36} \Rightarrow -\frac{3}{2}x = \sqrt{y^2 - 36} \Rightarrow \frac{9}{4}x^2 = y^2 - 36 \Rightarrow \frac{y^2}{36} - \frac{x^2}{16} = 1.$
Since $x \leq 0$ in the original equation,

its graph is the left halves of the branches of the hyperbola.

[59] Their equations are $\frac{x^2}{25} - \frac{y^2}{9} = 1$ and $\frac{x^2}{25} - \frac{y^2}{9} = -1$,

or, equivalently, $\frac{y^2}{9} - \frac{x^2}{25} = 1$.

Conjugate hyperbolas have the same asymptotes and exchange transverse and conjugate axes.

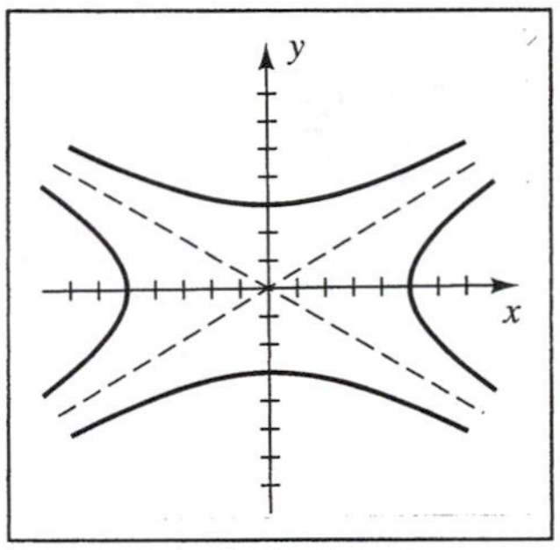

*Figure 59*

[61] The path is a hyperbola with $V(\pm 3, 0)$ and $W(0, \pm\frac{3}{2})$.

An equation is $\frac{x^2}{(3)^2} - \frac{y^2}{(\frac{3}{2})^2} = 1$ or equivalently, $x^2 - 4y^2 = 9$.

If only the right branch is considered, then $x = \sqrt{9 + 4y^2}$ is an equation of the path.

[63] Set up a coordinate system like the one in Example 6. Let the origin be located on the shoreline halfway between $A$ and $B$ and let $P(x, y)$ denote the coordinates of the ship. The coordinates of $A$ and $B$ (which can be thought of as the foci of the hyperbola) are $(-100, 0)$ and $(100, 0)$, respectively. Hence, $c = 100$. As in Exercise 45, the difference in distances is a constant—that is,

$$d(P, A) - d(P, B) = 2a \Rightarrow 160 = 2a \Rightarrow a = 80.$$

$b^2 = c^2 - a^2 = 100^2 - 80^2 \Rightarrow b = 60$. An equation of the hyperbola is $\frac{x^2}{80^2} - \frac{y^2}{60^2} = 1$.

Now, $y = 100 \Rightarrow$

$$\frac{x^2}{80^2} = 1 + \frac{100^2}{60^2} \Rightarrow x^2 = 80^2 \cdot \frac{13{,}600}{60^2} \Rightarrow x = 80 \cdot \frac{10}{60}\sqrt{136} = \frac{80}{3}\sqrt{34}.$$

The ship's coordinates are $(\frac{80}{3}\sqrt{34}, 100) \approx (155.5, 100)$.

**65** $\dfrac{(y-0.1)^2}{1.6}-\dfrac{(x+0.2)^2}{0.5}=1 \;\Rightarrow\; \dfrac{(y-0.1)^2}{1.6}=1+\dfrac{(x+0.2)^2}{0.5} \;\Rightarrow$

$(y-0.1)^2=1.6[1+(x+0.2)^2/0.5] \;\Rightarrow\; y-0.1=\pm\sqrt{1.6[1+(x+0.2)^2/0.5]} \;\Rightarrow$

$$y=0.1\pm\sqrt{1.6[1+(x+0.2)^2/0.5]}.$$

$\dfrac{(y-0.5)^2}{2.7}-\dfrac{(x-0.1)^2}{5.3}=1 \;\Rightarrow\; \dfrac{(y-0.5)^2}{2.7}=1+\dfrac{(x-0.1)^2}{5.3} \;\Rightarrow$

$(y-0.5)^2=2.7[1+(x-0.1)^2/5.3] \;\Rightarrow\; y-0.5=\pm\sqrt{2.7[1+(x-0.1)^2/5.3]} \;\Rightarrow$

$$y=0.5\pm\sqrt{2.7[1+(x-0.1)^2/5.3]}.$$

From the graph, the point of intersection in the first quadrant is approximately (0.741, 2.206).

[−15, 15] by [−10, 10]

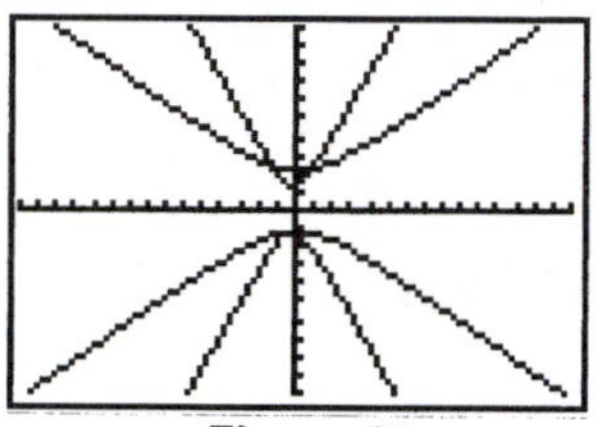

*Figure 65*

[−15, 15, 2] by [−10, 10, 2]

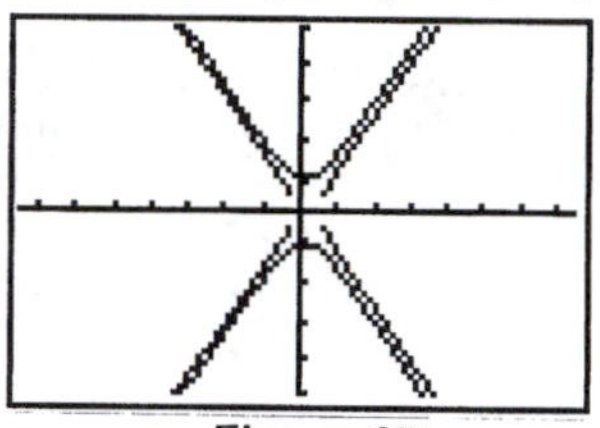

*Figure 67*

**67** $\dfrac{(x-0.3)^2}{1.3}-\dfrac{y^2}{2.7}=1 \;\Rightarrow\; \dfrac{(x-0.3)^2}{1.3}-1=\dfrac{y^2}{2.7} \;\Rightarrow\; y^2=2.7[-1+(x-0.3)^2/1.3] \;\Rightarrow$

$$y=\pm\sqrt{2.7[-1+(x-0.3)^2/1.3]}.$$

$\dfrac{y^2}{2.8}-\dfrac{(x-0.2)^2}{1.2}=1 \;\Rightarrow\; \dfrac{y^2}{2.8}=1+\dfrac{(x-0.2)^2}{1.2} \;\Rightarrow\; y^2=2.8[1+(x-0.2)^2/1.2] \;\Rightarrow$

$$y=\pm\sqrt{2.8[1+(x-0.2)^2/1.2]}.$$

The two graphs nearly intersect in the second and fourth quadrants,

but there are no points of intersection.

**69** (a) The comet's path is hyperbolic with $a^2=26\times10^{14}$ and $b^2=18\times10^{14}$.

$c^2=a^2+b^2=26\times10^{14}+18\times10^{14}=44\times10^{14} \;\Rightarrow\; c\approx6.63\times10^7$.

The coordinates of the sun are approximately $(6.63\times10^7,\ 0)$.

(b) The minimum distance between the comet and the sun will be

$$c-a=\sqrt{44\times10^{14}}-\sqrt{26\times10^{14}}=1.53\times10^7 \text{ mi.}$$

Since $r$ must be in meters, $1.53\times10^7$ mi $\times$ 1610 m/mi $\approx 2.47\times10^{10}$ m. At this distance, $v$ must be greater than $\sqrt{\dfrac{2k}{r}}\approx\sqrt{\dfrac{2(1.325\times10^{20})}{2.47\times10^{10}}}\approx 103{,}600$ m/sec.

## *Chapter 8 Review Exercises*

*Note:* Let the notation be the same as in §8.1–8.3.

1 $y^2 = 64x \Rightarrow x = \frac{1}{64}y^2 \Rightarrow a = \frac{1}{64}.$ $p = \frac{1}{4(\frac{1}{64})} = 16.$ $V(0,\ 0)$; $F(16,\ 0)$; $l\colon x = -16$

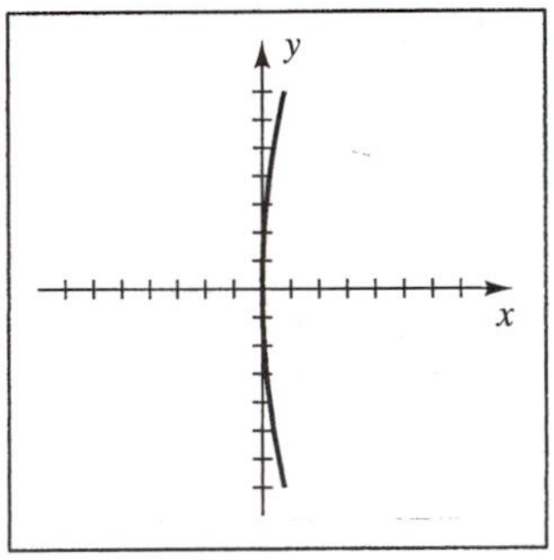


*Figure 1*

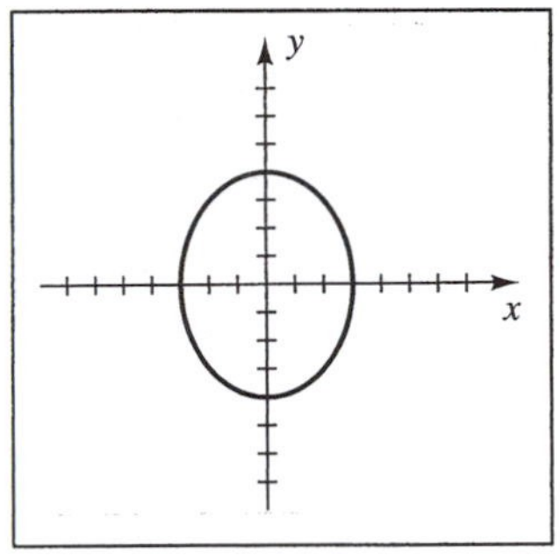


*Figure 3*

3 $9y^2 = 144 - 16x^2 \Rightarrow 16x^2 + 9y^2 = 144 \Rightarrow \frac{16x^2}{144} + \frac{9y^2}{144} = \frac{144}{144} \Rightarrow \frac{x^2}{9} + \frac{y^2}{16} = 1.$

$c^2 = 16 - 9 \Rightarrow c = \pm\sqrt{7}.$ $V(0,\ \pm 4)$; $F(0,\ \pm\sqrt{7})$; $M(\pm 3,\ 0)$

5 $x^2 - y^2 - 4 = 0 \Rightarrow \frac{x^2}{4} - \frac{y^2}{4} = 1.$ $c^2 = 4 + 4 \Rightarrow c = \pm\sqrt{8} \Rightarrow c = \pm 2\sqrt{2}.$

$V(\pm 2,\ 0)$; $F(\pm 2\sqrt{2},\ 0)$; $W(0,\ \pm 2)$; $y = \pm x$

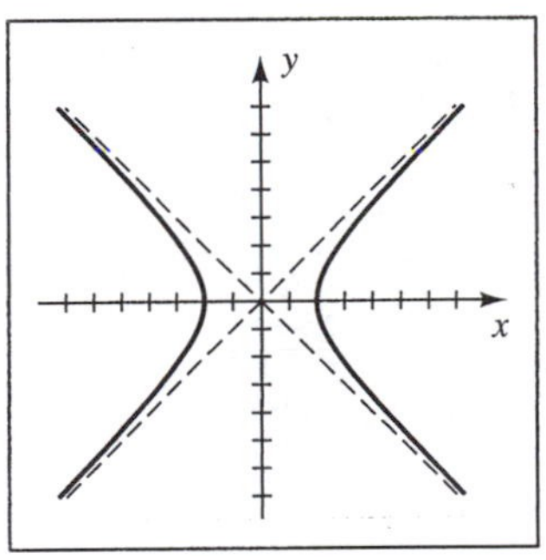


*Figure 5*

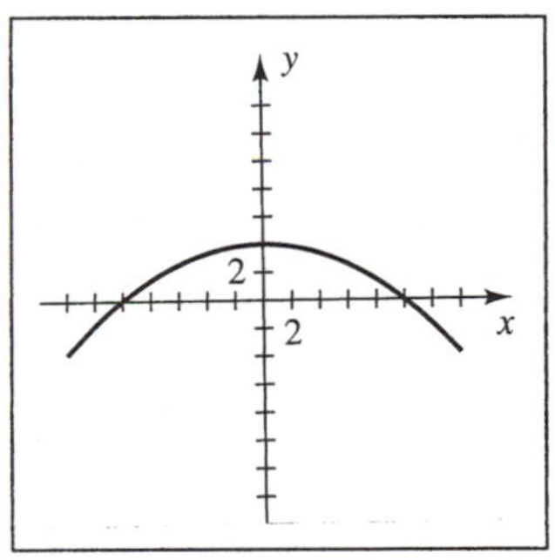


*Figure 7*

7 $25y = 100 - x^2 \Rightarrow y = 4 - \frac{1}{25}x^2 \Rightarrow a = -\frac{1}{25}.$ $p = \frac{1}{4(-\frac{1}{25})} = -\frac{25}{4}.$

The vertex is at $(0,\ 4)$ and the focus is $\frac{25}{4}$ units below the vertex. $4 - \frac{25}{4} = -\frac{9}{4}.$

The directrix is $\frac{25}{4}$ units above the vertex. $4 + \frac{25}{4} = \frac{41}{4}.$

$V(0,\ 4)$; $F(0,\ -\frac{9}{4})$; $l\colon y = \frac{41}{4}$

[9] $x^2 - 9y^2 + 8x + 90y - 210 = 0 \Rightarrow$

$(x^2 + 8x + \underline{16}) - 9(y^2 - 10y + \underline{25}) = 210 + \underline{16} - \underline{225} \Rightarrow$

$(x+4)^2 - 9(y-5)^2 = 1 \Rightarrow \dfrac{(x+4)^2}{1} - \dfrac{(y-5)^2}{\frac{1}{9}} = 1.\ c^2 = 1 + \frac{1}{9} \Rightarrow c = \pm\frac{1}{3}\sqrt{10}.$

$C(-4,\ 5);\ V(-4 \pm 1,\ 5);\ F(-4 \pm \frac{1}{3}\sqrt{10},\ 5);\ W(-4,\ 5 \pm \frac{1}{3});\ (y-5) = \pm\frac{1}{3}(x+4)$

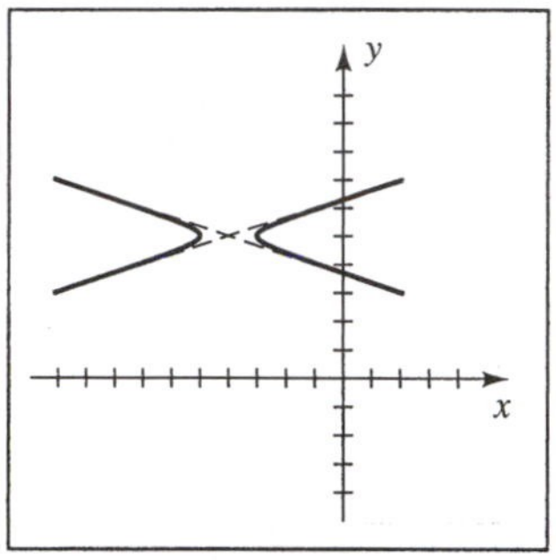


*Figure 9*

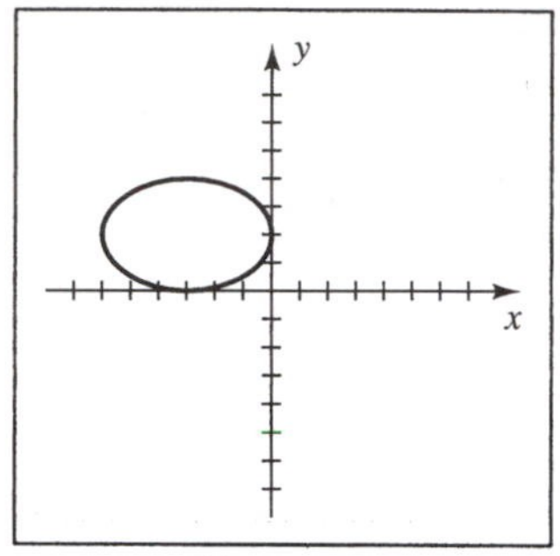


*Figure 11*

[11] $4x^2 + 9y^2 + 24x - 36y + 36 = 0 \Rightarrow$

$4(x^2 + 6x + \underline{9}) + 9(y^2 - 4y + \underline{4}) = -36 + \underline{36} + \underline{36} \Rightarrow$

$4(x+3)^2 + 9(y-2)^2 = 36 \Rightarrow \dfrac{(x+3)^2}{9} + \dfrac{(y-2)^2}{4} = 1.\ c^2 = 9 - 4 \Rightarrow c = \pm\sqrt{5}.$

$C(-3,\ 2);\ V(-3 \pm 3,\ 2);\ F(-3 \pm \sqrt{5},\ 2);\ M(-3,\ 2 \pm 2)$

[13] $y^2 - 8x + 8y + 32 = 0 \Rightarrow x = \frac{1}{8}y^2 + y + 4 \Rightarrow a = \frac{1}{8}.\ p = \dfrac{1}{4(\frac{1}{8})} = 2.$

$V(2,\ -4);\ F(4,\ -4);\ l\colon x = 0$

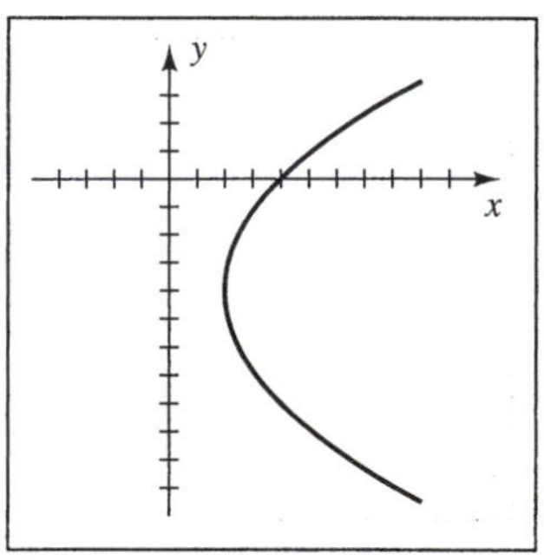


*Figure 13*

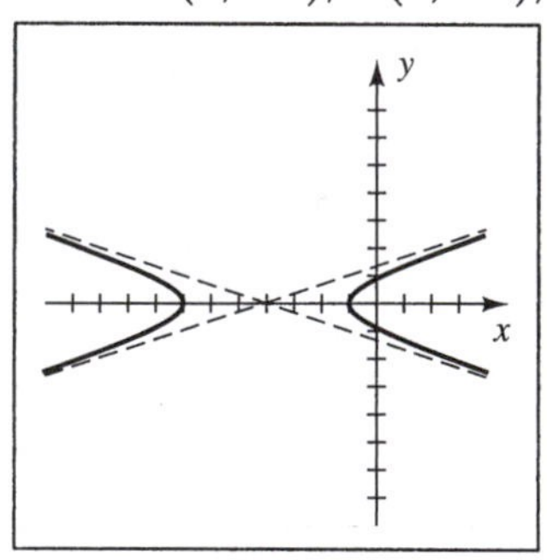


*Figure 15*

[15] $x^2 - 9y^2 + 8x + 7 = 0 \Rightarrow$

$(x^2 + 8x + \underline{16}) - 9(y^2) = -7 + \underline{16} \Rightarrow (x+4)^2 - 9(y^2) = 9 \Rightarrow$

$\dfrac{(x+4)^2}{9} - \dfrac{y^2}{1} = 1.\ c^2 = 9 + 1 \Rightarrow c = \pm\sqrt{10}.$

$C(-4,\ 0);\ V(-4 \pm 3,\ 0);\ F(-4 \pm \sqrt{10},\ 0);\ W(-4,\ 0 \pm 1);\ y = \pm\frac{1}{3}(x+4)$

[17] The vertex is halfway between the $x$-intercepts, so it is of the form $V(-7,\ k)$.

$y = a(x+10)(x+4)$ and $x = 0,\ y = 80 \Rightarrow 80 = a(10)(4) \Rightarrow a = 2.$

$x = -7 \Rightarrow y = 2(3)(-3) = -18.$ Hence, $y = 2(x+7)^2 - 18.$

**19** An equation of a hyperbola with vertices $V(0, \pm 7)$ and endpoints of the conjugate axis $(\pm 3, 0)$ is $\frac{y^2}{7^2} - \frac{x^2}{3^2} = 1$ or $\frac{y^2}{49} - \frac{x^2}{9} = 1$.

**21** $F(0, -10)$ and $l: y = 10 \Rightarrow p = -10$ and $V(0, 0)$.

An equation is $(x - 0)^2 = [4(-10)](y - 0)$, or $x^2 = -40y$.

**23** $V(0, \pm 10)$ and $F(0, \pm 5) \Rightarrow b^2 = 10^2 - 5^2 = 100 - 25 = 75$.

An equation is $\frac{x^2}{75} + \frac{y^2}{10^2} = 1$ or $\frac{x^2}{75} + \frac{y^2}{100} = 1$.

**25** Hyperbola, with vertices $V(0, \pm 6)$ and asymptotes $y = \pm 9x$ • The equations of the asymptotes are $y = \pm \frac{a}{b}x$ and $a = 6$, so $9 = \frac{6}{b} \Rightarrow b = \frac{6}{9} = \frac{2}{3}$.

An equation is $\frac{y^2}{6^2} - \frac{x^2}{(\frac{2}{3})^2} = 1$ or $\frac{y^2}{36} - \frac{x^2}{\frac{4}{9}} = 1$.

**26** $F(\pm 2, 0) \Rightarrow c^2 = 4$. Now $\frac{x^2}{a^2} + \frac{y^2}{b^2} = 1$ can be written as $\frac{x^2}{a^2} + \frac{y^2}{a^2 - 4} = 1$ since $b^2 = a^2 - c^2$. Substituting $x = 2$ and $y = \sqrt{2}$ into that equation yields

$\frac{4}{a^2} + \frac{2}{a^2 - 4} = 1 \Rightarrow 4a^2 - 16 + 2a^2 = a^4 - 4a^2 \Rightarrow a^4 - 10a^2 + 16 = 0 \Rightarrow$

$(a^2 - 2)(a^2 - 8) = 0 \Rightarrow a^2 = 2, 8$. Since $a > c$, $a^2$ must be 8 and $b^2$ is equal to 4.

An equation is $\frac{x^2}{8} + \frac{y^2}{4} = 1$.

**27** $M(\pm 5, 0) \Rightarrow b = 5$. $e = \frac{c}{a} = \frac{\sqrt{a^2 - b^2}}{a} = \frac{\sqrt{a^2 - 25}}{a} = \frac{2}{3} \Rightarrow \frac{2}{3}a = \sqrt{a^2 - 25} \Rightarrow$

$\frac{4}{9}a^2 = a^2 - 25 \Rightarrow \frac{5}{9}a^2 = 25 \Rightarrow a^2 = 45$. An equation is $\frac{x^2}{25} + \frac{y^2}{45} = 1$.

**29** (a) Substituting $x = 2$ and $y = -3$ in $Ax^2 + 2y^2 = 4 \Rightarrow A(2)^2 + 2(-3)^2 = 4 \Rightarrow$

$4A + 18 = 4 \Rightarrow 4A = -14 \Rightarrow A = -\frac{7}{2}$.

(b) The equation is $-\frac{7}{2}x^2 + 2y^2 = 4 \Rightarrow -\frac{7}{8}x^2 + \frac{2}{4}y^2 = 1 \Rightarrow \frac{y^2}{2} - \frac{7x^2}{8} = 1$,

a hyperbola.

**31** The focus is a distance of $p = \frac{1}{4a} = \frac{1}{4 \cdot \frac{1}{8}} = \frac{1}{\frac{1}{2}} = 2$ units from the origin. So the center of the circle is $(0, 2)$ and since the circle passes through the origin, the radius of the circle must be 2. An equation of the circle is $x^2 + (y - 2)^2 = 2^2 = 4$.

## Chapter 8 Discussion Exercises

[1] For $y = ax^2$, the horizontal line through the focus is $y = p$. Since $a = 1/(4p)$, we have $p = (1/(4p))x^2 \Rightarrow x^2 = 4p^2 \Rightarrow x = 2|p|$. Doubling this value for the width gives us $w = 4|p|$.

[3] Refer to Figure 2 and the derivation on text page 591.

[5] $P(x, y)$ is a distance of $(2 + d)$ from $(0, 0)$ and a distance of $d$ from $(4, 0)$. The difference of these distances is $(2 + d) - d = 2$, a positive constant. By the definition of a hyperbola, $P(x, y)$ lies on the right branch of the hyperbola with foci $(0, 0)$ and $(4, 0)$. The center of the hyperbola is halfway between the foci, that is, $(2, 0)$. The vertex is halfway from $(2, 0)$ to $(4, 0)$ since the distance from the circle to $P$ equals the distance from $P$ to $(4, 0)$.

Thus, the vertex is $(3, 0)$ and $a = 1$.

$b^2 = c^2 - a^2 = 2^2 - 1^2 = 3$ and

an equation of the right branch of the hyperbola is

$\dfrac{(x-2)^2}{1} - \dfrac{y^2}{3} = 1,\ x \geq 3$ or $x = 2 + \sqrt{1 + \dfrac{y^2}{3}}$.

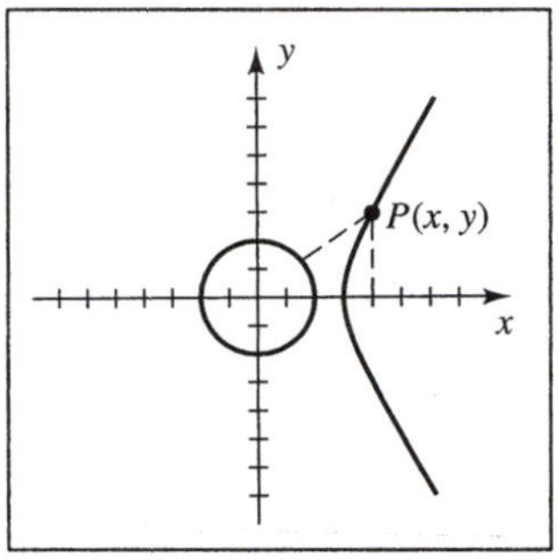

Figure 5